T0178547

Springer Optimization and Its Applications

VOLUME 122

Aims and Scope
Optimization has been expanding in all directions at an astonishing rate during the last few decades. New algorithmic and theoretical techniques have been developed, the diffusion into other disciplines has proceeded at a rapid pace, and our knowledge of all aspects of the field has grown even more profound. At the same time, one of the most striking trends in optimization is the constantly increasing emphasis on the interdisciplinary nature of the field. Optimization has been a basic tool in all areas of applied mathematics, engineering, medicine, economics and other sciences.

The series *Springer Optimization and Its Applications* publishes undergraduate and graduate textbooks, monographs and state-of-the-art expository works that focus on algorithms for solving optimization problems and also study applications involving such problems. Some of the topics covered include nonlinear optimization (convex and nonconvex), network flow problems, stochastic optimization, optimal control, discrete optimization, multi-objective programming, description of software packages, approximation techniques and heuristic approaches.

More information about this series at http://www.springer.com/series/7393

A.M. Mathai · H.J. Haubold

Fractional and Multivariable Calculus

Model Building and Optimization Problems

 Springer

A.M. Mathai
Centre for Mathematical and Statistical
  Sciences
Peechi Campus, Kerala
India

and

Department of Mathematics and Statistics
McGill University
Montreal, QC
Canada

H.J. Haubold
Vienna International Centre
Office for Outer Space Affairs
United Nations, Vienna
Austria

and

Centre for Mathematical and Statistical
  Sciences
Peechi Campus, Kerala
India

ISSN 1931-6828   ISSN 1931-6836   (electronic)
Springer Optimization and Its Applications
ISBN 978-3-319-86754-0   ISBN 978-3-319-59993-9   (eBook)
DOI 10.1007/978-3-319-59993-9

Mathematics Subject Classification (2010): 15B57, 26A33, 60B20, 62E15, 33C60, 40C05

Printed on acid-free paper

This Springer imprint is published by Springer Nature
The registered company is Springer International Publishing AG
The registered company address is: Gewerbestrasse 11, 6330 Cham, Switzerland

Preface

In SERC School, or Science and Engineering Research Council (SERC) of the Department of Science and Technology, Government of India, New Delhi (DST), research orientation program is a regular feature at the Centre for Mathematical and Statistical Sciences India (CMSS). It is a four- to five-week intensive program run in April–May of every year at CMSS. Lecture notes, in the form of research level books, are produced every year. The summary of the notes from the first five schools at CMSS was printed by Springer, New York, in 2008 under the title *"Special Functions for Applied Scientists."* The theme of those five schools was Special Functions and Their Applications.

The theme of the five SERC Schools from 2008 to 2012 was "Multivariable and matrix variable calculus with applications in model building, optimization, fractional calculus, statistical distributions, and astrophysics problems." The summary of the notes from the main lecturers in the five SERC Schools is put in the form of a book. This represents the present manuscript.

Chapter 1 starts with the properties of Mittag-Leffler and Wright's functions, which are associated with the solutions of fractional order differential and integral equations. Then, an exposure into the essentials of fractional calculus is given with reference to Cauchy problem, signaling problem, reaction-diffusion problems, relaxation, etc. The leading researchers in the area such as the late Prof. Dr. Anatoly A. Kilbas of Belarus, Prof. Dr. Rudolf Gorenflo of Germany, Prof. Dr. Francesco Mainardi of Italy, and Prof. Dr. R.K. Saxena of India have given lectures on this topic at CMSS in the SERC Schools in various years. The material in Chapter 1 is taken from the lectures of Prof. Dr. Francesco Mainardi of Italy.

Chapter 2 gives a basic introduction to multivariable calculus geared to applicable analysis. Some three people had lectured on this topic in different years at CMSS SERC Schools. We have included the lecture notes from Prof. Dr. D.V. Pai of IIT (Indian Institute of Technology) Bombay and IIT Gandhinagar.

Chapter 3 deals with deterministic models. Some three people had given lectures in different years on deterministic models, linear and nonlinear analysis, etc. We have included the introductory materials from the lectures of Prof. Dr. A.M. Mathai, the Director of SERC Schools, and Emeritus Professor of Mathematics and Statistics at McGill University, Canada.

Chapter 4 gives details of some non-deterministic models or models catering to random phenomena. Here, regression type prediction models are given extensive treatment. Concepts of regression, correlation, and the bases of prediction models are explained and illustrated properly. Some matrix-variate calculus in the form of vector and matrix derivatives and their applications into optimization problems leading into popular multivariate techniques such as principal component analysis, canonical correlation analysis, linear and nonlinear least squares, etc, are also described in this chapter. The material here is based on the lectures of Prof. A.M. Mathai.

Chapter 5 leads into some optimal designs in the area of Design of Experiments and Analysis of Variance. The material is based on the lectures of Prof. Dr. Stratis Kounias of the University of Athens, Greece.

Other topics covered in the SERC Schools of 2008–2012 include order statistics, time series analysis, optimization and game theory, astrophysics, stellar and solar models, and wavelet analysis. Since there were no extensive or deeper coverage of these topics, they are not included in the present manuscript. It is hoped that some of the topics presented in this manuscript will be of use to researchers working in the interface between different disciplines or in the area of applicable mathematics or applications of mathematical techniques to other disciplines such as differential equations, fractional calculus, statistical distributions, stochastic processes, theoretical physics, reaction-diffusion, input–output type problems, and model building in general.

The financial support for this project was available from Dr. A.M. Mathai's Project: No. SR/S4/MS:287/05 of the Department of Science and Technology, Government of India, New Delhi (DST). The authors would like to place on record their deep appreciation and thanks to DST for the timely help, especially to Dr. P.K. Malhotra and Dr. Ashok K. Singh of the Mathematical Sciences Division of DST. Dr. B.D. Acharya (now retired) and Dr. H.K.N. Trivedi (now retired) were instrumental in getting the original project sanctioned. Only due to Dr. B.D. Acharya's encouragement, the authors tried to bring out the summary of the first five schools in the form of the book: *Special Functions for Applied Scientists* (Springer, New York 2008), and the present manuscript, to give exposure to the contents of the SERC School notes to outside world. The authors express their heartfelt thanks to Dr. B.D. Acharya.

A lot of people have contributed in making this publication in this final printed form. Sini Devassy, Girija, R. of CMS office, and printing unit personnel of CMSS looked after the printing process. All the 18 JRFs/SRF and the various types of faculty and the Liaison Officer at CMSS all contributed directly or indirectly for bringing out this publication in this final form. The authors would like to thank each and every one who directly or indirectly contributed to bring out this publication in this final form.

CMSS India   A.M. Mathai
March 2017   H.J. Haubold

Symbols

$\hat{f}(k)$	Fourier transform, Section 1.2, 2
$\tilde{f}(s)$	Laplace transform, Section 1.2, 2
$f^*(s)$	Mellin transform, Section 1.2, 3
$\Gamma(z)$	Gamma function, Section 1.3, 5
$E_\alpha(x), E_{\alpha,\beta}(z)$	Mittag-Leffler functions, Section 1.3, 5
$(a)_n$	Pochhmmer symbol, Section 1.4, 11
$W_{\lambda,\mu}(x)$	Whittaker function, Section 1.4, 13
$J_t^\alpha f$	Fractional integral, Section 1.5, 15
D_t^μ	Fractional derivative, Section 1.5, 16
$^c_aD_t^\beta$	Caputo derivative, Section 1.7, 24
$\|(\cdot)\|$	Norm of $(\cdot)$, Section 2.4, 42
$E(\cdot)$	Expected value, Chapter 4 Notes, 118
$\mathrm{Var}(x)$	Variance, Section 4.4, 134

Contents

1 Essentials of Fractional Calculus 1
  1.1 Introduction 1
  1.2 Notions and Notations 2
  1.3 The Mittag-Leffler Functions 5
  1.4 The Wright Functions............................... 8
  1.5 Essentials of Fractional Calculus 15
  1.6 Fractional Differential Equations of Relaxation
      and Oscillation Type............................... 19
  1.7 Fractional Differential Equations of Diffusion-Wave Type 23
      1.7.1 Derivation of the Fundamental Solutions............ 25
  References...................................... 29

2 Multivariable Calculus 39
  2.1 Review of Euclidean n-Space R^n 39
  2.2 Geometric Approach: Vectors in 2-Space and 3-Space........ 39
      2.2.1 Components and the Norm of a Vector............. 40
      2.2.2 Position Vector 40
      2.2.3 Vectors as Ordered Triplets of Real Numbers........ 40
  2.3 Analytic Approach 41
      2.3.1 Properties of Addition and Scalar Multiplication...... 41
  2.4 The Dot Product or the Inner Product.................... 41
      2.4.1 The Properties of Dot Product.................... 42
      2.4.2 Properties of the Norm 43
      2.4.3 Geometric Interpretation of the Dot Product 43
  2.5 Multivariable Functions 44
      2.5.1 Scalar and Vector Functions and Fields............. 45
      2.5.2 Visualization of Scalar-Valued Functions 47
  2.6 Limits and Continuity.............................. 48
      2.6.1 Limits 48
      2.6.2 Continuity 50

2.7   Partial Derivatives and Differentiability   52
       2.7.1   Partial Derivatives   52
       2.7.2   Differentiability   53
       2.7.3   Chain Rule   55
       2.7.4   Gradient and Directional Derivatives   56
       2.7.5   Tangent Plane and Normal Line   58
       2.7.6   Differentiability of Vector-Valued Functions,
                Chain Rule   59
       2.7.7   Particular Cases   60
       2.7.8   Differentiation Rules   62
       2.7.9   Taylor's Expansion   63
2.8   Introduction to Optimization   64
       2.8.1   Unconstrained and Constrained Extremizers:
                Conditions for Local Extremizers   66
2.9   Classical Approximation Problems: A Relook   73
       2.9.1   Input–Output Process   73
       2.9.2   Approximation by Algebraic Polynomials   73
2.10   Introduction to Optimal Recovery of Functions   74
       2.10.1   Some Motivating Examples   76
       2.10.2   General Theory   81
       2.10.3   Central Algorithms   84
       2.10.4   Notes   87
References   88

3   **Deterministic Models and Optimization**   89
    3.1   Introduction   89
    3.2   Deterministic Situations   89
    3.3   Differential Equations   90
    3.4   Algebraic Models   92
    3.5   Computer Models   93
    3.6   Power Function Models   94
    3.7   Input–Output Models   94
    3.8   Pathway Model   95
           3.8.1   Optimization of Entropy and Exponential Model   96
           3.8.2   Optimization of Mathai's Entropy   97
    3.9   Fibonacci Sequence Model   98
    3.10   Fractional Order Integral and Derivative Model   101
    References   106

4   **Non-deterministic Models and Optimization**   107
    4.1   Introduction   107
           4.1.1   Random Walk Model   108
           4.1.2   Branching Process Model   109
           4.1.3   Birth-and-Death Process Model   109

		4.1.4	Time Series Models	110
		4.1.5	Regression Type Models	110
	4.2	Some Preliminaries of Statistical Distributions		111
		4.2.1	Minimum Mean Square Prediction	126
	4.3	Regression on Several Variables		130
	4.4	Linear Regression		133
		4.4.1	Correlation Between x_1 and Its Best Linear Predictor	140
	4.5	Multiple Correlation Coefficient $\rho_{1.(2...k)}$		142
		4.5.1	Some Properties of the Multiple Correlation Coefficient	143
	4.6	Regression Analysis Versus Correlation Analysis		150
		4.6.1	Multiple Correlation Ratio	150
		4.6.2	Multiple Correlation as a Function of the Number of Regressed Variables	152
	4.7	Residual Effect		155
	4.8	Canonical Correlations		157
		4.8.1	First Pair of Canonical Variables	164
		4.8.2	Second an Subsequent Pair of Canonical Variables	167
	4.9	Estimation of the Regression Function		170
		4.9.1	Linear Regression of y on x	171
		4.9.2	Linear Regression of y on $x_1, ..., x_k$	177
5	**Optimal Regression Designs**			183
	5.1	Introduction		183
	5.2	The General Linear Model		183
	5.3	Estimable Linear Functions		184
		5.3.1	The Gauss–Markov Theorem	186
	5.4	Correlated Errors		186
		5.4.1	Estimation of Some of the Parameters	187
	5.5	Other Distance Measures		188
	5.6	Means and Covariances of Random Vectors		188
		5.6.1	Quadratic Forms and Generalized Inverses	189
		5.6.2	Distribution of Quadratic Forms	189
		5.6.3	Normal Errors	191
		5.6.4	Rank and Nullity of a Matrix	191
	5.7	Exact or Discrete Optimal Designs		195
		5.7.1	Optimality Criteria	198
		5.7.2	Two-Level Factorial Designs of Resolution III	199
		5.7.3	Optimal Block Designs	201
		5.7.4	Approximate or Continuous Optimal Designs	202
		5.7.5	Algorithms	204
	References			207

6   Pathway Models 209
  6.1   Introduction 209
  6.2   Power Transformation and Exponentiation of a Type 1 Beta
         Density.. 209
         6.2.1   Asymmetric Models 212
  6.3   Power Transformation and Exponentiation on a Type 2 Beta
         Model.. 213
         6.3.1   Asymmetric Version of the Models............... 216
  6.4   Laplace Transforms and Moment Generating Functions 217
  6.5   Models with Appended Bessel Series..................... 219
  6.6   Models with Appended Mittag-Leffler Series 220
  6.7   Bayesian Outlook...................................... 221
  6.8   Pathway Model in the Scalar Case 222
  6.9   Pathway Model in the Matrix-Variate Case 224
         6.9.1   Arbitrary Moments 227
         6.9.2   Some Special Cases........................... 228
  References... 228

Index 231

Author Index... 233

Summary of SERC School Notes

Centre for Mathematical and Statistical Sciences India

SERC Schools are annual research orientation programs at the Centre for Mathematical and Statistical Sciences India (CMSS). The present manuscript is a summary of the lecture notes of five SERC Schools. These Schools usually run for five weeks, Monday–Friday, with each day starting at 08.30 hours with lectures from 08.30 to 10.30 hours and 14.00 to 16.00 hours, followed by problem-solving sessions from 10.30 to 13.00 hours and 16.00 to 18.00 hours with short coffee breaks at 10.30 hours and 16.00 hours. The participants are selected from all over India, M.Sc. to Ph.D. graduates, young college and university teachers and others who wish to pursue a research or teaching career.

The Department of Science and Technology, Government of India (DST), was kind enough to finance these SERC Schools as well as other programs at CMSS. Under the various programs, which DST, Delhi, has supported, CMSS has trained students, faculty and others at various levels, mainly focussing on quality education and training. CMSS would like to thank Dr. B.D. Acharya (now retired), Dr. H.K.N. Trivedi (now retired), Dr. P.K. Malhotra and Dr. Ashok K. Singh of the DST, Delhi, who took the initiative to release the necessary funds in time to CMSS so that the programs could be successfully completed as scheduled, and DST, Delhi, for the financial assistance for the SERC Schools and for other programs of CMSS.

CMSS Peechi

March 2017

A.M. Mathai

Director

Chapter 1
Essentials of Fractional Calculus

1.1 Introduction

In recent decades, the field of fractional calculus has attracted interest of researchers in several areas including mathematics, physics, chemistry, engineering, and even finance and social sciences.

In these lectures, we revisit the fundamentals of fractional calculus in the framework of the most simple time-dependent processes. These processes concern the basic phenomena of relaxation, diffusion, oscillations, and wave propagation which are of great relevance in physics; from a mathematical point of view, they are known to be governed by simple differential equations of orders 1 and 2 in time. The introduction of derivatives of non-integer order leads to processes that, in mathematical physics, we may refer to as fractional phenomena.

The objective of these lectures is to provide a general description of such phenomena adopting a mathematical approach to fractional calculus that is as simple as possible. The analysis carried out by the Laplace, Fourier, and Mellin transforms leads to special functions in one variable, which generalize in a straightforward way the characteristic functions of the basic phenomena, namely the exponential and the Gaussian functions. These special functions are shown to be of the Mittag-Leffler and Wright type, or, more generally, of the Fox type. However, for the sake of simplicity, we refrain to be too general in order to capture the interest of applied scientists.

The treatment mainly reflects the research activity and style of the author. The contents of the lecture notes are as follows.

In Section 1.2, we provide a list of preliminary notions and notations for the integral transforms and for the Eulerian functions. Section 1.3 and Section 1.4 are devoted to Mittag-Leffler and Wright functions, respectively. Here we provide the necessary formulas that will be used later.

This chapter is based on the lectures by Professor Francesco Mainardi of the University of Bologna, Italy.

© Springer International Publishing AG 2017
A.M. Mathai and H.J. Haubold, *Fractional and Multivariable Calculus*,
Springer Optimization and Its Applications 122, DOI 10.1007/978-3-319-59993-9_1

   In Section 1.5, we recall the essentials of fractional calculus limiting our analysis on the Riemann–Liouville approach for well-behaved functions in $\mathbb{R}^+$. We devote particular attention to the regularized notion of fractional derivative known nowadays as Caputo derivative, that is more useful for applications.

   Section 1.6 is devoted to phenomena of relaxation–oscillation type governed fractional differential equations of order less than 2 in time. Here we show the fundamental role of the Mittag-Leffler functions.

   Section 1.7 is devoted to phenomena of diffusion-wave type governed by partial differential equations of the second order in space and fractional order less than 2. Here we show the fundamental role of two auxiliary functions of the Wright type.

   Problems are suggested inside the text. Only seldom does the text explicitly give references to the literature: The references are reported in the extended Bibliography.

1.2   Notions and Notations

Integral Transforms

Since in what follows we shall meet only real- or complex-valued functions of a real variable that are defined and continuous in a given open interval $\mathcal{I} = (a, b)$, $-\infty \leq a < b \leq +\infty$, except, possibly, at isolated points where these functions can be infinite, we restrict our presentation of the integral transforms to the class of functions for which the Riemann improper integral on $\mathcal{I}$ absolutely converges. In so doing we follow Marichev [113], and we denote this class by $L^c(\mathcal{I})$ or $L^c(a, b)$.

The Fourier transform. Let

$$\widehat{f}(\kappa) = \mathcal{F}\{f(x); \kappa\} = \int_{-\infty}^{+\infty} e^{+i\kappa x} f(x)\, dx\,, \quad \kappa \in \mathbb{R}\,, \tag{1.2.1}$$

be the Fourier transform of a function $f(x) \in L^c(\mathbb{R})$, and let

$$f(x) = \mathcal{F}^{-1}\{\widehat{f}(\kappa); x\} = \frac{1}{2\pi} \int_{-\infty}^{+\infty} e^{-i\kappa x}\, \widehat{f}(\kappa)\, d\kappa\,, \quad x \in \mathbb{R}\,, \tag{1.2.2}$$

be the inverse Fourier transform.[1]

The Laplace transform. Let

$$\widetilde{f}(s) = \mathcal{L}\{f(t); s\} = \int_0^\infty e^{-st} f(t)\, dt\,, \quad \Re(s) > a_f\,, \tag{1.2.3}$$

be the Laplace transform of a function $f(t) \in \mathcal{L}^c(0, T)$, $\forall T > 0$ and let

---

[1]If $f(x)$ is piecewise differentiable, then the formula (1.2.2) holds true at all points where $f(x)$ is continuous and the integral in it must be understood in the sense of the Cauchy principal value.

$$f(t) = \mathcal{L}^{-1}\{\tilde{f}(s); t\} = \frac{1}{2\pi i} \int_{\gamma-i\infty}^{\gamma+i\infty} e^{st}\,\tilde{f}(s)\,ds\,, \quad \Re(s) = \gamma > a_f\,, \quad (1.2.4)$$

with $t > 0$, be the inverse Laplace transform[2]

The Mellin transform. Let

$$\mathcal{M}\{f(r); s\} = f^*(s) = \int_0^{+\infty} f(r)\,r^{s-1}\,dr\,, \quad \gamma_1 < \Re(s) < \gamma_2 \qquad (1.2.5)$$

be the Mellin transform of a sufficiently well-behaved function $f(r)$, and let

$$\mathcal{M}^{-1}\{f^*(s); r\} = f(r) = \frac{1}{2\pi i} \int_{\gamma-i\infty}^{\gamma+i\infty} f^*(s)\,r^{-s}\,ds \qquad (1.2.6)$$

be the inverse Mellin transform,[3] where $r > 0$, $\gamma = \Re(s)$, $\gamma_1 < \gamma < \gamma_2$.

The Eulerian Functions

Here we consider the so-called Eulerian functions (called after Euler), namely the Gamma and the Beta functions. We recall the main representations and properties of these functions.

The Gamma function. We take as its definition the *integral formula*

$$\Gamma(z) := \int_0^\infty u^{z-1}\,e^{-u}\,du\,, \quad \Re(z) > 0\,. \qquad (1.2.7)$$

This integral representation is the most common for $\Gamma(z)$, even if it is valid only in the right half plane of $\mathbb{C}$. The analytic continuation to the left half plane is provided by.

$$\Gamma(z) = \sum_{n=0}^\infty \frac{(-1)^n}{n!(z+n)} + \int_1^\infty e^{-u}\,u^{z-1}\,du\,, \quad z \in D_\Gamma\,. \qquad (1.2.8)$$

---

[2] A sufficient condition of the existence of the Laplace transform is that the original function is of exponential order as $t \to \infty$. This means that some constant a_f exists such that the product $e^{-a_f t}|f(t)|$ is bounded for all t greater than some T. Then $\tilde{f}(s)$ exists and is analytic in the half plane $\Re(s) > a_f$. If $f(t)$ is piecewise differentiable, then the formula (1.2.4) holds true at all points where $f(t)$ is continuous and the (complex) integral in it must be understood in the sense of the Cauchy principal value.

[3] For the existence of the Mellin transform and the validity of the inversion formula, we need to recall the following theorems *TM1, TM2* adapted from Marichev's [113] treatise, *TM1* Let $f(r) \in L^c(\epsilon, E)$, $0 < \epsilon < E < \infty$, be continuous in the intervals $(0, \epsilon]$, $[E, \infty)$, and let $|f(r)| \le M r^{-\gamma_1}$ for $0 < r < \epsilon$, $|f(r)| \le M r^{-\gamma_2}$ for $r > E$, where M is a constant. Then for the existence of a strip in the s-plane in which $f(r)\,r^{s-1}$ belongs to $L^c(0, \infty)$, it is sufficient that $\gamma_1 < \gamma_2$. When this condition holds, the Mellin transform $f^*(s)$ exists and is analytic in the vertical strip $\gamma_1 < \gamma = \Re(s) < \gamma_2$. *TM2* If $f(t)$ is piecewise differentiable, and $f(r)\,r^{\gamma-1} \in L^c(0, \infty)$, then the formula (1.2.6) holds true at all points where $f(r)$ is continuous and the (complex) integral in it must be understood in the sense of the Cauchy principal value.

so the *domain of analyticity* D_Γ of Γ turns out to be

$$D_\Gamma = \mathbb{C} - \{0, -1, -2, \ldots, \}. \tag{1.2.9}$$

The more common formulae are

$$\Gamma(z+1) = z\,\Gamma(z), \tag{1.2.10}$$

$$\Gamma(z+n) = z\,(z+1)\,\ldots\,(z+n-1)\,\Gamma(z), \quad n \in \mathbb{N}. \tag{1.2.11}$$

$$\Gamma(n+1) = n!, \quad n = 0, 1, 2, \ldots . \tag{1.2.12}$$

$$\Gamma(z) = 2\int_0^\infty e^{-v^2}\,v^{2z-1}dv, \quad \Re(z) > 0, \tag{1.2.13}$$

$$\Gamma\left(\frac{1}{2}\right) = \int_{-\infty}^{+\infty} e^{-v^2}\,dv = \sqrt{\pi} \approx 1.77245, \tag{1.2.14}$$

$$\Gamma\left(n+\frac{1}{2}\right) = \int_{-\infty}^{+\infty} e^{-v^2}\,v^{2n}\,dv = \Gamma\left(\frac{1}{2}\right)\frac{(2n-1)!!}{2^n} = \sqrt{\pi}\,\frac{(2n)!}{2^{2n}\,n!}. \tag{1.2.15}$$

$$\Gamma(z)\,\Gamma(1-z) = \frac{\pi}{\sin \pi z}. \tag{1.2.16}$$

$$\Gamma(nz) = (2\pi)^{(1-n)/2}\,n^{nz-1/2}\prod_{k=0}^{n-1}\Gamma(z+\frac{k}{n}), \quad n = 2, 3, \ldots, \tag{1.2.17}$$

which reduces, for $n = 2$, to *Legendre's Duplication Formula*

$$\Gamma(2z) = \frac{1}{\sqrt{2\pi}}\,2^{2z-1/2}\,\Gamma(z)\,\Gamma(z+\frac{1}{2}), \tag{1.2.18}$$

$$(z)_n := z\,(z+1)\,(z+2)\ldots(z+n-1) = \frac{\Gamma(z+n)}{\Gamma(z)}, \quad n \in \mathbb{N}. \tag{1.2.19}$$

$$\frac{1}{\Gamma(z)} = \frac{1}{2\pi i}\int_{Ha_-}\frac{e^\zeta}{\zeta^z}\,d\zeta, \quad z \in \mathbb{C}, \tag{1.2.20}$$

where Ha_- denotes the Hankel path defined as a contour that begins at $\zeta = -\infty - ia$ ($a > 0$), encircles the branch cut that lies along the negative real axis, and ends up at $\zeta = -\infty + ib$ ($b > 0$). Of course, the branch cut is present when z is non-integer because ζ^{-z} is a multi-valued function. When z is an integer, the contour can be taken to be simply a circle around the origin, described in the counterclockwise direction.

If a, b denote two positive constants, we have as $z \to \infty$ with $|\arg z| < \pi$

$$\Gamma(z + a) \simeq \sqrt{2\pi}\, e^{-z}\, (z)^{z+a-1/2}\,, \quad \frac{\Gamma(z+a)}{\Gamma(z+b)} \simeq z^{a-b}\,. \tag{1.2.21}$$

We conclude with two relevant formulas involving the Gamma function in two basic Laplace and Mellin transform pairs:

$$\mathcal{L}\{t^{\alpha}; s\} = \frac{\Gamma(\alpha+1)}{s^{\alpha+1}}\,, \quad \Re(\alpha) > -1\,, \ \Re(s) > 0\,, \tag{1.2.22}$$

$$\mathcal{M}\{e^{-r}; s\} = \Gamma(s)\,, \quad \Re(s) > 0\,. \tag{1.2.23}$$

The students are invited to derive the more common formulas listed above during the **Problem Session**.

1.3 The Mittag-Leffler Functions

In this section, we provide a survey of the Mittag-Leffler functions $E_\alpha(z)$, $E_{\alpha,\beta}(z)$, usually referred to as the standard and the generalized Mittag-Leffler functions, respectively. Both functions are known to play fundamental roles in various applications of the fractional calculus.

The Mittag-Leffler functions are so named after the great Swedish mathematician Gosta Mittag-Leffler, who introduced and investigated them at the beginning of the 20th century in a sequence of notes.

Series Representations The standard Mittag-Leffler function, denoted as $E_\alpha(z)$ with $\alpha > 0$ is defined by its power series, which converges in the whole complex plane,

$$E_\alpha(z) := \sum_{n=0}^{\infty} \frac{z^n}{\Gamma(\alpha n + 1)}\,, \quad \alpha > 0\,, \ z \in \mathbb{C}\,. \tag{1.3.1}$$

$E_\alpha(z)$ turns out to be an *entire function* of order $\rho = 1/\alpha$ and type 1. This property is still valid but with $\rho = 1/\Re(\alpha)$, if $\alpha \in \mathbb{C}$ with *positive real part*, as formerly noted by Mittag-Leffler himself.

The Mittag-Leffler function provides a simple generalization of the exponential function to which it reduces for $\alpha = 1$. Other particular cases of (1.3.1) from which elementary functions are recovered are:

$$E_2\left(+z^2\right) = \cosh z\,, \quad E_2\left(-z^2\right) = \cos z\,, \quad z \in \mathbb{C}\,, \tag{1.3.2}$$

and

$$E_{1/2}(\pm z^{1/2}) = e^z \left[1 + \mathrm{erf}\left(\pm z^{1/2}\right)\right] = e^z\, \mathrm{erfc}\left(\mp z^{1/2}\right)\,, \quad z \in \mathbb{C}\,, \tag{1.3.3}$$

where erf (erfc) denotes the (complementary) error function defined as

$$\text{erf} (z) := \frac{2}{\sqrt{\pi}} \int_0^z e^{-u^2} du \,, \quad \text{erfc} (z) := 1 - \text{erf} (z) \,, \quad z \in \mathbb{C} \,. \tag{1.3.4}$$

In (1.3.4) by $z^{1/2}$, we mean the principal value of the square root of z in the complex plane cut along the negative real semi-axis. With this choice, $\pm z^{1/2}$ turns out to be positive/negative for $z \in \mathbb{R}^+$.

A straightforward generalization of the Mittag-Leffler function is obtained by replacing the additive constant 1 in the argument of the Gamma function in (1.3.1) by an arbitrary complex parameter β. For the generalized Mittag-Leffler function, we agree to use the notation

$$E_{\alpha,\beta}(z) := \sum_{n=0}^{\infty} \frac{z^n}{\Gamma(\alpha n + \beta)} \,, \quad \alpha > 0, \ \beta \in \mathbb{C}, \ z \in \mathbb{C} \,. \tag{1.3.5}$$

Particular simple cases are:

$$E_{1,2}(z) = \frac{e^z - 1}{z} \,, \quad E_{2,2}(z) = \frac{\sinh (z^{1/2})}{z^{1/2}} \,. \tag{1.3.6}$$

We note that $E_{\alpha,\beta}(z)$ is still an entire function of order $\rho = 1/\alpha$ and type 1.

Integral Representations and Asymptotic Expansions Many of the important properties of $E_\alpha(z)$ follow from Mittag-Leffler's *integral representation*

$$E_\alpha(z) = \frac{1}{2\pi i} \int_{Ha} \frac{\zeta^{\alpha-1} e^\zeta}{\zeta^\alpha - z} d\zeta \,, \quad \alpha > 0, \ z \in \mathbb{C} \,, \tag{1.3.7}$$

where the path of integration Ha (the *Hankel path*) is a loop which starts and ends at $-\infty$ and encircles the circular disk $|\zeta| \leq |z|^{1/\alpha}$ in the positive sense: $-\pi \leq \arg \zeta \leq \pi$ on Ha. To prove (1.3.7), expand the integrand in negative powers of ζ, integrate term-by-term, and use Hankel's integral (1.2.20) for the reciprocal of the Gamma function.

The integrand in (1.3.7) has a branch point at $\zeta = 0$. The complex ζ-plane is cut along the negative real semi-axis, and in the cut plane the integrand is single-valued: the principal branch of ζ^α is taken in the cut plane. The integrand has poles at the points $\zeta_m = z^{1/\alpha} e^{2\pi i m/\alpha}$, m integer, but only those of the poles lie in the cut plane for which $-\alpha \pi < \arg z + 2\pi m < \alpha \pi$. Thus, the number of the poles inside Ha is either $[\alpha]$ or $[\alpha + 1]$, according to the value of arg z.

The most interesting properties of the Mittag-Leffler function are associated with its asymptotic developments as $z \to \infty$ in various sectors of the complex plane. These properties can be summarized as follows: For the case $0 < \alpha < 2$, we have

$$E_\alpha(z) \sim \frac{1}{\alpha} \exp\left(z^{1/\alpha}\right) - \sum_{k=1}^{\infty} \frac{z^{-k}}{\Gamma(1 - \alpha k)}, \quad |z| \to \infty, \quad |\arg z| < \alpha \pi/2,$$

$$(1.3.8a)$$

$$E_\alpha(z) \sim -\sum_{k=1}^{\infty} \frac{z^{-k}}{\Gamma(1 - \alpha k)}, \quad |z| \to \infty, \quad \alpha \pi/2 < \arg z < 2\pi - \alpha \pi/2.$$

$$(1.3.8b)$$

For the case $\alpha \geq 2$, we have

$$E_\alpha(z) \sim \frac{1}{\alpha} \sum_m \exp\left(z^{1/\alpha} e^{2\pi i m/\alpha}\right) - \sum_{k=1}^{\infty} \frac{z^{-k}}{\Gamma(1 - \alpha k)}, \quad |z| \to \infty, \quad (1.3.9)$$

where m takes all integer values such that $-\alpha\pi/2 < \arg z + 2\pi m < \alpha\pi/2$, and $\arg z$ can assume any value from $-\pi$ to $+\pi$. From the asymptotic properties (1.3.8) and (1.3.9) and the definition of the order of an entire function, we infer that the Mittag-Leffler function is an *entire function of order* $1/\alpha$ for $\alpha > 0$; in a certain sense, each $E_\alpha(z)$ is the simplest entire function of its order.

Finally, the integral representation for the generalized Mittag-Leffler function reads

$$E_{\alpha,\beta}(z) = \frac{1}{2\pi i} \int_{Ha} \frac{\zeta^{\alpha-\beta} e^\zeta}{\zeta^\alpha - z} d\zeta, \quad \alpha > 0, \quad \beta \in \mathbb{C}, \quad z \in \mathbb{C}. \quad (1.3.10)$$

The Laplace transform pairs related to the Mittag-Leffler functions The Mittag-Leffler functions are connected to the Laplace integral through the equation,

$$\int_0^\infty e^{-u} E_\alpha(u^\alpha z) \, du = \frac{1}{1 - z}, \quad \alpha > 0. \quad (1.3.11)$$

The integral at the L.H.S. was evaluated by Mittag-Leffler who showed that the region of its convergence contains the unit circle and is bounded by the line $\Re(z^{1/\alpha}) = 1$. Putting in (1.3.11) $u = st$ and $u^\alpha z = -a t^\alpha$ with $t \geq 0$ and $a \in \mathbb{C}$, and using the sign $\overset{\mathcal{L}}{\leftrightarrow}$ for the juxtaposition of a function depending on t with its Laplace transform depending on s, we get the following Laplace transform pairs

$$E_\alpha(-a t^\alpha) \overset{\mathcal{L}}{\leftrightarrow} \frac{s^{\alpha-1}}{s^\alpha + a}, \quad \Re(s) > |a|^{1/\alpha}. \quad (1.3.12)$$

More generally one can show

$$\int_0^\infty e^{-u} u^{\beta-1} E_{\alpha,\beta}(u^\alpha z) \, du = \frac{1}{1 - z}, \quad \alpha, \beta > 0, \quad (1.3.13)$$

and

$$t^{\beta-1} E_{\alpha,\beta}(a\,t^{\alpha}) \overset{\mathcal{L}}{\leftrightarrow} \frac{s^{\alpha-\beta}}{s^{\alpha}-a}, \quad \Re(s) > |a|^{1/\alpha}. \tag{1.3.14}$$

We note that the results (1.3.12) and (1.3.14) can be used to obtain a number of functional relations satisfied by $E_{\alpha}(z)$ and $E_{\alpha,\beta}(z)$. The students are invited to produce plots of the Mittag-Leffler functions during the Problem Session.

1.4   The Wright Functions

The *Wright* function, that we denote by $\Phi_{\lambda,\mu}(z)$, $z \in \mathbb{C}$, with the parameters $\lambda > -1$ and $\mu \in \mathbb{C}$, is so named after the British mathematician E. Maitland Wright, who introduced and investigated it between 1933 and 1940. We note that originally Wright considered such a function restricted to $\lambda \geq 0$ in connection with his investigations in the asymptotic theory of partitions. Only later, in 1940, he extended to $-1 < \lambda < 0$.

Like for the Mittag-Leffler functions, a description of the most important properties of the Wright functions (with relevant references up to the fifties) can be found in the third volume of the Bateman Project in the chapter *XVIII* on the *Miscellaneous Functions*. However, probably for a misprint, there λ is restricted to be positive.

The special cases $\lambda = -\nu$, $\mu = 0$ and $\lambda = -\nu$, $\mu = 1 - \nu$ with $0 < \nu < 1$ and z replaced by $-z$ provide the Wright type functions, $F_{\nu}(z)$ and $M_{\nu}(z)$, respectively, that have been so denoted and investigated by Mainardi starting since 1993. We refer to them as the *auxiliary functions of the Wright type*.

The series representation of the Wright function The Wright function is defined by the power series convergent in the whole complex plane,

$$\Phi_{\lambda,\mu}(z) := \sum_{n=0}^{\infty} \frac{z^{n}}{n!\,\Gamma(\lambda n + \mu)}, \quad \lambda > -1, \quad \mu \in \mathbb{C}, \quad z \in \mathbb{C}. \tag{1.4.1}$$

The case $\lambda = 0$ is trivial since the Wright function is reduced to the exponential function with the constant factor $1/\Gamma(\mu)$, which turns out to vanish identically for $\mu = -n$, $n = 0, 1, \ldots$ In general, it is proved that the Wright function for $\lambda > -1$ and $\mu \in \mathbb{C}$ ($\mu \neq -n$, $n = 0, 1, \ldots$ if $\lambda = 0$) is an entire function of finite order ρ and type σ given by,

$$\rho = \frac{1}{1+\lambda}, \quad \sigma = (1+\lambda)\,|\lambda|^{\lambda/(1+\lambda)}. \tag{1.4.2}$$

In particular, the Wright function turns out to be of *exponential type* if $\lambda \geq 0$.

The Wright integral representation and asymptotic expansions The integral representation of the Wright function reads

$$\Phi_{\lambda,\mu}(z) = \frac{1}{2\pi i} \int_{Ha} e^{\zeta + z\zeta^{-\lambda}} \frac{d\zeta}{\zeta^\mu}, \quad \lambda > -1, \quad \mu \in \mathbb{C}, \quad z \in \mathbb{C}. \quad (1.4.3)$$

Here Ha denotes an arbitrary Hankel path, namely a contour consisting of pieces of the two rays $\arg\zeta = \pm\phi$ extending to infinity, and of the circular arc $\zeta = \epsilon\, e^{i\theta}$, $|\theta| \le \phi$, with $\phi \in (\pi/2, \pi)$, and $\epsilon > 0$, arbitrary. The identity between the integral and series representations is obtained by using the Hankel representation of the reciprocal of the gamma function.

The complete picture of the asymptotic behavior of the Wright function for large values of z was given by Wright himself by using the method of steepest descent on the integral representation (1.4.3). Wright's results have been summarized by Gorenflo, Luchko, and Mainardi [56, 58]. Furthermore, Wong & Zhao [191, 192] have provided a detailed asymptotic analysis of the Wright function in the cases $\lambda > 0$ and $-1 < \lambda < 0$, respectively, achieving a uniform "exponentially improved" expansion with a smooth transition across the Stokes lines. The asymptotics of zeros of the Wright function has been investigated by Luchko [85].

Here we limit ourselves to recall Wright's result in the case $\lambda = -\nu \in (-1, 0)$, $\mu > 0$ where the following asymptotic expansion is valid in a suitable sector about the negative real axis

$$\Phi_{-\nu,\mu}(z) = Y^{1/2-\mu}\, e^{-Y} \left(\sum_{m=0}^{M-1} A_m\, Y^{-m} + O(|Y|^{-M}) \right), \quad |z| \to \infty, \quad (1.4.4)$$

with $Y = (1 - \nu)\, (-\nu^\nu z)^{1/(1-\nu)}$, where the A_m are certain real numbers.

The Wright functions as generalization of the Bessel functions For $\lambda = 1$ and $\mu = \nu + 1$, the Wright function turns out to be related to the well-known Bessel functions J_ν and I_ν

$$(z/2)^\nu\, \Phi_{1,\nu+1}\left(\mp z^2/4\right) = \begin{cases} J_\nu(z) \\ I_\nu(z) \end{cases}. \quad (1.4.5)$$

In view of this property, some authors refer to the Wright function as the *Wright generalized Bessel function* (misnamed also as the *Bessel–Maitland function*) and introduce the notation

$$J_\nu^{(\lambda)}(z) := \left(\frac{z}{2}\right)^\nu \sum_{n=0}^{\infty} \frac{(-1)^n (z/2)^{2n}}{n!\, \Gamma(\lambda n + \nu + 1)}; \quad J_\nu^{(1)}(z) := J_\nu(z). \quad (1.4.6)$$

As a matter of fact, the Wright function appears as the natural generalization of the entire function known as *Bessel–Clifford function*, see, e.g., Kiryakova [78] and referred by Tricomi, as the *uniform Bessel function*

$$T_\nu(z) := z^{-\nu/2} \, J_\nu(2\sqrt{z}) = \sum_{n=0}^{\infty} \frac{(-1)^n z^n}{n! \, \Gamma(n + \nu + 1)} = \Phi_{1,\nu+1}(-z) \,. \qquad (1.4.7)$$

Some of the properties which the Wright functions share with the popular Bessel functions were enumerated by Wright himself. Hereafter, we quote two relevant relations from the Bateman Project which can easily be derived from (1.4.1) or (1.4.3).

$$\lambda z \, \Phi_{\lambda,\lambda+\mu}(z) = \Phi_{\lambda,\mu-1}(z) + (1 - \mu) \, \Phi_{\lambda,\mu}(z) \,, \qquad (1.4.8)$$

$$\frac{d}{dz} \, \Phi_{\lambda,\mu}(z) = \Phi_{\lambda,\lambda+\mu}(z) \,. \qquad (1.4.9)$$

The auxiliary functions of the Wright type In our treatment of the time fractional diffusion-wave equation, we find it convenient to introduce two *auxiliary functions* $F_\nu(z)$ and $M_\nu(z)$, where z is a complex variable and ν a real parameter $0 < \nu < 1$. Both functions turn out to be analytic in the whole complex plane, i.e., they are entire functions. Their respective *integral representations* read,

$$F_\nu(z) := \frac{1}{2\pi i} \int_{Ha} e^{\zeta - z\zeta^\nu} \, d\zeta \,, \quad 0 < \nu < 1 \,, \quad z \in \mathbb{C} \,, \qquad (1.4.10)$$

$$M_\nu(z) := \frac{1}{2\pi i} \int_{Ha} e^{\zeta - z\zeta^\nu} \, \frac{d\zeta}{\zeta^{1-\nu}} \,, \quad 0 < \nu < 1 \,, \quad z \in \mathbb{C} \,. \qquad (1.4.11)$$

From a comparison of (1.4.10)–(1.4.11) with (1.4.3), we easily recognize that

$$F_\nu(z) = \Phi_{-\nu,0}(-z) \,, \qquad (1.4.12)$$

and

$$M_\nu(z) = \Phi_{-\nu,1-\nu}(-z) \,. \qquad (1.4.13)$$

From (1.4.8) and (1.4.12)–(1.4.13), we find the relation

$$F_\nu(z) = \nu \, z \, M_\nu(z) \,. \qquad (1.4.14)$$

This relation can be obtained directly from (1.4.10)–(1.4.11) via an integration by parts also. The *series representations* for our auxiliary functions turn out to be respectively,

$$F_\nu(z) := \sum_{n=1}^{\infty} \frac{(-z)^n}{n! \, \Gamma(-\beta n)} = -\frac{1}{\pi} \sum_{n=1}^{\infty} \frac{(-z)^n}{n!} \, \Gamma(\nu n + 1) \, \sin(\pi \nu n) \,, \qquad (1.4.15)$$

$$M_\nu(z) := \sum_{n=0}^{\infty} \frac{(-z)^n}{n! \, \Gamma[-\nu n + (1-\nu)]} = \frac{1}{\pi} \sum_{n=1}^{\infty} \frac{(-z)^{n-1}}{(n-1)!} \Gamma(\nu n) \sin(\pi \nu n) \,.$$

(1.4.16)

The series at the R.H.S. have been obtained by using the well-known reflection formula for the Gamma function (1.2.16). Furthermore, we note that $F_\nu(0) = 0$, $M_\nu(0) = 1/\Gamma(1-\nu)$ and that the relation (1.4.14) can be derived also from (1.4.15)–(1.4.16).

Explicit expressions of $F_\nu(z)$ and $M_\nu(z)$ in terms of known functions are expected for some particular values of ν. Mainardi & Tomirotti [100] have shown that for $\nu = 1/q$, with an integer $q \geq 2$, the auxiliary function $M_\nu(z)$ can be expressed as a sum of $(q-1)$ simpler entire functions, namely,

$$M_{1/q}(z) = \frac{1}{\pi} \sum_{h=1}^{q-1} c(h,q) \, G(z;h,q)$$

(1.4.17)

with

$$c(h,q) = (-1)^{h-1} \, \Gamma(h/q) \, \sin(\pi h/q) \,,$$

(1.4.18)

$$G(z;h,q) = \sum_{m=0}^{\infty} (-1)^{m(q+1)} \left(\frac{h}{q}\right)_m \frac{z^{qm+h-1}}{(qm+h-1)!} \,.$$

(1.4.19)

Here $(a)_m$, $m = 0, 1, 2, \ldots$ denotes Pochhammer's symbol

$$(a)_m := \frac{\Gamma(a+m)}{\Gamma(a)} = a(a+1)\ldots(a+m-1) \,.$$

We note that $(-1)^{m(q+1)}$ is equal to $(-1)^m$ for q even and $+1$ for q odd. In the particular cases $q = 2$, $q = 3$, we find, respectively

$$M_{1/2}(z) = \frac{1}{\sqrt{\pi}} \exp\left(-z^2/4\right) \,,$$

(1.4.20)

$$M_{1/3}(z) = 3^{2/3} \, \mathrm{Ai}\left(z/3^{1/3}\right) \,,$$

(1.4.21)

where Ai denotes the *Airy function*.

Furthermore, it can be proved that $M_{1/q}(z)$ (for integer ≥ 2) satisfies the differential equation of order $q-1$,

$$\frac{d^{q-1}}{dz^{q-1}} M_{1/q}(z) + \frac{(-1)^q}{q} z \, M_{1/q}(z) = 0 \,,$$

(1.4.22)

subjected to the $q-1$ initial conditions at $z = 0$, derived from the series expansion in (1.4.17)–(1.4.19),

$$M_{1/q}^{(h)}(0) = \frac{(-1)^h}{\pi} \Gamma[(h+1)/q] \sin[\pi (h+1)/q], \quad h = 0, 1, \ldots q - 2.$$
$$(1.4.23)$$

We note that, for $q \geq 4$, Eq. (1.4.22) is akin to the *hyper-Airy* differential equation of order $q - 1$, see, e.g., Bender & Orszag [8]. In the limiting case $\nu = 1$, we get $M_1(z) = \delta(z - 1)$, i.e., the M function degenerates into a generalized function of Dirac type.

From our purposes (time-fractional diffusion processes), it is relevant to consider the M_ν function for a positive (real) argument, that will be denoted by r. Later, by using its Laplace transform with the Bernstein theorem, we shall prove that $M_\nu(r) > 0$ for $r > 0$.

We note that for $0 < \nu \leq 1/2$, the function is monotonically decreasing, while for $1/2 < \nu < 1$ it exhibits a maximum whose position tends to $r = 1$ as $\nu \to 1^-$ (consistently with $M_1(r) = \delta(r - 1)$). The students are invited to produce plots of the auxiliary functions i the Problem Session.

The asymptotic representation of $M_\nu(r)$, as $r \to \infty$, can be obtained by using the ordinary saddle-point method. Choosing as a variable r/ν, rather than r, the computation is easier and yields, see Mainardi & Tomirotti [100]:

$$M_\nu(r/\nu) \sim a(\nu) \, r^{(\nu-1/2)/(1-\nu)} \exp\left[-b(\nu) \, r^{(1/(1-\nu))}\right], \quad r \to +\infty, \quad (1.4.24)$$

where $a(\nu) = 1/\sqrt{2\pi (1 - \nu)} > 0$, and $b(\nu) = (1 - \nu)/\nu > 0$.

The above asymptotic representation is consistent with the first term of the asymptotic expansion (1.4.4) obtained by Wright for $\Phi_{-\nu,\mu}(-r)$. In fact, taking $\mu = 1 - \nu$ so $1/2 - \mu = \nu - 1/2$, we obtain

$$M_\nu(r) \sim A_0 \, Y^{\nu-1/2} \exp(-Y), \quad r \to \infty, \quad (1.4.25)$$

where

$$A_0 = \frac{1}{\sqrt{2\pi} \, (1 - \nu)^\nu \, \nu^{2\nu-1}}, \quad Y = (1 - \nu) \, (\nu^\nu \, r)^{1/(1-\nu)}. \quad (1.4.26)$$

Because of the above exponential decay, any moment of order $\delta > -1$ for $M_\nu(r)$ is finite. In fact,

$$\int_0^\infty r^\delta \, M_\nu(r) \, dr = \frac{\Gamma(\delta + 1)}{\Gamma(\nu\delta + 1)}, \quad \delta > -1, \quad 0 < \nu < 1. \quad (1.4.27)$$

In particular we get the normalization property in $\mathbb{R}^+$, $\int_0^\infty M_\nu(r) \, dr = 1$.

Similarly, we can compute any moment of order $\delta > -1$ of $\Phi_{-\nu,\mu}(-r)$ in view of its exponential decay (1.4.4), obtaining

$$\int_0^\infty r^\delta \, \Phi_{-\nu,\mu}(-r) \, dr = \frac{\Gamma(\delta+1)}{\Gamma(\nu\delta+\nu+\mu)}, \quad \delta > -1, \quad 0 < \nu < 1, \quad \mu > 0.$$

$$(1.4.28)$$

We also quote an interesting formula derived by Stankovič [179], which provides a relation between the Whittaker function $W_{-1/2,1/6}$ and the Wright function $\Phi_{-2/3,0} = F_{2/3}$,

$$F_{2/3}\left(x^{-2/3}\right) = -\frac{1}{2\sqrt{3\pi}} \exp\left(-\frac{2}{27x^2}\right) W_{-1/2,1/6}\left(-\frac{4}{27x}\right). \qquad (1.4.29)$$

We recall that the generic Whittaker function $W_{\lambda,\mu}(x)$ satisfies the differential equation, see, e.g., Abramowitz & Stegun [1]

$$\frac{d^2}{dx^2} W_{\lambda,\mu}(x) + \left(-\frac{1}{4} + \frac{\lambda}{x} + \frac{\mu^2}{4x^2}\right) W_{\lambda,\mu}(x) = 0, \quad \lambda, \mu \in \mathbb{R}. \qquad (1.4.30)$$

Laplace transform pairs related to the Wright function Let us now consider some Laplace transform pairs related to the Wright functions. We continue to denote by r a positive variable.

In the case $\lambda > 0$, the Wright function is an entire function of order less than 1 and consequently, being of exponential type, its Laplace transform can be obtained by transforming term-by-term its Taylor expansion (1.4.1) in the origin, see, e.g., Doetsch [32] As a result we get,

$$\Phi_{\lambda,\mu}(\pm r) \overset{\mathcal{L}}{\leftrightarrow} \frac{1}{s} \sum_{k=0}^\infty \frac{(\pm s^{-1})^k}{\Gamma(\lambda k + \mu)} = \frac{1}{s} E_{\lambda,\mu}(\pm s^{-1}), \quad \lambda > 0, \quad \mu \in \mathbb{C}, \qquad (1.4.31)$$

with $0 < \epsilon < |s|$, ϵ arbitrarily small. Here $E_{\alpha,\beta}(z)$ denotes the generalized Mittag-Leffler function (1.3.5). In this case, the resulting Laplace transform turns out to be analytic for $s \neq 0$, vanishing at infinity and exhibiting an essential singularity at $s = 0$.

For $-1 < \lambda < 0$, the just applied method cannot be used since then the Wright function is an entire function of order greater than one. In this case, setting $\nu = -\lambda$, the existence of the Laplace transform of the function $\Phi_{-\nu,\mu}(-t)$, $t > 0$, follows from (1.4.4), which says us that the function $\Phi_{-\nu,\mu}(z)$ is exponentially small for large z in a sector of the plane containing the negative real semi-axis. To get the transform in this case, we can use the idea given in Mainardi [93], based on the integral representation (1.4.3). We have

$$\Phi_{-\nu,\mu}(-r) \overset{\mathcal{L}}{\leftrightarrow} \frac{1}{2\pi i} \int_{Ha} \frac{e^\zeta \zeta^{-\mu}}{s + \zeta^\nu} \, d\zeta = E_{\nu,\mu+\nu}(-s), \qquad (1.4.32)$$

where we have used the integral representation (1.3.10) of the generalized Mittag-Leffler function.

The relation (1.4.32) was given in Dzherbashian [34], $\mu \geq 0$ as a representation of the generalized Mittag-Leffler function in the whole complex plane as a Laplace integral of an entire function but without identifying this function as the Wright function. They also gave (in slightly different notations) the more general representation

$$E_{\alpha_2,\beta_2}(z) = \int_0^\infty E_{\alpha_1,\beta_1}(zr^{\alpha_1}) \, r^{\beta_1-1} \, \Phi_{-\nu,\gamma}(-r) \, dr \,, \qquad (1.4.33)$$

with $0 < \alpha_2 < \alpha_1$, $\beta_1, \beta_2 > 0$, and $0 < \nu = -\alpha_2/\alpha_1 < 1$, $\gamma = \beta_2 - \beta_1 \alpha_2/\alpha_1$.

An important particular case of the Laplace transform pair (1.4.32) is given for $\mu = 1 - \nu$ to yield, see also Mainardi [93],

$$M_\nu(r) \overset{\mathcal{L}}{\leftrightarrow} E_\nu(-s), \quad 0 < \nu < 1. \qquad (1.4.34)$$

As a further particular case, we recover the well-known Laplace transform pair, see, e.g., Doetsch [32]

$$M_{1/2}(r) = \frac{1}{\sqrt{\pi}} \exp\left(-r^2/4\right) \overset{\mathcal{L}}{\leftrightarrow} E_{1/2}(-s) := \exp\left(s^2\right) \mathrm{erfc}\,(s). \qquad (1.4.35)$$

We also note that transforming term-by-term the Taylor series of $M_\nu(r)$ (not being of exponential order) yields a series of negative powers of s, which represents the asymptotic expansion of $E_\nu(-s)$ as $s \to \infty$ in a sector around the positive real semi-axis.

Using the relation

$$\int_0^\infty r^n \, f(r) \, dr = \lim_{s \to 0} (-1)^n \frac{d^n}{ds^n} \mathcal{L}\{f(r); s\},$$

the Laplace transform pair (1.4.32) and the series representation of the generalized Mittag-Leffler function (1.3.5), we can compute all the moments of integer order for the Wright function $\Phi_{-\nu,\mu}(-r)$ with $0 < \nu < 1$ in $\mathbb{R}^+$:

$$\int_0^\infty r^n \Phi_{-\nu,\mu}(-r) \, dr = \frac{n!}{\Gamma(\nu n + \mu + \nu)}, \quad n \in \mathbb{N}_0. \qquad (1.4.36)$$

This formula is consistent with the more general formula (1.4.28) valid when the moments are of arbitrary order $\delta > -1$.

We can now obtain other Laplace transform pairs related to our auxiliary functions. Indeed, following Mainardi and using the integral representations (1.4.10)–(1.4.11), we get

$$\frac{1}{r} F_\nu\left(cr^{-\nu}\right) = \frac{c\nu}{r^{\nu+1}} M_\nu\left(cr^{-\nu}\right) \overset{\mathcal{L}}{\leftrightarrow} \exp\left(-cs^\nu\right), \quad 0 < \nu < 1, \, c > 0. \qquad (1.4.37)$$

The Laplace inversion in Eq. (1.4.37) was properly carried out by Pollard [143], based on a formal result by Humbert [72] and by Mikusiński [118]. A formal series inversion was carried out by Buchen & Mainardi [11] albeit unaware of the previous results.

By applying the formula for differentiation of the image of the Laplace transform to Eq. (1.4.37), we get a Laplace transform pair useful for our further discussions, namely

$$\frac{1}{r^{\nu}} M_{\nu}(cr^{-\nu}) \overset{\mathcal{L}}{\leftrightarrow} s^{\nu-1} \exp(-cs^{\nu}), \quad 0 < \nu < 1, \quad c > 0. \tag{1.4.38}$$

As particular cases of Eqs. (1.4.37)–(1.4.38), we recover the well-known pairs, see, e.g., Doetsch [32]

$$\frac{1}{2r^{3/2}} M_{1/2}(1/r^{1/2}) = \frac{1}{2\sqrt{\pi}} r^{-3/2} \exp\left(-1/(4r^2)\right) \overset{\mathcal{L}}{\leftrightarrow} \exp\left(-s^{1/2}\right), \tag{1.4.39}$$

$$\frac{1}{r^{1/2}} M_{1/2}(1/r^{1/2}) = \frac{1}{\sqrt{\pi}} r^{-1/2} \exp\left(-1/(4r^2)\right) \overset{\mathcal{L}}{\leftrightarrow} s^{-1/2} \exp\left(-s^{1/2}\right). \tag{1.4.40}$$

More generally, using the same method as in (1.4.37), we get, see Stanković [179], the Laplace transform pair

$$r^{\mu-1} \Phi_{-\nu,\mu}(-cr^{-\nu}) \overset{\mathcal{L}}{\leftrightarrow} s^{-\mu} \exp(-cs^{\nu}), \quad 0 < \nu < 1, \quad c > 0. \tag{1.4.41}$$

The Fourier and Mellin transforms related to the Wright functions Limiting ourselves to the M-Wright function we have,

$$M_{\nu}(|x|) \overset{\mathcal{F}}{\leftrightarrow} 2E_{2\nu}(-\kappa^2), \quad 0 < \nu < 1. \tag{1.4.42}$$

$$M_{\nu}(r) \overset{\mathcal{M}}{\leftrightarrow} \frac{\Gamma(s)}{\Gamma(\nu(s-1)+1)}, \quad 0 < \nu < 1. \tag{1.4.43}$$

1.5 Essentials of Fractional Calculus

For a sufficiently well-behaved function $f(t)$ (with $t \in \mathbb{R}^+$), we may define the derivative of a positive non-integer order in two different senses, that we refer here as to *Riemann–Liouville* (R-L) derivative and *Caputo* (C) derivative, respectively. Both derivatives are related to the so-called Riemann–Liouville fractional integral. For any $\alpha > 0$, this fractional integral is defined as

$$J_t^{\alpha} f(t) := \frac{1}{\Gamma(\alpha)} \int_0^t (t-\tau)^{\alpha-1} f(\tau) \, d\tau, \tag{1.5.1}$$

where $\Gamma(\alpha) := \int_0^\infty e^{-u} u^{\alpha-1} \, du$ denotes the Gamma function. For existence of the integral, it is sufficient that the function $f(t)$ is locally integrable in $\mathbb{R}^+$ and for $t \to 0$ behaves like $O(t^{-\nu})$ with a number $\nu < \alpha$.

For completion we define $J_t^0 = I$ (Identity operator).

We recall the semigroup property,

$$J_t^\alpha J_t^\beta = J_t^\beta J_t^\alpha = J_t^{\alpha+\beta}, \quad \alpha, \beta \geq 0. \tag{1.5.2}$$

Furthermore we note that for $\alpha \geq 0$,

$$J_t^\alpha t^\gamma = \frac{\Gamma(\gamma+1)}{\Gamma(\gamma+1+\alpha)} t^{\gamma+\alpha}, \quad \gamma > -1. \tag{1.5.3}$$

The fractional derivative of order $\mu > 0$ in the *Riemann–Liouville* sense is defined as the operator D_t^μ which is the left inverse of the Riemann–Liouville integral of order μ (in analogy with the ordinary derivative), that is

$$D_t^\mu J_t^\mu = I, \quad \mu > 0. \tag{1.5.4}$$

If m denotes the positive integer such that $m - 1 < \mu \leq m$, we recognize from Eqs. (1.5.2) and (1.5.4):

$$D_t^\mu f(t) := D_t^m J_t^{m-\mu} f(t), \quad m - 1 < \mu \leq m. \tag{1.5.5}$$

In fact, using the semigroup property (1.5.2), we have

$$D_t^\mu J_t^\mu = D_t^m J_t^{m-\mu} J_t^\mu = D_t^m J_t^m = I.$$

Thus (1.5.5) implies,

$$D_t^\mu f(t) = \begin{cases} \dfrac{d^m}{dt^m} \left[\dfrac{1}{\Gamma(m-\mu)} \displaystyle\int_0^t \dfrac{f(\tau) \, d\tau}{(t-\tau)^{\mu+1-m}} \right], & m - 1 < \mu < m; \\ \dfrac{d^m}{dt^m} f(t), & \mu = m. \end{cases} \tag{1.5.5'}$$

For completion, we define $D_t^0 = I$.

On the other hand, the fractional derivative of order μ in the *Caputo* sense is defined as the operator $_*D_t^\mu$ such that

$$_*D_t^\mu f(t) := J_t^{m-\mu} D_t^m f(t), \quad m - 1 < \mu \leq m. \tag{1.5.6}$$

This implies,

$$
{}_*D_t^\mu f(t) = \begin{cases} \dfrac{1}{\Gamma(m-\mu)} \displaystyle\int_0^t \dfrac{f^{(m)}(\tau)\,d\tau}{(t-\tau)^{\mu+1-m}}\,, & m-1 < \mu < m\,; \\[4mm] \dfrac{d^m}{dt^m} f(t)\,, & \mu = m\,. \end{cases} \tag{1.5.6$'$}
$$

Thus, when the order is not integer, the two fractional derivatives differ in that the standard derivative of order m does not generally commute with the fractional integral. Of course the Caputo derivative (1.5.6$'$) needs higher regularity conditions of $f(t)$ than the Riemann–Liouville derivative (1.5.5$'$).

We point out that the *Caputo* fractional derivative satisfies the relevant property of being zero when applied to a constant, and, in general, to any power function of nonnegative integer degree less than m, if its order μ is such that $m-1 < \mu \le m$. Furthermore, we note for $\mu \ge 0$,

$$
D_t^\mu t^\gamma = \frac{\Gamma(\gamma+1)}{\Gamma(\gamma+1-\mu)}\, t^{\gamma-\mu}\,, \quad \gamma > -1\,. \tag{1.5.7}
$$

It is instructive to compare Eqs. (1.5.3), (1.5.7).

In Gorenflo & Mainardi [47], we have shown the essential relationships between the two fractional derivatives for the same non-integer order,

$$
{}_*D_t^\mu f(t) = \begin{cases} D_t^\mu \left[f(t) - \displaystyle\sum_{k=0}^{m-1} f^{(k)}(0^+)\,\dfrac{t^k}{k!} \right], & \\[4mm] D_t^\mu f(t) - \displaystyle\sum_{k=0}^{m-1} \dfrac{f^{(k)}(0^+)\,t^{k-\mu}}{\Gamma(k-\mu+1)}\,, & \end{cases} m-1 < \mu < m\,. \tag{1.5.8}
$$

In particular we have from (1.5.6$'$) and (1.5.8)

$$
\begin{aligned}
{}_*D_t^\mu f(t) &= \frac{1}{\Gamma(1-\mu)} \int_0^t \frac{f^{(1)}(\tau)}{(t-\tau)^\mu}\,d\tau \\
&= D_t^\mu \left[f(t) - f(0^+) \right] = D_t^\mu f(t) - f(0^+)\,\frac{t^{-\mu}}{\Gamma(1-\mu)}\,,
\end{aligned} \quad 0 < \mu < 1\,. \tag{1.5.9}
$$

The *Caputo* fractional derivative represents a sort of regularization in the time origin for the *Riemann–Liouville* fractional derivative. We note that for its existence, all the limiting values $f^{(k)}(0^+) := \lim_{t\to 0^+} D_t^k f(t)$ are required to be finite for $k = 0, 1, 2, \ldots, m-1$. In the special case $f^{(k)}(0^+) = 0$ for $k = 0, 1, m-1$, the two fractional derivatives coincide.

We observe the different behavior of the two fractional derivatives at the end points of the interval $(m-1, m)$ namely when the order is any positive integer, as it can be noted from their definitions (1.5.5), (1.5.6). In fact, whereas for $\mu \to m^-$ both derivatives reduce to D_t^m, as stated in Eqs. (1.5.5$'$), (1.5.6$'$), due to the fact that the operator $J_t^0 = I$ commutes with D_t^m, for $\mu \to (m-1)^+$ we have,

$$\mu \to (m-1)^+ : \begin{cases} D_t^\mu f(t) \to D_t^m J_t^1 f(t) = D_t^{(m-1)} f(t) = f^{(m-1)}(t), \\ {}_* D_t^\mu f(t) \to J_t^1 D_t^m f(t) = f^{(m-1)}(t) - f^{(m-1)}(0^+). \end{cases}$$
(1.5.10)

As a consequence, roughly speaking, we can say that D_t^μ is, with respect to its order μ, an operator continuous at any positive integer, whereas ${}_* D_t^\mu$ is an operator only left-continuous.

The above behaviors have induced us to keep for the Riemann–Liouville derivative the same symbolic notation as for the standard derivative of integer order, while for the Caputo derivative to decorate the corresponding symbol with subscript $*$.

We also note, with $m - 1 < \mu \leq m$, and c_j arbitrary constants,

$$D_t^\mu f(t) = D_t^\mu g(t) \iff f(t) = g(t) + \sum_{j=1}^m c_j t^{\mu-j},$$
(1.5.11)

$$ {}_* D_t^\mu f(t) = {}_* D_t^\mu g(t) \iff f(t) = g(t) + \sum_{j=1}^m c_j t^{m-j}.$$
(1.5.12)

Furthermore, we observe that in case of a non-integer order for both fractional derivatives, the semigroup property (of the standard derivative for integer order) does not hold for both fractional derivatives when the order is not integer.

We point out the major utility of the Caputo fractional derivative in treating initial-value problems for physical and engineering applications where initial conditions are usually expressed in terms of integer-order derivatives. This can be easily seen using the Laplace transformation. We recall that under suitable conditions, the Laplace transform of the m-derivative of $f(t)$ is given by

$$\mathcal{L}\left\{D_t^m f(t); s\right\} = s^m \widetilde{f}(s) - \sum_{k=0}^{m-1} s^{m-1-k} f^{(k)}(0^+), \quad f^{(k)}(0^+) := \lim_{t \to 0^+} D_t^k f(t).$$

For the Caputo derivative of order μ with $m - 1 < \mu \leq m$, we have,

$$\mathcal{L}\left\{{}_* D_t^\mu f(t); s\right\} = s^\mu \widetilde{f}(s) - \sum_{k=0}^{m-1} s^{\mu-1-k} f^{(k)}(0^+),$$
(1.5.13)

$$f^{(k)}(0^+) := \lim_{t \to 0^+} D_t^k f(t).$$

The corresponding rule for the Riemann–Liouville derivative of order μ is,

$$\mathcal{L}\left\{D_t^\mu f(t); s\right\} = s^\mu \tilde{f}(s) - \sum_{k=0}^{m-1} s^{m-1-k} g^{(k)}(0^+),$$

(1.5.14)

$$g^{(k)}(0^+) := \lim_{t \to 0^+} D_t^k g(t), \quad \text{where} \quad g(t) := J_t^{(m-\mu)} f(t).$$

Thus, it is more cumbersome to use the rule (1.5.14) than (1.5.13). The rule (1.5.14) requires initial values concerning an extra function $g(t)$ related to the given $f(t)$ through a fractional integral. However, when all the limiting values $f^{(k)}(0^+)$ for $k = 0, 1, 2, \ldots$ are finite and the order is not integer, we can prove that the corresponding $g^{(k)}(0^+)$ vanish so that formula (1.5.14) simplifies into,

$$\mathcal{L}\left\{D_t^\mu f(t); s\right\} = s^\mu \tilde{f}(s), \quad m - 1 < \mu < m.$$

(1.5.15)

For this proof, it is sufficient to apply the Laplace transform to the second equation, by recalling that $\mathcal{L}\{t^\alpha; s\} = \Gamma(\alpha + 1)/s^{\alpha+1}$ for $\alpha > -1$, and then to compare (1.5.13) with (1.5.14).

1.6 Fractional Differential Equations of Relaxation and Oscillation Type

Generally speaking, we consider the following differential equation of fractional order $\alpha > 0$,

$$D_*^\alpha u(t) = D^\alpha \left(u(t) - \sum_{k=0}^{m-1} \frac{t^k}{k!} u^{(k)}(0^+) \right) = -u(t) + q(t), \quad t > 0,$$

(1.6.1)

where $u = u(t)$ is the field variable and $q(t)$ is a given function. Here m is a positive integer uniquely defined by $m - 1 < \alpha \le m$, which provides the number of the prescribed initial values $u^{(k)}(0^+) = c_k$, $k = 0, 1, 2, \ldots, m - 1$. Implicit in the form of (1.6.1) is our desire to obtain solutions $u(t)$ for which the $u^{(k)}(t)$ are continuous for positive t and right-continuous at the origin $t = 0$ for $k = 0, 1, 2, \ldots, m - 1$. In particular, the cases of *fractional relaxation* and *fractional oscillation* are obtained for $0 < \alpha < 1$ and $1 < \alpha < 2$, respectively.

The application of the Laplace transform through the Caputo formula (1.5.13) yields

$$\tilde{u}(s) = \sum_{k=0}^{m-1} c_k \frac{s^{\alpha-k-1}}{s^\alpha + 1} + \frac{1}{s^\alpha + 1} \tilde{q}(s).$$

(1.6.2)

Then, using the Laplace transform pair (1.3.12) for the Mittag-Leffler function, we put for $k = 0, 1, \ldots, m - 1$,

$$u_k(t) := J^k e_\alpha(t) \div \frac{s^{\alpha-k-1}}{s^\alpha + 1}, \quad e_\alpha(t) := E_\alpha(-t^\alpha) \div \frac{s^{\alpha-1}}{s^\alpha + 1}, \qquad (1.6.3)$$

and, from inversion of the Laplace transforms in (1.6.2), we find

$$u(t) = \sum_{k=0}^{m-1} c_k u_k(t) - \int_0^t q(t - \tau) u_0'(\tau) \, d\tau. \qquad (1.6.4)$$

In particular, the above formula (1.6.4) encompasses the solutions for $\alpha = 1, 2$, since $e_1(t) = \exp(-t)$, $e_2(t) = \cos t$. When α is not integer, namely for $m - 1 < \alpha < m$, we note that $m - 1$ represents the integer part of α (usually denoted by $[\alpha]$) and m the number of initial conditions necessary and sufficient to ensure the uniqueness of the solution $u(t)$. Thus, the m functions $u_k(t) = J^k e_\alpha(t)$ with $k = 0, 1, \ldots, m - 1$ represent those particular solutions *fractional relaxation and fractional oscillation* of the *homogeneous* equation which satisfy the initial conditions $u_k^{(h)}(0^+) = \delta_{kh}$, $h, k = 0, 1, \ldots, m - 1$, and therefore they represent the *fundamental solutions* of the fractional equation (1.6.1), in analogy with the case $\alpha = m$. Furthermore, the function $u_\delta(t) = -u_0'(t) = -e_\alpha'(t)$ represents the *impulse-response solution*.

We invite the reader to derive the relevant properties of the basic functions $e_\alpha(t)$ directly from their representation as a Laplace inverse integral

$$e_\alpha(t) = \frac{1}{2\pi i} \int_{Br} e^{st} \frac{s^{\alpha-1}}{s^\alpha + 1} \, ds, \qquad (1.6.5)$$

in detail for $0 < \alpha \le 2$, without detouring on the general theory of Mittag-Leffler functions in the complex plane. In (1.6.5), Br denotes the Bromwich path, i.e., a line $\Re(s) = \sigma$ with a value $\sigma \ge 1$ and $\operatorname{Im} s$ running from $-\infty$ to $+\infty$.

For transparency reasons, we separately discuss the cases (a) $0 < \alpha < 1$ and (b) $1 < \alpha < 2$, recalling that in the limiting cases $\alpha = 1, 2$, we know $e_\alpha(t)$ as elementary function, namely $e_1(t) = e^{-t}$ and $e_2(t) = \cos t$. For α not integer the power function s^α is uniquely defined as $s^\alpha = |s|^\alpha e^{i \arg s}$, with $-\pi < \arg s < \pi$, that is in the complex s-plane cut along the negative real axis.

The essential step consists in decomposing $e_\alpha(t)$ into two parts according to $e_\alpha(t) = f_\alpha(t) + g_\alpha(t)$, as indicated below. In case (a) the function $f_\alpha(t)$, in case (b) the function $-f_\alpha(t)$ is *completely monotone*; in both cases, $f_\alpha(t)$ tends to zero as t tends to infinity, from above in case (a), from below in case (b). The other part, $g_\alpha(t)$, is identically vanishing in case (a), but of *oscillatory* character with exponentially decreasing amplitude in case (b).

For the oscillatory part, we obtain

$$g_\alpha(t) = \frac{2}{\alpha} e^t \cos(\pi/\alpha) \cos\left[t \sin\left(\frac{\pi}{\alpha}\right)\right], \quad \text{if } 1 < \alpha < 2. \qquad (1.6.6)$$

We note that this function exhibits oscillations with circular frequency $w(\alpha) = \sin(\pi/\alpha)$ and with an exponentially decaying amplitude with rate $\lambda(\alpha) = |\cos(\pi/\alpha)|$.

For the monotonic part, we obtain

$$f_\alpha(t) := \int_0^\infty e^{-rt} K_\alpha(r)\, dr\,, \tag{1.6.7}$$

with

$$K_\alpha(r) = -\frac{1}{\pi}\, \text{Im}\left\{\left.\frac{s^{\alpha-1}}{s^\alpha + 1}\right|_{s=r\,e^{i\pi}}\right\} = \frac{1}{\pi}\,\frac{r^{\alpha-1}\,\sin(\alpha\pi)}{r^{2\alpha} + 2r^\alpha\,\cos(\alpha\pi) + 1}\,. \tag{1.6.8}$$

This function $K_\alpha(r)$ vanishes identically if α is an integer, it is positive for all r if $0 < \alpha < 1$, negative for all r if $1 < \alpha < 2$. In fact in (1.6.8), the denominator is, for α not integer, always positive being $> (r^\alpha - 1)^2 \geq 0$. Hence $f_\alpha(t)$ has the aforementioned monotonicity properties, decreasing toward zero in case (a), increasing toward zero in case (b). We also note that, in order to satisfy the initial condition $e_\alpha(0^+) = 1$, we find $\int_0^\infty K_\alpha(r)\, dr = 1$ if $0 < \alpha < 1$, $\int_0^\infty K_\alpha(r)\, dr = 1 - 2/\alpha$ if $1 < \alpha < 2$.

We invite the students to exhibit the plots of $K_\alpha(r)$, that we denote as the *basic spectral function*, for some values of α in the intervals (a) $0 < \alpha < 1$, (b) $1 < \alpha < 2$.

In addition to the basic fundamental solutions, $u_0(t) = e_\alpha(t)$, we need to compute the impulse-response solutions $u_\delta(t) = -D^1 e_\alpha(t)$ for cases (a) and (b) and, only in case (b), the second fundamental solution $u_1(t) = J^1 e_\alpha(t)$.

For this purpose, we note that in general it turns out that

$$J^k f_\alpha(t) = \int_0^\infty e^{-rt} K_\alpha^k(r)\, dr\,, \tag{1.6.9}$$

with

$$K_\alpha^k(r) := (-1)^k\, r^{-k}\, K_\alpha(r) = \frac{(-1)^k}{\pi}\,\frac{r^{\alpha-1-k}\,\sin(\alpha\pi)}{r^{2\alpha} + 2r^\alpha\,\cos(\alpha\pi) + 1}\,, \tag{1.6.10}$$

where $K_\alpha(r) = K_\alpha^0(r)$, and

$$J^k g_\alpha(t) = \frac{2}{\alpha}\, e^{t\,\cos(\pi/\alpha)} \cos\left[t\,\sin\left(\frac{\pi}{\alpha}\right) - k\frac{\pi}{\alpha}\right]. \tag{1.6.11}$$

This can be done in direct analogy to the computation of the functions $e_\alpha(t)$, the Laplace transform of $J^k e_\alpha(t)$ being given by (1.6.3). For the impulse-response solution, we note that the effect of the differential operator D^1 is the same as that of the virtual operator J^{-1}.

In conclusion, we can resume the solutions for the fractional relaxation and oscillation equations as follows:
(a) $0 < \alpha < 1$,

$$u(t) = c_0\, u_0(t) + \int_0^t q(t - \tau)\, u_\delta(\tau)\, d\tau\,, \tag{1.6.12a}$$

where

$$\begin{cases} u_0(t) = \int_0^\infty e^{-rt}\, K_\alpha^0(r)\, dr\,, \\ u_\delta(t) = -\int_0^\infty e^{-rt}\, K_\alpha^{-1}(r)\, dr\,, \end{cases} \tag{1.6.13a}$$

with $u_0(0^+) = 1$, $u_\delta(0^+) = \infty$;
(b) $1 < \alpha < 2$,

$$u(t) = c_0\, u_0(t) + c_1\, u_1(t) + \int_0^t q(t - \tau)\, u_\delta(\tau)\, d\tau\,, \tag{1.6.12b}$$

where

$$\begin{cases} u_0(t) = \int_0^\infty e^{-rt}\, K_\alpha^0(r)\, dr + \frac{2}{\alpha} e^{t\,\cos(\pi/\alpha)}\, \cos\left[t\, \sin\left(\frac{\pi}{\alpha}\right) \right], \\ u_1(t) = \int_0^\infty e^{-rt}\, K_\alpha^1(r)\, dr + \frac{2}{\alpha} e^{t\,\cos(\pi/\alpha)}\, \cos\left[t\, \sin\left(\frac{\pi}{\alpha}\right) - \frac{\pi}{\alpha} \right], \\ u_\delta(t) = -\int_0^\infty e^{-rt}\, K_\alpha^{-1}(r)\, dr - \frac{2}{\alpha} e^{t\,\cos(\pi/\alpha)}\, \cos\left[t\, \sin\left(\frac{\pi}{\alpha}\right) + \frac{\pi}{\alpha} \right], \end{cases} \tag{1.6.13b}$$

with $u_0(0^+) = 1$, $u_0'(0^+) = 0$, $u_1(0^+) = 0$, $u_1'(0^+) = 1$, $u_\delta(0^+) = 0$, $u_\delta'(0^+) = +\infty$.

The students are invited to exhibit the plots of the basic fundamental solution for the following cases: (a) $\alpha = 0.25$, 0.50, 0.75, 1, and (b) $\alpha = 1.25$, 1.50, 1.75, 2, obtained from the first formula in (1.6.13a) and (1.6.13b), respectively.

Of particular interest is the case $\alpha = 1/2$ where we recover a well-known formula of the Laplace transform theory, see, e.g., Doetsch [32]

$$e_{1/2}(t) := E_{1/2}(-\sqrt{t}) = e^t\, \mathrm{erfc}(\sqrt{t}) \div \frac{1}{s^{1/2}\,(s^{1/2} + 1)}\,, \tag{1.6.14}$$

where erfc denotes the *complementary error* function. Explicitly we have

$$E_{1/2}(-\sqrt{t}) = e^t\, \frac{2}{\sqrt{\pi}} \int_{\sqrt{t}}^\infty e^{-u^2}\, du\,. \tag{1.6.15}$$

We now desire to point out that in both the cases (a) and (b) (in which α is just not integer), i.e., for *fractional relaxation* and *fractional oscillation*, all the fundamental and impulse-response solutions exhibit an *algebraic decay* as $t \to \infty$, as discussed below. This algebraic decay is the most important effect of the non-integer derivative in our equations, which dramatically differs from the *exponential decay* present in the ordinary relaxation and damped-oscillation phenomena.

Let us start with the asymptotic behavior of $u_0(t)$. To this purpose, we first derive an asymptotic series for the function $f_\alpha(t)$, valid for $t \to \infty$. We then consider the spectral representation (1.6.7)–(1.6.8) and expand the spectral function for small r. Then the Watson lemma yields,

$$f_\alpha(t) = \sum_{n=1}^{N} (-1)^{n-1} \frac{t^{-n\alpha}}{\Gamma(1 - n\alpha)} + O\left(t^{-(N+1)\alpha}\right), \quad \text{as} \ \ t \to \infty. \quad (1.6.16)$$

We note that this asymptotic expansion coincides with that for $u_0(t) = e_\alpha(t)$, having assumed $0 < \alpha < 2$ ($\alpha \ne 1$). In fact the contribution of $g_\alpha(t)$ is identically zero if $0 < \alpha < 1$ and exponentially small as $t \to \infty$ if $1 < \alpha < 2$.

The asymptotic expansions of the solutions $u_1(t)$ and $u_\delta(t)$ are obtained from (1.6.16) integrating or differentiating term by term with respect to t. Taking the leading term of the asymptotic expansions, we obtain the asymptotic representations of the solutions $u_0(t)$, $u_1(t)$, and $u_\delta(t)$ as $t \to \infty$,

$$u_0(t) \sim \frac{t^{-\alpha}}{\Gamma(1-\alpha)}, \quad u_1(t) \sim \frac{t^{1-\alpha}}{\Gamma(2-\alpha)}, \quad u_\delta(t) \sim -\frac{t^{-\alpha-1}}{\Gamma(-\alpha)}, \quad (1.6.17)$$

that point out the algebraic decay.

We would like to remark the difference between fractional relaxation governed by the Mittag-Leffler type function $E_\alpha(-at^\alpha)$ and stretched relaxation governed by a stretched exponential function $\exp(-bt^\alpha)$ with $\alpha, a, b > 0$ for $t \geq 0$. A common behavior is achieved only in a restricted range $0 \leq t \ll 1$ where we can have

$$E_\alpha(-at^\alpha) \simeq 1 - \frac{a}{\Gamma(\alpha+1)} t^\alpha = 1 - bt^\alpha \simeq e^{-bt^\alpha}, \quad \text{if} \ \ b = \frac{a}{\Gamma(\alpha+1)}. \quad (1.6.18)$$

Finally, the students are invited to investigate the relevant aspects of the solutions through plots, as outlined by Gorenflo & Mainardi [47].

1.7 Fractional Differential Equations of Diffusion-Wave Type

In this section, we analyze some boundary value problems for partial differential equations of fractional order in time. These are fundamental for understanding phenomena of anomalous diffusion or intermediate between diffusion and wave propagation. Typical phenomena of the first kind are based on special stochastic processes that generalize the classical Brownian motion, including the random walks with memory investigated by Gorenflo & Mainardi [52], and the Grey Brownian Motion, introduced by Schneider [172] and generalized by Mura & Mainardi [129]. Typical phenomena of the second kind are related to the one-dimensional propagation of stress pulses in a linear viscoelastic medium with constant Q, of great interest in seismology, as outlined by Mainardi & Tomirotti [101].

The Evolution Equations of Fractional Order

It is known that the standard partial differential equations governing the basic phenomena of diffusion and wave propagation are the Fourier diffusion equation and the D'Alembert wave equation, respectively. Denoting as usual x, t the space and time variables, and $r = r(x, t)$ the response variable, these equations read:

$$\frac{\partial r}{\partial t} = d \frac{\partial^2 r}{\partial x^2}, \tag{1.7.1}$$

$$\frac{\partial^2 r}{\partial t^2} = c^2 \frac{\partial^2 r}{\partial x^2}. \tag{1.7.2}$$

In Eq. (1.7.1), the constant d denotes a diffusivity coefficient, whereas in Eq. (1.7.2) c denotes a characteristic velocity. In this chapter, we consider the family of evolution equations

$$\frac{\partial^\beta r}{\partial t^\beta} = a \frac{\partial^2 r}{\partial x^2}, \quad 0 < \beta \le 2, \tag{1.7.3}$$

where *the time derivative of order β is intended in the Caputo sense*, namely is the operator ${}^*_0 D^\beta_t$, introduced in this chapter 1, Eq. (1.5.12), and a is a positive constant of dimension $L^2 T^{-\beta}$. We must distinguish the cases $0 < \beta \le 1$ and $1 < \beta \le 2$. We have

$$\frac{\partial^\beta r}{\partial t^\beta} := \begin{cases} \dfrac{1}{\Gamma(1-\beta)} \displaystyle\int_0^t \left[\dfrac{\partial}{\partial \tau} r(x, \tau)\right] \dfrac{d\tau}{(t-\tau)^\beta}, & 0 < \beta < 1, \\[3mm] \dfrac{\partial r}{\partial t}, & \beta = 1; \end{cases} \tag{1.7.4a}$$

$$\frac{\partial^\beta r}{\partial t^\beta} := \begin{cases} \dfrac{1}{\Gamma(2-\beta)} \displaystyle\int_0^t \left[\dfrac{\partial^2}{\partial \tau^2} r(x, \tau)\right] \dfrac{d\tau}{(t-\tau)^{\beta-1}}, & 1 < \beta < 2, \\[3mm] \dfrac{\partial^2 r}{\partial t^2}, & \beta = 2. \end{cases} \tag{1.7.4b}$$

It should be noted that in both cases $0 < \beta \le 1, 1 < \beta \le 2$, the time fractional derivative in the L.H.S. of Eq. (1.7.3) can be removed by a suitable fractional integration,[4]

---

[4] We apply to Eq. (1.7.3) the fractional integral operator of order β, namely ${}_0 I^\beta_t$. For $\beta \in (0, 1]$ we have:

$${}_0 I^\beta_t \circ {}^*_0 D^\beta_t r(x, t) = {}_0 I^\beta_t \circ {}_0 I^{1-\beta}_t D^1_t r(x, t) = {}_0 I^1_t D^1_t r(x, t) = r(x, t) - r(x, 0^+).$$

leading to alternative forms where the necessary initial conditions at $t = 0^+$ explicitly appear. As a matter fact, we get the integro-differential equations: if $0 < \beta \le 1$:

$$r(x, t) = r(x, 0^+) + \frac{a}{\Gamma(\beta)} \int_0^t \left(\frac{\partial^2 r}{\partial x^2}\right) (t - \tau)^{\beta-1} \, d\tau; \qquad (1.7.5a)$$

if $1 < \beta \le 2$:

$$r(x, t) = r(x, 0^+) + t \frac{\partial}{\partial t} r(x, t)|_{t=0^+} + \frac{a}{\Gamma(\beta)} \int_0^t \left(\frac{\partial^2 r}{\partial x^2}\right) (t - \tau)^{\beta-1} \, d\tau. \quad (1.7.5b)$$

1.7.1 Derivation of the Fundamental Solutions

Green Functions for the Cauchy and Signaling Problems

In order to guarantee the existence and the uniqueness of the solution, we must equip (1.7.1) with suitable data on the boundary of the space–time domain. The basic boundary-value problems for diffusion are the so-called Cauchy and Signaling problems. In the Cauchy problem, which concerns the space–time domain $-\infty < x < +\infty$, $t \ge 0$, the data are assigned at $t = 0^+$ on the whole space axis (initial data). In the Signaling problem, which concerns the space–time domain $x \ge 0$, $t \ge 0$, the data are assigned both at $t = 0^+$ on the semi-infinite space axis $x > 0$ (initial data) and at $x = 0^+$ on the semi-infinite time axis $t > 0$ (boundary data); here, as mostly usual, the initial data are assumed to vanish.

Denoting by $f(x)$, $x \in \mathbb{R}$ and $h(t)$, $t \in \mathbb{R}^+$ sufficiently well-behaved functions, the basic problems are thus formulated as following, assuming $0 < \beta \le 1$,

a) Cauchy problem

$$r(x, 0^+) = f(x), \quad -\infty < x < +\infty; \quad r(\mp\infty, t) = 0, \quad t > 0; \qquad (1.7.6a)$$

b) Signaling problem

$$r(x, 0^+) = 0, \quad x > 0; \quad r(0^+, t) = h(t), \quad r(+\infty, t) = 0, \quad t > 0. \qquad (1.7.6b)$$

If $1 < \beta < 2$, we must add into (1.7.6a) and (1.7.6b) the initial values of the first time derivative of the field variable, $r_t(x, 0^+)$, since in this case the fractional derivative is expressed in terms of the second-order time derivative. To ensure the continuous

---

(Footnote 4 continued)
For $\beta \in (1, 2]$ we have:

$$_0I_t^\beta \circ {}_0^* D_t^\beta r(x, t) = {}_0I_t^\beta \circ {}_0I_t^{2-\beta} D_t^2 r(x, t) = {}_0I_t^2 D_t^2 r(x, t) = r(x, t) - r(x, 0^+) - r_t(x, 0^+).$$

dependence of our solution with respect to the parameter β also in the transition from $\beta = 1^-$ to $\beta = 1^+$, we agree to assume

$$\frac{\partial}{\partial t} r(x, t)|_{t=0^+} = 0, \text{ for } 1 < \beta \leq 2, \tag{1.7.7}$$

as it turns out from the integral forms (1.7.5a)–(1.7.5b).

In view of our subsequent analysis, we find it convenient to set

$$\nu := \beta/2, \quad \text{so} \quad \begin{cases} 0 < \nu \leq 1/2, & \Longleftrightarrow \ 0 < \beta \leq 1, \\ 1/2 < \nu \leq 1, & \Longleftrightarrow \ 1 < \beta \leq 2, \end{cases} \tag{1.7.8}$$

and from now on to add the parameter ν to the independent space–time variables x, t in the solutions, writing $r = r(x, t; \nu)$.

For the Cauchy and Signaling problems, we introduce the so-called Green functions $\mathcal{G}_c(x, t; \nu)$ and $\mathcal{G}_s(x, t; \nu)$, which represent the respective fundamental solutions, obtained when $f(x) = \delta(x)$ and $h(t) = \delta(t)$. As a consequence, the solutions of the two basic problems are obtained by a space or time convolution according to

$$r(x, t; \nu) = \int_{-\infty}^{+\infty} \mathcal{G}_c(x - \xi, t; \nu) f(\xi) d\xi, \tag{1.7.9a}$$

$$r(x, t; \nu) = \int_{0^-}^{t^+} \mathcal{G}_s(x, t - \tau; \nu) h(\tau) d\tau. \tag{1.7.9b}$$

It should be noted that in (1.7.9a) $\mathcal{G}_c(x, t; \nu) = \mathcal{G}_c(|x|, t; \nu)$ because the Green function of the Cauchy problem turns out to be an even function of x. According to a usual convention, in (1.7.9b) the limits of integration are extended to take into account for the possibility of impulse functions centered at the extremes.

Reciprocity relation and auxiliary functions for the standard diffusion and wave equations First, let us consider the particular cases $\nu = 1/2$ and $\nu = 1$, which correspond to the standard diffusion equation (1.7.1) with $a = d$ and to the standard wave equation (1.7.3) with $a = c^2$. For these cases, the two Green functions for the Cauchy and Signaling problems are usually derived in classical texts of mathematical physics by using the techniques of Fourier transforms in space and Laplace transforms in time, respectively.

Then, using the notation $\mathcal{G}_{c,s}^d(x, t) := \mathcal{G}_{c,s}(x, t; 1/2)$, the two Green functions of the *standard diffusion equation* read:

$$\mathcal{G}_c^d(x, t) = \frac{1}{2\sqrt{\pi a}} t^{-1/2} e^{-x^2/(4 a t)} \overset{\mathcal{F}}{\leftrightarrow} e^{-a t \kappa^2}, \tag{1.7.10a}$$

$$\mathcal{G}_s^d(x, t) = \frac{x}{2\sqrt{\pi a}} t^{-3/2} e^{-x^2/(4 a t)} \overset{\mathcal{L}}{\leftrightarrow} e^{-(x/\sqrt{a})s^{1/2}}. \tag{1.7.10b}$$

From the explicit expressions (1.7.10a)–(1.7.10b), we recognize *the reciprocity relation* between the two Green functions, for $x > 0$, $t > 0$:

$$x \, \mathcal{G}_c^d(x, t) = t \, \mathcal{G}_s^d(x, t) = F^d(\xi) = \frac{1}{2} \, \xi \, M^d(\xi) \,, \qquad (1.7.11)$$

where

$$M^d(\xi) = \frac{1}{\sqrt{\pi}} \, e^{-\xi^2/4}\,, \quad \xi = \frac{x}{\sqrt{a} \, t^{1/2}} > 0 \,. \qquad (1.7.12)$$

The variable ξ plays the role of *similarity variable* for the standard diffusion, whereas the two functions $F^d(\xi)$ and $M^d(\xi)$ can be considered the *auxiliary functions* for the diffusion equation itself because each of them provides the fundamental solutions through (1.7.11). We note that the function $M^d(\xi)$ satisfies the normalization condition $\int_0^\infty M^d(\xi) \, d\xi = 1$.

For the *standard wave equation*, using the notation $\mathcal{G}_{c,s}^w(x, t) := \mathcal{G}_{c,s}(x, t; 1)$, the two Green functions read

$$\mathcal{G}_c^w(x, t) = \frac{1}{2}\left[\delta(x - \sqrt{a}t) + \delta(x + \sqrt{a}t)\right] \overset{\mathcal{F}}{\leftrightarrow} \frac{1}{2}\left[e^{+i\sqrt{a}t\kappa} + e^{-i\sqrt{a}t\kappa}\right],$$
$$(1.7.13a)$$

$$\mathcal{G}_s^w(x, t) = \delta(t - x/\sqrt{a}) \overset{\mathcal{L}}{\leftrightarrow} e^{-(x/\sqrt{a})s} \,. \qquad (1.7.13b)$$

From the explicit expressions (1.7.13a)–(1.7.13b), we recognize *the reciprocity relation* between the two Green functions, for $x > 0$, $t > 0$:

$$2x \, \mathcal{G}_c^w(x, t) = t \, \mathcal{G}_s^w(x, t) = F^w(\xi) = \xi \, M^w(\xi) \,, \qquad (1.7.14)$$

where

$$M^w(\xi) = \delta(1 - \xi) \,, \quad \xi = \frac{x}{\sqrt{a} \, t} > 0 \,. \qquad (1.7.15)$$

Even if ξ does not appear as a similarity variable in the ordinary sense, we attribute to ξ and to $\{F^w(\xi), M^w(\xi)\}$ the roles of similarity variable and auxiliary functions of the standard wave equation.

Reciprocity relation and auxiliary functions for the fractional diffusion and wave equations We now properly extend the previous results to the general case $0 < \nu \leq 1$ by determining the two Green functions through the technique of Laplace transforms.

We show how this technique allows us to obtain the transformed functions $\widetilde{\mathcal{G}}_c(x, s; \nu), \widetilde{\mathcal{G}}_s(x, s; \nu)$, by solving ordinary differential equations of the second order in x and then, by inversion, the required Green functions $\mathcal{G}_c(x, t; \nu)$ and $\mathcal{G}_s(x, t; \nu)$ in the space–time domain.

For the Cauchy problem (1.7.6a) with $f(x) = \delta(x)$, the application of the Laplace transform to Eqs. (1.7.3)–(1.7.4) with $r(x, t) = \mathcal{G}_c(x, t; \nu)$ and $\nu = \beta/2$ leads to the non-homogeneous differential equation satisfied by the image of the Green function, $\widetilde{\mathcal{G}}_c(x, s; \nu)$,

$$a \frac{d^2\widetilde{\mathcal{G}}_c}{dx^2} - s^{2\nu}\, \widetilde{\mathcal{G}}_c = -\,\delta(x)\, s^{2\nu-1}\,, \quad -\infty < x < +\infty. \tag{1.7.16}$$

Because of the singular term $\delta(x)$, we have to consider the above equation separately in the two intervals $x < 0$ and $x > 0$, imposing the boundary conditions at $x = \mp\infty$, $\mathcal{G}_c(\mp\infty, t; \nu) = 0$, and the necessary matching conditions at $x = 0^\pm$. We obtain

$$\widetilde{\mathcal{G}}_c(x, s; \nu) = \frac{1}{2\sqrt{a}\, s^{1-\nu}}\, e^{-(|x|/\sqrt{a})s^\nu}\,, \quad -\infty < x < +\infty. \tag{1.7.17}$$

A different strategy to derive $\mathcal{G}_c(x, t; \nu)$ is to apply the Fourier transform to Eqs. (1.7.3)–(1.7.4) as illustrated in some papers by Mainardi et al, including Mainardi, Luchko, and Pagnini [104], Mainardi & Pagnini [97], Mainardi, Pagnini, and Gorenflo [105], Mainardi, Mura, and Pagnini [112], to which the interested reader is referred for details.

For the Signaling problem (1.7.6b) with $h(t) = \delta(t)$, the application of the Laplace transform to Eqs. (1.7.3)–(1.7.4) with $r(x, t) = \mathcal{G}_s(x, t; \nu)$ and $\nu = \beta/2$, leads to the homogeneous differential equation

$$a \frac{d^2\widetilde{\mathcal{G}}_s}{dx^2} - s^{2\nu}\, \widetilde{\mathcal{G}}_s = 0\,, \quad x \geq 0. \tag{1.7.18}$$

Imposing the boundary conditions $\mathcal{G}_s(0^+, t; \nu) = h(t) = \delta(t)$ and $\mathcal{G}_s(+\infty, t; \nu) = 0$, we obtain

$$\widetilde{\mathcal{G}}_s(x, s; \nu) = e^{-(x/\sqrt{a})s^\nu}\,, \quad x \geq 0. \tag{1.7.19}$$

The transformed solutions provided by Eqs. (1.7.17) and (1.7.19) must be inverted to provide the requested Green functions in the space–time domain. For this purpose, we recall the Laplace transform pairs related to the transcendental functions $F_\nu(r)$, $M_\nu(r)$ of the Wright type, discussed in Section 1.4, see Eqs. (1.4.37)–(1.4.38), where r stands for the actual time coordinate t:

$$\frac{1}{\nu}\, F_\nu\left(1/t^\nu\right) = \frac{1}{t^\nu}\, M_\nu\left(1/t^\nu\right) \div \frac{e^{-s^\nu}}{s^{1-\nu}}\,, \quad 0 < \nu < 1. \tag{1.7.20}$$

$$\frac{1}{t}\, F_\nu\left(1/t^\nu\right) = \frac{\nu}{t^{\nu+1}}\, M_\nu\left(1/t^\nu\right) \div e^{-s^\nu}\,, \quad 0 < \nu < 1, \tag{1.7.21}$$

so that these formulas can be used to invert the transforms in Eqs. (1.7.17) and (1.7.19). Then, introducing for $x > 0$, $t > 0$, the *similarity variable*

$$\xi := x/(\sqrt{a}\, t^\nu) > 0 \tag{1.7.22}$$

and recalling the rules of scale-change in the Laplace transform pairs; after some manipulation, we obtain the Green functions in the space–time domain in the form:

$$\mathcal{G}_c(x, t; \nu) = \frac{1}{2\nu x} F_\nu(\xi) = \frac{1}{2\sqrt{a}\, t^\nu} M_\nu(\xi), \qquad (1.7.23a)$$

$$\mathcal{G}_s(x, t; \nu) = \frac{1}{t} F_\nu(\xi) = \frac{\nu x}{\sqrt{a}\, t^{1+\nu}} M_\nu(\xi). \qquad (1.7.23b)$$

We also recognize the following *reciprocity relation* for the original Green functions,

$$2\nu x\, \mathcal{G}_c(x, t; \nu) = t\, \mathcal{G}_s(x, t; \nu) = F_\nu(\xi) = \nu\xi\, M_\nu(\xi). \qquad (1.7.24)$$

Now $F_\nu(\xi)$, $M_\nu(\xi)$ are the *auxiliary functions* for the general case $0 < \nu \leq 1$, which generalize those for the standard diffusion given in Eqs. (1.7.11)–(1.7.12) and for the standard wave equation given in Eqs. (1.7.14)–(1.7.15). In fact, for $\nu = 1/2$ and for $\nu = 1$, we recover the expressions of $M^d(\xi)$ and $M^w(\xi)$, respectively, given by (1.7.12) and (1.7.15), as it can be easily verified using the formulas provided in Section 1.4.

Equation (1.7.24) along with Eqs. (1.7.22) and (1.7.23a), (1.7.23b) allows us to note the exponential decay of $\mathcal{G}_c(x, t; \nu)$ as $x \to +\infty$ (at fixed t) and the algebraic decay of $\mathcal{G}_s(x, t; \nu)$ as $t \to +\infty$ (at fixed x), for $0 < \nu < 1$. In fact, we get

$$\mathcal{G}_c(x, t; \nu) \sim A(t)\, x^{(\nu-1/2)/(1-\nu)}\, e^{-B(t)x^{1/(1-\nu)}}, \quad x \to \infty, \qquad (1.7.25a)$$

$$\mathcal{G}_s(x, t; \nu) \sim C(x)\, t^{-(1+\nu)}, \quad t \to \infty, \qquad (1.7.25b)$$

where $A(t)$, $B(t)$, and $C(x)$ are positive functions to be determined.

The students are invited to investigate the relevant aspects of the solutions as outlined by Mainardi [91, 93] and Mainardi & Tomirotti [101], including the scaling properties, through plots in the Problem Session.

References

1. Abramowitz, M., & Stegun, I. A. (1965). *Handbook of mathematical functions*. New York: Dover.
2. Achar, B. N. N., Hanneken, J. W., & Clarke, T. (2004). Damping characteristics of a fractional oscillator. *Physica A, 339*, 311–319.
3. Agarwal, R. P. (1953). A propos d'une note de M. Pierre Humbert. *C.R. Acad. Sci. Paris, 236*, 2031–2032.
4. Anh, V. V., & Leonenko, N. N. (2001). Spectral analysis of fractional kinetic equations with random data. *Journal Statistical Physics, 104*, 1349–1387.
5. Atanackovic, T. M. (2004). Applications of fractional calculus in mechanics. Lecture Notes at the National Technical University of Athens (pp. 100).

6. Balescu, R. (2007). V-Langevin equations, continuous time random walks and fractional diffusion. *Chaos, Solitons and Fractals, 34*, 62–80.

7. Barret, J. H. (1954). Differential equations of non-integer order. *Canadian Journal of Mathematics, 6*, 529–541.

8. Bender, C. M., & Orszag, S. A. (1987). *Advanced mathematical methods for scientists and engineers*. Singapore: McGraw-Hill.

9. Berberan-Santos, M. N. (2005). Properties of the Mittag-Leffler relaxation function. *Journal of Mathematical Chemistry, 38*, 629–635.

10. Blank, L. (1997). Numerical treatment of differential equations of fractional order. *Non-linear World, 4*(4), 473–491.

11. Buchen, P. W., & Mainardi, F. (1975). Asymptotic expansions for transient viscoelastic waves. *Journal de Mécanique, 14*, 597–608.

12. Butzer, P., & Westphal, U. (2000). Introduction to fractional calculus. In H. Hilfer (Ed.), *Fractional calculus, applications in physics* (pp. 1–85). Singapore: World Scientific.

13. Cafagna, D. (2007). Fractional calculus: A mathematical tool from the past for present engineers. *IEEE Industrial Electronics Magazine, 1*, 35–40.

14. Camargo, R. F., Chiacchio, A. O., Charnet, R. & Capelas de Oliveira, E. (2009). Solution of the fractional Langevin equation and the Mittag-Leffler functions. *Journal of Mathematical Physics, 50*, 063507/1-8.

15. Caputo, M. (1966). Linear models of dissipation whose Q is almost frequency independent. *Annali di Geofisica, 19*, 383–393.

16. Caputo, M. (1967). Linear models of dissipation whose Q is almost frequency independent part II. *Geophysical Journal of the Royal Astronomical Society, 13*, 529–539.

17. Caputo, M. (1969). *Elasticità e Dissipazione*. Bologna: Zanichelli.

18. Caputo, M. (1973). Elasticity with dissipation represented by a simple memory mechanism, *Atti Accad. Naz. Lincei, Rend. Classe Scienze (Ser.8), 55*, 467–470.

19. Caputo, M. (1976). Vibrations of an infinite plate with a frequency independent Q. *Journal of the Acoustical Society of America, 60*, 634–639.

20. Caputo, M. (1979). A model for the fatigue in elastic materials with frequency independent Q. *Journal of the Acoustical Society of America, 66*, 176–179.

21. Caputo, M. (1996). The Green function of the diffusion in porous media with memory, *Rend. Fis. Acc. Lincei (Ser.9), 7*, 243–250.

22. Caputo, M. (1999). Diffusion of fluids in porous media with memory. *Geothermics, 28*, 113–130.

23. Caputo, M., & Mainardi, F. (1971). A new dissipation model based on memory mechanism. *Pure and Applied Geophysics (PAGEOPH), 91*, 134–147. [Reprinted in *Fractional Calculus and Applied Analysis, 10*(3), 309–324 (2007)]

24. Caputo, M., & Mainardi, F. (1971). Linear models of dissipation in anelastic solids. *Riv. Nuovo Cimento (Ser. II), 1*, 161–198.

25. Carcione, J. M., Cavallini, F., Mainardi, F., & Hanyga, A. (2002). Time-domain seismic modelling of constant-Q wave propagation using fractional derivatives. *Pure and Applied Geophysics (PAGEOPH), 159*, 1719–1736.

26. Carpinteri, A., & Cornetti, P. (2002). A fractional calculus approach to the description of stress and strain localization in fractal media. *Chaos, Solitons and Fractals, 13*, 85–94.

27. Chin, R. C. Y. (1980). Wave propagation in viscoelastic media. In A. Dziewonski & E. Boschi (Eds.), *Physics of the earth's interior* (pp. 213–246). Amsterdam: North-Holland [Enrico Fermi International School, Course 78].

28. Christensen, R. M. (1982). *Theroy of viscoelasticity*. New York: Academic Press (1st ed. (1972)).

29. Davis, H. T. (1936). *The theory of linear operators*. Bloomington: The Principia Press.

30. Diethelm, K. (2008). An investigation of some no-classical methods for the numerical approximation of Caputo-type fractional derivatives. *Numerical Algorithms, 47*, 361–390.

31. Diethelm, K. (2010). *The analysis of fractional differential equations* (Vol. 2004). Lecture notes in mathematics. Berlin: Springer.

32. Doetsch, G. (1974). *Introduction to the theory and application of the Laplace transformation*. Berlin: Springer.
33. Dzherbashyan, M. M. (1966). Integral transforms and representations of functions in the complex plane, Nauka, Moscow. [in Russian]. There is also the transliteration as Djrbashyan.
34. Dzherbashyan, M. M. (1993). *Harmonic analysis and boundary value problems in the complex domain*. Basel: Birkhäuser Verlag.
35. Eidelman, S. D., & Kochubei, A. N. (2004). Cauchy problem for fractional diffusion equations. *Journal of Differential Equations, 199*, 211–255.
36. Engler, H. (1997). Similarity solutions for a class of hyperbolic integro-differential equations. *Differential Integral Equations, 10*, 815–840.
37. Erdélyi, A., Magnus, W., Oberhettinger, F., & Tricomi, F. G. (1953–1955). *Higher transcendental functions*, 3 volumes. New York: McGraw-Hill [Bateman Project].
38. Feller, W. (1952). On a generalization of Marcel Riesz' potentials and the semigroups generated by them. *Meddelanden Lunds Universitets Matematiska Seminarium* (Comm. Sém. Mathém. Université de Lund), Tome suppl. dédié a M. Riesz, Lund (pp. 73–81).
39. Feller, W. (1971). *An introduction to probability theory and its applications* (2nd ed., Vol. II). New York: Wiley [First edition (1966)].
40. Fujita, Y. (1990). Integro-differential equation which interpolates the heat equation and the wave equation I, II. *Osaka Journal of Mathematics, 27*(309–321), 797–804.
41. Fujita, Y. (1990). Cauchy problems of fractional order and stable processes. *Japan Journal of Applied Mathematics, 7*, 459–476.
42. Gawronski, W. (1984). On the bell-shape of stable distributions. *Annals of Probability, 12*, 230–242.
43. Gel'fand, I. M., & Shilov, G. E. (1964). *Generalized functions* (Vol. 1). New York: Academic Press.
44. Giona, M., & Roman, H. E. (1992). Fractional diffusion equation for transport phenomena in random media. *Physica A, 185*, 82–97.
45. Gonsovskii, V. L., & Rossikhin, Yu. A. (1973). Stress waves in a viscoelastic medium with a singular hereditary kernel. *Zhurnal Prikladnoi Mekhaniki Tekhnicheskoi Fiziki, 4*, 184–186 [Translated from the Russian by Plenum Publishing Corporation, New Yorki (1975)].
46. Gorenflo, R. (1997). Fractional calculus: Some numerical methods. In A. Carpinteri & F. Mainardi (Eds.), *Fractals and fractional calculus in continuum mechanics* (pp. 277–290). Wien: Springer. http://www.fracalmo.org.
47. Gorenflo, R., & Mainardi, F. (1997). Fractional calculus: Integral and differential equations of fractional order. In A. Carpinteri & F. Mainardi (Eds.), *Fractals and fractional calculus in continuum mechanics* (pp. 223–276). Wien: Springer [E-print: arXiv:0805.3823].
48. Gorenflo, R., & Mainardi, F. (1998). Fractional calculus and stable probability distributions. *Archives of Mechanics, 50*, 377–388.
49. Gorenflo, R., & Mainardi, F. (1998). Random walk models for space-fractional diffusion processes. *Fractional Calculus and Applied Analysis, 1*, 167–191.
50. Gorenflo, R., & Mainardi, F. (1998). Signalling problem and Dirichlet-Neumann map for time-fractional diffusion-wave equations. *Matimyás Matematika, 21*, 109–118.
51. Gorenflo, R., & Mainardi, F. (2008). Continuous time random walk, Mittag-Leffler waiting time and fractional diffusion: Mathematical aspects. In R. Klages, G. Radons, & I. M. Sokolov (Eds.), *Anomalous transport: Foundations and applications* (pp. 93–127). Weinheim: Wiley-VCH [E-print arXiv:0705.0797].
52. Gorenflo, R., & Mainardi, F. (2009). Some recent advances in theory and simulation of fractional diffusion processes. *Journal of Computational and Applied Mathematics, 229*(2), 400–415 [E-print: arXiv:0801.0146].
53. Gorenflo, R., & Rutman, R. (1994). On ultraslow and intermediate processes. In P. Rusev, I. Dimovski, & V. Kiryakova (Eds.), *Transform methods and special functions, Sofia 1994* (pp. 171–183). Singapore: Science Culture Technology.
54. Gorenflo, R., & Vessella, S. (1991). *Abel integral equations: Analysis and applications* (Vol. 1461). Lecture notes in mathematics. Berlin: Springer.

55. Gorenflo, R., Luchko, Yu., & Rogosin, S. V. (1997). Mittag-Leffler type functions: Notes on growth properties and distribution of zeros, Preprint No A-97-04, Fachbereich Mathematik und Informatik, Freie Universität Berlin, Serie Mathematik (pp. 23) [E-print: http://www.math.fu-berlin.de/publ/index.html].

56. Gorenflo, R., Luchko, Yu., & Mainardi, F. (1999). Analytical properties and applications of the Wright function. *Fractional Calculus and Applied Analysis, 2*, 383–414.

57. Gorenflo, R., Iskenderov, A., & Luchko, Yu. (2000). Mapping between solutions of frational diffusion-wave equations. *Fractional Calculus and Applied Analysis, 3*, 75–86.

58. Gorenflo, R., Luchko, Yu., & Mainardi, F. (2000). Wright functions as scale-invariant solutions of the diffusion-wave equation. *Journal of Computational and Applied Mathematics, 118*, 175–191.

59. Gorenflo, R., Loutchko, J., & Luchko, Yu. (2002). Computation of the Mittag-Leffler function $E_{\alpha,\beta}(z)$ and its derivatives. *Fractional Calculus and Applied Analysis, 5*, 491–518.

60. Graffi, D. (1982). Mathematical models and waves in linear viscoelasticity. In F. Mainardi (Ed.), *Wave propagation in viscoelastic media* (Vol. 52, pp. 1–27). Research notes in mathematics. London: Pitman.

61. Gross, B. (1947). On creep and relaxation. *Journal of Applied Physics, 18*, 212–221.

62. Gupta, I. S., & Debnath, L. (2007). Some properties of the Mittag-Leffler functions. *Integral Transforms and Special Functions, 18*(5), 329–336.

63. Hanneken, J. W., Achar, B. N. N., Puzio, R., & Vaught, D. M. (2009). Properties of the Mittag-Leffler function for negative α. *Physica Scripta, T136*, 014037/1-5.

64. Hanyga, A. (2002). Multi-dimensional solutions of time-fractional diffusion-wave equation. *Proceedings of the Royal Society of London, 458*, 933–957.

65. Haubold, H. J., & Mathai, A. M. (2000). The fractional kinetic equation and thermonuclear functions. *Astrophysics and Space Science, 273*, 53–63.

66. Haubold, H. J., Mathai, A. M., & Saxena, R. K. (2007). Solution of fractional reaction-diffusion equations in terms of the H-function. *Bulletin of the Astronomical Society of India, 35*, 681–689.

67. Haubold, H. J., Mathai, A. M., & Saxena, R. K. (2009). Mittag-Leffler functions and their applications (pp. 49). arXiv:0909.0230.

68. Haubold, H. J., Mathai, A. M., & Saxena, R. K. (2011). Mittag-Leffler functions and their applications. *Journal of Applied Mathematics, 2011*, Article ID 298628, 51 p. Hindawi Publishing Corporation [E-Print: arXiv:0909.0230].

69. Hilfer, R. (2000). Fractional time evolution. In R. Hilfer (Ed.), *Applications of fractional calculus in physics* (pp. 87–130). Singapore: World Scientific.

70. Hilfer, R., & Seybold, H. J. (2006). Computation of the generalized Mittag-Leffler function and its inverse in the complex plane. *Integral Transforms and Special Functions, 17*(9), 637–652.

71. Hille, E., & Tamarkin, J. D. (1930). On the theory of linear integral equations. *Annals of Mathematics, 31*, 479–528.

72. Humbert, P. (1945). Nouvelles correspondances symboliques. *Bull. Sci. Mathém. (Paris, II ser.), 69*, 121–129.

73. Humbert, P. (1953). Quelques résultats relatifs à la fonction de Mittag-Leffler. *C.R. Acad. Sci. Paris, 236*, 1467–1468.

74. Humbert, P., & Agarwal, R. P. (1953). Sur la fonction de Mittag-Leffler et quelques-unes de ses généralisations. *Bull. Sci. Math (Ser. II), 77*, 180–185.

75. Kilbas, A. A., & Saigo, M. (1996). On Mittag-Leffler type functions, fractional calculus operators and solution of integral equations. *Integral Transforms and Special Functions, 4*, 355–370.

76. Kilbas, A. A., Saigo, M., & Trujillo, J. J. (2002). On the generalized Wright function. *Fractional Calculus and Applied Analysis, 5*(4), 437–460.

77. Kilbas, A. A., Srivastava, H. M., & Trujillo, J. J. (2006). *Theory and applications of fractional differential equations* (Vol. 204). North-Holland series on mathematics studies. Amsterdam: Elsevier.

78. Kiryakova, V. (1994). *Generalized fractional calculus and applications* (Vol. 301). Pitman research notes in mathematics. Harlow: Longman.
79. Kiryakova, V. (1997). All the special functions are fractional differintegrals of elementary functions. *Journal of Physics A: Mathematical and General, 30*, 5085–5103.
80. Kochubei, A. N. (1989). A Cauchy problem for evolution equations of fractional order. *Differential Equations, 25*, 967–974 [English translation from the Russian Journal *Differentsial'nye Uravneniya*].
81. Kochubei, A. N. (1990). Fractional order diffusion. *Differential Equations, 26*, 485–492 [English translation from the Russian Journal *Differentsial'nye Uravneniya*].
82. Kolsky, H. (1956). The propagation of stress pulses in viscoelastic solids. *Philosophical Magazine (Series 8), 2*, 693–710.
83. Kreis, A., & Pipkin, A. C. (1986). Viscoelastic pulse propagation and stable probability distributions. *Quarterly of Applied Mathematics, 44*, 353–360.
84. Luchko, Yu. (1999). Operational method in fractional calculus. *Fractional Calculus and Applied Analysis, 2*, 463–488.
85. Luchko, Yu. (2000). Asymptotics of zeros of the Wright function. *Zeit. Anal. Anwendungen, 19*, 583–595.
86. Luchko, Yu. (2001). On the distribution of zeros of the Wright function. *Integral Transforms and Special Functions, 11*, 195–200.
87. Luchko, Yu. (2008). Algorithms for evaluation of the Wright function for the real arguments' values. *Fractional Calculus and Applied Analysis, 11*, 57–75.
88. Magin, R. L. (2006). *Fractional calculus in bioengineering*. Connecticut: Begell House Publishers.
89. Mainardi, F. (1994). On the initial value problem for the fractional diffusion-wave equation. In S. Rionero & T. Ruggeri (Eds.), *Waves and stability in continuous media* (pp. 246–251). Singapore: World Scientific.
90. Mainardi, F. (1995). The time fractional diffusion-wave equation. *Radiophysics and Quantum Electronics, 38*(1–2), 20–36 [English translation from the Russian of *Radiofisika*].
91. Mainardi, F. (1996). The fundamental solutions for the fractional diffusion-wave equation. *Applied Mathematics Letters, 9*(6), 23–28.
92. Mainardi, F. (1996). Fractional relaxation-oscillation and fractional diffusion-wave phenomena. *Chaos, Solitons and Fractals, 7*, 1461–1477.
93. Mainardi, F. (1997). Fractional calculus: Some basic problems in continuum and statistical mechanics. In A. Carpinteri & F. Mainardi (Eds.), *Fractals and fractional calculus in continuum mechanics* (pp. 291–348). Wien: Springer. http://www.fracalmo.org.
94. Mainardi, F. (2010). *Fractional calculus and waves in linear viscoelasticity*. London: Imperial College Press.
95. Mainardi, F., & Gorenflo, R. (2000). On Mitag-Leffler type functions in fractional evolution processes. *Journal of Computational and Applied Mathematics, 118*, 283–299.
96. Mainardi, F., & Gorenflo, R. (2007). Time-fractional derivatives in relaxation processes: A tutorial survey. *Fractional Calculus and Applied Analysis, 10*, 269–308 [E-print: arXiv:0801.4914].
97. Mainardi, F., & Pagnini, G. (2003). The Wright functions as solutions of the time-fractional diffusion equations.
98. Mainardi, F., & Paradisi, P. (2001). Fractional diffusive waves. *Journal of Computational Acoustics, 9*, 1417–1436.
99. Mainardi, F., & Spada, G. (2011). Creep, relaxation and viscosity properties for basic fractional models in rheology. *The European Physical Journal, Special Topics, 193*, 133–160.
100. Mainardi, F., & Tomirotti, M. (1995). On a special function arising in the time fractional diffusion-wave equation. In P. Rusev, I. Dimovski, & V. Kiryakova (Eds.), *Transform methods and special functions, Sofia 1994* (pp. 171–183). Singapore: Science Culture Technology Publications.
101. Mainardi, F., & Tomirotti, M. (1997). Seismic pulse propagation with constant Q and stable probability distributions. *Annali di Geofisica, 40*, 1311–1328.

102. Mainardi, F., & Turchetti, G. (1975). Wave front expansion for transient viscoelastic waves. *Mechanics Research Communications, 2*, 107–112.

103. Mainardi, F., Raberto, M., Gorenflo, R., & Scalas, E. (2000). Fractional calculus and continuous-time finance II: The waiting-time distribution. *Physica A, 287*(3–4), 468–481.

104. Mainardi, F., Luchko, Yu., & Pagnini, G. (2001). The fundamental solution of the space-time fractional diffusion equation. *Fractional Calculus and Applied Analysis, 4*, 153–192 [E-print arXiv:cond-mat/0702419].

105. Mainardi, F., Pagnini, G., & Gorenflo, R. (2003). Mellin transform and subordination laws in fractional diffusion processes. *Fractional Calculus and Applied Analysis, 6*(4), 441–459 [E-print: http://arxiv.org/abs/math/0702133].

106. Mainardi, F., Gorenflo, R., & Scalas, E. (2004). A fractional generalization of the Poisson processes. *Vietnam Journal of Mathematics, 32* SI, 53–64 [E-print arXiv:math/0701454].

107. Mainardi, F., Pagnini, G., & Saxena, R. K. (2005). Fox H-functions in fractional diffusion. *Journal of Computational and Applied Mathematics, 178*, 321–331.

108. Mainardi, F., Gorenflo, R., & Vivoli, A. (2005). Renewal processes of Mittag-Leffler and Wright type. *Fractional Calculus and Applied Analysis, 8*, 7–38 [E-print arXiv:math/0701455].

109. Mainardi, F., Gorenflo, R., & Vivioli, A. (2007). Beyond the Poisson renewal process: A tutorial survey. *Journal of Computational and Applied Mathematics, 205*, 725–735.

110. Mainardi, F., Mura, A., Gorenflo, R., & Stojanovic, M. (2007). The two forms of fractional relaxation of distributed order. *Journal of Vibration and Control, 13*(9–10), 1249–1268 [E-print arXiv:cond-mat/0701131].

111. Mainardi, F., Mura, A., Pagnini, G., & Gorenflo, R. (2008). Time-fractional diffusion of distributed order. *Journal of Vibration and Control, 14*(9–10), 1267–1290 [arxiv.org/abs/cond-mat/0701132].

112. Mainardi, F., Mura, A., & Pagnini, G. (2009). The M-Wright function in time-fractional diffusion processes: A tutorial survey. *International Journal of Differential Equations*.

113. Marichev, O. I. (1983). *Handbook of integral transforms of higher transcendental functions, theory and algorithmic tables*. Chichester: Ellis Horwood.

114. Mathai, A. M., & Haubold, H. J. (2008). *Special functions for applied scientists*. New York: Springer.

115. Mathai, A. M., Saxena, R. K., & Haubold, H. J. (2010). *The H-function: Theory and applications*. New York: Springer.

116. Meshkov, S. I., & Rossikhin, Yu. A. (1970). Sound wave propagation in a viscoelastic medium whose hereditary properties are determined by weakly singular kernels. In Yu. N. Rabotnov (Kishniev) (Ed.), *Waves in inelastic media* (pp. 162–172) [in Russian].

117. Metzler, R., Glöckle, W. G., & Nonnenmacher, T. F. (1994). Fractional model equation for anomalous diffusion. *Physica A, 211*, 13–24.

118. Mikusiński, J. (1959). On the function whose Laplace transform is exp $(-s^\alpha)$. *Studia Math., 18*, 191–198.

119. Miller, K. S. (1993). The Mittag-Leffler and related functions. *Integral Transforms and Special Functions, 1*, 41–49.

120. Miller, K. S. (2001). Some simple representations of the generalized Mittag-Leffler functions. *Integral Transforms and Special Functions, 11*(1), 13–24.

121. Miller, K. S., & Ross, B. (1993). *An Introduction to the fractional calculus and fractional differential equations*. New York: Wiley.

122. Miller, K. S., & Samko, S. G. (1997). A note on the complete monotonicity of the generalized Mittag-Leffler function. *Real Analysis Exchange, 23*(2), 753–755.

123. Miller, K. S., & Samko, S. G. (2001). Completely monotonic functions. *Integral Transforms and Special Functions, 12*, 389–402.

124. Mittag-Leffler, M. G. (1903). Une généralisation de l'intégrale de Laplace-Abel. *C.R. Acad. Sci. Paris* (Ser. II), *137*, 537–539.

125. Mittag-Leffler, M. G. (1903). Sur la nouvelle fonction $E_\alpha(x)$. *C.R. Acad. Sci. Paris* (Ser. II), *137*, 554–558.

126. Mittag-Leffler, M. G. (1904). Sopra la funzione $E_\alpha(x)$. *Rendiconti R. Accademia Lincei* (Ser. V), *13*, 3–5.

127. Mittag-Leffler, M. G. (1905). Sur la représentation analytique d'une branche uniforme d'une fonction monogène. *Acta Mathematica, 29*, 101–181.

128. Mura, A. (2008). Non-Markovian stochastic processes and their applications: From anomalous diffusion to time series analysis. Ph.D. thesis, University of Bologna (Supervisor: Professor F. Mainardi). Now available by Lambert Academic Publishing (2011).

129. Mura, A., & Mainardi, F. (2009). A class of self-similar stochastic processes with stationary increments to model anomalous diffusion in physics. *Integral Transforms and Special Functions, 20*(3-4), 185–198. E-print: arXiv:0711.0665.

130. Mura, A., & Pagnini, G. (2008). Characterizations and simulations of a class of stochastic processes to model anomalous diffusion. *Journal of Physics A: Mathematical and Theoretical, 41*(28), 285002/1-22. E-print arXiv:0801.4879.

131. Mura, A., Taqqu, M. S., & Mainardi, F. (2008). Non-Markovian diffusion equations and processes: Analysis and simulation. *Physica A, 387*, 5033–5064.

132. Nigamatullin, R. R. (1986). The realization of the generalized transfer equation in a medium with fractal geometry. *Physica Status Solidi B, 133*, 425–430 [English translation from the Russian].

133. Nonnenmacher, T. F., & Glöckle, W. G. (1991). A fractional model for mechanical stress relaxation. *Philosophical Magazine Letters, 64*, 89–93.

134. Nonnenmacher, T. F., & Metzler, R. (1995). On the Riemann-Liouville fractional calculus and some recent applications. *Fractals, 3*, 557–566.

135. Oldham, K. B., & Spanier, J. (1974). *The fractional calculus*. New York: Academic Press.

136. Pagnini, G. (2012). Erdeélyi-Kober fractional diffusion. *Fractional Calculus and Applied Analysis, 15*(1), 117–127.

137. Pillai, R. N. (1990). On Mittag-Leffler functions and related distributions. *Annals of the Institute of Statistical Mathematics, 42*, 157–161.

138. Pipkin, A. C. (1986). *Lectures on viscoelastic theory* (pp. 56–76). New York: Springer. [1st edition 1972].

139. Podlubny, I. (1999). *Fractional differential equations* (Vol. 198). Mathematics in science and engineering. San Diego: Academic Press.

140. Podlubny, I. (2002). Geometric and physical interpretation of fractional integration and fractional differentiation. *Fractional Calculus and Applied Analysis, 5*, 367–386.

141. Podlubny, I. (2006). Mittag-Leffler function, WEB Site of MATLAB Central. http://www.mathworks.com/matlabcentral/fileexchange.

142. Pollard, H. (1946). The representation of $\exp(-x^\lambda)$ as a Laplace integral. *Bulletin of the American Mathematical Society, 52*, 908–910.

143. Pollard, H. (1948). The completely monotonic character of the Mittag-Leffler function $E_\alpha(-x)$. *Bulletin of the American Mathematical Society, 54*, 1115–1116.

144. Prabhakar, T. R. (1971). A singular integral equation with a generalized Mittag-Leffler function in the kernel. *The Yokohama Mathematical Journal, 19*, 7–15.

145. Prudnikov, A. P., Brychkov, Y. A., & Marichev, O. I. (1986). *Integrals and series* (Vol. I, II, III). New York: Gordon and Breach.

146. Prüsse, J. (1993). *Evolutionary integral equations and applications*. Basel: Birkhauser Verlag.

147. Pskhu, A. V. (2003). Solution of boundary value problems for the fractional diffusion equation by the Green function method. *Differential Equations, 39*(10), 1509–1513 [English translation from the Russian Journal *Differentsial'nye Uravneniya*].

148. Pskhu, A. V. (2005). *Partial differential equations of fractional order*. Moscow: Nauka [in Russian].

149. Pskhu, A. V. (2009). The fundamental solution of a diffusion-wave equation of fractional order. *Izvestiya: Mathematics, 73*(2), 351–392.

150. Rangarajan, G., & Ding, M. Z. (2000). Anomalous diffusion and the first passage time problem. *Physical Review E, 62*, 120–133.

151. Rangarajan, G., & Ding, M. Z. (2000). First passage time distribution for anomalous diffusion. *Physics Letters A, 273*, 322–330.

152. Robotnov, Yu. N. (1969). *Creep problems in structural members*. Amsterdam: North-Holland [English translation of the 1966 Russian edition].

153. Ross, B. (Ed.). (1975). *Fractional calculus and its applications* (Vol. 457). Lecture notes in mathematics. Berlin: Springer.

154. Ross, B. (1977). The development of fractional calculus 1695–1900. *Historia Mathematica, 4*, 75–89.

155. Rossikhin, Yu. A., & Shitikova, M. V. (1997). Application of fractional calculus to dynamic problems of linear and nonlinear hereditary mechanics of solids. *Applied Mechanics Reviews, 50*, 15–67.

156. Rossikhin, Yu. A., & Shittikova, M. V. (2007). Comparative analysis of viscoelastic models involving fractional derivatives of different orders. *Fractional Calculus and Applied Analysis, 10*(2), 111–121.

157. Rossikhin, Yu. A., & Shittikova, M. V. (2010). Applications of fractional calculus to dynamic problems of solid mechanics: Novel trends and recent results. *Applied Mechanics Reviews, 63*, 010801/1-52.

158. Saichev, A., & Zaslavsky, G. (1997). Fractional kinetic equations: Solutions and applications. *Chaos, 7*, 753–764.

159. Saigo, M., & Kilbas, A. A. (1998). On Mittag-Leffler type function and applications. *Integral Transforms Special Functions, 7*(1–2), 97–112.

160. Saigo, M., & Kilbas, A. A. (2000). Solution of a class of linear differential equations in terms of functions of Mittag-Leffler type. *Differential Equations, 36*(2), 193–200.

161. Samko, S. G., Kilbas, A. A., & Marichev, O. I. (1993). *Fractional integrals and derivatives, theory and applications*. Amsterdam: Gordon and Breach [English translation from the Russian, Nauka i Tekhnika, Minsk, 1987].

162. Sansone, G., & Gerretsen, J. (1960). *Lectures on the theory of functions of a complex variable* (Vol. I). Holomorphic functions. Groningen: Nordhoff.

163. Saxena, R. K., Mathai, A. M., & Haubold, H. J. (2004). On generalized fractional kinetic equations. *Physica A, 344*, 657–664.

164. Saxena, R. K., Mathai, A. M., & Haubold, H. J. (2004). Unified fractional kinetic equations and a fractional diffusion. *Astrophysics and Space Science, 290*, 299–310.

165. Saxena, R. K., Mathai, A. M., & Haubold, H. J. (2006). Fractional reaction-diffusion equations. *Astrophysics and Space Science, 305*, 289–296.

166. Saxena, R. K., Mathai, A. M., & Haubold, H. J. (2006). Reaction-diffusion systems and nonlinear waves. *Astrophysics and Space Science, 305*, 297–303.

167. Saxena, R. K., Mathai, A. M., & Haubold, H. J. (2006). Solution of generalized fractional reaction-diffusion equations. *Astrophysics and Space Science, 305*, 305–313.

168. Saxena, R. K., Mathai, A. M., & Haubold, H. J. (2006). Solution of fractional reaction-diffusion equation in terms of Mittag-Leffler functions. *International Journal of Science and Research, 15*, 1–17.

169. Saxena, R. K., Mathai, A. M., & Haubold, H. J. (2008). Solutions of certain fractional kinetic equations a fractional diffusion equation. *International Journal of Science and Research, 17*, 1–8.

170. Scalas, E., Gorenflo, R., & Mainardi, F. (2000). Fractional calculus and continuous-time finance. *Physica A, 284*, 376–384.

171. Scalas, E., Gorenflo, R., & Mainardi, F. (2004). Uncoupled continuous-time random walks: Solution and limiting behavior of the master equation. *Physical Review E,69*, 011107/1-8.

172. Schneider, W. R. (1990). Grey noise. In S. Albeverio, G. Casati, U. Cattaneo, D. Merlini, & R. Moresi (Eds.), *Stochastic processes, physics and geometry* (pp. 676–681). Singapore: World Scientific.

173. Schneider, W. R. (1996). Completely monotone generalized Mittag-Leffler functions. *Expositiones Mathematicae, 14*, 3–16.

174. Schneider, W. R., & Wyss, W. (1989). Fractional diffusion and wave equations. *Journal of Mathematical Physics, 30,* 134–144.

175. Scott-Blair, G. W. (1949). *Survey of general and appplied rheology.* London: Pitman.

176. Srivastava, H. M. (1968). On an extension of the Mittag-Leffler function. *The Yokohama Mathematical Journal, 16,* 77–88.

177. Srivastava, H. M., & Saxena, R. K. (2001). Operators of fractional integration and their applications. *Applied Mathematics and Computation, 118,* 1–52.

178. Srivastava, H. M., Gupta, K. C., & Goyal, S. P. (1982). *The H-functions of one and two variables with applications.* New Delhi and Madras: South Asian Publishers.

179. Stankovič, B., (1970). On the function of E.M. Wright. Publ. de l'Institut Mathèmatique. *Beograd, Nouvelle Sèr., 10,* 113–124.

180. Stankovič, B., (2002). Differential equations with fractional derivatives and nonconstant coefficients. *Integral Transforms and Special Functions, 6,* 489–496.

181. Strick, E. (1970). A predicted pedestal effect for pulse propagation in constant-Q solids. *Geophysics, 35,* 387–403.

182. Strick, E. (1982). Application of linear viscoelasticity to seismic wave propagation. In F. Mainardi (Ed.), *Wave propagation in viscoelastic media* (Vol. 52, pp. 169–193). Research notes in mathematics. London: Pitman.

183. Strick, E., & Mainardi, F. (1982). On a general class of constant Q solids. *Geophysical Journal of the Royal Astronomical Society, 69,* 415–429.

184. Temme, N. M. (1996). *Special functions: An introduction to the classical functions of mathematical physics.* New York: Wiley.

185. Uchaikin, V. V. (2003). Relaxation processes and fractional differential equations. *International Journal of Theoretical Physics, 42,* 121–134.

186. Uchaikin, V. V. (2008). *Method of fractional derivatives.* Ulyanovsk: ArteShock-Press [in Russian].

187. Uchaikin, V. V., & Zolotarev, V. M. (1999). *Chance and stability: Stable distributions and their applications.* Utrecht: VSP.

188. West, B. J., Bologna, M., & Grigolini, P. (2003). *Physics of fractal operators.* Institute for nonlinear science. New York: Springer.

189. Wiman, A. (1905). Über den Fundamentalsatz der Theorie der Funkntionen $E_\alpha(x)$. *Acta Mathematica, 29,* 191–201.

190. Wiman, A. (1905). Über die Nullstellen der Funkntionen $E_\alpha(x)$. *Acta Mathematica, 29,* 217–234.

191. Wong, R., & Zhao, Y.-Q. (1999). Smoothing of Stokes' discontinuity for the generalized Bessel function. *Proceedings of the Royal Society of London A, 455,* 1381–1400.

192. Wong, R., & Zhao, Y.-Q. (1999). Smoothing of Stokes' discontinuity for the generalized Bessel function II. *Proceedings of the Royal Society of London A, 455,* 3065–3084.

193. Wong, R., & Zhao, Y.-Q. (2002). Exponential asymptotics of the Mittag-Leffler function. *Constructive Approximation, 18,* 355–385.

194. Wright, E. M. (1933). On the coefficients of power series having exponential singularities. *Journal of the London Mathematical Society, 8,* 71–79.

195. Wright, E. M. (1935). The asymptotic expansion of the generalized Bessel function. *Proceedings of the London Mathematical Society (Series II), 38,* 257–270.

196. Wright, E. M. (1935). The asymptotic expansion of the generalized hypergeometric function. *Journal of the London Mathematical Society, 10,* 287–293.

197. Wright, E. M. (1940). The generalized Bessel function of order greater than one. *The Quarterly Journal of Mathematics, Oxford Series, 11,* 36–48.

Chapter 2
Multivariable Calculus

2.1 Review of Euclidean n-Space R^n

2.2 Geometric Approach: Vectors in 2-Space and 3-Space

A *vector* is a quantity that is determined by both its magnitude and its direction: thus, it is a directed line segment (Figure 2.1).

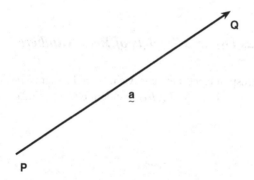

Fig. 2.1 A vector

The length of a vector **a** is also called its *norm* (Euclidean norm), denoted by $\|a\|$.

This chapter is based on the lectures of Professor D.V. Pai, IIT Bombay, and IIT Gandhinagar, Gujarat, India.

2.2.1   Components and the Norm of a Vector

For a vector $\mathbf{a}$ with the initial point $P : (u_1, u_2, u_3)$ and the terminal point $Q :$
(v_1, v_2, v_3), its *components* are the three numbers:

$$a_1 = v_1 - u_1, \quad a_2 = v_2 - u_2, \quad a_3 = v_3 - u_3,$$

and we write

$$\mathbf{a} = (a_1, a_2, a_3).$$

The norm of $\mathbf{a}$ is the number

$$\|a\| = \sqrt{a_1^2 + a_2^2 + a_3^2}.$$

2.2.2   Position Vector

Given a Cartesian coordinate system, each point $P : (x_1, x_2, x_3)$ is determined by its
position vector $\mathbf{r} = (x_1, x_2, x_3)$ with the initial point origin and the terminal point P.
The origin is determined by the null vector or *zero-vector* $\mathbf{O} = (0, 0, 0)$ with length
0 and no direction.

2.2.3   Vectors as Ordered Triplets of Real Numbers

There is a 1-1 correspondence between vectors and ordered triplets of R. Given
vectors $\mathbf{a} = (a_1, a_2, a_3)$, $\mathbf{b} = (b_1, b_2, b_3)$, we define (Figure 2.2)

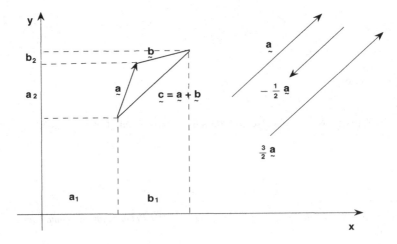

Fig. 2.2 Scalar multiplication and addition of vectors

$\mathbf{a} = \mathbf{b} \Leftrightarrow a_1 = b_1, a_2 = b_2, a_3 = b_3$ *(equality)*
$\mathbf{a} + \mathbf{b} = (a_1 + b_1, a_2 + b_2, a_3 + b_3)$ *(vector addition)*
$\alpha \mathbf{a} = (\alpha a_1, \alpha a_2, \alpha a_3)$ for all $\alpha \in R$; *(scalar multiplication)*.

2.3 Analytic Approach

Given $n \in N$, let $R^n := \{\mathbf{a} = (a_1, a_2, \ldots, a_n) : a_i \in R, \ i = 1, 2, \ldots n\}$ denote the *Cartesian n-space*. The elements of R^n are called *vectors* (or more precisely *n-vectors*).

Definition 2.1. Given two vectors $\mathbf{a} = (a_1, \ldots, a_n)$, $\mathbf{b} = (b_1, \ldots, b_n)$, we define:
(equality) $\mathbf{a} = \mathbf{b}$ iff $a_i = b_i$, $i = 1, 2, \ldots, n$.
(addition) $\mathbf{a} + \mathbf{b} = (a_1 + b_1, \ldots, a_n + b_n)$.
(scalar multiplication) $c\mathbf{a} = (ca_1, \ldots, ca_n)$ for all $c \in R$.

2.3.1 Properties of Addition and Scalar Multiplication

For all $\mathbf{a}, \mathbf{b}, \mathbf{c}$ in R^n and α, β in R, we have

(i) $\mathbf{a} + \mathbf{b} = \mathbf{b} + \mathbf{a}$ *(commutativity)*,
(ii) $(\mathbf{a} + \mathbf{b}) + \mathbf{c} = \mathbf{a} + (\mathbf{b} + \mathbf{c})$ *(associativity)*,
(iii) $\mathbf{a} + \mathbf{O} = \mathbf{O} + \mathbf{a} = \mathbf{a}$ *(zero element)*,
(iv) $\mathbf{a} + (-\mathbf{a}) = \mathbf{O}$ *(inverse element)*,
(v) $\alpha(\mathbf{a} + \mathbf{b}) = \alpha\mathbf{a} + \alpha\mathbf{b}$ *(distributivity)*,
(vi) $(\alpha + \beta)\mathbf{a} = \alpha\mathbf{a} + \beta\mathbf{a}$ *(distributivity)*,
(vii) $\alpha(\beta\mathbf{a}) = (\alpha\beta)\mathbf{a}$,
(viii) $1\mathbf{a} = \mathbf{a}$.

The above properties of addition and scalar multiplication are used as axioms for defining a general *vector space*.

2.4 The Dot Product or the Inner Product

Definition 2.2. For $\mathbf{a}, \mathbf{b}$ in R^n, their *dot* (or *inner*) product, written $\mathbf{a} \cdot \mathbf{b}$ (or $(\mathbf{a}, \mathbf{b})$) is the number

$$\mathbf{a} \cdot \mathbf{b} = \sum_{i=1}^{n} a_i b_i. \tag{2.4.1}$$

2.4.1   The Properties of Dot Product

For all $\mathbf{a}, \mathbf{b}, \mathbf{c}$ in R^n and α in R, we have:

(i) $\mathbf{a}.\mathbf{a} \geq 0$,
(ii) $\mathbf{a}.\mathbf{a} = 0$ iff $\mathbf{a} = \mathbf{O}$ $((i)\&(ii) \Rightarrow positivity)$,
(iii) $\mathbf{a}.\mathbf{b} = \mathbf{b}.\mathbf{a}$ $(commutativity)$,
(iv) $\mathbf{a} \cdot (\mathbf{b} + \mathbf{c}) = \mathbf{a}.\mathbf{b} + \mathbf{a}.\mathbf{c}$
(v) $\alpha(\mathbf{a}.\mathbf{b}) = (\alpha\mathbf{a}) \cdot \mathbf{b} = \mathbf{a} \cdot (\alpha\mathbf{b})$ $((iv)\&(v) \Rightarrow bilinearity)$.

Theorem 2.1. (The Cauchy–Schwarz inequality) *For all $\mathbf{a}, \mathbf{b}$ in R^n, we have*

$$(\mathbf{a}.\mathbf{b})^2 \leq (\mathbf{a}.\mathbf{a})(\mathbf{b}.\mathbf{b}). \tag{2.4.2}$$

Equality holds in (2.4.2) if and only if $\mathbf{a}$ is a scalar multiple of $\mathbf{b}$, or one of them is $\mathbf{O}$.

Proof. If either $\mathbf{a}$ or $\mathbf{b}$ equals $\mathbf{O}$, then (2.4.2) holds trivially. So let $\mathbf{a} \neq \mathbf{O}$ and $\mathbf{b} \neq \mathbf{O}$. Let $\mathbf{x} = \alpha\mathbf{a} - \beta\mathbf{b}$ where $\alpha = \mathbf{b}.\mathbf{b}$ and $\beta = \mathbf{a}.\mathbf{b}$, then

$$\begin{aligned} 0 \leq \mathbf{x}.\mathbf{x} &= (\alpha\mathbf{a} - \beta\mathbf{b}) \cdot (\alpha\mathbf{a} - \beta\mathbf{b}) \\ &= \alpha^2(\mathbf{a}.\mathbf{a}) - 2\alpha\beta(\mathbf{a}.\mathbf{b}) + \beta^2(\mathbf{b}.\mathbf{b}) \\ &= (\mathbf{b}.\mathbf{b})^2(\mathbf{a}.\mathbf{a}) - 2(\mathbf{b}.\mathbf{b})(\mathbf{a}.\mathbf{b})^2 + (\mathbf{a}.\mathbf{b})^2(\mathbf{b}.\mathbf{b}) \\ &= (\mathbf{b}.\mathbf{b})^2(\mathbf{a}.\mathbf{a}) - (\mathbf{a}.\mathbf{b})^2(\mathbf{b}.\mathbf{b}). \end{aligned} \tag{2.4.3}$$

Since $\mathbf{b}.\mathbf{b} \neq 0$, we may divide by $(\mathbf{b}.\mathbf{b})$ to obtain

$$(\mathbf{a}.\mathbf{a})(\mathbf{b}.\mathbf{b}) - (\mathbf{a}.\mathbf{b})^2 \geq 0,$$

which is (2.4.2). Next, if the equality $(\mathbf{a}.\mathbf{b})^2 = (\mathbf{a}.\mathbf{a})(\mathbf{b}.\mathbf{b})$ holds in (2.4.2), then by (2.4.3), $\mathbf{x}.\mathbf{x} = 0$ and so $\mathbf{x} = \mathbf{O}$, that is, $(\mathbf{b}.\mathbf{b})\mathbf{a} = (\mathbf{a}.\mathbf{b})\mathbf{b}$. Thus, $\mathbf{a}$ equals the scalar multiple $\frac{(\mathbf{a}.\mathbf{b})}{(\mathbf{b}.\mathbf{b})}\mathbf{b}$ of $\mathbf{b}$ in case $\mathbf{b} \neq \mathbf{O}$. On the other hand, if either one of $\mathbf{a}, \mathbf{b}$ equals $\mathbf{O}$ or $\mathbf{a} = k\mathbf{b}$ for some $k \in R$, then it is easily seen that

$$(\mathbf{a}.\mathbf{b})^2 = (\mathbf{a}.\mathbf{a})(\mathbf{b}.\mathbf{b}).$$

Definition 2.3. For a vector $\mathbf{a} \in R^n$, the length or *norm* of $\mathbf{a}$ is the number

$$\|\mathbf{a}\| = (\mathbf{a}.\mathbf{a})^{\frac{1}{2}}. \tag{2.4.4}$$

This enables us to rewrite the Cauchy–Schwarz inequality (2.4.2) as

$$|(\mathbf{a}.\mathbf{b})| \leq \|\mathbf{a}\| \, \|\mathbf{b}\|. \tag{2.4.5}$$

2.4.2 Properties of the Norm

For all $\mathbf{a}, \mathbf{b}$ in R^n and $\alpha \in R$, we have

(*i*) $\|\mathbf{a}\| \geq 0$,
(*ii*) $\|\mathbf{a}\| = 0$ iff $\mathbf{a} = \mathbf{O}$ ((*i*)&(*ii*) $\Rightarrow$ *positivity*),
(*iii*) $\|\alpha\mathbf{a}\| = |\alpha| \|\mathbf{a}\|$ (*homogeneity*),
(*iv*) $\|\mathbf{a} + \mathbf{b}\| \leq \|\mathbf{a}\| + \|\mathbf{b}\|$ (*triangle inequality*).

Remark 2.1. (i) The equality holds in the triangle inequality

$$\|\mathbf{a} + \mathbf{b}\| = \|\mathbf{a}\| + \|\mathbf{b}\| \tag{2.4.6}$$

iff $\mathbf{a}$ is a positive scalar multiple of $\mathbf{b}$, or one of them is $\mathbf{O}$.
(ii) The *Pythagorean identity*

$$\|\mathbf{a} + \mathbf{b}\|^2 = \|\mathbf{a}\|^2 + \|\mathbf{b}\|^2$$

holds iff $\mathbf{a}.\mathbf{b} = 0$.

Definition 2.4. Two vectors a, b in R^n are said to be *orthogonal* if $\mathbf{a}.\mathbf{b} = 0$.

2.4.3 Geometric Interpretation of the Dot Product

The Cauchy–Schwarz inequality (2.4.2) shows that for any two non-null vectors $\mathbf{a}, \mathbf{b}$ in R^n, we have

$$-1 \leq \frac{\mathbf{a}.\mathbf{b}}{\|\mathbf{a}\| \|\mathbf{b}\|} \leq 1.$$

Thus, there is exactly one real θ in the interval $0 \leq \theta \leq \pi$ such that

$$\frac{\mathbf{a}.\mathbf{b}}{\|\mathbf{a}\| \|\mathbf{b}\|} = \cos \theta.$$

This defines the *angle* θ between the vectors $\mathbf{a}$ and $\mathbf{b}$. With this definition of θ, we obtain

$$\mathbf{a}.\mathbf{b} = \|\mathbf{a}\| \|\mathbf{b}\| \cos \theta. \tag{2.4.7}$$

Recall that for the vectors in 3-space, (2.4.7) is, in fact, taken as the definition of the dot product for the geometric approach.

Given two vectors $\mathbf{x}, \mathbf{y}$ in R^n, the *distance* between them is defined by

$$d(\mathbf{x}, \mathbf{y}) = \|\mathbf{x} - \mathbf{y}\| = \left[\sum_{i=1}^{n} |x_i - y_i|^2 \right]^{\frac{1}{2}} \quad \text{(Euclidean distance)}.$$

For developing the basic notions of multivariable calculus, just as in the univariate case, the notion of *neighborhood* of a point is crucial to us.

Definition 2.5. Let $\mathbf{x}^0 = (x_1^0, \ldots, x_n^0) \in R^n$ and $r > 0$ be given. A convenient neighborhood of $\mathbf{x}^0$ is the set of all vectors $\mathbf{x} \in R^n$ such that

$$d(\mathbf{x}, \mathbf{x}^0) = \|\mathbf{x} - \mathbf{x}^0\| < r.$$

This is called the *open ball* with center $\mathbf{x}^0$ and radius r, denoted by $B(\mathbf{x}^0, r)$. Thus,

$$B(\mathbf{x}^0, r) = \{\mathbf{x} \in R^n : \|\mathbf{x} - \mathbf{x}^0\| < r\}.$$

As in the case of univariate calculus, we may prefer to begin with the notion of convergence of a sequence.

Definition 2.6. Given a sequence $\{\mathbf{a}^k\}_{k \in N}$ of vectors in R^n, we say that this sequence converges to a vector $\mathbf{a} \in R^n$, provided the sequence of real numbers $\{\|\mathbf{a}^k - \mathbf{a}\| : k \in N\}$ converges to 0. Put differently, $\mathbf{a}^k \to \mathbf{a}$ provided for every $\epsilon > 0$, there corresponds a number $K \in N$, such that

$$\|\mathbf{a}^k - \mathbf{a}\| < \epsilon \text{ for all } k \geq K.$$

This is the same as saying that for every $\epsilon > 0$, $\mathbf{a}^k$ belongs to the open ball $B(\mathbf{a}, \epsilon)$ for k large enough.

Remark 2.2. If $\mathbf{a}^{(k)} = (a_1^{(k)}, \ldots, a_n^{(k)}), k \in N$, and $\mathbf{a} = (a_1, \ldots, a_n), k \in N$, then clearly from the above definition, we have

$$\lim_{k \to \infty} \mathbf{a}^{(k)} = \mathbf{a} \iff \lim_{k \to \infty} a_i^{(k)} = a_i, \ i = 1, 2, \ldots, n.$$

2.5 Multivariable Functions

Definition 2.7. Let $D \subset R^n$. By a function $\mathbf{F}$ on D to R^m, we mean a correspondence that assigns a unique vector

$$\mathbf{y} = \mathbf{F}(\mathbf{x})$$

in R^m to each element $\mathbf{x} = (x_1, \ldots, x_n)$ in D. We write $\mathbf{F} : D \subset R^n \to R^m$ to signify that the set D is the *domain* of $\mathbf{F}$ and R^m is the target space. The *range* of $\mathbf{F}$ denoted $\mathcal{R}_{\mathbf{F}}$ is the set $\{\mathbf{F}(\mathbf{x}) : \mathbf{x} \in D\}$ of all images of elements of D. Here, $\mathbf{x}$ is called an *input vector* and $\mathbf{y}$ is called an *output vector*. If $m = 1$, the function $\mathbf{F}$ is *real-valued* or *scalar-valued* which we simply denote by F. If $m > 1$, the function is called *vector-valued*. If $n > 1$, we call such a function a *function of several variables* or a *multivariate function*.

Remark 2.3. Note that for $m > 1$, each vector-valued function $\mathbf{F} : D \subset R^n \to R^m$ can be written in terms of its components

$$\mathbf{F}(\mathbf{x}) = (F_1(\mathbf{x}), \ldots, F_m(\mathbf{x})), \quad \mathbf{x} \in D. \tag{2.5.1}$$

Here, if $\mathbf{y} = (y_1, \ldots, y_m) = \mathbf{F}(\mathbf{x})$, then we define

$$F_i(\mathbf{x}) = F_i(x_1, \ldots, x_n) = y_i, \quad \mathbf{x} \in D, \quad i = 1, \ldots, m, \tag{2.5.2}$$

called the ith component of $\mathbf{F}$. Conversely, given m scalar-valued functions $F_i : D \subset R^n \to R$, $i = 1, \ldots, m$, we can get a vector-valued function $\mathbf{F} : D \subset R^n \to R^m$ defined by (2.5.1).

2.5.1 Scalar and Vector Functions and Fields

If to each point P of a set $D \subset R^3$ is assigned a scalar $f(P)$, then a *scalar field* is said to be defined in D and the function $f : D \to R$ is called a *scalar function* (or a *scalar field* itself). Likewise, if to each point P in D is assigned a vector $\mathbf{F}(P) \in R^3$ then a *vector field* is said to be defined in D and the vector-valued function $\mathbf{F} : D \to R^3$ is called a *vector function* (or a *vector field* itself).

If we introduce Cartesian coordinates x, y, z, then instead of $f(P)$ we can write $f(x, y, z)$ and

$$\mathbf{F}(x, y, z) = (F_1(x, y, z), F_2(x, y, z), F_3(x, y, z))$$

where F_1, F_2, F_3 are the components of $\mathbf{F}$.

Remark 2.4. (i) A scalar field or a vector field arising from geometric or physical considerations must depend only on the points P where it is defined and not on the particular choice of Cartesian coordinates.

(ii) More generally, if D is a subset of R^n, then a *scalar field* in D is a function $f : D \to R$ and a *vector field* in D is a function $\mathbf{F} : D \to R^n$. In the latter case,

$$\mathbf{F}(x_1, \ldots, x_n) = (F_1(x_1, \ldots, x_n), \ldots, F_n(x_1, \ldots, x_n))$$

where $F_1, \ldots, F_n$ are the components of $\mathbf{F}$. If $n = 2$, $f(resp.\mathbf{F})$ is called a *scalar* (resp. *vector*) *field in the plane*. If $n = 3$, $f(resp.\mathbf{F})$ is a *scalar* (resp. *vector*) *field in space*.

Example 2.1. (*Euclidean distance*) Let $D = R^3$ and $f(P) = \| \overrightarrow{PP_0} \|$ the distance of point P from a fixed point P_0 in space. $f(P)$ defines a scalar field in space. If we introduce a Cartesian coordinate system in which $P_0 : (x_0, y_0, z_0)$, then

$$f(P) = f(x, y, z) = \|(x - x_0, y - y_0, z - z_0)\|$$
$$= \sqrt{(x - x_0)^2 + (y - y_1)^2 + (z - z_0)^2}.$$

Note that the value of $f(P)$ does not depend on the particular choice of Cartesian coordinate system.

Example 2.2. (*Thermal field*) In a region D of R^3, one may be required to specify steady state temperature distribution function $u : D \to R, (x, y, z) \in D \to u(x, y, z)$. In this case, D becomes a scalar field.

Example 2.3. (*Gravitational force field*) Place the origin of a coordinate system at the center of the earth (assumed spherical). By Newton's law of gravity, the force of attraction of the earth (assumed to be of mass M) on a mass m situated at point P is given by

$$\mathbf{F} = -\frac{c}{r^3}\mathbf{r}, \quad c = GMm, \quad G = \text{ the gravitational constant.}$$

Here $\mathbf{r}$ denotes the position vector of point P. If we introduce Cartesian coordinates and $P : (x, y, z)$, then

$$\mathbf{F}(x, y, z) = (-\frac{c}{r^3}x, -\frac{c}{r^3}y, -\frac{c}{r^3}z) = -\frac{c}{r^3}(x\mathbf{i} + y\mathbf{j} + z\mathbf{k}),$$

where $r = \|\mathbf{r}\|$.

Example 2.4. (*Electrostatic force field*) According to Coulomb's law, the force acting on a charge e at a position $\mathbf{r}$ due to a charge Q at the origin is given by

$$\mathbf{F} = \frac{\epsilon Q e}{r^3}\mathbf{r},$$

ϵ being a constant. For $Qe > 0$ (like charges) the force is repulsive and for $Qe < 0$ (unlike charges) the force is attractive (Figure 2.3).

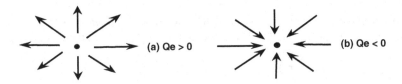

Fig. 2.3 Electrostatic force field

Example 2.5. One may be required to specify the reaction rate of a solution consisting of say five reacting chemicals $C_1, C_2, \ldots, C_5$. This requires a scalar function $\phi : D \subset R^5 \to R$ where $\phi(x_1, x_2, \ldots, x_5)$ gives the rate when the chemicals are in the indicated proportion. Again, in this case, D becomes a scalar field.

Example 2.6. In medical diagnostics, for carrying out a *stress test*, it may be required to specify the *cardiac vector* (the vector giving the magnitude and direction of electric current flow in the heart) as it depends on time. This requires a vector-valued function $\mathbf{r} : R \to R^3$ (which is *not* a vector field).

2.5.2 Visualization of Scalar-Valued Functions

In view of Remark 2.3, we first consider a scalar-valued function $f : R^n \to R$ of n real variables. The *natural domain* $\mathcal{D}_f$ of such a function is the set of all vectors $\mathbf{x} \in R^n$ such that $f(\mathbf{x}) \in R$.

Example 2.7. Let $f(x, y, z) = \sqrt{x^2 + y^2 + z^2}$. Here, $\mathcal{D}_f = R^3$ and $\mathcal{R}_f = \{x \in R : x \geq 0\}$.

Example 2.8. Let $g(x, y, z) = \sqrt{25 - x^2 - y^2 - z^2}$. Here, $\mathcal{D}_g = \{(x, y, z) : x^2 + y^2 + z^2 \leq 25\}$ and $\mathcal{R}_g = [0, 5]$.

Example 2.9. Let $h(x, y, z) = \sin(\frac{1}{xyz})$. Here, $\mathcal{D}_h = \{(x, y, z) \in R^3 : xyz \neq 0\}$ and $\mathcal{R}_h = [-1, 1]$.

Definition 2.8. (i) If $f : D \subset R^n \to R$ is a given function, then its *graph* $G(f)$ is the set

$$\{(x_1, \ldots, x_n, f(x_1, \ldots, x_n)) \in R^{n+1} : \mathbf{x} = (x_1, \ldots, x_n) \in D\}$$

in R^{n+1}. For $n = 2$, the graph of the function $f(x, y)$ of two variables is the surface consisting of all the points (x, y, z) such that $(x, y) \in D$ and $z = f(x, y)$.
(ii) Given a function $f : R^n \to R$, its *level set at height* α is the set

$$\{\mathbf{x} = (x_1, \ldots, x_n) : f(\mathbf{x}) = \alpha\}.$$

For $n = 2$, it is called a *level curve* and for $n = 3$, it is called a *level surface*.

Example 2.10. (i) For the function $f(x, y) = x + y + 1$, its graph is the plane $\{(x, y, z = x + y + 1) : (x, y) \in R^2\}$, and its level curves are the straight lines $x + y = c$.
(ii) For the function $f(x, y) = x^2 + y^2$, its graph is the paraboloid of revolution

$$\{(x, y, z = x^2 + y^2) : (x, y) \in R^2\}$$

and its level curves are circles $x^2 + y^2 = c^2$ in the xy-plane.
(iii) For the function $f(x, y) = \sqrt{4 - x^2 - y^2}$, its graph is the upper hemisphere of radius 2, and its level curves are circles $x^2 + y^2 = 4 - c^2$, $|c| \leq 2$.

2.6 Limits and Continuity

We first consider scalar-valued function of multivariables. For simplicity, we confine ourselves to functions of two variables. Let $(x_0, y_0) \in R^2$ and $r > 0$. Recall that the *open ball* $B_r(x_0, y_0)$ of center (x_0, y_0) and radius r (for the Euclidean distance) is the set

$$\{(x, y) \in R^2 : \|(x, y) - (x_0, y_0)\| = \sqrt{(x - x_0)^2 + (y - y_0)^2} < r\},$$

which is, in fact, the open disk of center (x_0, y_0) and radius r. Let us also denote by $\tilde{B}_r(x_0, y_0)$ the set

$$\{(x, y) \in R^2 : |x - x_0| < r \text{ and } |y - y_0| < r\},$$

which is, in fact, the open square centered at (x_0, y_0) of side $2r$. It is easy to see that $\tilde{B}_r(x_0, y_0)$ is the open ball of center (x_0, y_0) and radius r for the metric

$$d_\infty((x_1, y_1), (x_2, y_2)) = \max\{|x_1 - x_2|, |y_1 - y_2|\}, (x_1, y_1), (x_2, y_2) \in R^2.$$

Observe that
$$B_r(x_0, y_0) \subset \tilde{B}_r(x_0, y_0) \subset B_{\sqrt{2}r}(x_0, y_0).$$

This inclusion of open balls makes it possible to choose either of the two sets $B_r(x_0, y_0), \tilde{B}_r(x_0, y_0)$ as a neighborhood of the point (x_0, y_0).

2.6.1 Limits

The notion of the limit $\lim_{(x,y)\to(x_0,y_0)} f(x, y)$ is defined analogously as in the univariate case. Let us recall that a point $(x_0, y_0) \in R^2$ is called a *limit point* of the set $D \subset R^2$ if for every $r > 0$, the neighborhood $B_r(x_0, y_0)$ of (x_0, y_0) contains a point of D other than (x_0, y_0). Throughout in this subsection, we will make one of the following assumptions:
(i) The natural domain $\mathcal{D}_f$ of f contains a neighborhood $B_r(x_0, y_0)$ of (x_0, y_0) except possibly (x_0, y_0) itself;
(ii) $f : D \subset R^2 \to R$ is a given function and (x_0, y_0) is a limit point of D.

Definition 2.9. Assume either (i) or (ii) above holds. Given a number $L \in R$, one says

$$f(x, y) \to L \text{ as } (x, y) \to (x_0, y_0),$$

written $\lim_{(x,y)\to(x_0,y_0)} f(x, y) = L$ if for every $\epsilon > 0$, there corresponds a $\delta > 0$, such that

$$(x, y) \in D \cap B_\delta(x_0, y_0), (x, y) \neq (x_0, y_0) \Rightarrow |f(x, y) - L| < \epsilon.$$

In case, we are assuming (i), the above condition can be simply replaced by

$$(x, y) \in B_\delta(x_0, y_0), (x, y) \neq (x_0, y_0) \Rightarrow |f(x, y) - L| < \epsilon.$$

Remarks 2.5. (i) In the above definition, $B_\delta(x_0, y_0)$ can be replaced by $\tilde{B}_\delta(x_0, y_0)$.
(ii) Intuitively, the definition simply says that $f(x, y)$ comes arbitrarily close to L whenever (x, y) is sufficiently close to (x_0, y_0).
(iii) Generalization of this definition to a function $f(x_1, \ldots, x_n)$ of n variables is clear. Given a vector $\mathbf{x}^0 = (x_1^0, \ldots, x_n^0) \in R^n$ and $r > 0$, we take the open ball $B_r(\mathbf{x}^0) = \{\mathbf{x} \in R^n : \|\mathbf{x} - \mathbf{x}^0\| < r\}$ as a neighborhood of the vector $\mathbf{x}^0$. Under assumptions analogous to (i) and (ii), we proceed exactly as before to define the notion of the limit $\lim_{\mathbf{x} \to \mathbf{x}^0} f(\mathbf{x}) = L$.

Example 2.11. Let $f : R^3 \to R$ be defined by

$$f(x, y, z) = \begin{cases} \frac{xyz}{\sqrt{x^2+y^2+z^2}} \sin(\frac{1}{xyz}), & \text{if } x \neq 0, y \neq 0 \text{ and } z \neq 0 \\ 0, & \text{if } x = 0 \text{ or } y = 0 \text{ or } z = 0. \end{cases}$$

The natural domain of the function is $\mathcal{D}_f = \{(x, y, z) : x \neq 0, y \neq 0 \text{ and } z \neq 0\}$. We have

$$|f(x, y, z) - 0| \leq \frac{|x|\,|y|\,|z|}{\sqrt{x^2 + y^2 + z^2}} \leq (x^2 + y^2 + z^2) \leq \epsilon,$$

whenever $0 < \sqrt{x^2 + y^2 + z^2} < \delta \leq \epsilon^{\frac{1}{2}}$. This shows that

$$\lim_{(x,y,z) \to (0,0,0)} f(x, y, z) = 0.$$

Theorem 2.2. *Assume conditions (i) or (ii) as in the definition of limit. Let $y = \phi(x), x \in [a, b]$ be a curve such that $x_0 \in (a, b)$ and $\lim_{x \to x_0} \phi(x) = y_0$. If*

$$\psi(x) = f(x, \phi(x)), x \in [a, b], \quad and \quad \lim_{(x,y) \to (x_0,y_0)} f(x, y) = L,$$

then $\lim_{x \to x_0} \psi(x) = L$.

Proof: Exercise.

The above theorem gives the following test for nonexistence of a limit: If there is some curve as in the last theorem, along which the limit does not exist or that the limit is different along two different curves as (x, y) approaches (x_0, y_0), then $\lim_{(x,y) \to (x_0,y_0)} f(x, y)$ does not exist.

Example 2.12. Let $f : R^2 \to R$ be defined by

$$f(x, y) = \begin{cases} \frac{xy^2}{x^2+y^4}, & (x, y) \neq (0, 0) \\ 0, (x, y) = (0, 0). \end{cases}$$

Then, $\lim_{(x,y)\to(0,0)} f(x, y)$ does not exist. Indeed, along the curve $x = my^2$, $y \neq 0$, the function has the constant value $f(my^2, y) = \frac{m}{1+m^2}$. Therefore,

$$\lim_{(x,y)\to(0,0) \text{ along } x=my^2} f(x, y) = \lim_{y\to 0} f(my^2, y) = \frac{m}{1 + m^2}.$$

The limit is different for curves with different values of m. Hence, $\lim_{(x,y)\to(0,0)}$ $f(x, y)$ does not exist.

It is sometimes convenient to use polar coordinates for examining the limit of a function of two variables as illustrated by the next example.

Example 2.13. Let $f : R^2 \to R$ be defined by

$$f(x, y) = \begin{cases} \frac{x^4 y - 2x^3 y^2 + 3x^2 y^3 + y^5}{(x^2+y^2)^2}, & (x, y) \neq (0, 0) \\ 0, & (x, y) = (0, 0). \end{cases}$$

We have

$$|f(r \cos \theta, r \sin \theta) - 0| \leq r(1 + 2 + 3 + 1) = 7r = 7\sqrt{x^2 + y^2}.$$

Therefore, $|f(x, y) - 0| < \epsilon$, whenever $0 < \sqrt{x^2 + y^2} < \delta < \epsilon/7$. This shows that $\lim_{(x,y)\to(0,0)} f(x, y) = 0$.

2.6.2 Continuity

The notion of continuity of an univariate function extends easily to multivariate functions. As before, we start with a real-valued function of two variables.

Definition 2.10. *(continuity)* Let $f : D \subset R^2 \to R$ be a function and $(x_0, y_0) \in D$. We say that f is *continuous* at (x_0, y_0) if for every $\epsilon > 0$, there exists a $\delta > 0$ such that

$$(x, y) \in B_\delta(x_0, y_0) \Rightarrow |f(x, y) - f(x_0, y_0)| < \epsilon.$$

Further, we say that f is *continuous on D* if it is continuous at each point of D.

Remarks 2.6. (i) If (x_0, y_0) is a limit point of D, then f is continuous at (x_0, y_0) $\Leftrightarrow$ $\lim_{(x,y)\to(x_0,y_0)} f(x, y)$ exists and equals $f(x_0, y_0)$.

(ii) If (x_0, y_0) is an isolated point of D, that is, for some $r > 0$, $B_r(x_0, y_0) \cap D = \{(x_0, y_0)\}$, then f is trivially continuous at (x_0, y_0).

(iii) Let $f : D \subset R^2 \to R$ be a function and $(x_0, y_0) \in D$. Then, f is continuous at $(x_0, y_0) \Leftrightarrow f$ is sequentially continuous at (x_0, y_0). Recall that f is sequentially continuous at (x_0, y_0) if whenever a sequence $\{(x_n, y_n)\}$ in D is such that $(x_n, y_n) \to (x_0, y_0)$, we have $f(x_n, y_n) \to f(x_0, y_0)$.

Theorem 2.3. *(Composition of continuous functions) (1):* $f : D \subset R^2 \to R$ is *continuous at a point* $(x_0, y_0) \in D$. *(2):* $E \subset R$ *is such that* $\mathcal{R}_f \subset E$. *(3):* $g : E \to R$ *is continuous at* $t_0 = f(x_0, y_0)$. *Then, the composed function* $g \circ f : D \to R$ *is continuous at* (x_0, y_0).

Proof: Prove this result as an exercise using the definition of norm in R^m.

We remark that the notions of limit and continuity extend easily to vector-valued functions $\mathbf{F} : D \subset R^n \to R^m$. In fact, let $\mathbf{x}^0 \in R^n$ be a limit point of D and let $\mathbf{L} \in R^m$. Then, we say

$$\lim_{\mathbf{x} \to \mathbf{x}^0} \mathbf{F}(\mathbf{x}) = \mathbf{L} \text{ provided } \lim_{\mathbf{x} \to \mathbf{x}^0} \|\mathbf{F}(\mathbf{x}) - \mathbf{L}\| = 0.$$

Remarks 2.7. If the functions $F_i : D \subset R^n \to R$, $i = 1, \ldots, m$ are the components of the function $\mathbf{F} : D \subset R^n \to R^m$:

$$\mathbf{F}(\mathbf{x}) = (F_1(\mathbf{x}), \ldots, F_m(\mathbf{x})), \ \mathbf{x} \in R^n,$$

then clearly

$$\lim_{\mathbf{x} \to \mathbf{x}^0} \mathbf{F}(\mathbf{x}) = \mathbf{L} \Leftrightarrow \lim_{\mathbf{x} \to \mathbf{x}^0} F_i(\mathbf{x}) = L_i, \ i = 1, \ldots, m \text{ where } \mathbf{L} = (L_1, \ldots, L_m).$$

For the sake of completeness, we mention that the following sandwich theorem also holds.

Theorem 2.4. *(Sandwich Theorem) Let* f, g, h *be functions defined on* $D \subset R^n$ *to* R. *Let* $\mathbf{x}^0$ *be a limit point of* D. *If*

$$g(\mathbf{x}) \leq f(\mathbf{x}) \leq h(\mathbf{x}), \ \mathbf{x} \in D,$$

and

$$\lim_{\mathbf{x} \to \mathbf{x}^0} g(\mathbf{x}) = L = \lim_{\mathbf{x} \to \mathbf{x}^0} h(\mathbf{x}),$$

then $\lim_{\mathbf{x} \to \mathbf{x}^0} f(\mathbf{x}) = L$.

Proof: Prove this as an exercise. (**Hint:** Let $H(\mathbf{x}) = h(\mathbf{x}) - g(\mathbf{x})$, $F(\mathbf{x}) = f(\mathbf{x}) - g(\mathbf{x})$. Then, $0 \leq F(\mathbf{x}) \leq H(\mathbf{x})$ and $\lim_{\mathbf{x} \to \mathbf{x}^0} H(\mathbf{x}) = 0$.)

It is now clear that it looks natural to define a function $\mathbf{F} : D \subset R^n \to R^m$ to be continuous at a point $\mathbf{x}^0 \in D$ if for every $\epsilon > 0$, there exists $\delta > 0$ such that

$$\mathbf{x} \in B_\delta(\mathbf{x}^0) \cap D \Rightarrow \|\mathbf{F}(\mathbf{x}) - \mathbf{F}(\mathbf{x}^0)\| < \epsilon.$$

Remark 2.8. (i) If $D \subset R^n$ and $\mathbf{x}^0 \in D$ is a limit point of D, then $\mathbf{F} : D \subset R^n \to R^m$ is continuous at $\mathbf{x}^0 \Leftrightarrow \lim_{\mathbf{x} \to \mathbf{x}^0} \mathbf{F}(\mathbf{x})$ exists and equals $\mathbf{F}(\mathbf{x}^0)$.
(ii) If $\mathbf{x}^0$ is an isolated point of D, then $\mathbf{F}$ is trivially continuous at $\mathbf{x}^0$.
(iii) $\mathbf{F} : D \subset R^n \to R^m$ is continuous at a point $\mathbf{x}^0 \in D$ if and only if each of its component function F_i is continuous at $\mathbf{x}^0$, $i = 1, \ldots, m$.

2.7 Partial Derivatives and Differentiability

2.7.1 Partial Derivatives

Let $f : D \subset R^2 \to R$ be a function, and let (x_0, y_0) be an **interior point** of D. By that we mean that there exists $\delta > 0$ such that $B_\delta(x_0, y_0) \subset D$. Let us recall that the partial derivative of f with respect to x at (x_0, y_0) denoted by $f_x(x_0, y_0)$ or $\frac{\partial f}{\partial x}(x_0, y_0)$ is this limit

$$\lim_{h \to 0} \frac{f(x_0 + h, y_0) - f(x_0, y_0)}{h},$$

if it exists. Geometrically, it is the slope of the tangent to the curve $z = f(x, y_0)$ obtained by intersecting the graph $(x, y, z = f(x_0, y_0))$, $(x, y) \in D$ of the function $f(x, y)$ with the plane $y = y_0$ at the point $(x_0, y_0, f(x_0, y_0, f(x_0, y_0))$. The other partial derivative $f_y(x_0, y_0)$ or $\frac{\partial f}{\partial y}(x_0, y_0)$ is defined analogously. More generally, if $f : D \subset R^n \to R$ and $\mathbf{x}^0 \in R^n$, we define the partial derivative of f with respect to x_i at $\mathbf{x}^0$ denoted by

$$D_i f(\mathbf{x}^0) \text{ or } \frac{\partial f}{\partial x_i}(\mathbf{x}^0)$$

as the limit

$$\lim_{h_i \to o} \frac{f(x_1^0, \ldots, x_i^0 + h_i, \ldots, x_n^0) - f(x_1^0, \ldots, x_n^0)}{h_i}$$

if it exists.

Example 2.14. (i) Let $f(x_1, x_2, x_3, x_4) = x_2 \sin(x_1 x_2) + e^{x_2} \cos(x_3) + x_4^2$. Then, $D_2 f = \sin(x_1 x_2) + x_1 x_2 \cos(x_1 x_2) + e^{x_2} \cos(x_3)$.

(ii) Let $f(x, y, z) = \sqrt{x^2 + y^2 + z^2}$. Then, for $(x_0, y_0, z_0) \neq (0, 0, 0)$, $f_x(x_0, y_0, z_0) = \frac{x_0}{\sqrt{x_0^2 + y_0^2 + z_0^2}}$ etc. However, it is easily seen that $f_x(0, 0, 0)$, $f_y(0, 0, 0)$, $f_z(0, 0, 0)$ do not exist.

2.7.2 Differentiability

In the case of an univariate function $f : (a, b) \to R$, one says that f is differentiable at a point $x_0 \in (a, b)$ provided the derivative $f'(x_0)$ exists and that it is differentiable in (a, b) if it is differentiable at each point of (a, b). Standard facts about univariate real-valued differentiable functions are that such functions are continuous and that the chain rule for differentiation applies to them. Going from univariate to multivariate functions, one may be tempted to believe that the existence of partial derivatives constitutes differentiability of such functions. The following example refutes such a belief.

Example 2.15. Let

$$f(x, y) = \begin{cases} 0, & \text{if } x = 0 \text{ or } y = 0 \\ 1, & \text{if } x \neq 0 \text{ and } y \neq 0. \end{cases}$$

Clearly, $f_x(0, 0) = f_y(0, 0) = 0$. However, $\lim_{(x,y)\to(0,0)} f(x, y)$ does not exist, and hence, f is not continuous at $(0, 0)$.

This example shows that existence of partial derivatives alone is not adequate for its differentiability at a point. In the univariate case, the second approach for differentiability of a function $f : (a, b) \to R$ at a point $x_0 \in (a, b)$ is the following. f is said to be differentiable at x_0, provided we can write

$$f(x_0 + h) = f(x_0) + hf'(x_0) + h\eta(h), \ |h| < \delta,$$

for a suitable $\delta > 0$ and a suitable function $\eta(h)$ defined in this range such that $\lim_{h\to 0} \eta(h) = 0$. Geometrically, this says f is differentiable at x_0, provided the tangent line approximation $L(x) = f(x_0) + (x - x_0)f'(x_0)$ is a good approximation to f in a neighborhood of the point x_0.

The next theorem is in the same spirit for a function of two variables.

Theorem 2.5. Let $f : D \subset R^2 \to R$ be a given function, and let (x_0, y_0) be an interior point of D. The following statements are equivalent: (i) There exist numbers $\alpha, \beta \in R$ such that

$$\lim_{(h,k)\to(0,0)} \frac{|f(x_0 + h, y_0 + k) - f(x_0, y_0) - \alpha h - \beta k|}{\sqrt{h^2 + k^2}} = 0. \qquad (2.7.1)$$

(ii) There exist numbers $\alpha, \beta \in R$ and functions $\epsilon_1(h, k), \epsilon_2(h, k)$ defined in an open ball $B_r(0, 0)$, for a suitable $r > 0$, such that

$$f(x_0 + h, y_0 + k) - f(x_0, y_0) = \alpha h + \beta k + \epsilon_1(h, k)h + \epsilon_2(h, k)k \qquad (2.7.2)$$

and

$$\lim_{(h,k)\to(0,0)} \epsilon_1(h,k) = 0 = \lim_{(h,k)\to(0,0)} \epsilon_2(h,k). \qquad (2.7.3)$$

Proof: $(i) \Rightarrow (ii)$: Suppose (2.7.1) holds. Define

$$\eta(h,k) := \begin{cases} \frac{f(x_0+h,y_0+k)-f(x_0,y_0)-\alpha h-\beta k}{\sqrt{h^2+k^2}}, & (h,k) \neq (0,0) \\ 0, \text{ if } (h,k) = (0,0). \end{cases}$$

By (i), $\lim_{(h,k)\to(0,0)} \eta(h,k) = 0$, and

$$f(x_0+h, y_0+k) - f(x_0, y_0) = \alpha h + \beta k + \sqrt{h^2+k^2}\, \eta(h,k).$$

We write

$$\sqrt{h^2+k^2}\, \eta(h,k) = \frac{h^2+k^2}{\sqrt{h^2+k^2}}\, \eta(h,k),$$

and let

$$\epsilon_1(h,k) = \frac{h}{\sqrt{h^2+k^2}}\, \eta(h,k), \ \ \epsilon_2(h,k) = \frac{k}{\sqrt{h^2+k^2}}\, \eta(h,k).$$

Then clearly,

$$\lim_{(h,k)\to(0,0)} \epsilon_1(h,k) = 0 = \lim_{(h,k)\to(0,0)} \epsilon_2(h,k)$$

and (2.7.3) holds. $(ii) \Rightarrow (i)$: Note that (ii) implies

$$\frac{|f(x_0+h, y_0+k) - f(x_0, y_0) - \alpha h - \beta k|}{\sqrt{h^2+k^2}} = \epsilon_1 \frac{h}{\sqrt{h^2+k^2}} + \epsilon_2 \frac{k}{\sqrt{h^2+k^2}},$$

which tends to 0 as $(h,k) \to (0,0)$.

Definition 2.11. Let (x_0, y_0) be an interior point of $D \subset R^2$ and $f : D \to R$ be a function. The function f is said to be *differentiable at* (x_0, y_0) if f satisfies condition (ii) of the last theorem. If D is an open set, then f is said to be *differentiable in D* if it is differentiable at each point of D.

Theorem 2.6. Let $f : D \subset R^2 \to R$ be a function and let (x_0, y_0) be an interior point of D. If f is differentiable at (x_0, y_0), then (a): f is continuous at (x_0, y_0); (b): both the partial derivatives of f exist at (x_0, y_0). In fact $\alpha = f_x(x_0, y_0)$ and $\beta = f_y(x_0, y_0)$ where α, β are as in (ii) of the last theorem.

Proof: This is an easy exercise.

Theorem 2.7. *(Increment Theorem) Let $D \subset R^2$ and (x_0, y_0) be an interior point of D. If $f : D \to R$ is such that the partial derivatives f_x, f_y exist at all points in an open ball $B(x_0, y_0)$ around (x_0, y_0) and these are continuous at (x_0, y_0), then f is differentiable at (x_0, y_0).*

Proof: Proof using MVT is left as an exercise.

Remarks 2.9. The conditions in the Increment Theorem are only sufficient but not necessary for differentiability of f at (x_0, y_0). As an example, consider

$$f(x, y) = \begin{cases} (x^2 + y^2) \sin \left(\frac{1}{x^2+y^2} \right), & (x, y) \neq (0, 0) \\ 0, \text{ if } (x, y) = (0, 0). \end{cases}$$

Here $f_x(0, 0) = f_y(0, 0) = 0$ and f is differentiable at $(0, 0)$. In fact, we have

$$\epsilon_1(h, k) = h \sin \left(\frac{1}{h^2 + k^2} \right), \quad \epsilon_2(h, k) = k \sin \left(\frac{1}{h^2 + k^2} \right),$$

and both $\epsilon_1(h, k)$, $\epsilon_2(h, k) \to 0$ as $(h, k) \to (0, 0)$. However,

$$f_x(x, y) = 2x \sin \left(\frac{1}{x^2 + y^2} \right) - \frac{2x}{x^2 + y^2} \cos \left(\frac{1}{x^2 + y^2} \right), \quad (x, y) \neq (0, 0),$$

and it is easily seen that along $y = 0$, $f_x(x, 0) \to -\infty$ as $x \to 0^+$ and $f_x(x, 0) \to \infty$ as $x \to 0^-$.

2.7.3 Chain Rule

We can easily extend the notion of differentiability to a function $f : D \subset R^n \to R$. More precisely, let $\mathbf{x}^0$ be an interior point of D. By the discussion in the preceding subsection, we can define f to be differentiable at $\mathbf{x}^0$ if the partial derivatives $D_i f(\mathbf{x}^0)$, $i = 1, \ldots, m$ exist and we have

$$f(x_1^0 + h_1, \ldots, x_n^0 + h_n) - f(x_1^0, \ldots, x_n^0) = h_1 D_1 f(\mathbf{x}^0) + \ldots$$
$$+ h_n D_n f(\mathbf{x}^0) + \epsilon_1(\mathbf{h})h_1 + \ldots + \epsilon_n(\mathbf{h})h_n \quad (2.7.4)$$

where $\lim_{\mathbf{h} \to \mathbf{0}} \epsilon_i(\mathbf{h}) = 0$, $i = 1, \ldots, m$. Writing

$$\mathbf{D}f(\mathbf{x}^0) = (D_1 f, \ldots, D_n f)_{\mathbf{x}^0} \quad (2.7.5)$$

as the *total derivative* of f at $\mathbf{x}^0$, we can write (2.7.4) in compact form as

$$f(\mathbf{x}^0 + \mathbf{h}) - f(\mathbf{x}^0) = \mathbf{D}f(\mathbf{x}^0) \cdot \mathbf{h} + \epsilon(\mathbf{h}) \cdot \mathbf{h} \quad (2.7.6)$$

where $\epsilon(\mathbf{h}) = (\epsilon_1(\mathbf{h}), \ldots, \epsilon_n(\mathbf{h})) \to \mathbf{O}$ as $\mathbf{h} \to \mathbf{O}$. Frequently, in calculus, one writes $w = f(x_1, \ldots, x_n)$ and (2.7.4) is written in the form

$$\Delta w = \Delta x_1 D_1 f(\mathbf{x}^0) + \ldots + \Delta x_n D_n f(\mathbf{x}^0) + \epsilon_1 \Delta x_1 + \ldots \epsilon_n \Delta x_n, \qquad (2.7.7)$$

by taking $h_i = \Delta x_i$, $i = 1, \ldots, n$. We now consider the first version of the chain rule for the case under consideration.

Theorem 2.8. *(1): $D \subset R^n$ and $\mathbf{x}^0$ is an interior point of D. (2): $f : D \subset R^n \to R$ is differentiable at $\mathbf{x}^0$. (3): $x_1 = x_1(t), \ldots, x_n = x_n(t)$ are functions defined from (a, b) to R, which are differentiable at $t_0 \in (a, b)$. (4): $(x_1(t_0), \ldots, x_n(t_0)) = \mathbf{x}^0$ and $(x_1(t), \ldots, x_n(t)) \in D$ for $t \in (a, b)$. Then, $W = f(x_1(t), \ldots, x_n(t))$ is differentiable at t_0 and*

$$W'(t_0) = \sum_{i=1}^{n} D_i'(\mathbf{x}^0) x_i'(t_0). \qquad (2.7.8)$$

Proof: As t changes from t_0 to $t_0 + \Delta t$, the function x_i changes to $x_i + \Delta x_i$ where $\Delta x_i = x_i(t_0 + \Delta t) - x_i(t_0)$, $i = 1, \ldots, n$. Differentiability of f entails

$$f(x_1^0 + \Delta x_1, \ldots, x_n^0 + \Delta x_n) - f(x_1^0, \ldots, x_n^0) = \sum_{i=1}^{n} D_i f(\mathbf{x}^0) \Delta x_i + \epsilon(\Delta \mathbf{x}) \cdot \Delta \mathbf{x}.$$

Divide both sides by Δt and let $\Delta t \to 0$, to obtain (2.7.8).

2.7.4 Gradient and Directional Derivatives

Let $D \subset R^n$ and $\mathbf{x}^0$ be an interior point of D. Let $f : D \subset R^n \to R$ be given. If the partial derivatives $D_i f(\mathbf{x}^0)$ all exist, then the vector $(D_1 f(\mathbf{x}^0), \ldots, D_n f(\mathbf{x}^0))$ is called the *gradient* of f at $\mathbf{x}^0$. It is denoted by $\nabla f(\mathbf{x}^0)$ or $\mathrm{grad} f(\mathbf{x}^0)$. Now, fix up any direction $\mathbf{u} = (u_1, \ldots, u_n)$ in R^n. The requirement that $\|\mathbf{u}\| = 1$ is not essential. The line through $\mathbf{x}^0$ in the direction $\mathbf{u}$ has the equation: $\mathbf{x} = \mathbf{x}^0 + t\mathbf{u}, t \in R$. This gives rise to the parametric equations

$$x_i = x_i(t) = x_i^0 + t u_i, \ i = 1, \ldots, n.$$

Definition 2.12. The directional derivative of f at $\mathbf{x}^0$ in the direction $\mathbf{u}$ is the limit

$$\lim_{t \to 0} \frac{f(\mathbf{x}^0 + t\mathbf{u}) - f(\mathbf{x}^0)}{t}$$

if it exists. It is denoted variously by $D_{\mathbf{u}} f(\mathbf{x}^0)$, $\frac{\partial f}{\partial \mathbf{u}}(\mathbf{x}^0)$, or $f'(\mathbf{x}^0, \mathbf{u})$.

The notion of directional derivative extends the notion of partial derivative: If $\mathbf{e}_i = (0, \ldots, 1, 0, \ldots, 0)$, then $D_{\mathbf{e}_i} f(\mathbf{x}^0)$, $i = 1, \ldots, n$. It is clear from the definition that if $D_{\mathbf{u}} f(\mathbf{x}^0)$ exists, then $D_{-\mathbf{u}} f(\mathbf{x}^0)$ also exists, and $D_{-\mathbf{u}} f(\mathbf{x}^0) = -D_{\mathbf{u}} f(\mathbf{x}^0)$. The next theorem is a crucial link between differentiability and the existence of directional derivatives of a function.

Theorem 2.9. *(1): $D \subset R^n$ and $\mathbf{x}^0$ is an interior point of D. (2): $f : D \subset R^n \to R$ is differentiable at $\mathbf{x}^0$. Then, in every direction $\mathbf{u} \in R^n$, the directional derivative $D_{\mathbf{u}} f(\mathbf{x}^0)$ exists and is equal to*

$$\nabla f(\mathbf{x}^0) \cdot \mathbf{u} = D_1 f(\mathbf{x}^0) u_1 + \ldots + D_n F(\mathbf{x}^0) u_n.$$

Proof: Indeed, the differentiability of f at $\mathbf{x}^0$ entails

$$f(\mathbf{x}^0 + t\mathbf{u}) - f(\mathbf{x}^0) = (tu_1) D_1 f(\mathbf{x}^0) + \ldots + (tu_n) D_n f(\mathbf{x}^0) + t\mathbf{u} \cdot \epsilon(t\mathbf{u})$$

where $\epsilon(t\mathbf{u}) \to \mathbf{O}$ as $t \to 0$. Dividing both sides by t and letting $t \to 0$, we conclude that $D_{\mathbf{u}} f(\mathbf{x}^0)$ exists and is equal to $\nabla f(\mathbf{x}^0) \cdot \mathbf{u}$.

Remarks 2.10. (i): Note that existence of all partial derivatives need not imply existence of directional derivative $D_{\mathbf{u}} f(\mathbf{x}^0)$ in every direction. By way of an example, let

$$f(x, y) = \begin{cases} x + y, & \text{if } x = 0 \text{ or } y = 0 \\ 1, & \text{otherwise.} \end{cases}$$

Clearly, $f_x(0, 0) = f_y(0, 0) = 1$. Nevertheless, if we take $\mathbf{u} = (u_1, u_2)$, $u_1 \neq 0$, and $u_2 \neq 0$, then

$$\frac{f(\mathbf{O} + t\mathbf{u}) - f(\mathbf{O})}{t} = \frac{f(t\mathbf{u})}{t} = \frac{1}{t},$$

and this does not tend to a limit as $t \to 0$. (ii): The converse of the preceding theorem is false. For example, let

$$f(x, y) = \begin{cases} \frac{xy^2}{x^2 + y^4}, & \text{if } x \neq 0 \\ 0, & \text{if } x = 0. \end{cases}$$

It is easily seen that f is not continuous at $(0, 0)$; hence, it is, a fortiori, not differentiable at $(0, 0)$. Let $\mathbf{u} = (u_1, u_2)$. Then,

$$\frac{f(0 + tu_1, 0 + tu_2) - f(0, 0)}{t} = \frac{f(tu_1, tu_2)}{t} = \frac{u_1 u_2^2}{u_1^2 + t^2 u_2^4}.$$

Therefore,

$$D_{\mathbf{u}} f(0, 0) = \begin{cases} \frac{u_2^2}{u_1}, & \text{if } u_1 \neq 0 \\ 0, & \text{if } u_1 = 0. \end{cases}$$

The next example shows that the directional derivative $D_{\mathbf{u}} f(\mathbf{x}^0)$ may exist in every direction and f may be continuous, and yet f may be non-differentiable at $\mathbf{x}^0$.

Example 2.16. Let

$$f(x, y) = \begin{cases} \frac{y}{|y|}\sqrt{x^2 + y^2}, & \text{if } y \neq 0 \\ 0, & \text{if } y = 0. \end{cases}$$

Clearly, $|f(x, y)| = \sqrt{x^2 + y^2}$, which implies f is continuous at $(0, 0)$. Fix up $\mathbf{u} = (u_1, u_2) \in R^2$ such that $u_1^2 + u_2^2 = 1$. Then,

$$\frac{f(tu_1, tu_2) - f(0, 0)}{t} = \frac{\frac{tu_2}{|tu_2|}\sqrt{t^2 u_1^2 + t^2 u_2^2}}{t} = \frac{u_2}{|u_2|}, \quad u_2 \neq 0.$$

Therefore,

$$D_{\mathbf{u}} f(0, 0) = \begin{cases} \frac{u_2}{|u_2|}, & \text{if } u_2 \neq 0 \\ 0, & \text{if } u_2 = 0. \end{cases}$$

Also, it is easily seen that $f_x(0, 0) = 0$, $f_y(0, 0) = 1$, and grad $f \cdot \mathbf{u} \neq D_{\mathbf{u}} f(0, 0) = \frac{u_2}{|u_2|}$ if $u_2 \neq 0$ and $|u_2| \neq 1$. This shows that f is not differentiable at $(0, 0)$.

2.7.5   Tangent Plane and Normal Line

We consider a differentiable function $f : D \subset R^3 \to R$ defined in an open set D. If $C : \mathbf{r} = \dot{\mathbf{r}}(t) = (x(t), y(t), z(t)), a \leq t \leq b$ is a smooth curve (i.e., we are assuming $\dot{\mathbf{r}}(t)$ is continuous on $[a, b]$) on the level surface $S : f(x, y, z) = c$ of f, then $f(x(t), y(t), z(t)) = c$. Differentiating both sides of this equation with respect to t, by chain rule, we get

$$\frac{\partial f}{\partial x}\frac{dx}{dt} + \frac{\partial f}{\partial y}\frac{dy}{dt} + \frac{\partial f}{\partial z}\frac{dz}{dt} = 0.$$

Equivalently, we can write $\nabla f \cdot \dot{\mathbf{r}} = 0$, which says that at every point along the curve C, ∇f is orthogonal to the tangent vector $\dot{\mathbf{r}}$ to C. Thus, if we fix up a point P_0 on S and consider all possible curves on S passing through P_0, all the tangent lines to these curves are orthogonal to the vector $\nabla f(P_0)$. This motivates us to define

Definition 2.13. The *tangent plane* at the point $P_0 : (x_0, y_0, z_0)$ to the level surface $f(x, y, z) = c$ is the plane through P_0 which is orthogonal to the vector $\nabla f(P_0)$. The *normal line* to the surface at P_0 is the line through P_0 parallel to $\nabla f(P_0)$.

The vector equations of the tangent plane and normal line are, respectively,

$$(\mathbf{r} - \mathbf{r}_0) \cdot \nabla \mathbf{f}(P_0) = 0, \ \mathbf{r} = \mathbf{r}_0 + t \nabla \mathbf{f}(P_0), \ \mathbf{r}_0 = (x_0, y_0, z_0).$$

The corresponding scalar equations are, respectively,

$$f_x(P_0)(x - x_0) + f_y(P_0)(y - y_0) + f_z(P_0)(z - z_0) = 0,$$
$$x = x_0 + t f_x(P_0), \ y = y_0 + t f_y(P_0), \ z = z_0 + t f_z(P_0).$$

If we are given a differentiable function $f : D \subset R^2 \to R$ where D is an open set, then its graph $\{(x, y, z = f(x, y)) : (x, y) \in D\}$ is the level surface $F(x, y, z) = 0$ of the function

$$F(x, y, z) = f(x, y) - z,$$

whose partial derivatives are $F_x = f_x$, $F_y = f_y$, $F_z = -1$, respectively. In this case, the equations of the tangent plane and the normal line become, respectively,

$$f_x(P_0)(x - x_0) + f_y(P_0)(y - y_0) - (z - z_0),$$
$$x = x_0 + t f_x(P_0), \ y = y_0 + t f_y(P_0), \ z = z_0 - t.$$

2.7.6 Differentiability of Vector-Valued Functions, Chain Rule

We are now ready for the general definition of differentiability of functions from R^n to R^m. Let $\mathbf{F} : \Omega \subset R^n \to R^m$ be a given function and let $\mathbf{x}^0$ be an interior point of Ω. As before, we write

$$\mathbf{F}(\mathbf{x}) = (F_1(\mathbf{x}), \ldots, F_m(\mathbf{x})), \ \mathbf{x} \in \Omega \tag{2.7.9}$$

where each F_i maps Ω to R, so that F_i's are the components of $\mathbf{F}$. Let us continue to denote by $D_j F_i$, the jth partial derivative $\frac{\partial F_i}{\partial x_j}$ of F_i, $i = 1, \ldots, m$, $j = 1, \ldots, n$.

Definition 2.14. The *derivative matrix* of $\mathbf{F}$ at $\mathbf{x}^0$ is the matrix

$$\mathbf{DF}(\mathbf{x}^0) = [D_j F_i]_{1 \leq i \leq m, 1 \leq j \leq n} \tag{2.7.10}$$

where the partial derivatives $D_j F_i$ evaluated at $\mathbf{x}^0$ are assumed to exist. We say that the vector-valued function $\mathbf{F}$ is differentiable at $\mathbf{x}^0$, provided these partial derivatives exist at $\mathbf{x}^0$ and if

$$\lim_{\mathbf{h} \to \mathbf{0}} \frac{\|\mathbf{F}(\mathbf{x}^0 + \mathbf{h}) - \mathbf{F}(\mathbf{x}^0) - \mathbf{DF}(\mathbf{x}^0)\mathbf{h}\|}{\|\mathbf{h}\|} = 0. \tag{2.7.11}$$

In the last equation, we regard $\mathbf{h} = [h_1, \ldots, h_n]^T$ as a n-column vector, so that it can be multiplied by the $m \times n$ matrix $\mathbf{DF}(\mathbf{x}^0)$. In case $\mathbf{F}$ is differentiable at $\mathbf{x}^0$, the derivative matrix of $\mathbf{F}$ at $\mathbf{x}^0$ is sometimes called the *total derivative* of $\mathbf{F}$ at $\mathbf{x}^0$.

Remark 2.11. Applying Remark 2.7 to equation (2.7.11), it is immediately clear that $\mathbf{F} : \Omega \subset R^n \to R^m$ is differentiable at an interior point $\mathbf{x}^0$ of Ω if and only if each component function $F_i : \Omega \subset R^n \to R$, $i = 1, \ldots, m$ is differentiable at $\mathbf{x}^0$.

2.7.7 Particular Cases

(i) Let $m = 1$. Here, $F : \Omega \subset R^n \to R$ and

$$\mathbf{DF}(\mathbf{x}^0) = [D_1 F, \ldots, D_n F]_{\mathbf{x}^0},$$

as already seen earlier.

(ii) Let $n = 1$ and $m > 1$. Here, $\mathbf{F} : (a, b) \subset R \to R^m$, written

$$\mathbf{F}(t) = (F_1(t), \ldots, F_m(t)), \quad t \in (a, b),$$

which is a vector-valued function encountered frequently for parametrizing a *curve* in R^m. Clearly,

$$\mathbf{DF}(t) = [F_1'(t), \ldots, F_m'(t)]^T$$

which is written frequently as $\mathbf{F}'(t)$ or $\dot{\mathbf{F}}(t)$ and it represents tangent vector to the curve $C : \mathbf{F} = \mathbf{F}(t), t \in (a, b)$ in R^m.

For the sake of completeness, we state below without proof the following analogue of the *Increment Theorem* for the general case.

Theorem 2.10. *Let* $\mathbf{F} : \Omega \subset R^n \to R^m$ *and* $\mathbf{x}^0$ *be an interior point of* Ω. *Writing* $\mathbf{F}$ *in terms of its components*

$$\mathbf{F}(\mathbf{x}) = (F_1(\mathbf{x}), \ldots, F_m(\mathbf{x})), \mathbf{x} \in \Omega,$$

if the partial derivatives

$$D_j F_i, \; i = i, \ldots, m; \; j = 1, \ldots, n$$

all exist in an open ball $B_\delta(\mathbf{x}^0)$ *around* $\mathbf{x}^0$ *and are continuous at* $\mathbf{x}^0$, *then* $\mathbf{F}$ *is differentiable at* $\mathbf{x}^0$.

Remark 2.12. It follows from Remark 2.9 and the example therein that the conditions in the above theorem are only sufficient but not necessary for the differentiability of $\mathbf{F}$ at $\mathbf{x}^0$.

Next, we state, without proof, the general form of the chain rule.

Theorem 2.11. *(Chain Rule)(1):* $\mathbf{G} : \Omega \subset R^p \to R^n$ *is differentiable at an interior point* $\mathbf{a}$ *of* Ω. *(2):* $\mathbf{F} : \Omega_1 \subset R^n \to R^m$ *is differentiable at* $\mathbf{b} = \mathbf{G}(\mathbf{a})$, *which is an interior point of* Ω_1. *Then, the composite function* $\mathbf{H} : \Omega \subset R^p \to R^m$ *defined by*

$$\mathbf{H} = \mathbf{F} \circ \mathbf{G} : \mathbf{H}(\mathbf{x}) = \mathbf{F}(\mathbf{G}(\mathbf{x})), \quad \mathbf{x} \in R^p$$

is differentiable at $\mathbf{a}$ *and we have*

$$\mathbf{DH}(\mathbf{a}) = \mathbf{DF}(\mathbf{b}) \cdot \mathbf{DG}(\mathbf{a}). \tag{2.7.12}$$

Note that the matrix on the left-hand side is of order $m \times p$, which is a product of an $m \times n$ matrix with an $n \times p$ matrix. The following special case is encountered frequently in multivariable calculus: $F : \Omega_1 \subset R^n \to R$, and $\mathbf{G} : \Omega \subset R^p \to R^n$. Here, $m = 1$, and $H : \Omega \to R$. We have

$$\mathbf{DH}(\mathbf{a}) = [D_1 H, \dots, D_p H]_{\mathbf{a}}, \mathbf{DF}(\mathbf{b}) = [D_1 F, \dots, D_n F]_{\mathbf{b}},$$

and $\mathbf{DG}(\mathbf{a}) = [D_j G_i]_{i=1,\dots,n; \ j=1,\dots,p}(\mathbf{a})$. Hence, from (2.7.12),

$$[D_1 H, \dots, D_p H]_{\mathbf{a}} = [D_1 F, \dots, D_n F]_{\mathbf{b}}[D_j G_i](\mathbf{a}).$$

This gives the familiar formulae using chain rule:

$$D_j H(\mathbf{a}) = \sum_{i=1}^{n} D_i F(\mathbf{b}) D_j G_i(\mathbf{a}), \quad j = 1, \dots, p. \tag{2.7.13}$$

For example, let $f : R^3 \to R, \mathbf{g} : R^3 \to R^3$ be differentiable. Write

$$g(u, v, w) = (x(u, v, w), y(u, v, w), z(u, v, w))$$

and define $h : R^3 \to R$ by setting

$$h(u, v, w) = f(x(u, v, w), y(u, v, w), z(u, v, w)).$$

Then,

$$\left[\frac{\partial h}{\partial u} \frac{\partial h}{\partial v} \frac{\partial h}{\partial w} \right] = \left[\frac{\partial f}{\partial x} \frac{\partial f}{\partial y} \frac{\partial f}{\partial z} \right] \begin{bmatrix} \frac{\partial x}{\partial u} & \frac{\partial x}{\partial v} & \frac{\partial x}{\partial w} \\ \frac{\partial y}{\partial u} & \frac{\partial y}{\partial v} & \frac{\partial y}{\partial w} \\ \frac{\partial z}{\partial u} & \frac{\partial z}{\partial v} & \frac{\partial z}{\partial w} \end{bmatrix}.$$

which gives the familiar chain rule for three intermediate and three independent variables.

2.7.8 Differentiation Rules

Here and in the sequel, it would be convenient for us to regard R^n as the space of column n-vectors $\mathbf{x} = [x_1, \ldots, x_n]^T$ with real entries. Let us recall that if the function $f : R^n \to R$ is differentiable, then the function $\nabla : R^n \to R^n$ called the *gradient* of f is defined by

$$\nabla f(\mathbf{x}) = \mathbf{D}f(\mathbf{x})^T = [\frac{\partial f}{\partial x_1}(\mathbf{x}), \ldots, \frac{\partial f}{\partial x_n}(\mathbf{x})]^T.$$

Given $f : R^n \to R$, if ∇f is differentiable, then f is said to be *twice differentiable*, and we write the derivative of ∇f as

$$\mathbf{D}^2 f = \begin{bmatrix} \frac{\partial^2 f}{\partial x_1^2} & \cdots & \frac{\partial^2 f}{\partial x_n \partial x_1} \\ \cdots & \cdots & \cdots \\ \frac{\partial^2 f}{\partial x_1 \partial x_n} & \cdots & \frac{\partial^2 f}{\partial x_n \partial x_n} \end{bmatrix}.$$

The matrix $\mathbf{D}^2 f(\mathbf{x})$ is called the *Hessian matrix* of f at $\mathbf{x}$. Let us also recall that a function $\mathbf{F} : \Omega \to R^m$, where Ω is an open subset of R^n is said to be continuously differentiable in Ω if it is differentiable, and $\mathbf{DF} : \Omega \to R^{m \times n}$ is continuous. Here, $R^{m \times n}$ denotes the space of all $m \times n$ matrices with real entries. (This amounts to saying that the components of $\mathbf{F}$ have continuous first partial derivatives.) In this case, we write $\mathbf{F} \in \mathcal{C}^{(1)}(\Omega.)$

Theorem 2.12. *Let Ω be an open subset of R^n, $g : \Omega \to R$ be differentiable in Ω, and let $\mathbf{f} : (a, b) \to \Omega$ be differentiable in (a, b). Then, $h = g \circ \mathbf{f} : (a, b) \to R$ defined by $h(t) = g(\mathbf{f}(t)), t \in (a, b)$ is differentiable in (a, b) and*

$$h'(t) = Dg(\mathbf{f}(t))D\mathbf{f}(t) = \nabla g(\mathbf{f}(t))^T [f_1'(t), \ldots, f_n'(t)]^T, \ t \in [a, b].$$

Theorem 2.13. *Let $\mathbf{f} : R^n \to R^m$, $\mathbf{g} : R^n \to R^m$ be two differentiable functions. Let $h : R^n \to R$ be defined by*

$$h(\mathbf{x}) = \mathbf{f}(\mathbf{x})^T \mathbf{g}(\mathbf{x}) = < f(\mathbf{x}), g(\mathbf{x}) > . \tag{2.7.14}$$

Then, h is differentiable and

$$Dh(\mathbf{x}) = \mathbf{f}(\mathbf{x})^T Dg(\mathbf{x}) + \mathbf{g}(\mathbf{x})^T Df(\mathbf{x}). \tag{2.7.15}$$

Remarks 2.13. (Some useful formulae)
Let $\mathbf{A} \in R^{m \times n}$ and $\mathbf{y} \in R^m$ be given. Then, we have:

(i):

$$D(\mathbf{y}^T \mathbf{A}\mathbf{x}) = \mathbf{y}^T \mathbf{A}, \ \mathbf{x} \in R^n. \tag{2.7.16}$$

(ii):

$$D(\mathbf{x}^T A \mathbf{x}) = \mathbf{x}^T (A + A^T), \ \mathbf{x} \in R^n. \tag{2.7.17}$$

2.7.9 Taylor's Expansion

Let us recall the following order symbols. Let $g : B \to R$ be a function defined in an open ball B around $\mathbf{0} \in R^n$ such that $g(\mathbf{x}) \neq 0$ for $\mathbf{x} \neq \mathbf{0}$, and let $f : C \to R^m$ be defined in a set $C \subseteq R^n$ that contains $\mathbf{0} \in R^n$. Then, (a) : the symbol

$$\mathbf{f}(\mathbf{x}) = O(g(\mathbf{x}))$$

means that $\frac{\|\mathbf{f}(\mathbf{x})\|}{|g(\mathbf{x})|}$ is bounded near $\mathbf{0}$. More precisely, there are numbers $K > 0$ and $\delta > 0$ such that

$$x \in C \text{ and } \|x\| < \delta \Rightarrow \frac{\|\mathbf{f}(\mathbf{x})\|}{|g(\mathbf{x})|} \leq K.$$

(b) : The symbol $f(\mathbf{x}) = o(g(\mathbf{x}))$ means that

$$\lim_{\mathbf{x}\to 0, \mathbf{x}\in C} \frac{\|\mathbf{f}(\mathbf{x})\|}{|g(\mathbf{x})|} = 0.$$

Theorem 2.14. *(Taylor's Theorem) Let $f : R \to R$ be in $C^m[a, b]$, and $0 \leq h \leq b - a$. Then,*

$$f(a + h) = f(a) + \frac{h}{1!}f'(a) + \dots + \frac{h^{m-1}}{(m-1)!}f^{(m-1)}(a) + R_m \tag{2.7.18}$$

where $R_m = \frac{h^m}{m!}f^{(m)}(a + \theta h)$, for a suitable $\theta \in (0, 1)$.

Remarks 2.14. Note that since $f \in C^m[a, b]$, $f^{(m)}(a + \theta h) = f^{(m)} + o(1)$. Thus, if $f \in C^{(m)}$, then we can write Taylor's formula as

$$f(a + h) = f(a) + \frac{h}{1!}f'(a) + \dots + \frac{h^m}{m!}f^m(a) + o(h^m). \tag{2.7.19}$$

In addition, if we assume that $f \in C^{(m+1)}[a, b]$, then $R_{m+1} = \frac{h^{m+1}}{(m+1)!}f^{(m+1)}(a + \theta h)$, and since $f^{(m+1)}$ is bounded, we can conclude that $R_{m+1} = O(h^{(m+1)})$. Thus, in this case, we can write Taylor's formula as

$$f(a + h) = f(a) + \frac{h}{1!}f'(a) + \dots + \frac{h^m}{m!}f^m(a) + O(h^{m+1}). \tag{2.7.20}$$

Theorem 2.15. *Let $f : \Omega \to R$, where Ω is an open subset of R^n, be a function in class $C^2(\Omega)$. Let $\mathbf{x_0} \in \Omega$ and $\mathbf{x}$ be in an open ball around $\mathbf{x_0}$ contained in Ω. Let $\mathbf{h} = \mathbf{x} - \mathbf{x_0}$. Then,*

$$f(\mathbf{x}_0 + \mathbf{h}) = f(\mathbf{x}_0) + \frac{1}{1!}\mathbf{D}f(\mathbf{x}_0)\mathbf{h} + \frac{1}{2!}\mathbf{h}^T\mathbf{D}^2f(\mathbf{x}_0)\mathbf{h} + o(\|\mathbf{h}\|^2). \qquad (2.7.21)$$

Moreover, if $f \in C^3(\Omega)$, then we have:

$$f(\mathbf{x}_0 + \mathbf{h}) = f(\mathbf{x}_0) + \frac{1}{1!}\mathbf{D}f(\mathbf{x}_0)\mathbf{h} + \frac{1}{2!}\mathbf{h}^T\mathbf{D}^2f(\mathbf{x}_0)\mathbf{h} + O(\|\mathbf{h}\|^2). \qquad (2.7.22)$$

Proof: This is left as an exercise by considering the function $\mathbf{y}(t) = \mathbf{x}_0 + t\left(\frac{\mathbf{x}-\mathbf{x}_0}{\|\mathbf{x}-\mathbf{x}_0\|}\right)$ for $0 \leq t \leq \|\mathbf{x} - \mathbf{x}_0\|$. Let $h : R \to R$ be defined by $h(t) = f(\mathbf{y}(t))$. Use chain rule and the univariate Taylor theorem, Theorem 2.15 to complete the proof.

2.8 Introduction to Optimization

Let us begin by recalling the so-called *Max-Min Theorem* which asserts that a continuous function $f : D \subset R^n \to R$ defined on a *compact* subset D of R^n is *bounded* and that it attains its (global) maximum and its (global) minimum at some points of D. Since the topology of R^n that we are using is *metrizable*, by Heine–Borel theorem, saying that D is compact is equivalent to saying that D is closed and bounded. More importantly, it is equivalent to saying that every sequence $\mathbf{x}^{(n)}$ in D has a convergent subsequence $\mathbf{x}^{(n_k)}$ converging in D.

We intend to give here a slightly more general result than the above stated result. For this purpose, let us observe that since $\max_D f = -\min_D(-f)$, the problem of maximizing f is equivalent to the problem of minimizing $-f$. Thus, without loss of generality, we may confine ourself to the minimization problem. We need the following definitions.

Definition 2.15. Let X be a normed linear space. A function $f : X \to R \cup \{\infty\}$ is said to be (i): *inf-compact*, if for each $\alpha \in R$, the sublevel set of f at height α:

$$lev_\alpha f = \{x \in X : f(x) \leq \alpha\}$$

is compact. It is said to be (ii): *lower semi-continuous*(lsc), if $lev_\alpha f$ is closed for each $\alpha \in R$. Furthermore, f is said to be (iii): *coercive*, if $\lim_{\|x\| \to +\infty} f(x) = +\infty$.

Remarks 2.15. It is clear from the definitions that for a function $f : X \to R \cup \{\infty\}$ defined on a normed space X, f is coercive if and only if f is *inf-bounded*, that is to say that the sublevel set $lev_\alpha f$ of f at height α is bounded for each $\alpha \in R$. As a result, we see that if $X = R^n$ with the usual topology, then for a function $f : R^n \to R \cup \{\infty\}$ which is lsc, f is coercive if and only if f is inf-compact.

One main reason for bringing in extended real-valued functions in optimization is that these provide a flexible modelization of minimization problems with constraints. Most minimization problems can be formulated as

$$\min\{f_0(x) : x \in \Omega\} \tag{2.8.1}$$

where $f_0 : X \to R$ is a real-valued function, and $\Omega \subseteq X$ is the so-called *constraint set* or *feasible set*, X being some vector space, usually R^n. A natural way of dealing with such a problem is to apply penalization to it. For example, introduce a distance d on X and for any positive real number λ, let us consider the minimization problem

$$\min\{f_0(x) + \lambda \, dist(x, \Omega) : x \in X\} \tag{2.8.2}$$

where

$$dist(x, \Omega) = \inf\{d(x, y) : y \in \Omega\} \tag{2.8.3}$$

is the distance function from x to Ω. Let us note that the penalization is equal to zero if $x \in \Omega$ (that is if the constraint is satisfied), and when $x \notin \Omega$ it takes larger and larger values with λ (when the constraint is violated). Notice that the approximated problem (2.8.2) can be written as

$$\min\{f_\lambda(x) : x \in X\} \tag{2.8.4}$$

where

$$f_\lambda(x) = f_0(x) + \lambda dist(x, \Omega) \tag{2.8.5}$$

is a real-valued function. Thus, the approximated problems (2.8.4) are unconstrained problems. As $\lambda \to +\infty$, the (generalized) sequence of functions (2.8.5) increases to the function $f : X \to R \cup \{\infty\}$, which is equal to

$$f(x) = \begin{cases} f_0(x), & \text{if } x \in \Omega, \\ +\infty, & \text{otherwise.} \end{cases} \tag{2.8.6}$$

Thus, we are led to minimization of an extended real-valued function f:

$$\min\{f(x) : x \in X\}$$

where f is given by (2.8.6). Let us note that if we introduce the *indicator function* δ_Ω of the set Ω:

$$\delta_\Omega(x) = \begin{cases} 0, & \text{if } x \in \Omega, \\ +\infty, & \text{otherwise,} \end{cases} \tag{2.8.7}$$

then we have the convenient expression $f = f_0 + \delta_\Omega$. We now come to the following generalization of the *Min-Max Theorem* called the extended Weierstrass theorem.

Theorem 2.16. *Let X be a normed linear space and let $f : X \to R \cup \{+\infty\}$ be an extended real-valued function which is lower semi-continuous and inf-compact. Then, $\inf_X f > -\infty$, and there exists some $\hat{x} \in X$ which minimizes f on X:*

$$f(\hat{x}) \le f(x) \, \textit{for all } x \in X.$$

Proof: Given a function $f : X \to R \cup \{+\infty\}$, by the definition of $\inf_X f$, the infimum of f on X, we can construct a *minimizing sequence*, that is, a sequence $\{x_n\}$ such that $f(x_n) \to \inf_X f$ as $n \to +\infty$. Indeed, if $\inf_X f > -\infty$, pick $\{x_n\}$ such that

$$\inf_X f \le f(x_n) \le inf_X f + \frac{1}{n};$$

and if $\inf_X f = -\infty$, pick $\{x_n\}$ such that $f(x_n) \le -n$. Without any restriction, we may assume that f is proper (that is finite somewhere), and hence, $\inf_X f < +\infty$. Thus, for $n \in N$,

$$f(x_n) \le \max\{\inf_X f + 1/n, -n\} \le \max\{\inf_X f + 1, -1\} := \alpha_0.$$

Since, $\alpha_0 > \inf_X f$, $lev_{\alpha_0}(f) \ne \emptyset$, and $x_n \in lev_{\alpha_0}(f)$, $n \in N$. By inf-compactness of f, this sublevel set in which the sequence $\{x_n\}$ is contained is compact. Hence, we can extract a subsequence $\{x_{n_k}\}$ converging to some element $\hat{x} \in X$. By lower semi-continuity of f, we have

$$f(\hat{x}) \le \lim_k f(x_{n_k}) = \inf_X f.$$

This proves that $\inf_X f > -\infty$, since $f : X \to R \cup \{+\infty\}$, and

$$f(\hat{x}) \le f(x) \quad \text{for all } x \in X,$$

which completes the proof.

Remarks 2.16. The set of all the minimizers of f on X is usually denoted by $\arg\min_X(f)$. The above theorem gives the conditions under which $\arg\min_X(f) \ne \emptyset$. As a corollary of the preceding theorem, we have the following:

Corollary 2.1. (Weierstrass theorem) *Let X be a normed linear space and K be a compact subset of X. If $f : X \to R \cup \{+\infty\}$ is lower semi-continuous, then $\arg\min_X(f) \ne \emptyset$.*

2.8.1 Unconstrained and Constrained Extremizers: Conditions for Local Extremizers

We consider here the minimization problem (D, f):

$$\min\{f(\mathbf{x}) : \mathbf{x} \in D\}.$$

Here $f : R^n \to R$ is a given function called the *objective function* or *cost function*, the vector $\mathbf{x}$ is called the vector of *decision variables* $x_1, \ldots, x_n$, and the set D is a subset of R^n, called the *constraint set* or *feasible set*. The above problem is a general *constrained* minimization problem. In case $D = R^n$, one refers to the problem as *unconstrained* minimization problem. Constrained and unconstrained maximization problems are defined analogously.

Definition 2.16. Let $f : D \subseteq R^n \to R$ be a given function and let $\mathbf{x}^0$ be an interior point of D. Then, (i): $\mathbf{x}^0$ is called a *local minimizer* of f, if there is some $\delta > 0$ such that

$$f(\mathbf{x}) \geq f(\mathbf{x}^0), \quad \text{for all } \mathbf{x} \in B_\delta(\mathbf{x}^0).$$

In this case, $f(\mathbf{x}^0)$ is called a *local minimum value* of f. (ii): A *local maximizer* $\mathbf{x}^0$ of f and *local maximum value* $f(\mathbf{x}^0)$ of f are defined analogously. (iii): A point $\mathbf{x}^0 \in D$ is called a *global minimizer* of f if

$$f(\mathbf{x}^0) \leq f(\mathbf{x}), \quad \forall \mathbf{x} \in D.$$

A *global maximizer* of f is defined analogously. A local (resp.global)minimizer or maximizer of f is called a *local (resp.global)extremizer* of f.

Let us consider the constrained minimization problem (D, f) where D, f are as given before. A minimizer $\mathbf{x}^0$ of problem (D, f) can be either an interior point or a boundary point of D. The following definition is useful in this respect.

Definition 2.17. A vector $\mathbf{u} \in R^n$, $\mathbf{u} \neq \mathbf{0}$ is called a *feasible direction* (f.d.) at $\mathbf{x}^0 \in D$, if there exists $\delta > 0$ such that $\mathbf{x}^0 + \lambda \mathbf{u} \in D$ for all $\lambda \in [0, \delta]$.

Let us recall Definition 2.16 of the directional derivative $f'(\mathbf{x}^0; \mathbf{u})$, which is also sometimes written as $\frac{df}{du}$, given by

$$f'(\mathbf{x}^0 : \mathbf{u}) = \mathbf{f} \nabla f(\mathbf{x}^0)^T \mathbf{u} = < \nabla f(\mathbf{x}^0), \mathbf{u} > = \mathbf{u}^T \nabla f(\mathbf{x}^0),$$

in case f is differentiable at $\mathbf{x}^0$. Note that if $\mathbf{u}$ is a unit vector, then

$$|f'(\mathbf{x}^0; \mathbf{u})| = | < \nabla f(\mathbf{x}^0), \mathbf{u} > | \leq |\nabla f(\mathbf{x}^0)|.$$

Theorem 2.17. *(First-Order Necessary Condition) Let D be a subset of R^n and $f : D \to R$ be a $C^{(1)}$ function. If $\mathbf{x}^0 \in D$ is a local minimizer of f, then for every f.d. $\mathbf{u}$ at $\mathbf{x}^0$, we have*

$$\nabla f(\mathbf{x}^0)^T \mathbf{u} \geq 0.$$

Proof: Note that $\mathbf{u}$ is a f.d. at $\mathbf{x}^0$. $\Rightarrow \exists \delta > 0$ such that $\mathbf{x}^0 + \lambda \mathbf{u} \in D$, $\forall \lambda \in [0, \delta]$. Let us define

$$\phi(\lambda) = f(\mathbf{x_0} + \lambda \mathbf{u}), \quad \lambda \in [0, \delta].$$

By Taylor's theorem, we have

$$f(\mathbf{x}^0 + \lambda\mathbf{u}) - f(\mathbf{x}^0) = \phi(\lambda) - \phi(0)$$
$$= \lambda\phi'(0) + o(\lambda)$$
$$= \lambda\nabla f(\mathbf{x}^0)^T\mathbf{u} + o(\lambda).$$

Since $\mathbf{x}^0$ is a local minimizer of f, $\phi(\lambda) - \phi(0) \geq 0$ for sufficiently small values of $\lambda > 0$. This implies that $\nabla f(\mathbf{x}^0)^T\mathbf{u} \geq 0$.

Remarks 2.16. (i): The theorem holds under the weaker assumption: f is differentiable at $\mathbf{x}^0$.
(ii): An alternative way to express the first-order necessary condition is $f'(\mathbf{x}^0; \mathbf{u}) \geq 0$ for every f.d. $\mathbf{u}$ at $\mathbf{x}^0$.

Corollary 2.2. *(Interior point case) Let $D \subseteq R^n$, $\mathbf{x}^0$ be an interior point of D, and $f : D \to R$ be differentiable at $\mathbf{x}^0$. If $\mathbf{x}^0$ is a local minimizer of f, then we have*

$$\nabla f(\mathbf{x}^0) = \mathbf{0}.$$

Proof: Indeed, if $\mathbf{x}^0$ is an interior point of D, then the set of f.d.s at $\mathbf{x}^0$ is the whole of R^n. Thus, for any $\mathbf{u} \in R^n$, both $\nabla f(\mathbf{x}^0)^T\mathbf{u} \geq 0$ as well as $\nabla f(\mathbf{x}^0)^T(-\mathbf{u}) \geq 0$ hold. Thus, $\nabla f(\mathbf{x}^0) = \mathbf{0}$.

Theorem 2.18. *Let $D \subseteq R^n$ and $f : D \to R$ be a C^2 function. If $\mathbf{x}^0 \in D$ is a local minimizer of f and $\mathbf{u}$ is a f.d. at $\mathbf{x}^0$ such that $\nabla f(\mathbf{x}^0)^T\mathbf{u} = 0$, then*

$$\mathbf{u}^T\mathbf{H}_f(\mathbf{x}^0)\mathbf{u} \geq 0.$$

Here, $\mathbf{H}_f(\mathbf{x}^0)$ denotes the Hessian of f at $\mathbf{x}^0$.

Proof: This is left as an exercise to the reader. One makes use of Theorem 2.11.

Remark 2.18. Theorem holds under the weaker assumption: f is twice differentiable at $\mathbf{x}^0$.

Corollary 2.3. *Let $\mathbf{x}^0$ be an interior point of $D \subseteq R^n$, $f : D \to R$ be a given function. If f is twice differentiable at $\mathbf{x}^0$ and $\mathbf{x}^0$ is a local minimizer of f, then $\nabla f(\mathbf{x}^0) = \mathbf{0}$, and $\mathbf{H}_f(\mathbf{x}^0)$ is positive semi-definite: $\mathbf{u}^T\mathbf{H}_f(\mathbf{x}^0)\mathbf{u} \geq 0$, $\forall \mathbf{u} \in R^n$.*

Proof: Note that $\mathbf{x}^0$ is an interior point of D implies all directions $\mathbf{u} \in R^n$ are feasible. Hence, the result follows from the preceding theorem and the last corollary.

Exercises I

2.1. Find the natural domains of the following functions of two variables.

$(i) \dfrac{xy}{x^2 - y^2}$, (ii) $\ln(x^2 + y^2 + z^2)$, (iii) $\sqrt{25 - x^2 - y^2 - z^2}$, (iv) $\dfrac{1}{xyz}$.

2.2. (a): Use level curves to sketch graphs of the following functions.

$$(i) \ f(x, y) = x^2 + y^2, \quad (ii) \ f(x, y) = x^2 + y^2 - 4x - 6y + 13.$$

(b): Sketch some level surfaces of the following functions.

$$(i) \ f(x, y, z) = 4x^2 + y^2 + 9z^2, \quad (ii) \ f(x, y, z) = x^2 + y^2.$$

2.3. Using definition, examine the following functions for continuity at $(0, 0)$. The expressions below give the value at $(x, y) \neq (0, 0)$. At $(0, 0)$, the value should be taken as zero.

$$(i) \ \frac{x^3 y}{x^6 + y^2}, \quad (ii) \ \frac{x^2 y}{x^2 + y^2}, \quad (iii) \ xy \frac{x^2 - y^2}{x^2 + y^2},$$

$$(iv) \ |\,|x| - |y|\,| - |x| - |y|, \quad (v) \ \frac{\sin^2(x + y)}{|x| + |y|}.$$

2.4. Using definition, examine the following functions for continuity at $(0, 0, 0)$. The expressions below give the value at $(x, y, z) \neq (0, 0, 0)$. At $(0, 0, 0)$, the value should be taken as zero.

$$(i) \ \frac{xyz}{\sqrt{x^2 + y^2 + z^2}} \sin(\frac{1}{x^2 + y^2 + z^2}), \quad (ii) \ \frac{2xy}{x^2 - 3z^2}, \quad (iii) \ \frac{xy + yz + zx}{\sqrt{x^2 + y^2 + z^2}}.$$

2.5. Suppose $f, g : R \to R$ are continuous functions. Show that each of the following functions of $(x, y) \in R^2$ is continuous.

$(i) \ f(x) \pm g(y), \quad (ii) \ f(x)g(y), \quad (iii) \ \max\{f(x), g(y)\}, \quad (iv) \ \min\{f(x), g(y)\}.$

2.6. Using the results from the above problem, show that $f(x, y) = x + y$ and $g(x, y) = xy$ are continuous in R^2. Deduce that every polynomial function in two variables is continuous in R^2.

2.7. Examine each of the following functions for continuity.

$$(i) \ f(x, y) = \begin{cases} \frac{y}{|y|}\sqrt{x^2 + y^2}, & \text{if } y \neq 0 \\ 0, & \text{if } y = 0. \end{cases}$$

$$(ii) \ f(x, y) = \begin{cases} x \sin\frac{1}{x} + y \sin\frac{1}{y}, & \text{if } x \neq 0, \ y \neq 0 \\ x \sin\frac{1}{x}, & \text{if } x \neq 0, \ y = 0 \\ y \sin\frac{1}{y}, & \text{if } x = 0, \ y \neq 0 \\ 0, & \text{if } x = 0, \ y = 0. \end{cases}$$

2.8. Let $f(x, y) = \frac{x^2 y^2}{x^2 y^2 + (x-y)^2}$, for $(x, y) \neq (0, 0)$. Show that the iterated limits $\lim_{x \to 0}[\lim_{y \to 0} f(x, y)]$ and $\lim_{y \to 0}[\lim_{x \to 0} f(x, y)]$ exist and both are equal to 0, but $\lim_{(x,y) \to (0,0)} f(x, y)$ does not exist.

2.9. Express the definition of $\lim_{(x,y) \to (0,0)} f(x, y)$ in terms of polar coordinates and analyze it for the following functions:

(i) $f(x, y) = \frac{x^3 - xy^2}{x^2 + y^2}$. (ii) $f(x, y) = \tan^{-1}\left(\frac{|x| + |y|}{x^2 + y^2}\right)$.

(iii) $f(x, y) = \frac{y^2}{x^2 + y^2}$. (iv) $f(x, y) = \frac{x^4 y - 3x^2 y^3 + y^5}{(x^2 + y^2)^2}$.

Exercises II

2.10. Examine the following functions for the existence of partial derivatives at $(0, 0)$. The expressions below give the value at $(x, y) \neq (0, 0)$. At $(0, 0)$, the value should be taken as zero.

(i) $\frac{x^3 y}{x^6 + y^2}$, (ii) $xy \frac{x^3}{x^2 + y^2}$, (iii) $\frac{x^2 y}{x^4 + y^2}$

(iv) $xy \frac{x^2 - y^2}{x^2 + y^2}$, (v) $||x| - |y|| - |x| - |y|$, (vi) $\frac{\sin^2(x+y)}{|x| + |y|}$.

2.11. Let $D \subset R^2$ be an open disk centered at (x_0, y_0) and $f : D \to R$ be such that both f_x and f_y exist and are bounded in D. Prove that f is continuous at (x_0, y_0). Conclude that f is continuous everywhere in D.

2.12. Let $f(0, 0) = 0$ and $f(x, y) = (x^2 + y^2) \sin \frac{1}{x^2 + y^2}$ for $(x, y) \neq (0, 0)$. Show that f is continuous at $(0, 0)$, and the partial derivatives of f exist but are not bounded in any disk (howsoever small) around $(0, 0)$.

2.13. Let $f : R^2 \to R$ be defined by $f(0, 0) = 0$ and $f(x, y) = \frac{xy}{x^2 + y^2}$ if $(x, y) \neq (0, 0)$. Show that f_x and f_y exist at every $(x_0, y_0) \in R^2$, but f is not continuous at $(0, 0)$.

2.14. Let $f(0, 0) = 0$ and

$$f(x, y) = \begin{cases} x \sin(\frac{1}{x}) + y \sin(\frac{1}{y}), & \text{if } x \neq 0, \ y \neq 0 \\ x \sin(\frac{1}{x}), & \text{if } x \neq 0, \ y = 0 \\ y \sin(\frac{1}{y}), & \text{if } y \neq 0, \ x = 0. \end{cases}$$

Show that none of the partial derivatives of f exist at $(0, 0)$ although f is continuous at $(0, 0)$.

2.15. Suppose the implicit equation $F(x, y, z) = 0$ determines z as a function of x and y, that is, there exists a function w of two variables such that $F(x, y, w(x, y)) = 0$. Assume that F_x, F_y, F_z exist and are continuous, $F_z \neq 0$ and w_y exists. Show that $w_y = -F_y/F_z$.

2.16. Suppose $F : R^3 \to R$ has the property that there exists $n \in N$ such that $F(tx, ty, tz) = t^n F(x, y, z)$ for all $t \in R$ and $(x, y, z) \in R^3$. [Such a function is said to be *homogeneous* of degree n.] If the first-order partial derivatives of f exist and are continuous, then show that $x\frac{\partial F}{\partial x} + y\frac{\partial F}{\partial y} + z\frac{\partial F}{\partial z} = nF$. [This result is sometimes called *Euler's Theorem*.]

2.17. Examine the following functions for the existence of directional derivatives and differentiability at $(0, 0)$. The expressions below give the value at $(x, y) \neq (0, 0)$. At $(0, 0)$, the value should be taken as zero.

(i) $xy\frac{x^2-y^2}{x^2+y^2}$, (ii) $\frac{x^3}{x^2+y^2}$, (iii) $(x^2 + y^2)\sin\frac{1}{x^2+y^2}$.

2.18. Let $f(x, y) = 0$ if $y = 0$ and $f(x, y) = \frac{y}{|y|}\sqrt{x^2 + y^2}$ if $y \neq 0$. Show that f is continuous at $(0, 0)$, $D_{\mathbf{u}} f(0, 0)$ exists for every vector $\mathbf{u}$, yet f is not differentiable at $(0, 0)$.

2.19. Assume that f_x and f_y exist and are continuous in some disk centered at $(1, 2)$. If the directional derivative of f at $(1, 2)$ toward $(2, 3)$ is $2\sqrt{2}$ and toward $(1, 0)$ is -3, then find $f_x(1, 2)$, $f_y(1, 2)$ and the directional derivative of f at $(1, 2)$ toward $(4, 6)$.

Exercises III

2.20. Given $z = x^2 + 2xy$, $x = u \cos v$ and $y = u \sin v$, find $\frac{\partial z}{\partial u}$ and $\frac{\partial z}{\partial v}$.

2.21. Given $\sin(x + y) + \sin(y + z) = 1$, find $\frac{\partial^2 z}{\partial x \partial y}$, provided $\cos(y + z) \neq 0$.

2.22. If $f(0, 0) = 0$ and $f(x, y) = xy\frac{x^2-y^2}{x^2+y^2}$ for $(x, y) \neq (0, 0)$, show that both f_{xy} and f_{yx} exist at $(0, 0)$, but they are not equal. Are f_{xy} and f_{yx} continuous at $(0, 0)$?

2.23. Show that the following functions have local minima at the indicated points.

(i) $f(x, y) = x^4 + y^4 + 4x - 32y - 7$, $(x_0, y_0) = (-1, 2)$
(ii) $f(x, y) = x^3 + 3x^2 - 2xy + 5y^2 - 4y^3$, $(x_0, y_0) = (0, 0)$.

2.24. Analyze the following functions for local maxima, local minima, and saddle points:

(i) $f(x, y) = (x^2 - y^2)e^{-\frac{1}{2}(x^2+y^2)}$, (ii) $f(x, y) = x^3 - 3xy^2$,
(iii) $f(x, y) = 6x^2 - 2x^3 + 3y^2 + 6xy$, (iv) $f(x, y) = x^3 + y^3 - 3xy + 15$,
(v) $f(x, y) = (x^2 + y^2)\cos(x + 2y)$.

2.25. Find the absolute minimum and the absolute maximum of $f(x, y) = 2x^2 - 4x + y^2 - 4y + 1$ on the closed triangular plate bounded by the lines $x = 0$, $y = 2$ and $y = 2x$.

2.26. Find the absolute maximum and the absolute minimum of $f(x, y) = (x^2 - 4x)\cos y$ over the region $1 \le x \le 3$, $-\frac{\pi}{4} \le y \le \frac{\pi}{4}$.

2.27. The temperature at a point (x, y, z) in 3-space is given by $T(x, y, z) = 400xyz$. Find the highest temperature on the unit sphere $x^2 + y^2 + z^2 = 1$.

2.28. Find the point nearest to the origin on the surface defined by the equation $z = xy + 1$

2.29. Find the absolute maximum and minimum of $f(x, y) = \frac{1}{2}x^2 + \frac{1}{2}y^2$ in the elliptic region D defined by $\frac{1}{2}x^2 + y^2 \le 1$.

2.30. A space probe in the shape of the ellipsoid $4x^2 + y^2 + 4z^2 = 16$ enters the earth's atmosphere and its surface begins to heat. After one hour, the temperature at the point (x, y, z) on the surface of the probe is given by $T(x, y, z) = 8x^2 + 4yz - 16z + 600$. Find the hottest and the coolest points on the surface of the probe.

2.31. Maximize the quantity xyz subject to the constraints $x + y + z = 40$ and $x + y = z$.

2.32. Minimize the quantity $x^2 + y^2 + z^2$ subject to the constraints $x + 2y + 3z = 6$ and $x + 3y + 4z = 9$.

2.33. Consider the minimization problem: minimize $x_1^2 + \frac{1}{2}x_2^2 + 3x_2 + \frac{9}{2}$ subject to $x_1, x_2 \ge 0$. Answer the following questions: (In the following, we abbreviate first-order necessary condition by f.o.n.c.)

(a) Is the f.o.n.c. for a local minimizer satisfied at $x_0 = [1, 3]^T$?
(b) Is the f.o.n.c. for a local minimizer satisfied at $x_0 = [0, 3]^T$?
(c) Is the f.o.n.c. for a local minimizer satisfied at $x_0 = [1, 0]^T$?
(d) Is the f.o.n.c. for a local minimizer satisfied at $x_0 = [0, 0]^T$?

2.34. Consider the quadratic function $f : R^2 \to R$ given below:

$$f(\mathbf{x}) = \mathbf{x}^T \begin{pmatrix} 1 & 2 \\ 4 & 7 \end{pmatrix} \mathbf{x} + \mathbf{x}^T [3, 5]^T + 6$$

(a) Find the gradient and Hessian of f at the point $[1, 1]^T$.
(b) Find the directional derivative of f at $[1, 1]^T$ with respect to a unit vector in the direction in which the function decreases most rapidly.
(c) Find a point that satisfies the f.o.n.c.(interior case) for f. Does this point satisfy the second-order necessary condition(s.o.n.c.)(for a minimizer)?

2.35. Consider the problem: minimize $f(\mathbf{x})$ subject to $\mathbf{x} \in \Omega$, where $f : R^2 \to R$ is given by $f(\mathbf{x}) = 5x_2$ with $\mathbf{x} = [x_1, x_2]^T$, and $\Omega = \{\mathbf{x} = [x_1, x_2]^T : x_1^2 + x_2 \ge 1\}$. Answer each of the following questions giving justification.

(a) Does the point $x_0 = [0, 1]^T$ satisfy the f.o.n.c.?
(b) Does the point $x_0 = [0, 1]^T$ satisfy the s.o.n.c.?
(c) Is the point $x_0 = [0, 1]^T$ a local minimizer?

2.9 Classical Approximation Problems: A Relook

2.9.1 Input–Output Process

In, many practical situations of interest, such as in *economic forecasting, weather prediction*, etc, one is required to construct a model for what is generally called an *input-output process*. For example, one may be interested in the price of a stock 5 years from now. The *rating industry* description of a stock typically lists such indicators as, the increase in the price over the last 1 year or the increase in the price over the last 2 years/ 3 years, the life of the stock, P/E ratio, α, β risk factors etc. The investor is expected to believe that the price of the stock depends on these parameters. Of course, no one knows precisely a formula to compute this price as a closed form function of these parameters. (Else, one would strike it rich quickly!) Typically, this is an example of an input-output process.

$$\text{input data } \mathbf{x} \in R^n \longrightarrow \text{ actual process } \longrightarrow \text{ output } f(\mathbf{x})$$

where $f(\mathbf{x})$ is the so-called *target function*. In practice, the computed model is as follows:

$$\text{finite data } \longrightarrow \text{ computed model } \longrightarrow P_f(\mathbf{x}).$$

Here, the finite data may consist of the values of the function or the values of some of its derivatives sampled at some points or possibly it may consist of the Fourier coefficients or coefficients in some other series associated with f, etc.

Broadly speaking, there are two kinds of errors in using the model $P_f(\mathbf{x})$ as a predictor for the target function $f(\mathbf{x})$.

(a) **Noise:** This comes from the fact that the observations on which the model is based are usually subjected to errors-human errors, machine errors, interference from nature, etc. Also, this could arise from faulty assumptions about what one is modeling. Statistics is mainly concerned with the *reliability* of the model by eliminating or controlling the noise.

(b) **Intrinsic Error:** This comes from the fact that one is computing a *model* rather than the *actual function*. Approximation Theory is concerned for the most part with "intrinsic error."

2.9.2 Approximation by Algebraic Polynomials

Under the setting of an input-output process in Section 2.9.1, the problems of *Approximation Theory* can be broadly classified into four categories.

Let I denote the compact interval $[a, b]$ of R, and let $X = C(I)$ denote the space of functions

$$\{f : I \to R : f \text{ continuous}\}$$

normed by

$$\|f\| = \max\{|f(t)| : t \in I\}, f \in \mathcal{C}(I).$$

Let us assume that our target function f be in the class $\mathcal{C}(I)$. The *model* for such a function is typically a real algebraic polynomial

$$P_n(t) = \sum_{k=0}^{n} a_k x^k, a_k \in R, k = 0, \dots, n.$$

The integer n is called the degree of the polynomial and we denote by Π_n the class of all real algebraic polynomials of degree $\leq n$. The first problem that we encounter in this context is the *density problem*: to decide whether it is feasible to approximate the target function arbitrarily closely by choosing more and more complex models. More precisely, let

$$E_n(f) := \inf_{P \in \Pi_n} \|f - P\|$$

denote the *degree of approximation* of f from Π_n. The density problem is to decide whether $E_n(f) \to 0$ as $n \to \infty$. It is the classical Weierstrass approximation theorem that answers this question affirmatively.

The next problem is the so-called *complexity problem* which is concerned with estimating the rate at which $E_n(f) \to 0$. Frequently in applications, the target function is usually unknown, but one may assume that $f \in \mathcal{F}$, where $\mathcal{F}$ is a certain class of functions. For example, $\mathcal{F}$ may consist of functions f such that f' exists, is continuous, and satisfies $\|f'\| \leq 1$. Typically, then, one is interested in estimating $\sup_{f \in \mathcal{F}} E_n(f)$.

The next problem is addressed in the *theory of best approximation*. This deals with the existence, uniqueness, characterization, and computability of elements $P \in \Pi_n$ such that

$$\|f - P\| = E_n(f).$$

Lastly, the *theory of good approximation* deals with the approximation capabilities of different procedures for computing the approximants based on the values of the function, its derivatives, its Fourier coefficients, etc. It is to be emphasized here that these approximants are sometimes more interesting than the best approximant, P, because of the *relative ease of its computability* or because of some desirable properties such as *shape preservation*. For a more detailed exposition of ideas in this direction, the reader may consult the book by Mhaskar and Pai [3].

2.10   Introduction to Optimal Recovery of Functions

Let us begin by considering what is usually called the problem of *best simultaneous approximation* in the literature in Approximation Theory. Here, the main concern is with *simultaneous approximation* which is best approximation in some sense of sets, rather than that of just single elements. One of the motivations for treating this problem is the following practical situation involving *optimal estimation*. We assume that in the mathematical modeling of a physical process an entity E is represented by an unknown element x_E of some normed linear space X. Mostly, X is a function space such as $\mathcal{C}([a, b], R)$, $H^m[a, b]$, etc. By means of suitable experiments, observations are obtained for E which give rise to *limited information* concerning x_E. For instance, the information could be the values of x_E and/some of its derivatives sampled on a discrete set of points, or it could be the Fourier coefficients of x_E etc. In addition, the information could be error contaminated due to experimental inaccuracies. We assume that the information so gathered is incomplete to specify x_E completely; it only identifies a certain subset $F \subset X$ called the *data set*. Our estimation problem then is to find the best estimate of x_E given only that $x_E \in F$ (Figure 2.4).

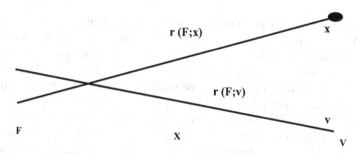

Fig. 2.4 Optimal estimation

We pick an element $x \in X$ (usually, a normed linear space) and ask how bad it is from the point of view of representing the data set F. The measure of *worstness* of x as a representer of F is given by the quantity

$$r(F; x) := \sup\{\|x - y\| : y \in F\} \tag{2.10.1}$$

(in a *worst-case scenario*). In order that this quantity be finite, we must assume that F be bounded. The *intrinsic error* in our estimation problem is then determined by the number

$$\mathrm{rad}(F) := \inf\{r(F; x) : x \in X\},$$

called the *Chebyshev radius* of F. It is impossible for the worstness of x as a representer of F to fall below this number. An element $x_0 \in X$ will be a best representer (or a global approximator) of the data set F if it minimizes the measure of worstness:

$$r(F; x_0) = \min\{r(F; x) : x \in X\}. \tag{2.10.2}$$

An element $x_0 \in X$ satisfying (2.10.2) is called a *Chebyshev center* of F and Cent(F) denotes (the possibly void) set of all Chebyshev centers of F.

Practical reasons may require us to restrict our search of a best representer of the data set F to another set V which may perhaps be a subspace or a convex set obtained by taking the intersection of a subspace with a set determined by affine constraints, etc. In this case, the intrinsic error in our estimation problem will be determined by the number

$$\text{rad}_V(F) := \inf \{r(F; v) : v \in V\} \tag{2.10.3}$$

called the *(restricted) radius* of F in V and a best representer (or global approximator) of F in V will be an element $v_0 \in V$ called *(restricted) center* of F in V satisfying

$$r(F; v_0) = \text{rad}_V(F). \tag{2.10.4}$$

For a more detailed exposition of ideas in this direction, the reader may consult Chapter VIII of the book by Mhaskar and Pai [3]. Let us observe that the above stated problem of *Optimal Estimation* is closely related to the so-called problem of *optimal recovery* of functions. Following Micchelli and Rivlin [4, 5], by *optimal recovery* we will mean the problem of estimating some required feature of a function, known to belong to some class of functions prescribed *a priori*, from limited and possibly error-contaminated information about it, as effectively as possible. This problem is again subsumed by what goes on under *information-based complexity* [7] and it has some rich connections with the problem of *image reconstruction* or *image recovery* which, for instance, is important in mathematical studies of computer-assisted tomography. We begin by looking at some simple examples given below, motivating the general theory.

2.10.1 Some Motivating Examples

Example 2.17. Let $X = \mathcal{C}[0, 1]$ and let

$$\mathcal{K} := \left\{x \in \mathcal{C}^{(n)}[0, 1] : \|x^{(n)}\| \le 1\right\}.$$

Suppose we are given:
(i): $x \in \mathcal{K}$, and (ii): the values $x(t_1), x(t_2), \ldots, x(t_n)$ sampled at distinct points $t_1, t_2, \ldots, t_n$ in $[0, 1]$. Then, we ask the following:

(I): Given $t_0 \in [0, 1] \setminus \{t_1, \ldots, t_n\}$, what is the *best possible* estimate of $x(t_0)$, based solely on the information (i) and (ii)? (II): What is the *best possible* estimate of x itself based solely on the information (i) and (ii)?

Here, (I) is a problem of *optimal interpolation* and (II) is a problem of *optimal approximation*. We proceed to answer Question (I) first. Specially, let

$$\mathcal{I} := \{(x(t_1), \ldots, x(t_n)) : x \in \mathcal{K}\}.$$

An *algorithm* A is any function of $\mathcal{I}$ into R. Then, the *error* in the algorithm is given by

$$E_A(\mathcal{K}) := \sup_{x \in \mathcal{K}} |x(t_0) - A(x(t_1), \ldots, x(t_n))|,$$

and

$$E(\mathcal{K}) := \inf_A E_A(\mathcal{K})$$

denotes the *intrinsic error* in our recovery problem. An algorithm $\hat{A}$, if one such exists, satisfying $E_{\hat{A}}(\mathcal{K}) = E(\mathcal{K})$ is called an *optimal algorithm*, which is said to effect the *optimal recovery* of $x(t_0)$. It is easily seen that the polynomial

$$\hat{P}(t) := \frac{(t - t_1) \ldots (t - t_n)}{n!} \in \mathcal{K}$$

and so also does the polynomial $-\hat{P}$. Thus, if A is any algorithm, then

$$|\hat{P}(t_0) - A(\hat{P}(t_1), \ldots, \hat{P}(t_n))| = |\hat{P}(t_0) - A(0, \ldots, 0)| \le E_A(\mathcal{K})$$

and also

$$|-\hat{P}(t_0) - A(0, \ldots, 0)| \le E_A(\mathcal{K}),$$

whence, by triangle inequality, $|\hat{P}(t_0)| \le E_A(\mathcal{K})$, and we obtain

$$|\hat{P}(t_0)| \le E(\mathcal{K}), \qquad (2.10.5)$$

which gives a lower bound for the intrinsic error. Now, consider the algorithm $\hat{A} : \mathcal{I} \to R$ given by

$$\hat{A}(x(t_1), \ldots, x(t_n)) = L_{t_1, t_2, \ldots, t_n}(x; t_0)$$

where $L_{t_1, t_2, \ldots, t_n}(x; t)$ denotes the unique Lagrange interpolant in $P_n(R)$ of x on the nodes t_is. Suppose we show that

$$|x(t_0) - \hat{A}(x(t_1), \ldots, x(t_n))| \le |\hat{P}(t_0)|,$$

for all $x \in \mathcal{K}$, then

$$E_{\hat{A}}(\mathcal{K}) \le |\hat{P}(t_0)| \le E(\mathcal{K}),$$

and $\hat{A}$ would be an optimal algorithm sought, with $E(\mathcal{K}) = |\hat{P}(t_0)|$. Assume the contrary that

$$x(t_0) - L_{t_1, t_2, \ldots, t_n}(x; t_0) = \alpha \hat{P}(t_0),$$

for some α, $|\alpha| > 1$. Then, the function

$$h(t) := x(t) - L_{t_1,\ldots,t_n}(x; t) - \alpha \hat{P}(t)$$

would have $n + 1$ distinct zeros $t_0, t_1, \ldots, t_n$. Since $h \in \mathcal{C}^{(n-1)}[0, 1]$, by Rolle's theorem, $h^{(n)} = x^{(n)} - \alpha$ would have at least one zeros $\xi : x^{(n)}(\xi) - \alpha = 0$. Hence,

$$\|x^{(n)}\| \leq 1,$$

would be contradicted. Thus $\hat{A}$ is an optimal algorithm sought.

Remarks 2.19. If our object is to find the best possible estimate of $x^{(m)}(t_j)$ for some fixed $m \leq n$ and j, $1 \leq j \leq n$, based solely on (i) $\wedge$ (ii), then one can show on the same lines as above that $|\hat{P}^{(m)}(t_j)| \leq E(\mathcal{K})$, and that the algorithm $\tilde{A} : \mathcal{I} \to R$ defined by

$$\tilde{A}(x(t_1), \ldots, x(t_n)) = L^{(m)}_{t_1, t_2, \ldots, t_n}(x; t_j)$$

is an optimal algorithm.

Next, we address Question (II). An algorithm now is any function $A : \mathcal{I} \longrightarrow X$, and

$$E_A(\mathcal{K}) := \sup \|x - A(x(t_1), \ldots, x(t_n))\|_\infty$$

is the error in algorithm A. By exactly the same reasoning as above, we obtain $\|\hat{P}\|_\infty \leq E_A(\mathcal{K})$, which yields $\|\hat{P}\|_\infty \leq E(\mathcal{K})$ as a lower bound for the intrinsic error. Consider the algorithm $A^* : \mathcal{I} \to X$ given by

$$A^*(x(t_1), \ldots, x(t_n)) = L_{t_1, t_2, \ldots, t_n}(x; t).$$

We claim that

$$\|x - A^*(x(t_1), \ldots, x(t_n))\|_\infty \leq \|\hat{P}\|_\infty, \text{ for all } x \in \mathcal{K}.$$

Indeed, if we assume the contrary, then

$$|x(t_0) - L_{t_1,\ldots,t_n}(x; t_0)| > |\hat{P}(t_0)|, \text{ for some } t_0 \in [0, 1],$$

which contradicts Rolle's theorem, exactly as before. Thus

$$E_{A^*}(\mathcal{K}) \leq \|\hat{P}\|_\infty \leq E(\mathcal{K}),$$

and we conclude that A^* is an optimal algorithm for the recovery problem involving optimal approximation. The problem of finding *optimal sampling nodes* in this case amounts to minimizing the quantity $E(\mathcal{K}) = \|\hat{P}\|_\infty$ over all distinct points $t_1, \ldots, t_n \in [0, 1]$. This problem is easily solved thanks to two well-known results in classical Approximation Theory, cf., e.g., Theorem 2.4.1 and Corollary 2.2.7, in the book Mhaskar and Pai [3]. Indeed,

$$\min \{E(\mathcal{K}) : t_1, \ldots, t_n \in [0, 1] \text{ distinct } \} = \left(\frac{1}{2^{n-1}} \right) \left(\frac{1}{n!} \right),$$

and if we denote by $\tilde{t}_k := \cos \left(\frac{2k-1}{2n} \right) \pi$, $k = 1, 2, \ldots, n$ the zeros of the nth Chebyshev polynomial $T_n(t) := \cos(n \cos^{-1}(t))$, then the optimal sampling nodes are precisely the Chebyshev nodes on $[0, 1]$,

$$\hat{t}_k = \frac{1}{2}[1 + \tilde{t}_k], \quad k = 1, 2, \ldots, n.$$

Example 2.18. Let T be a compact subset of R^m and let $X = L_\infty(T)$. Let

$$\mathcal{K} := \{x \in L_\infty(T) : |x(t_1) - x(t_2)| \leq \|t_1 - t_2\| \text{ for all } t_1, t_2 \in T\} \qquad (2.10.6)$$

denote the set of all *non-expansive functions* in X. We can ask analogous questions as in Example 2.17.

Question 1 (Optimal interpolation): With $\mathcal{K}$ as above and given $t_0 \in T$, what is the best possible estimate of $x(t_0)$ based solely on the information (i) and (ii) (cf. Example 2.17)?

Let $I : X \to R^n$ be defined by $Ix := (x(t_1), \ldots, x(t_n))$, and let

$$\mathcal{I} := I(\mathcal{K}) = \{(x(t_1), \ldots, x(t_n)) : x \in \mathcal{K}\}.$$

A lower bound on the intrinsic error of our estimation problem is easily seen to be given by

$$E(\mathcal{K}) \geq \sup \{|x(t_0)| : x \in \mathcal{K}, x|_\Delta = 0\} \qquad (2.10.7)$$

where $\Delta := \{t_1, t_2, \ldots, t_n\}$. Let

$$T_i := \left\{ t \in T : \min_j \|t - t_j\| = \|t - t_i\| \right\},$$

and let $\hat{q}(t) := \min_j \|t - t_j\|$. Clearly, $\hat{q} \in \mathcal{K}$ and $\hat{q}|_\Delta = 0$. Thus, if $t_0 \in T_k$, then $E(\mathcal{K}) \geq \hat{q}(t_0) = \|t_0 - t_k\|$. Define the algorithm $\tilde{A} : \mathcal{I} \to R$ by

$$\hat{A}(x(t_1), \ldots, x(t_n)) = x(t_k).$$

Then,

$$|x(t_0) - \hat{A}(Ix)| = |x(t_0) - x(t_k)| \leq \|t_0 - t_k\| = \hat{q}(t_0) \text{ for all } x \in \mathcal{K}.$$

Consequently, $E_{\hat{A}}(\mathcal{K}) = \hat{q}(t_0)$ and $\hat{A}$ is an optimal algorithm.

Question 2 (Optimal approximation): In the same setting as above with $\mathcal{K}$ as defined in (I), what is the best possible estimate of x itself based solely on the information (i) and (ii)?

Note that an algorithm now is any function $A : \mathcal{I} \to L_\infty(T)$. An analogue of (2.10.7) is:

$$E(\mathcal{K}) \geq \sup\{\|x\|_\infty : x \in \mathcal{K}, x|_\Delta = 0\}$$
$$\geq \|\hat{q}\|_\infty.$$

Let us denote by s the step function

$$s(t) = x(t_i), \ t \in \text{ int } T_i, \ i = 1, \ldots, n.$$

Then, $s \in L_\infty(T)$ and the algorithm $\tilde{A} : \mathcal{I} \to L_\infty(T)$ defined by $\tilde{A}(x(t_1), \ldots, x(t_n)) = s$ is an optimal algorithm. Indeed, if $t \in \text{ int } T_i$ and $x \in \mathcal{K}$ then

$$|x(t) - \tilde{A}(Ix)(t)| = |x(t) - x(t_i)| \leq \|t - t_i\| = \hat{q}(t).$$

Thus $\|x - \tilde{A}(Ix)\|_\infty \leq \|\hat{q}\|_\infty$, and we conclude that

$$E_{\tilde{A}}(\mathcal{K}) \leq \|\hat{q}\|_\infty.$$

Question 3 (Optimal integration): With the same setting and the same information as in the previous two questions, we look for an optimal recovery of $\int_T x \, dt$.

In this case, a lower bound for the intrinsic error is given by

$$E(\mathcal{K}) \geq \sup\left\{\left|\int_T x \, dt\right| : x \in \mathcal{K}, x|_\Delta = 0\right\}$$
$$\geq \int_T \hat{q} \, dt.$$

Consider the algorithm $A_0 : \mathcal{I} \to R$ given by

$$A_0(x(t_1), \ldots, x(t_n)) = \int_T s \, dt = \sum_{i=1}^{n} x(t_i) vol(T_i).$$

We have

$$\left|\int_T x \, dt - A_0(Ix)\right| = \left|\int_T x \, dt - \int_T s \, dt\right| \leq \int_T |x - s| \, dt \leq \int_T \hat{q} \, dt.$$

Thus $E_{A_0}(\mathcal{K}) \leq \int_T \hat{q} \, dt$ and we conclude that A_0 is an optimal algorithm.

2.10.2 General Theory

Let X be a linear space and Y, Z be normed linear spaces. Let K be a balanced convex subset of X. Let $T : X \to Z$ be a linear operator (the so-called *feature operator*). Our object is to estimate Tx for $x \in K$ using "limited information" about x. A linear operator $I : X \to Y$ called the *information operator* is prescribed and we assume Ix for $x \in K$ is known possibly with some error (Figure 2.5).

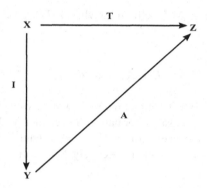

Fig. 2.5 Schematic representation

The algorithm A produces an error. Thus, while attempting to recover Tx for $x \in K$, we only know $y \in Y$ satisfying $\|Ix - y\| \leq \epsilon$ for some preassigned $\epsilon \geq 0$. An *algorithm* is any function—not necessarily a linear one—from $IK + \epsilon U(Y)$ into Z.

Schematically,

$$E_A(K, \epsilon) := \sup\{\|Tx - Ay\| : x \in K \text{ and } \|y - Ix\| \leq \epsilon\},$$

and

$$E(K, \epsilon) := \inf_A E_A(K, \epsilon)$$

is the *intrinsic error* in our estimation problem.

Any algorithm $\hat{A}$ satisfying $E_{\hat{A}}(K, \epsilon) = E(K, \epsilon)$ is called an *optimal algorithm*. When $\epsilon = 0$, which corresponds to the recovery problem with *exact information*, we simply denote $E_A(K, 0)$ and $E(K, 0)$ by $E_A(K)$, $E(K)$ respectively. A lower bound for $E(K, \epsilon)$ is given by the next proposition.

Theorem 2.19. *We have*

$$e(K, \epsilon) := \sup\{\|Tx\| : x \in K, \ \|Ix\| \leq \epsilon\} \leq E(K, \epsilon). \tag{2.10.8}$$

Proof. For every $x \in K$ such that $\|Ix\| \leq \epsilon$ and any algorithm A, we have

$$\|Tx - A\theta\| \le E_A(K, \epsilon),$$

as well as

$$\|T(-x) - A\theta\| = \|Tx + A\theta\| \le E_A(K, \epsilon),$$

whence $\|Tx\| \le E_A(K, \epsilon)$. This implies $e(K, \epsilon) \le E_A(K, \epsilon)$, which yields the result.

The next result, due to Micchelli and Rivlin (1977), gives an upper bound for $E(K, \epsilon)$. We omit its proof.

Theorem 2.20. *We have*

$$E(K, \epsilon) \le 2e(K, \epsilon). \tag{2.10.9}$$

It is important to observe that optimal recovery with error is, at least theoretically, equivalent to recovery with exact information. To see this, let $\hat{X} = X \times Y$ and in $\hat{X}$ let $\hat{K}$ denote the balanced convex set $K \times U(Y)$. We extend T and I to $\hat{X}$ by defining $\hat{T}(x, y) = Tx$, $\hat{I}(x, y) = Ix + \epsilon y$, respectively. Schematically, (Figure 2.6).

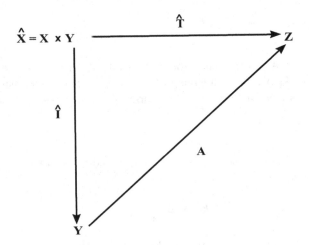

Fig. 2.6 Another schematic representation

Next, we observe that

$$E(\hat{K}) = \inf_{A} \sup_{(x,y)\in\hat{K}} \|\hat{T}(x, y) - A(\hat{I}(x, y))\|$$

$$= \inf_{A} \sup\{\|Tx - A(Ix + \epsilon y)\| : x \in K, y \in U(Y)\}$$

$$= \inf_A \sup\{\|Tx - Ay\| : x \in K, \|Ix - y\| \le \epsilon\}$$
$$= E(K, \epsilon).$$

This justifies the observation made before.

In many situations, it becomes possible to bridge the gap between Theorems 2.19 and 2.20 and thereby solve the optimal recovery problem. Specifically, let

$$K_0 := \{x \in K : Ix = \theta\}, \tag{2.10.10}$$

and let $e(K)$ denote $e(K, 0)$. Then, we have

Theorem 2.21. (Morozov and Grebbenikov) *Suppose there exists a transformation* $S : IX \to X$, *such that* $x - SIx \in K_0$ *for all* $x \in K$. *Then,* $e(K) = E(K)$ *and* $\hat{A} = TS$ *is an optimal algorithm.*

Proof: Indeed,

$$E_{\hat{A}}(K) = \sup_{x \in K} \|Tx - TS(Ix)\|$$
$$= \sup_{x \in K} \|T(x - SIx)\|$$
$$\le \sup_{x \in K_0} \|Tx\| = e(K) \le E(K).$$

The above proposition enables us to obtain a variant of Example 2.17 above.

Example 2.19. Let

$$X := W_\infty^n[0, 1] := \{x : [0, 1] \longrightarrow R : x^{(n-1)} \in AC[0, 1] \text{ and } x^{(n)} \in L_\infty[0, 1]\}$$

and $K := \{x \in X : \|x^{(n)}\|_\infty \le 1\}$. Let $I : X \to R^n$ be defined by $I(x) = (x(t_1), \ldots, x(t_n))$ where $t_1, t_2, \ldots, t_n$ in $[0, 1]$ are distinct points. Define $S : IX \to X$ by $S(Ix) = L_{t_1, t_2, \ldots, t_n}(x)$, the unique polynomial in $P_n(R)$ interpolating x at the points t_i's. We have

$$(x - S(Ix))^{(n)} = x^{(n)}, \|(x - S(Ix))^{(n)}\|_\infty \le 1 \text{ for all } x \in K.$$

Therefore, $x - S(Ix) \in K_0$ for all $x \in K$. Theorem 2.21 now reveals that $e(K) = E(K)$ and TS is an optimal algorithm. The problem of optimal interpolation which corresponds to $Z = R$, $Y = R^{n+r}$, $Tx = x(t_0)$ for some point $t_0 \in [0, 1] \setminus \{t_1, \ldots, t_{n+r}\}$, in case the number of sampling nodes $> n$ was treated by Micchelli, Rivlin, and Winograd [6] by making use of perfect spline for bounding the intrinsic error $E(K)$ from below.

Example 2.20. Let $X = W_2^n[0, 1] = Z$, $Y = R^{n+r}$, $r \ge 0$. Let $\Delta := \{t_i\}_1^{n+r}$ be a strictly increasing sequence of data nodes in $(0, 1)$. For $x \in X$, let

$$Ix = x|_\Delta = (x(t_1), \ldots, x(t_{n+r})),$$
$$K := \{x \in X : \|x^{(n)}\|_2 \leq 1\}.$$

Let us denote by $\hat{S}_{2n}(\Delta)$ the space of natural splines of order $2n$ with the knot sequence Δ. For $x \in X$, let $s(x)$ denote the unique element of $\hat{S}_{2n}(\Delta)$ interpolating x on $\Delta : x|_\Delta = s|_\Delta$ (cf. Theorem 6.3.3.3 in Mhaskar and Pai [3]). It follows from Theorem 6.3.3.6 of Mhaskar and Pai [3] that

$$\|x^{(n)} - (s(x))^{(n)}\|_2 \leq \|x^{(n)}\|_2, \text{ for all } x \in X.$$

Thus if we define $S : IX \to X$ by $S(Ix) = s(x)$, then $I(x - s(x)) = 0$ and $x - SIx \in K$ for all $x \in K$. Theorem 2.21 now reveals that natural spline interpolant is an optimal algorithm.

2.10.3 Central Algorithms

As remarked in the introduction, the problem of optimal recovery is closely related to the notion of Chebyshev center of a set. The reader may recall the notations in the introduction of Section 2.8. In particular, recall that for a bounded subset F of a normed linear space X, $\mathrm{rad}(F) := \inf_{x \in X} r(F; x)$ denotes the Chebyshev radius of F, and

$$\mathrm{Cent}(F) := \{x \in X : r(F; x) = \mathrm{rad}(F)\}$$

denotes the Chebyshev center of F (possibly void). The next two elementary results relate the Chebyshev radius of a bounded set F with its diameter denoted by $\mathrm{diam}(F)$.

Theorem 2.22. *If F is a bounded subset of a normed linear space X, then*

$$\frac{1}{2}\mathrm{diam}(F) \leq \mathrm{rad}(F) \leq \mathrm{diam}(F). \tag{2.10.11}$$

Proof: For $y_1, y_2 \in F$ and $x \in X$ we have

$$\|y_1 - y_2\| \leq \|y_1 - x\| + \|x - y_2\| \leq 2r(F; x).$$

Therefore, $\mathrm{diam}(F) \leq 2r(F; x)$ which entails $\frac{1}{2}\mathrm{diam}(F) \leq \mathrm{rad}(F)$. On the other hand, $\|y_1 - y_2\| \leq \mathrm{diam}(F) \Rightarrow r(F; y_1) \leq \mathrm{diam}(F)$. Hence, $\mathrm{rad}(F) \leq \mathrm{diam}(F)$.

Definition 2.18. A subset F of a linear space X is said to *symmetric about an element* $x_0 \in X$ if

$$z \in X, \ x_0 + z \in F \Rightarrow x_0 - z \in F.$$

Theorem 2.23. *If F is a bounded subset of a normed linear space X which is symmetric about an element $x_0 \in X$, then $x_0 \in \text{Cent}(F)$ and*

$$\text{rad}(F) = \frac{1}{2}\text{diam}(F). \qquad (2.10.12)$$

Proof: If x_0 were not to belong to $\text{Cent}(F)$, then there would exist $x \in F$ such that $r(F; x) < r(F; x_0)$ and one could pick $y_0 \in F$ such that $r(F; x) < \|x_0 - y_0\|$. Then, $\|x - y\| < \|x_0 - y_0\|$ for all $y \in F$. Let $x_0 - y_0 = z_0$. Then, $x_0 - z_0 \in F$ and since F is symmetric about x_0, $x_0 + z_0 \in F$. Therefore,

$$\begin{aligned} 2\|z_0\| &= \|(x_0 + z_0) - x - (x_0 - z_0) + x\| \\ &\leq \|(x_0 + z_0) - x\| + \|(x_0 - z_0) - x\| \\ &< 2\|x_0 - y_0\| = 2\|z_0\|, \text{ a contradiction.} \end{aligned}$$

Thus $x_0 \in \text{Cent}(F)$. Let $x \in X$ and $\delta > 0$. Pick $y_1 \in F$ such that $\|x - y_1\| > r(F; x) - \delta$ and put $x - y_1 = z$. Then, $y_1 = x - z \in F$ and symmetry of F about x_0 shows that $y_2 = x + z \in F$. Therefore,

$$\|y_1 - y_2\| = 2\|z\| > 2r(F; x) - 2\delta \geq 2\text{rad}(F) - 2\delta,$$

which yields $\text{diam}(F) \geq 2\text{rad}(F) - 2\delta$. In view of (2.10.11), we obtain

$$\text{rad}(F) = \frac{1}{2}\text{diam}(F).$$

The preceding theorem motivates the next definition.

Definition 2.19. A bounded set F is said to be *centerable* if (2.10.12) holds.

For $y \in IK$, let us denote by K_y the set

$$K_y := \{x \in K : Ix = y\} = I^{-1}\{y\} \cap K,$$

and let the "hypercircle" $H(y)$ be defined by

$$H(y) = T(K_y) = \{Tx : x \in K, Ix = y\}.$$

We assume that $H(y)$ is *bounded* for each $y \in IK$. Let us call

$$r_I(T, K; y) := \text{rad}\, H(y)$$

the *local radius of information* and

$$r_I(T, K) = \sup_{y \in IK} r_I(T, K; y)$$

the *(global) radius of information*. We have

Theorem 2.24. *Under the hypothesis as before,*

$$e(K) = \operatorname{rad} H(\theta) = r_I(T, K; \theta), \tag{2.10.13}$$

and

$$E(K) = \sup_{y \in IK} \operatorname{rad} H(y) = r_I(T, K). \tag{2.10.14}$$

Proof: We have

$$e(K) = e(K, 0) = \sup\{\|Tx\| : x \in K, Ix = \theta\}$$
$$= \sup\{\|z\| : z \in H(\theta)\} = r(H(\theta), \theta).$$

Observe that $H(\theta)$ is symmetric about θ. Indeed, if $z \in H(\theta)$, then $z = Tx$ for some $x \in K$ such that $Ix = \theta$. This implies $-z = T(-x)$, and since $-x \in K$ and $I(-x) = \theta$, we have $-z \in H(\theta)$. Thus $\theta \in \operatorname{Cent}(H(\theta))$ and $r(H(\theta), \theta) = \operatorname{rad} H(\theta)$, which proves (2.10.13).

To prove (2.10.14), observe that since for any algorithm A, we have

$$\|Tx - AIx\| \le E_A(K) \text{ for all } x \in K,$$
$$\sup_{z \in H(y)} \|z - Ay\| \le E_A(K),$$

which implies

$$\operatorname{rad} H(y) \le r(H(y), Ay) \le E_A(K) \text{ for all } y \in IK.$$

Thus $\sup_{y \in IK} \operatorname{rad}(H(y)) \le E(K)$. To reverse this inequality $\epsilon > 0$ given, for each $y \in IK$ pick $z_y \in Z$ such that $r(H(y), z_y) < \operatorname{rad} H(y) + \epsilon$. Then, for each $x \in K$ and $y \in Y$ such that $Ix = y$, we have

$$\|Tx - z_y\| < \operatorname{rad} H(y) + \epsilon \le \sup_{y \in IK} \operatorname{rad}(H(y)) + \epsilon.$$

Now, consider the algorithm $\tilde{A} : IK \to Z$ defined by $\tilde{A}(y) = z_y$, $y \in IK$. We have

$$E_{\tilde{A}}(K) = \sup\{\|Tx - \tilde{A}y\| : x \in K, Ix = y\}$$
$$= \sup\{\|Tx - z_y\| : x \in K, Ix = y\}$$
$$\leq \sup_{y \in IK} \text{rad}(H(y)) + \epsilon.$$

Thus $E(K) \leq \sup_{y \in IK} \text{rad}(H(y))$, which establishes (2.10.14).

Corollary 2.4. *Under the hypothesis as before, suppose Z admits centers for the sets $H(y)$, $y \in IK$. Pick $c_y \in \text{Cent}\,H(y)$ for each $y \in IK$, then $y \to c_y$ is an optimal algorithm.*

Lastly, aside from Theorem 2.24, we present one more case wherein $e(K) = E(K)$.

Theorem 2.25. *Assume $T(K)$ is bounded and that each set $H(y)$, $y \in IK$ is centerable, then $e(K) = E(K)$.*

Proof: For each $y \in IK$, we have $\text{diam}\,H(y) = 2\text{rad}\,H(y)$, and

$$\text{diam}\,H(y) = \sup\{\|Tx_1 - Tx_2\| : x_1, x_2 \in K \text{ and } Ix_1 = Ix_2 = y\}$$
$$= \sup\{\|T(x_1 - x_2)\| : x_1, x_2 \in K \text{ and } Ix_1 = Ix_2 = y\}$$
$$\leq \sup\{\|Tx\| : x \in 2K, Ix = \theta\}$$
$$= 2\sup\{\|Tx\| : x \in K, Ix = \theta\}$$
$$= 2e(K).$$

Thus $\text{rad}\,H(y) \leq e(K)$ for all $y \in IK$ and we conclude from (2.10.14) that $E(K) \leq e(K)$.

2.10.4 Notes

One of the first landmark articles devoted to optimal recovery is Golomb and Weinberger [1] and one of the first surveys in this area is Section 33 in Holmes [2]. A more extensive survey of the literature in this direction up to 1977 is contained in Micchelli and Rivlin [4]. An update of this survey has also been contributed by the same authors in 1985. This topic is now encompassed by a wider topic called *information-based complexity*. The interested reader is referred to the book of Traub and Wozniakowski [7] for more details in this direction.

Exercises IV

2.41. (Rivlin and Micchelli): For optimal approximation of non-expansive functions with $m = 1$ in Example II, let $T = [0, 1], 0 \leq t_1 < t_2 \ldots < t_n \leq 1$, $\xi_i := \frac{1}{2}(t_i +$

t_{i+1}), $i = 1, 2, \ldots, n-1$, $\xi_0 := 0$, $\xi_n := 1$. Then, $T_1 = [0, \xi_1]$, $T_n = [\xi_{n-1}, 1]$, and $T_i = [\xi_{i-1}, \xi_i]$, $i = 2, \ldots, n-1$. Show that with $\Delta := \max_{1 \le i \le n-1} \Delta_i$, $\Delta_i := t_{i+1} - t_i$, $E(\mathcal{K}) = \|\hat{q}\|_\infty = \max\{t_1, \frac{1}{2}\Delta, 1 - t_n\}$ and that an optimal approximant of $x \in \mathcal{K}$ is given by the step function

$$s(t) = x(t_i), \quad \xi_{i-1} < t < \xi_i, \quad i = 1, \ldots, n.$$

Show also that the optimal sampling nodes in this case are given by $t_i = \frac{2i-1}{2n}$, $i = 1, \ldots, n$ with

$$E(\mathcal{K}) = \frac{1}{2n} = \min\{E(\mathcal{K}) : t_1, \ldots, t_n \in [0, 1], \text{ distinct }\}.$$

Furthermore, show that for the optimal approximation problem as above another simple optimal algorithm is

$$\mathring{A} : (x(t_1), \ldots, x(t_n)) \longrightarrow p(t)$$

where $p(t)$ is the piecewise linear interpolant of x at the nodes t_is.

2.42. (Rivlin and Micchelli): For the optimal approximation of non-expansive functions with $m = 2$, $n = 3$ in Example II, take T to be a triangle with vertices $\mathbf{t}_1$, $\mathbf{t}_2$, $\mathbf{t}_3$ (not obtuse angled). Show that another optimal approximant of $x \in \mathcal{K}$ is given by $\mathring{A} = (x(\mathbf{t}_1), x(\mathbf{t}_2), x(\mathbf{t}_3)) \longrightarrow \ell$ where ℓ is the linear interpolant to x,

$$\ell(\mathbf{t}) = au + bv + c, \mathbf{t} = (u, v)$$

satisfying $\ell(\mathbf{t}_i) = x(\mathbf{t}_i)$, $i = 1, 2, 3$.

References

1. Golomb, M., & Weinberger, H. (1959). Optimal approximation and error bounds. In R. Langer (Ed.), *On numerical approximation* (pp. 117–190). Madison: University of Wisconsin Press.
2. Holmes, R. B. (1972). *A course on optimization and best approximation* (Vol. 257). Lecture notes. New York: Springer.
3. Mhaskar, H. N., & Pai, D. V. (2007). *Fundamentals of approximation theory*. New Delhi: Narosa Publishing House.
4. Micchelli, C. A., & Rivlin, T. J. (1977). A survey of optimal recovery. In C. A. Micchelli & T. J. Rivlin (Eds.), *Optimal estimation in approximation theory* (pp. 1–54). New York: Plenum Press.
5. Micchelli, C. A., & Rivlin, T. J. (1985). *Lectures on optimal recovery* (Vol. 1129, pp. 21–93). Lecture notes in mathematics. Berlin: Springer.
6. Micchelli, C. A., Rivlin, T. J., & Winograd, S. (1976). Optimal recovery of smooth functions. *Numerische Mathematik, 260*, 191–200.
7. Traub, J. F., & Wozniakowski, H. (1980). *A general theory of optimal algorithms*. New York: Academic Press.

Chapter 3
Deterministic Models and Optimization

3.1 Introduction

There are various types of models that one can construct for a given set of data. The types of model that is chosen depend upon the type of data for which the model is to be constructed. If the data are coming from a deterministic situation, then there may be already an underlying mathematical formula such as a physical law. Perhaps the law may not be known yet. Hence, one may try to fit a model to the data collected about the situation. When the physical law is known, then there is no need to fit a model, but for verification purposes, one may substitute data points into the speculated physical law.

3.2 Deterministic Situations

For example, a simple physical law for gases says that the pressure P multiplied by volume V is a constant under a constant temperature. Then, the physical law that is available is

$$PV = c$$

where c is a constant. When it is a mathematical relationship, then all pairs of observations on P and V must satisfy the equation $PV = c$. If V_1 is one observation on V and if the corresponding observation is P_1 for P, then $P_1 V_1 = c$ for the same constant c. Similarly, other pairs of observations $(P_2, V_2), \ldots$ will satisfy the equation $PV = c$. If there are observational errors in observing P and V, then the equation may not be satisfied exactly by a given observational pair. If the model proposed $PV = c$ is not true exactly, then of course the observational pairs (P_i, V_i) for some specific i may not satisfy the equation $PV = c$.

This chapter is based on the lectures of Professor A.M. Mathai of CMS.

© Springer International Publishing AG 2017
A.M. Mathai and H.J. Haubold, *Fractional and Multivariable Calculus*,
Springer Optimization and Its Applications 122, DOI 10.1007/978-3-319-59993-9_3

3.3 Differential Equations

Another way of describing a deterministic situation is through the help of differential equations. We may be able to place some plausible assumptions on the rate of change of one variable with respect to another variable. For example, consider the growth of a bacterial colony over time. The rate of change of the population size may be proportional to the existing or current size. For example, if new cells are formed by cell division, then the number of new cells will depend upon the number of cells already there. Growth of human population in a community is likely to be proportional to the number of fertile women in the community. In all these situations, the rate of change of the actual size, denoted by y, over time t, will be a constant multiple of y itself. That is,

$$\frac{dy}{dt} = ky \tag{3.1}$$

where k is a constant. The solution is naturally

$$y = me^{kt}$$

where m is an arbitrary constant. This constant m is available if we have an initial condition such as when $t = 0$, $y_0 = me^0 = m$ or $m = y_0$ or the model is

$$y = y_0 e^{kt}, \quad t \geq 0. \tag{3.2}$$

In this model, there is only one parameter k (unknown quantity sitting in the model), and this k is the constant of proportionality of the rate of change of the population size y over time t.

Example 3.1. Coconut oil (u = demand in kilograms) and palm oil (v = demand in kilograms) are two competing vegetable cooking oils that are sold in a supermarket. The shopkeeper found that if either coconut oil or palm oil is absent, then the demand sores exponentially. The rate of change of demand over time is found to be the following:

$$\frac{du}{dt} = 2u - v \text{ and } \frac{dv}{dt} = -u + v$$

$$\Rightarrow \frac{d}{dt}\begin{bmatrix} u \\ v \end{bmatrix} = \begin{bmatrix} 2 & -1 \\ -1 & 1 \end{bmatrix}\begin{bmatrix} u \\ v \end{bmatrix}$$

$$\Rightarrow \frac{d}{dt}W = AW, \ W = \begin{bmatrix} u \\ v \end{bmatrix}, \ A = \begin{bmatrix} 2 & -1 \\ -1 & 1 \end{bmatrix}.$$

Suppose that the starting stocks at $t = 0$ are $u = 100$ kilos and $v = 200$ kilos, time being measured in days. What will be the demands (1) on the 10th day, (2) eventually? [This is a special case of a general class of models known as *prey–predator models*].

Solution 3.1. Note that individually the demands vary exponentially, that is, when $v = 0$, $\frac{du}{dt} = 2u \Rightarrow u = e^{2t}$ and when $u = 0$, $\frac{dv}{dt} = v \Rightarrow v = e^{t}$. Hence, we may look for a general solution of the type

$$u = e^{\lambda t} x_1 \text{ and } v = e^{\lambda t} x_2$$

where λ, x_1, x_2 are unknown quantities. Then,

$$W = \begin{bmatrix} u \\ v \end{bmatrix} = \begin{bmatrix} e^{\lambda t} x_1 \\ e^{\lambda t} x_2 \end{bmatrix} = e^{\lambda t} X, \ X = \begin{bmatrix} x_1 \\ x_2 \end{bmatrix}.$$

Pre-multiplying both sides by A, we have

$$\frac{du}{dt} = \lambda e^{\lambda t} x_1, \ \frac{dv}{dt} = \lambda e^{\lambda t} x_2$$

$$\frac{d}{dt} W = AW = e^{\lambda t} AX \Rightarrow$$

$$\lambda e^{\lambda t} x_1 = 2 e^{\lambda t} x_1 - e^{\lambda t} x_2$$

$$\lambda e^{\lambda t} x_2 = -e^{\lambda t} x_1 + e^{\lambda t} x_2.$$

That is,

$$AX = \lambda X, \ A = \begin{bmatrix} 2 & -1 \\ -1 & 1 \end{bmatrix}, \ X = \begin{bmatrix} x_1 \\ x_2 \end{bmatrix}.$$

This means λ is an eigenvalue of A.

$$|A - \lambda I| = 0 \Rightarrow \left\| \begin{bmatrix} 2 & -1 \\ -1 & 1 \end{bmatrix} - \lambda \begin{bmatrix} 1 & 0 \\ 0 & 1 \end{bmatrix} \right\| = 0$$

$$\Rightarrow \begin{vmatrix} 2-\lambda & -1 \\ -1 & 1-\lambda \end{vmatrix} = 0$$

$$\Rightarrow \lambda^2 - 3\lambda + 1 = 0 \Rightarrow \lambda_1 = \frac{3+\sqrt{5}}{2}, \ \lambda_2 = \frac{3-\sqrt{5}}{2}.$$

What are the eigenvectors? Corresponding to $\lambda_1 = \frac{3+\sqrt{5}}{2}$ we have the second equation from $AX = \lambda_1 X$

$$-x_1 + \left[1 - \frac{3+\sqrt{5}}{2} \right] x_2 = 0 \Rightarrow x_1 = \frac{1-\sqrt{5}}{2}$$

for $x_2 = 1$ or one solution is

$$W_1 = \begin{bmatrix} x_1 \\ x_2 \end{bmatrix} = \begin{bmatrix} \frac{1-\sqrt{5}}{2} \\ 1 \end{bmatrix}.$$

For $\lambda_2 = \frac{3-\sqrt{5}}{2}$, substituting in $AX = \lambda_2 X$ then from the first equation

$$\left(2 - \frac{3-\sqrt{5}}{2}\right)x_1 - x_2 = 0 \Rightarrow x_2 = \frac{1+\sqrt{5}}{2}$$

for $x_1 = 1$. Thus, an eigenvector is

$$W_2 = \begin{bmatrix} x_1 \\ x_2 \end{bmatrix} = \begin{bmatrix} 1 \\ \frac{1+\sqrt{5}}{2} \end{bmatrix}.$$

Then the general solution is

$$W = c_1 W_1 + c_2 W_2$$

$$= c_1 e^{\lambda_1 t} \begin{bmatrix} \frac{1-\sqrt{5}}{2} \\ 1 \end{bmatrix} + c_2 e^{\lambda_2 t} \begin{bmatrix} 1 \\ \frac{1+\sqrt{5}}{2} \end{bmatrix}$$

where λ_1 and λ_2 are given above. From the initial conditions, by putting $t = 0$

$$W_0 = \begin{bmatrix} 100 \\ 200 \end{bmatrix} = c_1 \begin{bmatrix} \frac{1-\sqrt{5}}{2} \\ 1 \end{bmatrix} + c_2 \begin{bmatrix} 1 \\ \frac{1+\sqrt{5}}{2} \end{bmatrix} \Rightarrow$$

$$100 = \frac{(1-\sqrt{5})}{2}c_1 + c_2$$

$$200 = c_1 + \frac{(1+\sqrt{5})}{2}c_2.$$

This gives $c_1 = 75 - 25\sqrt{5}$ and $c_2 = 50\sqrt{5}$. Thus, the answers are the following:
(1): For $t = 10$

$$u = (75 - 25\sqrt{5})e^{(\frac{3+\sqrt{5}}{2})(10)}(\frac{1-\sqrt{5}}{2}) + 50\sqrt{5}e^{(\frac{3-\sqrt{5}}{2})(10)} \text{ and}$$

$$v = (75 - 25\sqrt{5})e^{(\frac{3+\sqrt{5}}{2})(10)} + 50\sqrt{5}e^{(\frac{3-\sqrt{5}}{2})(10)}(\frac{1+\sqrt{5}}{2}).$$

Both these tend to $+\infty$ when $t \to \infty$, and hence, the demand explodes eventually.

3.4   Algebraic Models

Another category of models for a deterministic situation is algebraic in nature. Suppose that we want to construct models for mass, pressure, energy generation, and luminosity in the Sun, then we have to speculate about the nature of matter density distribution in the Sun. The Sun can be taken as a gaseous ball in hydrostatic equilibrium. The matter density can be speculated to be more and more concentrated toward the center or core of the Sun and more and more thin when moving outward from the core to the surface. This density distribution is deterministic in nature. But due to lack of knowledge about the precise mechanism working inside the Sun, one has to speculate about the matter density distribution. If r is an arbitrary distance from the center of the Sun, then one can start with a simple linear model for the matter density $f(r)$ at a distance r from the center. Then, our assumption is that

$$f(r) = a(1 - x), \quad a > 0, \quad x = \frac{r}{R_o} \tag{3.3}$$

where R_o is the radius of the Sun. This model indicates a straight line decrease from the center of the Sun, going outward. From observational data, we can see that (3.3) is not a good model for the density distribution in the Sun. Another speculative solar model (model for the Sun) is of the form

$$f(r) = f_0(1 - x^\delta)^\rho \tag{3.4}$$

where δ and ρ are two parameters. It can be shown that we have specific choices of δ and ρ so that the model in (3.4) can be taken as a good model for the density distribution in the Sun. Once a model such as the one in (3.4) is proposed, then simple mathematical formulation from (3.4) will give models for mass $M(r)$, pressure $P(r)$, and luminosity $L(r)$, at arbitrary point r. Such models are called analytical models. Various such analytical solar and stellar models may be seen from Mathai and Haubold (1988: *Modern Problems in Nuclear and Neutrino Astrophysics*, Akademie-Verlag, Berlin) and later papers by the same authors and others.

3.5 Computer Models

Problems such as energy generation in the Sun or in a star depend on many factors. One such factor is the distribution of matter density in the star. A model based on just one variable such as matter density will not cover all possibilities. Hence, what scientists often do is to take into account all possible factors and all possible forms in which the factors enter into the scenario. This may result in dozens of differential equations to solve. Naturally, an analytical solution or exact mathematical formulation will not be possible. Then, the next best thing to do is to come up with some numerical solutions with the help of computers. Such models are called computer models. The standard solar model, which is being widely used now, is a computer model. One major drawback of a computer model is that the computer will give out some numbers as numerical solutions but very often it is not feasible to check and see whether the numbers given out by the computer are correct or not.

3.6 Power Function Models

There are a number of physical situations where the rate of change of an item is in the form of a power of the original item. In certain decaying situations such as certain radiation activity, energy production, reactions, diffusion processes, the rate of change over time is in the form of the power. If y is the original amount and if it is a decaying situation, then the law governing the process may be of the form

$$\frac{dy}{dt} = -y^\alpha, \text{ (decaying)} \tag{3.5}$$

$$\frac{dy}{dt} = y^\alpha, \text{ (growing)}. \tag{3.6}$$

Then a solution of the decaying model in (3.5) is of the form

$$y = [1 - a(1 - \alpha)t]^{\frac{1}{1-\alpha}} \Rightarrow \frac{dy}{dt} = -ay^\alpha, \ a > 0 \tag{3.7}$$

and the growth model in (3.6) has the solution

$$y = [1 + a(1 - \alpha)t]^{-\frac{1}{\alpha-1}} \Rightarrow \frac{dy}{dt} = ay^\alpha, \ a > 0. \tag{3.8}$$

In (3.7), the rate of decay of y is proportional to a power of y. For example, if $\alpha = \frac{1}{2}$, then the rate of change of y over time is proportional to the square root of y. The current hot topics of Tsallis statistics and q-exponential functions are associated with the differential equations in (3.5) and (3.6) with resulting solutions in (3.7) and (3.8), respectively. The power parameter α may be determined from some physical interpretation of the system or one may estimate from data. The behavior in (3.5) and (3.6) is also associated with some generalized entropies in Information Theory. Some details may be seen from Tsallis [12] and Mathai and Haubold [7–9].

3.7 Input–Output Models

There are a large number of situations in different disciplines where what is observed is the residual effect of an input activity and an output activity. Storage of water in a dam or storage of grain in a silo at any given time is the residual effect of the total amount of water flowed into the dam and the total amount of water taken out for irrigation, power generation, etc. Grain is stored in a silo during harvest seasons, and grain is sold out or consumed from time to time, and thus what is in the storage silo is the residual effect of the input quantity and output quantity. A chemical called melatonin is produced in every human body. The production starts in the evening, it attains the maximum peak by around 1 am, and then it comes down and by morning the level is back to normal. At any time in the night, the amount of melatonin seen in the body is what is produced, say x, minus what is consumed or converted or spent by the body, say y. Then, what is seen in the body is $z = x - y$, $x \geq y$. It is speculated that by monitoring z, one can determine the body age of a human being, because it is seen that the peak value of z grows as the person grows in age, and it attains a maximum during what we will call the middle age of the person, and the peak value of z starts coming down when the person crosses the middle age. Thus, it is of great importance in modeling this z, and lot of studies are there on z.

Another similar situation is the production of certain particles in the Sun called solar neutrinos. These particles are capable of passing through almost all media and practically inert, not reacting with other particles. There are experiments being conducted in the USA, Europe, and Japan, to capture these neutrinos. There is discrepancy between what is theoretically predicted from physical laws and what is captured through these experiments. This is known as the solar neutrino problem. Modeling the amount of captured neutrinos is a current activity to see whether there is any cyclic pattern, whether there is time variation, and so on. What is seen mathematically is an input–output type behavior. There are also several input–output type situations in economics, social sciences, industrial production, commercial activities, and so on. An analysis of the general input–output structure is given in Mathai [3], and some studies of the neutrino problem may be seen from Haubold and Mathai [1, 2]. In input–output modeling, the basic idea is to model

$$z = x - y$$

by imposing assumptions on the behaviors or types of x and y and assumption about whether x and y are independently varying or not.

3.8 Pathway Model

Another idea that was introduced recently, see Mathai [4], is to come up with a model which can switch around, assume different functional forms, or move from one functional form to another, thus creating a pathway to different functional forms. Such a model will be useful in describing transitional stages in the evolution of any item or describing chaos or chaotic neighborhoods where there is a stable solution. The pathway model is created on the space of rectangular matrices. For the very special case of real scalar positive variable x, the model is of the following form:

$$f_1(x) = c_1 x^\gamma [1 - a(1 - \alpha)x^\delta]^{\frac{1}{1-\alpha}}, \ 1 - a(1 - \alpha)x^\delta > 0, \ a > 0 \qquad (3.9)$$

where $\gamma, \delta \geq 0$, $a > 0$ are parameters and c_1 can behave as a normalizing constant if the total integral over $f(x)$ is to be made equal to 1 so that with $f(x) \geq 0$ for $x \geq 0$ this $f(x)$ can act as a density for a positive random variable x also. The parameter α is the pathway parameter. Note that for $\alpha < 1$ and $1 - a(1 - \alpha)x^\delta > 0$, we have (1.9) belonging to the generalized type-1 beta family of functions. When $\alpha > 1$ writing $1 - \alpha = -(\alpha - 1)$, we have

$$f_2(x) = c_1 x^\gamma [1 + a(\alpha - 1)x^\delta]^{-\frac{1}{\alpha-1}}, \ \alpha > 1, \ 0 \le x < \infty. \tag{3.10}$$

The function in (3.10) belongs to the generalized type-2 beta family of functions. But when $\alpha \to 1$, both the forms in (3.9) and (3.10) go to the exponential form

$$f_3(x) = c_3 x^\gamma e^{-ax^\delta}, \ \alpha \to 1, a > 0, \delta > 0, x \ge 0. \tag{3.11}$$

This $f_3(x)$ belongs to the generalized gamma family of functions. Thus, the basic function $f_1(x)$ in (3.9) can move to three different families of functions through the parameter α as α goes from $-\infty$ to 1, then from 1 to ∞, and finally the limiting case of $\alpha \to 1$. If the limiting form in (3.11) is the stable solution to describe a system, then unstable neighborhoods of this stable solution are described by (3.9) and (3.10). All the transitional stages can be modeled with the help of the parameter α. For example, as α moves from the left to 1, this function $f_1(x)$ moves closer and closer and finally to $f_3(x)$. Similarly as α moves to 1 from the right, $f_2(x)$ moves closer and closer and finally to $f_3(x)$.

The same model can create a pathway for certain classes of differential equations, a pathway for classes of statistical densities and a pathway for certain classes of generalized entropies. Thus, essentially there are three pathways in the model in (3.9), the entropic pathway, the differential pathway, and the distributional pathway. More details may be seen from Mathai and Haubold [8, 9]. The entropic pathway will be described in the next subsection.

3.8.1 Optimization of Entropy and Exponential Model

The exponential model of Section 3.2 can also be obtained by optimization of entropy or other optimization processes. Shannon's entropy, a measure of uncertainty in a scheme of events and the corresponding probabilities, for the continuous situation is the following:

$$S(f) = - \int_{-\infty}^\infty f(x) \ln f(x) \, dx \tag{3.12}$$

where $f(x) \ge 0$ for all x and $\int_{-\infty}^\infty f(x) \, dx = 1$, or $f(x)$ is a statistical density. If $S(f)$ is optimized over all functional $f(x) \ge 0$, subject to the condition that the first moment is fixed and that $f(x)$ is a density, that is

$$\int_{-\infty}^\infty f(x) \, dx = 1 \text{ and } \int_{-\infty}^\infty x f(x) \, dx = \mu = \text{ fixed}$$

then if we use calculus of variation and Lagrangian multipliers, the quantity to be optimized is

$$U(f) = -\int_{-\infty}^{\infty} f(x) \ln f(x)\, dx - \lambda_1 \left(\int_{-\infty}^{\infty} f(x) dx - 1 \right) - \lambda_2 \left(\int_{-\infty}^{\infty} x f(x) dx - \mu \right)$$

over all functional f. By using calculus of variation, the Euler equation reduces to the following:

$$\frac{\partial}{\partial f} [-f \ln f - \lambda_1 f - \lambda_2 x f] = 0 \Rightarrow$$

$$-\ln f - 1 - \lambda_1 - \lambda_2 x = 0 \Rightarrow$$

$$f = k e^{-kx}, \ x \geq 0, \ k > 0 \tag{3.13}$$

where k is a constant, which is the exponential density. The first moment μ being fixed can be given many types of physical interpretations in different fields.

3.8.2 Optimization of Mathai's Entropy

There are many types of generalizations of Shannon's entropy of (3.12). One such generalization is Mathai's entropy

$$M_\alpha(f) = \frac{\int_{-\infty}^{\infty} [f(x)]^{2-\alpha} dx - 1}{\alpha - 1}, \ \alpha \neq 1. \tag{3.14}$$

When $\alpha \to 1, M_\alpha(f) = S(f) =$ Shannon's entropy, and hence $M_\alpha(f)$ is an α-generalized Shannon's entropy $S(f)$. By optimizing $M_\alpha(f)$, under different constraints, we can obtain the power law of Section 3.5 and the pathway model of Section 3.7. For example, if we consider the constraints, $f(x) \geq 0$ for all x, $\int_{-\infty}^{\infty} f(x) dx = 1$ and $\int_{-\infty}^{\infty} x f(x) dx = \mu$ fixed, that is, $f(x)$ is a density and that the first moment is fixed, then from calculus of variation the Euler equation is the following:

$$\frac{\partial}{\partial f} \left[\frac{f^{2-\alpha}}{\alpha - 1} - \lambda_1 f + \lambda_2 x f \right] = 0 \Rightarrow$$

$$f = c_1 [1 - \frac{\lambda_2}{\lambda_1} x]^{\frac{1}{1-\alpha}}.$$

If we write $\frac{\lambda_2}{\lambda_1} = a(1 - \alpha), a > 0$, then we have the model

$$f = c_1 [1 - a(1 - \alpha) x]^{\frac{1}{1-\alpha}}, \ x \geq 0, \ a > 0 \tag{3.15}$$

where c_1 is the normalizing constant. This is Tsallis' model, which leads to the whole area of non-extensive statistical mechanics. Then, observe that $g = \frac{f}{c_1}$ for $a = 1$ is the power function model

$$\frac{d}{dx}g = -g^{\alpha}. \tag{3.16}$$

If $M_{\alpha}(f)$ is optimized by using two moment-like constraints

$$\int_{-\infty}^{\infty} x^{\gamma}f(x)dx = \text{fixed}, \quad \int_{-\infty}^{\infty} x^{\gamma+\delta}f(x)dx = \text{fixed}$$

then, over all functional $f(x) \geq 0$ such that $\int_{-\infty}^{\infty}f(x)dx < \infty$ we have the Euler equation

$$\frac{\partial}{\partial f}[\frac{f^{2-\alpha}}{\alpha - 1} - \lambda_1 x^{\gamma}f + \lambda_2 x^{\gamma+\delta}f] = 0.$$

This leads to the form

$$f = c_2 x^{\gamma}[1 - a(1 - \alpha)x^{\delta}]^{\frac{1}{1-\alpha}}, \ x \geq 0, \ a > 0, \ \delta > 0 \tag{3.17}$$

where c_2 can act as a normalizing constant. This f is the pathway model of Section 3.7. For $\gamma = 0, a = 1, \delta = 1$, we have Tsallis model, Tsallis [12] from (3.17). It is said that over 3000 papers are published on Tsallis' model during 1990–2010 period. For $\alpha > 1$ in (3.17) with $a = 1$ and $\delta = 1$, one has Beck–Cohen superstatistics model of Astrophysics. Dozens of articles are published on superstatistics since 2003.

3.9   Fibonacci Sequence Model

Consider the growth of a certain mechanism in the following fashion. It takes one unit of time to grow and one unit of time to reproduce. Reproduction means producing one more item or splitting into two items. For example, consider the growth of rabbit population in the certain locality. Suppose that we count only female rabbits. Suppose that every female rabbit gives birth to one female rabbit at every six month's period. After six months of giving birth, the mother rabbit can give birth again, one unit of time. But the daughter rabbit needs six months time to grow and then six months to reproduce, or two units of time. Then, at each six months, the population size will be of the form 1, 1, 2, 3, 5, 8, 13, ... or the sum of the two previous numbers is the next number. This sequence of numbers is called the Fibonacci sequence of numbers. Thus, any growth mechanism which takes one unit of time to grow and one unit of time to reproduce can be modeled by a Fibonacci sequence. Some application of these types of models in explaining many types of patterns that are seen in nature may be seen from Mathai and Davis [5].

Example 3.2. When a population is growing according to a Fibonacci sequence and if F_k denotes the population size at the k-th generation, with the initial conditions $F_0 = 0, F_1 = 1$, then the difference equation governing the growth process is $F_{k+2} = F_{k+1} + F_k$. Evaluate F_k for a general k.

Solution 3.2. This can be solved by a simple property of matrices. To this end, write one more equation that is an identity $F_{k+1} = F_{k+1}$ and write the system of equations:

$$F_{k+2} = F_{k+1} + F_k$$
$$F_{k+1} = F_{k+1} \tag{i}$$

Take

$$V_k = \begin{bmatrix} F_{k+1} \\ F_k \end{bmatrix}. \tag{ii}$$

Then the equation in (*ii*) can be written as

$$V_{k+1} = A V_k, \quad A = \begin{bmatrix} 1 & 1 \\ 1 & 0 \end{bmatrix}. \tag{iii}$$

But from the initial condition

$$V_0 = \begin{bmatrix} F_1 \\ F_0 \end{bmatrix} = \begin{bmatrix} 1 \\ 0 \end{bmatrix}.$$

From (*iii*)

$$V_1 = A V_0 \Rightarrow V_2 = A V_1 = A^2 V_0 \Rightarrow V_k = A^k V_0. \tag{iv}$$

Therefore, we need to compute only A^k because V_0 is already known. For example, if $k = 100$, then we need to multiply A by itself 100 times, which is not an easy process. Hence, we will make use of some properties of eigenvalues. Let us compute the eigenvalues of A.

$$|A - \lambda I| = 0 \Rightarrow \begin{vmatrix} 1 - \lambda & 1 \\ 1 & -\lambda \end{vmatrix} = 0$$
$$\Rightarrow \lambda^2 - \lambda - 1 = 0$$
$$\Rightarrow \lambda_1 = \frac{1 + \sqrt{5}}{2} \text{ and } \lambda_2 = \frac{1 - \sqrt{5}}{2}.$$

Therefore, the eigenvalues of A^k are λ_1^k and λ_2^k. Let us compute the eigenvectors. Let $X = \begin{pmatrix} x_1 \\ x_2 \end{pmatrix}$ be the eigenvector corresponding to λ_1. Then,

$$(A - \lambda_1 I)X = O \Rightarrow \begin{bmatrix} 1 - \frac{(1+\sqrt{5})}{2} & 1 \\ 1 & -\frac{(1+\sqrt{5})}{2} \end{bmatrix} \begin{pmatrix} x_1 \\ x_2 \end{pmatrix} = \begin{pmatrix} 0 \\ 0 \end{pmatrix}$$

$$\Rightarrow X_1 = \begin{pmatrix} x_1 \\ x_2 \end{pmatrix} = \begin{pmatrix} \frac{1+\sqrt{5}}{2} \\ 1 \end{pmatrix} = \begin{pmatrix} \lambda_1 \\ 1 \end{pmatrix}.$$

Similarly, the eigenvector corresponding to λ_2 is

$$X_2 = \begin{pmatrix} \lambda_2 \\ 1 \end{pmatrix}.$$

Then the matrix of eigenvectors is given by

$$Q = (X_1, X_2) = \begin{bmatrix} \lambda_1 & \lambda_2 \\ 1 & 1 \end{bmatrix} \Rightarrow Q^{-1} = \frac{1}{\lambda_1 - \lambda_2} \begin{bmatrix} 1 & -\lambda_2 \\ -1 & \lambda_1 \end{bmatrix}.$$

Then

$$A = QDQ^{-1} = \begin{bmatrix} \lambda_1 & \lambda_2 \\ 1 & 1 \end{bmatrix} \begin{bmatrix} \lambda_1 & 0 \\ 0 & \lambda_2 \end{bmatrix} \begin{bmatrix} \frac{1}{\lambda_1 - \lambda_2} \begin{pmatrix} 1 & -\lambda_2 \\ -1 & \lambda_1 \end{pmatrix} \end{bmatrix}.$$

Since λ and λ^k share the same eigenvectors,

$$A^k = \frac{1}{\lambda_1 - \lambda_2} \begin{bmatrix} \lambda_1 & \lambda_2 \\ 1 & 1 \end{bmatrix} \begin{bmatrix} \lambda_1^k & 0 \\ 0 & \lambda_2^k \end{bmatrix} \begin{bmatrix} 1 & -\lambda_2 \\ -1 & \lambda_1 \end{bmatrix}$$

$$= \frac{1}{\lambda_1 - \lambda_2} \begin{bmatrix} \lambda_1^{k+1} - \lambda_2^{k+1} & -\lambda_2 \lambda_1^{k+1} + \lambda_1 \lambda_2^{k+1} \\ \lambda_1^k - \lambda_2^k & -\lambda_2 \lambda_1^k + \lambda_1 \lambda_2^k \end{bmatrix}. \tag{v}$$

But

$$V_k = A^k V_0 = A^k \begin{pmatrix} 1 \\ 0 \end{pmatrix} \Rightarrow$$

$$V_k = \frac{1}{\lambda_1 - \lambda_2} \begin{bmatrix} \lambda_1^{k+1} - \lambda_2^{k+1} \\ \lambda_1^k - \lambda_2^k \end{bmatrix}.$$

That is,

$$F_k = \frac{1}{\lambda_1 - \lambda_2} (\lambda_1^k - \lambda_2^k)$$

$$= \frac{1}{\sqrt{5}} \left\{ \left(\frac{1 + \sqrt{5}}{2} \right)^k - \left(\frac{1 - \sqrt{5}}{2} \right)^k \right\}. \tag{vi}$$

Then F_k can be computed for any k by using (vi). We can make some interesting observations. Since $\left|\frac{1-\sqrt{5}}{2}\right| < 1$, $\lambda_k \to 0$ as $k \to \infty$. Hence, for large values of k,

$$F_k \approx \frac{1}{\sqrt{5}}\left(\frac{1+\sqrt{5}}{2}\right)^k \Rightarrow \lim_{k\to\infty}\frac{F_{k+1}}{F_k} = \frac{1+\sqrt{5}}{2}$$

$$\Rightarrow \lim_{k\to\infty}\frac{F_k}{F_{k+1}} = \frac{2}{1+\sqrt{5}} = \frac{\sqrt{5}-1}{2} \tag{3.18}$$

This $\frac{\sqrt{5}-1}{2}$ is known as the golden ratio which appears at many places in nature.

3.10 Fractional Order Integral and Derivative Model

Consider a simple reaction-diffusion type situation. Particles react with each other, and new particles are produced. If $N(t)$ is the number density of the reacting particles at time t, then the number of particles produced will be proportional to the number already in the system. That is, the rate of change in $N(t)$, over time, is a constant multiple of $N(t)$. Then, the simple reaction equation is given by

$$\frac{d}{dt}N(t) = \alpha\,N(t), \quad \alpha = \text{a constant} > 0,$$

whose solution is an exponential function, $N(t) = N_0 e^{\alpha t}$, where N_0 is the number at $t = 0$ or at the start of the observation period. If part of the number of particles is destroyed or if it is a diffusion- or decay-type process, then the equation governing such a situation is

$$\frac{d}{dt}N(t) = -\beta\,N(t), \quad \beta > 0 \text{ is a constant.}$$

If it is a resultant decay process that β dominates over α, then the equation could be of the form

$$\frac{d}{dt}N(t) = -\gamma\,N(t), \quad \gamma = \beta - \alpha > 0 \Rightarrow N(t) = N_0 e^{-\gamma t}. \tag{3.19}$$

In the above notation, $D = \frac{d}{dt}$ is the single integer derivative and $D^2 = \frac{d^2}{dt^2}$ is the second integer order derivative. Here D can measure velocity, and D^2 can measure acceleration. If we think of a situation between D and D^2, that is, the rate of change is more than the velocity D but less than the acceleration D^2, then we have something which can be denoted as D^α, $1 \le \alpha \le 2$ or D^α can be called a fractional order derivative. Then, $D^{-\alpha}$ can denote a fractional order integral.

In many reaction-diffusion type problems in physical situations, it is noted that the solution through a fractional differential or integral operator can describe the physical situation better, compared to integer order derivative or integrals. Currently, there is intensive activity in many applied areas to consider fractional order differential and integral equations rather than integer order derivatives and integrals, and the corresponding models will be fractional order derivative or integral models.

We will consider the simple problem in (3.19), and let us see what happens if we change the integer order derivative to a fractional derivative. The differential equation in (3.19) can also be converted into an integral equation of the type

$$N(t) - N_0 = -\gamma \int N(t) \mathrm{d}t. \tag{3.20}$$

Let us replace the integral in (3.20) by a fractional integral. There are various types of fractional integrals and fractional derivatives. The left-sided Riemann–Liouville fractional integral is denoted and defined as follows:

$$_0D_x^{-\alpha}f(x) = \frac{1}{\Gamma(\alpha)} \int_0^x (x - t)^{\alpha-1} f(t) \mathrm{d}t, \ \Re(\alpha) > 0. \tag{3.21}$$

If the total integral in (3.20) is replaced by the fractional integral in (3.21), then we have the fractional integral equation

$$N(t) - N_0 = -\gamma \, _0D_x^{-\alpha} N(t). \tag{3.22}$$

Let us see what will happen to the exponential solution in (3.20). In order to solve (3.22), we can appeal to the Laplace transform. Take the Laplace transform on both sides of (3.22), with Laplace parameter s. Let us denote the Laplace transform of $N(t)$ by $\tilde{N}(t)$. Then, we have,

$$\int_0^\infty e^{-st} N(t) \mathrm{d}t - N_0 \int_0^\infty e^{-st} \mathrm{d}t = -\gamma \int_0^\infty e^{-st} \{_0D_x^{-\alpha} N(t)\} \mathrm{d}x.$$

But, denoting the Laplace transform on the fractional derivative by L, we have

$$L = \int_0^\infty e^{-st} [_0D_x^{-\alpha} N(t)] \mathrm{d}x$$

$$= \int_0^\infty e^{-sx} \{\frac{1}{\Gamma(\alpha)} \int_0^x (x - t)^{\alpha-1} N(t) \mathrm{d}t\} \mathrm{d}x, \ \Re(\alpha) > 0.$$

Now, changing the order of integration, we have the following:

$$L = \frac{1}{\Gamma(\alpha)} \int_{t=0}^\infty \{\int_{x=t}^\infty e^{-sx} (x - t)^{\alpha-1} \mathrm{d}x\} N(t) \mathrm{d}t, \ \text{put } u = x - t$$

$$= s^{-\alpha} \int_0^\infty e^{-st} N(t) dt$$

$$= s^{-\alpha} \tilde{N}(s)$$

where $\tilde{N}(s)$ is the Laplace transform of $N(t)$. The equation (3.22) yields

$$\tilde{N}(s) - \frac{N_0}{s} = -\gamma s^{-\alpha} \tilde{N}(s) \Rightarrow$$

$$\tilde{N}(s) = N_0 \frac{1}{s[1 + \frac{\gamma}{s^\alpha}]}$$

$$= N_0 \sum_{k=0}^\infty (-1)^k \frac{\gamma^k}{s^{\alpha k+1}}, \quad |\frac{\gamma}{s^{\alpha k+1}}| < 1.$$

Taking the inverse Laplace transform by using the fact that

$$\int_0^\infty x^{\alpha k} e^{-sx} dx = \Gamma(1 + \alpha k) s^{-\alpha k - 1}, \quad \Re(s) > 0$$

we have

$$N(t) = N_0 \sum_{k=0}^\infty \frac{(-1)^k t^{\alpha k} \gamma^k}{\Gamma(1 + \alpha k)} = N_0 E_\alpha(-\gamma t^\alpha) \qquad (3.23)$$

$$= N_0 e^{-\gamma t} \text{ for } \alpha = 1 \qquad (3.24)$$

where $E_\alpha(\cdot)$ is the simplest form of a Mittag-Leffler function. Thus, the exponential solution in (3.24) changes to the Mittag-Leffler solution in (3.23) when we go from the total integral to a fractional integral. It can be seen that (3.23) has a thicker tail compared to the tail behavior in (3.24). A thicker-tailed model seems to be a better fit compared to the exponential-tailed model in many practical situations. Many models of this type, coming from various types of fractional differential equations, may be seen from the recent books: Mathai and Haubold [10]; Mathai, Saxena, and Haubold [11].

A general form of the Mittag-Leffler function is the following:

$$E_{\alpha,\beta}^\gamma(x) = \sum_{k=0}^\infty \frac{(\gamma)_k}{k!} \frac{x^k}{\Gamma(\beta + \alpha k)}, \quad \Re(\alpha) > 0, \ \Re(\beta) > 0$$

$$E_{\alpha,\beta}^1(x) = E_{\alpha,\beta}(x) = \sum_{k=0}^\infty \frac{x^k}{\Gamma(\beta + \alpha k)}, \quad \Re(\alpha) > 0, \ \Re(\beta) > 0$$

$$E_\alpha(x) = E_{\alpha,1}(x) = \sum_{k=0}^{\infty} \frac{x^k}{\Gamma(1+\alpha k)}, \quad \Re(\alpha) > 0.$$

$$E_1(x) = e^x = \sum_{k=0}^{\infty} \frac{x^k}{\Gamma(1+k)} = \sum_{k=0}^{\infty} \frac{x^k}{k!}.$$

The Pochhammer symbol is defined as follows:

$$(\gamma)_k = \gamma(\gamma+1)(\gamma+2)...(\gamma+k-1), \quad (\gamma)_0 = 1, \quad \gamma \neq 0.$$

Note that $E_{\alpha,\beta}^\gamma(x)$ is a particular case of a Wright's function, which is a particular case of a H-function. The theory and applications of H-function are available from Mathai, Saxena, and Haubold [11].

Exercises 3.

3.1. Do Example 3.1 if the matrix A is the following:

$$A = \begin{bmatrix} 2 & 1 \\ 1 & 2 \end{bmatrix}.$$

3.2. A biologist has found that the owl population u and the mice population v in a particular region are governed by the following system of differential equations, where $t = $ time being measured in months.

$$\frac{du}{dt} = 2u + 2v$$

$$\frac{dv}{dt} = u + 2v.$$

Initially there are two owls and 100 mice. Solve the system. What will be the populations of owl and mice after 5 months? What will be the populations eventually?

3.3. Suppose that 2 falcons are also in the region in Exercise 1.2. Thus, the initial populations of falcon f, owl u, and mince v are $f_0 = 2$, $u_0 = 2$, $v_0 = 100$. Suppose that the following are the differential equations:

$$\frac{df}{dt} = f + u + v$$

$$\frac{du}{dt} = -f + u + 2v$$

$$\frac{dv}{dt} = -f - u + 3v.$$

Solve for f, u, v and answer the same questions in Exercise 3.2, including falcons.

3.4. For the model of density in equation (3.3), evaluate the mass of the Sun at an arbitrary value r if the law is

$$\frac{dM(r)}{dr} = 4\pi r^2 f(r)$$

where $f(r)$ is the density at a distance r from the center.

3.5. If pressure in the Sun is given by the formula

$$\frac{dP}{dr} = -\frac{G\, M(r)f(r)}{r^2}$$

where G is a constant and $f(r)$ is the density, evaluate the pressure at arbitrary distance r from the center, from the model in (3.3).

3.6. For the pathway model in (3.9), evaluate c_1 if $f_1(x)$ is a density for $\alpha < 1$ and $1 - a(1 - \alpha)x^\delta > 0, a > 0, \delta > 0$.

3.7. For the pathway model in (3.10), evaluate c_2 if $f_2(x)$ is a density for $x \geq 0$, $a > 0, \delta > 0, \alpha > 1$.

3.8. For the pathway model in (3.11) evaluate c_3.

3.9. For the models in (3.9), (3.10), and (3.11), compute the Mellin transform of $f_i(x)$ or $\int_x x^{s-1} f_i(x) dx, i = 1, 1, 3$ and state the conditions for their existence.

3.10. Draw the graphs of the models in (3.9), (3.10), and (3.11) for various values of the pathway parameter α and for some fixed values of the other parameters and check the movement of the function.

3.11. Suppose that a system is growing in the following fashion. The first stage size plus 2 times the second stage plus 2 times the third stage equals the fourth stage. That is,

$$F_k + 2F_{k+1} + 2F_{k+2} = F_{k+3}.$$

Suppose that the initial conditions are $F_0 = 0, F_1 = 1, F_2 = 1$. Compute F_k for $k = 100$.

3.12. Suppose that a system is growing in the following fashion: The first stage size plus 3 times the second stage size plus the third stage size equals the fourth stage or

$$F_k + 3F_{k+1} + F_{k+} = F_{k+3}.$$

Compute F_k for $k = 50$ if $F_0 = 0, F_1 = 1, F_2 = 1$.

3.13. In a savannah, there are some lions and some deers. Lions eat out deers but deers do not eat lions. There are small games like rabbits in the area. If no deer is left, the lions cannot survive for long hunting small games. If no lion is there, then deer population will explode. It is seen that the lion population at the i-th stage, L_i, and the deer population at the i-th stage, D_i, are related by the system

$$L_{i+1} = 0.1L_i + 0.4D_i$$
$$D_{i+1} = 1.2D_i - pL_i$$

where p is some number. The initial population at $i = 0$ are $L_0 = 10, D_0 = 100$. Compute the population of lion and deer for $i = 1, 2, 3, 4$. For what value of p, (i) the deer population will be extinct eventually, (ii) the lion population will be extinct eventually, (iii) lion and deer population will explode. When a number falls below 1, it should be counted as extinction.

3.14. Show that every third Fibonacci number is odd, starting with $1, 1$.

3.15. Consider the sequence $F_k + F_{k+1} = F_{k+2}$ with $F_0 = 0, F_1 = a > 0$. Show that

$$\lim_{k \to \infty} \frac{F_k}{F_{k+1}} = \frac{\sqrt{5} - 1}{2} = \text{ golden ratio.}$$

References

1. Haubold, H. J., & Mathai, A. M. (1994). Solar nuclear energy generation and the chlorine solar neutrino experiment. *American Institute of Physics, Conference Proceedings, 320*, 102–116.
2. Haubold, H. J., & Mathai, A. M. (1995). A heuristic remark on the periodic variation in the number of solar neutrinos detected on Earth. *Astrophysics and Space Science, 228*, 113–134.
3. Mathai, A. M. (1993). The residual effect of growth-decay mechanism and the distributions of covariance structures. *The Canadian Journal of Statistics, 21*(3), 277–283.
4. Mathai, A. M. (2005). A pathway to matrix-variate gamma and normal densities. *Linear Algebra and Its Applications, 396*, 317–328.
5. Mathai, A. M., & Davis, T. A. (1974). Constructing the sunflower head. *Journal of Mathematical Biosciences, 20*, 117–133.
6. Mathai, A. M., & Haubold, H. J. (1988). *Modern problems in nuclear and neutrino astrophysics*. Berlin: Akademie-Verlag.
7. Mathai, A. M., & Haubold, H. J. (2007). Pathway model, superstatistics, Tsallis statistics and a generalized measure of entropy. *Physica A, 375*, 110–122.
8. Mathai, A. M., & Haubold, H. J. (2007a). On generalized entropy measures and pathways. *Physica A, 385*, 493–500.
9. Mathai, A. M., & Haubold, H. J. (2007b). On entropic, distributional, and differential pathways. *Bulletin of the Astronomical Society of India, 35*, 669–680.
10. Mathai, A. M., & Haubold, H. J. (2008). *Special functions for applied scientists*. New York: Springer.
11. Mathai, A. M., Saxena, R. K., & Haubold, H. J. (2010). *The H-function: Theory and applications*. New York: Springer.
12. Tsallis, C. (1988). Possible generalizations of Boltzmann-Gibbs statistics. *Journal of Statistical Physics, 52*, 479–487.

Chapter 4
Non-deterministic Models and Optimization

4.1 Introduction

We considered models which were governed by definite mathematical rules or fully deterministic in nature. There was no chance variation involved. But most of the practical situations, mostly in social sciences, economics, commerce, management, etc., as well as many physical phenomena, are non-deterministic in nature. Earthquake at a place cannot be predetermined, but with sophisticated prediction tools, we may be able to predict the occurrence to some extent. We know that Meenachil River will be flooded during monsoon season, but we cannot tell in advance what will be the flood level on July 1, 2020, in front of Pastoral Institute. Even though many factors about weight gain are known, we cannot state for sure how much will be the weight gain on a cow under a certain feed. A student who is going to write an examination cannot tell beforehand what exact grade she is going to get. She may be able to predict that it will be above 80% from her knowledge about the subject matter. But after writing the examination, she may be able to give a better prediction that she will get at least 95% if not 100%. She is able to improve her prediction by knowing additional information of having written the examination.

Situations described above and many other situations of the same type are not deterministic in nature. When chance variation is involved, prediction can be made by using properties of random variables or chance variables or measurement of chance or probabilities.

Since a lot of practical situations are random or non-deterministic in nature, when we talk about model building, people naturally think that we are trying to describe a random or non-deterministic situation by mathematical modeling. Attempts to describe random situations have given birth to different branches of science. Stochastic process is one such area where we study a collection of random variables. Time series analysis is an area where we study a collection of random variables over time. Regression is an area where we try to describe random situations by analyzing

---

This chapter is based on the lectures of Professor A.M. Mathai of CMS.

© Springer International Publishing AG 2017
A.M. Mathai and H.J. Haubold, *Fractional and Multivariable Calculus*,
Springer Optimization and Its Applications 122, DOI 10.1007/978-3-319-59993-9_4

conditional expectations. As can be seen, even to give a basic description of all the areas and disciplines where we build models, it will take hundreds of pages. Hence, what we will do here is to pick a few selected topics and give basic introduction to these topics.

4.1.1   Random Walk Model

Consider a simple practical situation of a drunkard coming out of the liquor bar, denoted by S in Figure 4.1.

_____
       x H   ↓ S   B |
_____

Fig. 4.1   Random walk

He tries to walk home. Since he is fully drunk, assume that he walks in the following manner. At every minute, he will either take a step to the right or to the left. Let us assume a straight line path. Suppose that he covers 1 foot (about a third of a meter) at each step. He came out from the bar or station S as indicated by the arrow. Then if his first step is to the right, then he is one foot closer to home, whereas if his first step is to the left, then he is farther away from home by one foot. At the next minute, he takes the next step either to his right or to his left. If he had taken the second step also to the left, then now he is farther away from home by two feet. One can associate a chance or probability for taking a step to the left or right at each stage. If the probabilities are $\frac{1}{2}$ each, then at each step there is 50% chance that the step will be to the left and 50% chance that the step will be to the right. If the probabilities of going to the left or right are 0.6 and 0.4, respectively, then there is a 60% chance that his first step will be to the left.

Some interesting questions to ask in this case are the following: What is the chance that eventually he will reach home? What is the chance that eventually he will get lost and walk away from home to infinity? Where is his position after n steps? There can be several types of modifications to the simple random walk on a line. In Figure 4.1, a point is marked as B. This B may be a barrier. This barrier may be such that once he hits the barrier, he falls down dead or the walk is finished, or the barrier may be such that if he hits the barrier, there is a certain positive chance of bouncing back to the previous position so that the random walk can continue and there may be a certain chance that the walk is finished, and so on.

An example of a two-dimensional random walk is the case of Mexican jumping beans. There is a certain variety of Mexican beans (lentils). If you place a dried bean on the tabletop, then after a few seconds, it jumps by itself in a random direction to another point on the table. This is due to an insect making the dry bean as its home and the insect moves around by jumping. This is an example of a two-dimensional random walk. We can also consider random walk in space and random walk in higher dimensions.

4.1.2 Branching Process Model

In nature, there are many species which behave in the following manner. There is a mother and the mother gives birth to n offspring once in her lifetime. After giving birth, the mother dies. The number of offspring could be $0, 1, 2, ..., k$, where k is a fixed number, not infinitely large. A typical example is the banana plant. One banana plant gives only one bunch of bananas. You cut the bunch, and the mother plant is destroyed. The next-generation offspring are the new shoots coming from the bottom. The number of shoots could be $0, 1, 2, 3, 4, 5$, usually a maximum of 5 shoots. These shoots are the next-generation plants. Each shoot, when mature, can produce one bunch of bananas each. Usually, after cutting the mother banana plant, the farmer will separate the shoots and plant all shoots, except one, elsewhere so that all have good chances of growing up into healthy banana plants.

Another example is pineapple. Again, one pineapple plant gives only one pineapple. The pineapple itself will have one shoot of plant at the top of the fruit and other shoots will be coming from the bottom. Again, the mother plant is destroyed when the pineapple is plucked. Another example is certain species of spiders. The mother carries the sack of eggs around and dies after the new offspring are born. Another example is salmon fish. From the wide ocean, the mother salmon enters into freshwater river, goes to the birthplace of the river, overcoming all types of obstacles on the way, and lays one bunch of eggs and then promptly dies. Young salmons come out of these eggs, and they find their way down river to the ocean. The life cycle is continued.

Assume that the last name of a person, for example "Rumfeld," is carried only by the sons and not by the daughters. It is assumed that the daughters take the last names of their husbands. Then, there is a chance that the father's last name will be extinct after a few generations. What is the chance that the name Rumfeld will disappear from Earth?

These are some examples for branching processes. Interesting questions to ask are the following: What is the chance that the species will be extinct eventually? This can happen if there is a positive probability of having no offspring in a given birth. What is the expected population size after n generations? Branching process is a subject matter by itself, and it is a special case of a general process known as birth and death processes.

4.1.3 Birth-and-Death Process Model

This can be explained with a simple example. Suppose that there is a good pool area in a river, a good fishing spot for fishermen. Fish move in and move out of the pool area. If $N(t)$ is the number of fish in the pool area at time t and if one fish moved out at the next unit of time, then $N(t + 1) = N(t) - 1$. On the other hand, if one fish moved into the pool area at the next time unit, then $N(t + 1) = N(t) + 1$. When one addition is there, then one can say that there is one birth, and when one deletion is there, we can say that there is one death. Thus, if we are modeling such a process where there is possibility of birth and death, then we call it a birth-and-death model.

4.1.4 Time Series Models

Suppose that we are monitoring the flood level at Meenachil River at the Pastoral Institute. If $x(t)$ denotes the flood level on the tth day, time $= t$ being measured in days, then at the zeroth day or starting of the observation period, the flood level is $x(0)$, the next day the flood level is $x(1)$, and so on. We are observing a phenomenon, namely flood level, over time. In this example, time is measured in discrete time units. Details of various types of processes and details of the various techniques available for time series modeling of data are available in SERC School Notes of 2005–2010. Since these are available to the students, the material will not be elaborated here. But some more details will be given in the coming chapters.

4.1.5 Regression Type Models

It is assumed that the participants have some exposure to probability and statistics. The first prediction problem that we are going to handle is associated with a concept called regression, and our models will be regression-type models. When Sreelakshmy was born, her doctor predicted by looking at the heights of parents and grand parents that she would be $5'5''$ at the age of 11. When she grew up, when she hit 11 years of age her height was only $5'2''$. Thus, the true or observed height was $5'2''$ against the predicted height of $5'5''$. Thus, the prediction was not correct and the error in the prediction = observed minus predicted = $5'2'' - 5'5'' = -3''$. Thus, the prediction was off by $3''$ in magnitude. [We could have also described error as predicted minus observed]. Of course, the guiding principle is that smaller the distance between the observed and predicted, better the prediction. Thus, we will need to consider some measures of distance between the observed and predicted. When random variables are involved, some measures of distances between the real scalar random variable x and a fixed point a are the following:

$$E|x - a| = \text{mean deviation of expected difference between } x \text{ and } a \qquad \text{(i)}$$

$$[E|x - a|^2]^{\frac{1}{2}} = \text{mean square deviation between } x \text{ and } a \qquad \text{(ii)}$$

$$[E|x - a|^r]^{\frac{1}{r}}, r \geq 1 \qquad \text{(iii)}$$

where E denotes the expected value. Many such measures can be proposed. Since we are going to deal with only mean deviation and mean square deviations mainly, we will not look into other measures of distance between x and a here. For the sake of curiosity, let us see what should be a, if a is an arbitrary constant, such that $E|x - a|$ is a minimum or what should be an arbitrary constant b such that $[E|x - b|^2]^{\frac{1}{2}}$ is a minimum?

4.2 Some Preliminaries of Statistical Distributions

We start with the concept of statistical densities and applications of the concept of *expected values*. It is assumed that the student is familiar with the ideas of random experiments, sample space, the definition of probability of an event, random variable, expected values, etc. A brief note on these items will be given here before we start with standard densities. We only consider real random variables in this chapter.

Note 4.1. A note on the concept of probability.

Definition 4.1 A random experiment: It is an experiment where the outcome is not predetermined.

If a person jumps off a cliff to see whether he flies upward or downward or fly straight horizontal, the outcome is predetermined. He will only go downward. If a granite piece is put in water to see whether it sinks or floats or fly out, then the outcome is predetermined because we know all factors producing the outcome and the granite piece will sink in water. If a box contains 10 identical marbles of which 6 are green and 4 are red and if a marble is picked blind-folded to see whether a red marble or a green marble is obtained, then the outcome is not predetermined. The marble picked could be red or green. If we speculate about the possibility of rain at CMS campus today at 4pm, then the outcome is not predetermined. It is rainy season, but at 4pm, there is a very good chance of having rain, but we cannot tell beforehand for sure. If the outcomes are not predetermined, then the experiment is called a random experiment.

Definition 4.2 Outcome set or sample space: The set of all possible outcomes, where no outcome can be subdivided in any sense, is called an outcome set or sample space of a random experiment, and it is usually denoted by S.

Suppose that a coin, with one side called head or H and the other side called tail or T, is tossed twice. Assuming that either H or T will turn up, that is, the coin will not stand on its edge, the possible outcomes are $S = \{(H, H), (H, T), (T, H), (T, T)\}$ where the first letter corresponds to the occurrence in the first trial and the second letter that of the second trial. Suppose that someone says that the possible outcomes are 0 head, 1 head, and 2 heads, thus exhausting all possibilities, then the outcome set can be denoted as $S_1 = \{0H, 1H, 2H\}$. In a sense, both S and S_1 cover all possible outcomes. But in S_1, $1H$ allows a subdivision as (H, T) or (T, H), and hence, we will take S as the sample space and not S_1.

Definition 4.3 Events: Any subset of the sample space S is called an event.

Consider the subset $A = \{(H, H)\} \subset S$ in the experiment of tossing a coin twice, then A is an event, and it can be called the event of getting exactly two heads. If $B = \{(H, H), (H, T), (T, H)\} \subset S$, then B is the event of getting at least one head in this random experiment. Let $C = \{(T, T)\}$, then C is the event of getting exactly two tails or exactly no head in this sample space. Then, the event of getting either exactly two heads or exactly two tails is then $\{(H, H), (T, T)\} = A \cup C$ or the union of two sets, and hence, $\cup$ can be interpreted as "*either or*". In this experiment, the event of getting exactly two heads and exactly two tails is impossible. That is, $A \cap C = \phi$ where a null set or an impossible event is denoted by ϕ or O. Here, $B \cup C$ is the event of getting at least one head or exactly two tails, which is sure to happen. Here, $B \cup C = S$ itself or S can be called a *sure event*. Intersection of two events or simultaneous occurrence of two events can be described as "*and* ." For example, occurrence of A and B is $A \cap B =$ the set of common elements in A and B, which in this case is $\{(H, H)\} = A$ itself. If the simultaneous occurrence of two events $A_1 \subset S$, $A_2 \subset S$ is impossible, that is, $A_1 \cap A_2 = \phi$, then they are also called *mutually exclusive events*, and the occurrence of one excludes the occurrence of the other or simultaneous occurrence is impossible. In our example, A and C are mutually exclusive, and B and C are mutually exclusive, whereas A and B are not mutually exclusive.

The probability of an event or the chance of occurrence of an event will be measured by a number associated with the event. The standard notations used are $P(A) =$ probability of the event A, $Pr\{\text{event}\} =$ probability of the event described within the brackets such as $Pr\{x \geq 50 \, kilos\}$ where x denotes the weight of 10th person who entered a particular shop on a particular day. Probability will be defined by using the following axioms or postulates [Postulates are assumptions that we make, which are mutually non-overlapping and which are desirable properties that we wish to have for the item being defined].

Definition 4.4 Probability of an event A **or** $P(A)$: It is defined as that number satisfying the following axioms or postulates:

(i): $0 \leq P(A) \leq 1$ [It is a number between 0 and 1]
(ii): $P(S) = 1$ [The sure event has probability 1]
(iii): $P(A_1 \cup A_2 \cup ...) = P(A_1) + P(A_2) + ...$ whenever $A_1, A_2, ...$ are mutually exclusive [Here all the events $A_1, A_2, ...$ belong to the same sample space S and they may be finite in number or countably infinite in number]

By using the same axioms, it is easy to prove that $P(\phi) = 0$ or the probability of an impossible event is zero [Hint: Take the mutually exclusive events S and ϕ. $S = S \cup \phi$ and $S \cap \phi = O$. Hence, $1 = P(S)$ (by axiom (ii)) $= P(S) + P(\phi)$ (by axiom (iii)) $= 1 + P(\phi)$, which gives $1 = 1 + P(\phi)$. Since the probability is a real number, $P(\phi) = 0$.] Also, it is easy to show that $A^c = \bar{A} =$ non-occurrence of A or the complement of A in S has the probability, $1 -$ the probability of A, or $P(\bar{A}) = 1 - P(A)$. [Hint: Consider the mutually exclusive events A and $\bar{A}$].

The definition given above is not sufficient to evaluate the probability of a given event. What is the probability of getting a head when a coin is tossed once? Assuming that the coin will not stand on its edge or it will fall head or tail, then the sample space is $S = \{H, T\}$. Let A be the event of getting a head, then $\bar{A}$ will be the event of getting a tail. But these are mutually exclusive and their union is S. Then, $S = A \cup \bar{A}$ and $A \cap \bar{A} = O$ which means that $1 = P(A) + P(\bar{A})$. We can only come up to this stage that the probability of getting a head plus the probability of getting a tail is 1. From here, there is no way of computing the probability of getting a head. It could be any number between 0 and 1, $0 \leq P(A) \leq 1$. We cannot conclude that it is $\frac{1}{2}$ because it does not depend upon the number of sample points in the sample space. Suppose that a housewife is drying her washed clothes in the terrace of a multi-storied building. What is the probability that she will jump off the building? There are two possibilities, either she jumps off or she does not jump off. It is foolish to conclude that, therefore, the probability is $\frac{1}{2}$. There is practically very little chance that she will jump off and the probability is not definitely $\frac{1}{2}$.

The axioms for defining probability of an event are not of much use in computing the probability of an event in a given situation. It needs further rules for *assigning probabilities* to individual events. There are several rules.

Method of symmetry for assigning probabilities to given events. This says that if the physical aspects of the experiment are such that there is complete symmetry in the outcomes and if there is a finite number of sample points in the sample space, then assign equal probabilities to the individual elementary events or the individual points in the sample space.

For example, in the case of tossing a coin twice the sample space $S = \{(H, H), (H, T), (T, H), (T, T)\}$. Suppose that with respect to the physical characteristics of the coin, the coin is balanced or there is no way of giving more weight to one side compared to the other. The toss is done in such a way that from the flight path, there is no way of preferring one side over the other. In such a case, we say that there is symmetry in the outcomes, and in this case, we assign equal probabilities to the elementary events $A_1 = \{(H, H)\}$, $A_2 = \{(H, T)\}$, $A_3 = \{(T, H)\}$, $A_4 = \{(T, T)\}$. Then, the probability of getting at least one head is $P(A_1 \cup A_2 \cup A_3) = P(A_1) + P(A_2) + P(A_3)$ because these are mutually exclusive. Each one is assigned a probability of $\frac{1}{4}$, and hence, the required probability $\frac{1}{4} + \frac{1}{4} + \frac{1}{4} = \frac{3}{4}$. Another way of saying that we assume symmetry is to say "a balanced" coin is tossed or "an unbiased coin" is tossed. Similarly, "a balanced die" is rolled once or a marble is taken "at random" or a card is taken "at random" from a well-shuffled deck of 52 playing cards etc. When symmetry is assumed, then the calculation of individual probabilities can be achieved with the help of permutations and combinations and this aspect is known as *combinatorial probability*.

Method of assigning probabilities proportional to length, area, volume. Suppose that a child cuts a string of length $10\,\text{cm}$ into two pieces. If one end is denoted by 0 and the distance from 0 to the point of cut as x, then what is the probability that this $x \leq 6$ or what is the probability that $2 \leq x \leq 7$? Here, the sample space consists of a continuum of points and not a space of a finite number of sample points. Hence, the previous method does not work. Here, we may assign probabilities proportional to the length. In such a case, we say that the string is "cut at random" so that every piece of the string having the same length will have the same probability of finding the cut on the respective interval. Then, what is the probability that the random cut is on the interval $[2, 7]$? The length of this interval is $7 - 2 = 5$ units, and hence, the required probability is $\frac{5}{10} = \frac{1}{2}$. In this case of random cut, what is the probability that the point of cut is between 3 and 3.001 or in the interval $[3, 3.001]$? The length of this interval is 0.001, and hence, if the cut is random, then the required probability is $\frac{0.001}{10} = 0.0001$. What is the probability that the cut is exactly at the point $x = 2$? Then it is $\frac{2-2}{10} = 0$? Hence, we have assigned the probability zero here because a point has no length. But we are not saying that the event is impossible. Assigning a probability zero by using some rule does not mean that the event is impossible. It is possible to cut the string exactly at the point $x = 2$, but by this rule of assigning probabilities proportional to the length, it is practically an impossible event.

The classical example is Buffon's "clean tile problem". A floor is paved with square tiles of length m units, and a circular coin of diameter d unit is thrown upward, assuming that $d < m$. What is the probability that the coin will "fall clean," falling clean means that the coin will not touch or cut any of the sides of any of the tiles. This problem is one of the starting problems in probability theory. Another is Buffon's needle problem. Place the coin touching any of the sides of a typical tile or square. Then, draw an inside square $\frac{d}{2}$ units away from the outer square. If the center of the coin is inside the inner square, then the coin will not cut any of the sides of the outer square. If we are prepared to assign probabilities proportional to the area, then the required probability is:

$$p = \frac{[m - \frac{d}{2} - \frac{d}{2}]^2}{m^2} = \frac{[m - d]^2}{m^2}.$$

We can think of a dust particle flying in a room of fixed length, width, and height. If we assign probabilities proportional to the volume, then the probability of finding this dust particle in a given region can be calculated if the particle is flying "at random." Here, we will assign probabilities proportional to the volume.

Example 4.1. A box contains 10 identical marbles of which 4 are green and 6 are red. Two marbles are picked at random one by one ["at random" means that when a marble is picked, every marble in the box has equal chances of being picked] (a) without replacement and (b) with replacement ["without replacement" means that the marble picked is set aside after noting the color, it is not returned to be box, whereas "with replacement" means after noting the color the marble is returned to be box]. What is the probability of getting (1) the first marble picked is red; (2) the second marble picked is red, given that the first marble picked is green; (3) the sequence red, green; (4) the second marble picked is red; (5) one red marble and one green marble? Compute these probabilities in each case of (a) and (b).

Solution 4.1. Consider the case (a) of without replacement. Since marbles are picked at random, we can assume symmetry in the outcomes. The answer to (1) is $\frac{6}{10}$ since there are 6 red marbles out of 10 marbles. In (2), one green marble is removed and there are 9 marbles left out of which 6 are red, and hence, the answer is $\frac{6}{9}$. For answering (3), we can look at the total number of sample points. First picking can be done in 10 ways, and for each such picking, the second picking can be done in 9 ways, and thus, the total number of sample points is $10 \times 9 = 90$. The red marble can be picked in 6 ways and the green in 4 ways, and hence, the number of points favorable to the event is $6 \times 4 = 24$. Then, the required probability is $\frac{24}{90}$. In (4), there are two possibilities, the sequences red red (RR) and green red (GR). These are mutually exclusive events. Hence, the answer is:

$$Pr\{(RR)\} + Pr\{(GR)\} = \frac{30}{90} + \frac{24}{90} = \frac{54}{90} = \frac{6}{10}.$$

It is equivalent to picking one red marble from the set of 10 marbles. In (5), since we need only one red and one green, there are two possible sequences (RG) or (GR) and these are mutually exclusive, and hence, the answer is:

$$Pr\{(RG)\} + Pr\{(GR)\} = \frac{6 \times 4}{10 \times 9} + \frac{4 \times 6}{10 \times 9} = 2 \times \frac{24}{90}. \tag{i}$$

This is also equivalent to picking two marbles together at random which results in one red and one green. If a subset of two marbles is picked, then picking at random means that every such subsets of two is given equal chance of being picked. The total number of such subsets is the total number of combinations of two items out of 10 items or $\binom{10}{2} = \frac{10 \times 9}{1 \times 2}$. The red marbles can be picked only from the set of red marbles, and this can be done in $\binom{6}{1}$ ways, and for every such choice, the green marbles can be chosen in $\binom{4}{1}$ ways, and hence, the required answer is:

$$\frac{\binom{6}{1} \times \binom{4}{1}}{\binom{10}{2}} = 2 \times \frac{6 \times 4}{10 \times 9}. \tag{ii}$$

Note that equations (i) and (ii) produce the same result.

Now let us consider the case (b) of sampling with replacement. The answer to (1) is evidently $\frac{6}{10}$. In (2), it does not make any difference whether the first marble is green or red because it is returned to the box, and hence, the answer is $\frac{6}{10}$. For (3), the total number of sample points is 10×10 and red green can come in 6×4 ways, and the required probability is $\frac{6 \times 4}{10 \times 10} = \frac{6}{10} \times \frac{4}{10}$. In (4), getting second green has nothing to do with what happened in the first trial because the marble is returned to the box, and hence, the answer is $\frac{6}{10}$. For (5), one red and one green can come either RG or GR, and hence, the answer is $2 \times (\frac{6}{10})(\frac{4}{10})$.

In Example 4.1 part (2), it is a conditional statement. If A denotes the event that the first marble picked is green and if B denotes the event that the second marble picked is red, then it is a statement of the form B given A which is denoted as $B|A$, B and then a vertical bar and then A, and the corresponding probability can be written as $P(B|A) = $ probability of the event B, given that the event A has already occurred. In this case, $A \cap B$ is the event of getting the sequence green red, and from the case (a)(ii) of Example 4.1, we have seen that $P(B|A)$ is available from $P(B \cap A)$ divided by $P(A)$. We will now define conditional probabilities formally.

Definition 4.5 Conditional probabilities. For any two events A and B in the same sample space S

$$P(A|B) = \frac{P(A \cap B)}{P(B)} \text{ if } P(B) \neq 0$$

$$P(B|A) = \frac{P(B \cap A)}{P(A)} \text{ if } P(A) \neq 0$$

$$P(A \cap B) = P(A|B)P(B) = P(B|A)P(A) \text{ if } P(A) \neq 0, \; P(B) \neq 0 \qquad (iii)$$

$$P(A \cap B) = P(A)P(B) \text{ if } P(A|B) = P(A), \; P(B|A) = P(B). \qquad (iv)$$

Definition 4.6 Independent events. If two event A and B in the same sample space S are such that $P(A|B) = P(A)$ and $P(B|A) = P(B)$ or $P(A \cap B) = P(A)P(B)$, then A and B are said to be *independent events*.

Here, the term "independent" is used in the sense that the product probability property (PPP) holds or the conditional statements are equivalent to the marginal statements or free of the conditions imposed or free of the fact that some other events have occurred.

Definition 4.7 A random variable. It is a variable, say x, which is a mathematical variable, and in addition, we can attach a probability statement of the type $Pr\{x \leq a\}$ for every real a; then, x will be called a real random variable.

Consider the experiment of throwing a balanced coin twice. Let x denote the number of heads in the outcomes. Then what is the probability that $x = 2$? It is the probability of getting exactly two heads in this sample space. Under the assumption of symmetry, the probability is $\frac{1}{4}$. Hence, if $f(a)$ denotes the probability that $x = a$ or $Pr\{x = a\}$, then we have

$$f(a) = \begin{cases} \frac{1}{4} \text{ for } a = 2 \\ \frac{2}{4} \text{ for } a = 1 \\ \frac{1}{4} \text{ for } a = 0 \\ 0, \text{ for all other real } a. \end{cases}$$

Can we compute a function of the form $F(a) = Pr\{x \leq a\}$ for every real a? Let us see what happens. If a is in the interval $(-\infty, 0)$, then the cumulative probability up to a is zero. $F(0) = \frac{1}{4}$. If $0 \leq a < 1$, then the cumulative probability up to this a is $0 + \frac{1}{4} + 0 = \frac{1}{4}$, and so on. Hence, the function is the following:

$$F(a) = Pr\{x \leq a\} = \begin{cases} 0 \text{ when } -\infty < a < 0 \\ \frac{1}{4} \text{ when } 0 \leq a < 1 \\ \frac{1}{4} + \frac{2}{4} = \frac{3}{4} \text{ when } 1 \leq a < 2 \\ 1 \text{ when } a \geq 2. \end{cases}$$

It is a step function, a function looking like steps. Here, x is a real random variable because we can construct $F(a) = Pr\{x \leq a\}$ for all real a.

Can we define any other random variable on the same sample space? The answer is that any number of random variables can be defined over a given sample space if one random variable can be defined there. This example is for a *discrete random variable*, a random variable which takes individually distinct values with nonzero probabilities. Our x is taking values 0, 1, 2 with nonzero probabilities and other values with zero probabilities.

Now, consider a sample space of continuum of points. Let us take the case of a random cut of a string of length 10 cm. Let x be the distance from one end (marked as 0) to the point of cut. Then, under the assumption of random cut

$$F(a) = Pr\{x \leq a\} = \frac{a}{10} = \int_0^a \frac{1}{10} dx.$$

This also means here that

$$\frac{d}{dx} F(x) = \frac{1}{10} \text{ or } dF(x) = f(x)dx$$

where

$$f(x) = \begin{cases} \frac{1}{10}, 0 \leq x \leq 10 \\ 0, \text{ elsewhere,} \end{cases}$$

is the probability function corresponding to this *continuous* random variable x [continuous in the sense that x can take a continuum of points with nonzero probabilities as opposed to *discrete* in the sense of taking individually distinct points with nonzero probabilities]. Probability function in the continuous case is usually called a *density function* and for the discrete case as *probability mass function* or simply *probability function*. Some authors do not make a difference and call both as either probability function or density function. We will use the style that in the continuous case, we will call density function and in the discrete and mixed cases (part of the probability mass is distributed on individually distinct points and the remaining part on a continuum of points), we will call probability function.

Irrespective of a random variable, we can define a probability function or density function by using the following conditions:

(1): $f(x) \geq 0$ for all real x, $-\infty < x < \infty$

(2):

$$\begin{cases} \sum_{-\infty}^{\infty} f(x) = 1 \text{ if discrete;} \\ \int_{-\infty}^{\infty} f(x)dx = 1 \text{ if continuous;} \end{cases} \qquad (v)$$

sum up over discrete points and integrate over the continuum in the mixed case. Any function satisfying the above conditions (1) and (2) can be called a probability/density function corresponding to some random variable x. Similarly, one can define a cumulative probability function $F(t) = Pr\{x \leq t\}$, which is also often called a *distribution function*, by using the following conditions:

(1): $F(-\infty) = 0$, $F(+\infty) = 1$

(2): $F(a) \leq F(b)$ for $a < b$

(3): $F(t)$ is right continuous.

Definition 4.8 Expected values. The standard notation used is $E[\psi(x)] = $ expected value of $\psi(x)$ a function of the random variable x. It is defined as follows:

$$E[\psi(x)] = \begin{cases} \int_{-\infty}^{\infty} \psi(x)f(x)dx = \int_{-\infty}^{\infty} \psi(x)dF(x) \text{ for } x \text{ continuous} \\ \sum_{-\infty}^{\infty} \psi(x)f(x) \text{ for } x \text{ discrete} \end{cases} \qquad (vi)$$

and in the mixed case, sum up over discrete points and integrate over the continuum of points, where $f(x)$ is the density/probability function and $F(x)$ is the cumulative function or the distribution function.

Example 4.2. Evaluate the following expected values, whenever they exist, for the following density/probability functions:

(1): $E(x)$ in

$$f(x) = \begin{cases} \binom{n}{x} p^x q^{n-x}, x = 0, 1, ..., n, 0 < p < 1, q = 1 - p \\ 0, \text{ elsewhere;} \end{cases}$$

[This probability function is known as the binomial probability law, coming from a sequence of independent Bernoulli trials].

(2): $E(x^2)$ in

$$f(x) = \begin{cases} \frac{\lambda^x}{x!} e^{-\lambda}, x = 0, 1, ..., \lambda > 0 \\ 0, \text{ elsewhere;} \end{cases}$$

[This probability function is known as the Poisson probability law].

(3): $E[2 + 3x]$ in

$$f(x) = \begin{cases} \frac{1}{\beta}e^{-\frac{x}{\beta}}, & 0 \le x < \infty, \ \beta > 0 \\ 0, & \text{elsewhere}; \end{cases}$$

[This is known as the exponential density].

(4): $E[x^h]$ in

$$f(x) = \begin{cases} \frac{x^{\alpha-1}e^{-\frac{x}{\beta}}}{\beta^\alpha \Gamma(\alpha)}, & 0 \le x < \infty, \ \alpha > 0, \ \beta > 0 \\ 0, & \text{elsewhere}; \end{cases}$$

[This is known as the gamma density].

Solution 4.2. (1): By definition

$$E(x) = \sum_{-\infty}^{\infty} x f(x) = 0 + \sum_{x=0}^{n} \binom{n}{x} p^x q^{n-x}$$

$$= \sum_{x=1}^{n} x \binom{n}{x} p^x q^{n-x} \text{ since at } x = 0 \text{ the term is zero}$$

$$= np \sum_{x=1}^{n} \frac{(n-1)!}{(x-1)!(n-x)!} p^{x-1} q^{n-x}, \ n - x = (n-1) - (x-1)$$

$$= np(q + p)^{n-1} = np \text{ since } q + p = 1.$$

Here, the expected value of x, which is also, in general, called the *mean value of x* is $E(x) = np$ where n and p are *parameters* in the binomial probability function. A *parameter* is an unknown quantity sitting in a probability/density function, such that irrespective of the values of the parameter in a parameter space, $f(x)$ is still a probability/density function.

(2): Here, we need $E[x^2]$. For convenience, we can write $x^2 = x(x - 1) + x$ since factorials are sitting in the denominator.

$$E[x^2] = 0 + \sum_{x=0}^{\infty} x^2 \frac{\lambda^x}{x!} e^{-\lambda}$$

$$= e^{-\lambda} \sum_{x=0}^{\infty} [x(x-1)\frac{\lambda^x}{x!} + x\frac{\lambda^x}{x!}]$$

$$= e^{-\lambda} [\sum_{x=2}^{\infty} x(x-1)\frac{\lambda^x}{x!} + \sum_{x=1}^{\infty} x\frac{\lambda^x}{x!}]$$

$$= e^{-\lambda}[\lambda^2 \sum_{x=2}^{\infty} \frac{\lambda^{x-2}}{(x-2)!} + \lambda \sum_{x=1}^{\infty} \frac{\lambda^{x-1}}{(x-1)!}]$$

$$= \lambda^2 + \lambda.$$

Here, λ is a parameter in the Poisson probability law. Here, the mean value or $E(x) = \lambda$. Another general concept called *variance* = $\text{Var}(x)$, which in general, is defined as $\text{Var}(x) = E[x - E(x)]^2 = E(x^2) - [E(x)]^2$. In the Poisson case, the variance is $(\lambda^2 + \lambda) - (\lambda)^2 = \lambda =$ the mean value itself.

(3): Here, we need

$$E[2 + 3x] = \int_{-\infty}^{\infty} [2 + 3x]f(x)dx \text{ in the continuous case}$$

$$= 2 \int_{-\infty}^{\infty} f(x)dx + 3 \int_{-\infty}^{\infty} x\, f(x)dx$$

$$= 2 + 3E(x)$$

provided the expected value $E(x)$ exists. This is a general property that when the expected value exists $E[a + bx] = a + bE(x)$ where a and b are constants. For the exponential density, we compute $E(x)$ first.

$$E(x) = 0 + \int_0^{\infty} x \frac{e^{-\frac{x}{\beta}}}{\beta}dx$$

$$= \beta \int_0^{\infty} ye^{-y}dy, \quad y = \frac{x}{\beta}$$

$$= \beta$$

by integrating by parts or by using gamma function. Hence $E[2 + 3x] = 2 + 3\beta$ in this case.

(4): Here for arbitrary h, we need $E[x^h]$. This is also known as *h*th *arbitrary moment of* x, and for $h = 0, 1, 2, \ldots$ are called the *integer moments of* x.

$$E(x^h) = 0 + \int_0^{\infty} x^h \frac{x^{\alpha-1}e^{-\frac{x}{\beta}}}{\beta^{\alpha}\Gamma(\alpha)}dx$$

$$= \beta^h \int_0^{\infty} y^{\alpha+h-1}e^{-y}dy, \quad y = \frac{x}{\beta}$$

$$= \beta^h \frac{\Gamma(\alpha + h)}{\Gamma(\alpha)} \text{ for } \Re(\alpha + h) > 0$$

where $\Re(\cdot)$ denotes the real part of $(\cdot)$. Thus, the arbitrary moments will exist even for complex values of h provided the real part of $\alpha + h$ is positive. The integral here is evaluated by using a gamma integral.

Definition 4.9 Joint density/joint probability function. If $f(x_1, ..., x_k)$ is the joint probability/density function of the random variables $x_1, ..., x_k$, then $f(x_1, ..., x_k)$ must satisfy the following conditions:

(1): $f(x_1, ..., x_k) \geq 0$ for all $x_1, ..., x_k$;

(2): $\sum_{-\infty}^{\infty} \cdots \sum_{-\infty}^{\infty} f(x_1, ..., x_k) = 1$ if all variables are discrete,
$\int_{-\infty}^{\infty} \cdots \int_{-\infty}^{\infty} f(x_1, ..., x_k)dx_1 \wedge ... \wedge dx_k = 1$ if all variables are continuous. Integrate out over the continuous cases and sum up over the discrete ones and the total probability or total volume under the surface $z = f(x_1, ..., x_k)$ must be one. Any function satisfying these two conditions is called a joint probability/density function.

As in the one variable or univariate case, the expected values can be defined in a similar way. Let $\psi(x_1, ..., x_k)$ be a function of $x_1, ..., x_k$. Then the expected value of $\psi(x_1, ..., x_k)$ denoted by $E[\psi(x_1, ..., x_k)]$ is defined as follows, whenever it exists:

Definition 4.10 Expected values.

$$E[\psi(x_1, ..., x_k)] = \sum_{-\infty}^{\infty} \psi(x_1, ..., x_k)f(x_1, ..., x_k) \text{ if all discrete}$$

$$= \int_{-\infty}^{\infty} \cdots \int_{-\infty}^{\infty} \psi(x_1, ..., x_k)f(x_1, ..., x_k)dx_1 \wedge ... \wedge dx_k \quad \text{(vii)}$$

if all variables are continuous. In the mixed case, integrate over the continuous variables and sum up over the discrete ones.

Let $f(x, y)$ be the density/probability function in a two-variable situation or in a bivariate case. Then, as in the case of events, we can have a decomposition into marginal and conditional densities/probability functions. That is,

$$f(x, y) = f_1(x)g_1(y|x) = f_2(y)g_2(x|y) \quad \text{(viii)}$$

$$g_1(y|x = a) = \frac{f(x, y)}{f_1(x)}|_{x=a} = \frac{f(a, y)}{f_1(a)} \text{ for } f_1(a) \neq 0 \quad \text{(ix)}$$

$$g_2(x|y = b) = \frac{f(x, y)}{f_2(y)}|_{y=b} = \frac{f(x, b)}{f_2(b)} \text{ for } f_2(b) \neq 0 \quad \text{(x)}$$

$$= f_1(x)f_2(y) \text{ when } x \text{ and } y \text{ are independently distributed} \quad \text{(xi)}$$

where $f_1(x)$ and $f_2(y)$ are the marginal densities/probability functions of x and y alone, and $g_1(y|x)$ and $g_2(x|y)$ are the conditional densities/probability functions of y given x and x given y, respectively.

Definition 4.11 Marginal and conditional densities. When $x_1, ..., x_k$ have a joint probability/density function (or have a joint distribution), then the joint probability/density function of a subset $x_1, ...x_r, r < k$ is called the marginal joint probability/density of $x_1, .., x_r$ for all $1 \le r \le k$. The individual density/probability function of x_j, say $f_j(x_j)$, is called the marginal density/probability function of x_j for $j = 1, ..., k$, from the joint density/probability function of $x_1, ..., x_k$. If the density/probability function of x_j is evaluated at given values of other variables, then it is called the conditional density/probability function of x_j.

Example 4.3. If a bivariate discrete probability mass function is given in the following table, evaluate (a) the marginal probability functions, (b) the conditional probability function of y given $x = 0$, (c) the conditional probability function of x given $y = 1$, (d) the conditional expectation of y, given $x = 0$:

$y \rightarrow$	-1	1	sum
$x \downarrow$			
0	$\frac{1}{10}$	$\frac{3}{10}$	$\frac{4}{10}$
1	$\frac{2}{10}$	$\frac{4}{10}$	$\frac{6}{10}$
sum	$\frac{3}{10}$	$\frac{7}{10}$	1

and $f(x, y) = 0$ elsewhere.

Solution 4.3. From the marginal sums, we have

$$f_1(x) = \begin{cases} \frac{4}{10}, & x = 0 \\ \frac{6}{10}, & x = 1 \\ 0, & \text{elsewhere,} \end{cases} \quad \text{and } f_2(y) = \begin{cases} \frac{3}{10}, & y = -1 \\ \frac{7}{10}, & y = 1 \\ 0, & \text{elsewhere.} \end{cases}$$

The conditional probability functions are the following:

$$g_1(y|x = 0) = \frac{f(0, y)}{f_1(0)} = \frac{4}{10} \begin{cases} \frac{1}{10}, & y = -1 \\ \frac{3}{10}, & y = 1 \end{cases}$$

$$= \begin{cases} \frac{1}{4}, & y = -1 \\ \frac{3}{4}, & y = 1 \\ 0, & \text{elsewhere} \end{cases}$$

and

$$g_2(x|y = -1) = \frac{f(x, -1)}{f_2(-1)} = \frac{10}{3} \begin{cases} \frac{1}{10}, & x = 0 \\ \frac{2}{10}, & x = 1 \end{cases}$$

$$= \begin{cases} \frac{1}{3}, & x = 0 \\ \frac{2}{3}, & x = 1 \\ 0, & \text{elsewhere.} \end{cases}$$

The conditional expectation of y, given $x = 0$, is the expected value in the conditional probability function of y, given $x = 0$, that is, the expectation in the probability function $g_1(y|x = 0)$. That is,

$$E(y|x = 0) = \sum_y y \, g_1(y|x = 0) = 0 + (-1) \times \frac{1}{4} + (1) \times \frac{3}{4}$$

$$= \frac{2}{4} = \frac{1}{2}.$$

Example 4.4. If the following is a bivariate density, then evaluate (a) the marginal density of x, (b) the conditional density of y given $x = \frac{1}{3}$, (c) the conditional expectation of y given $x = \frac{1}{3}$:

$$f(x, y) = \begin{cases} \frac{1}{1+2x} e^{-\frac{y}{1+2x}}, & 0 \le y < \infty, \ 0 \le x \le 1 \\ 0, & \text{elsewhere.} \end{cases}$$

Solution 4.4. The marginal density of x is available by integrating out y. That is,

$$f_1(x) = \int_0^\infty \frac{1}{1+2x} e^{-\frac{y}{1+2x}} dy, \quad u = \frac{y}{1+2x}$$

$$= \begin{cases} 1, & 0 \le x \le 1 \\ 0 & \text{elsewhere.} \end{cases}$$

Hence, the conditional density of y, given $x = \frac{1}{3}$, is given by

$$g_1(y|x = \frac{1}{3}) = \frac{f(x, y)}{f_1(x)}\Big|_{x=\frac{1}{3}}$$

$$= \frac{1}{1+2x} e^{-\frac{y}{1+2x}}\Big|_{x=\frac{1}{3}} = \begin{cases} \frac{3}{5} e^{-\frac{3}{5}y}, & 0 \le y < \infty \\ 0, & \text{elsewhere.} \end{cases}$$

Since the conditional density of y, given x, is an exponential density, the expected value there is the parameter itself. Hence, $E(y|x) = 1 + 2x$ and this evaluated at $x = \frac{1}{3}$ is $\frac{5}{3}$.

Some standard densities in current use, other than the ones listed in the notes above, are the following:

Gaussian or normal density

$$f_G(x) = \frac{1}{\sigma\sqrt{2\pi}} e^{-\frac{1}{2\sigma^2}(x-\mu)^2},$$

for $-\infty < x < \infty$, $-\infty < \mu < \infty$, $\sigma > 0$.

Type-1 beta density

$$f_{b1}(x) = \frac{\Gamma(\alpha+\beta)}{\Gamma(\alpha)\Gamma(\beta)} x^{\alpha-1}(1-x)^{\beta-1}, \ 0 \le x \le 1, \ \alpha > 0, \ \beta > 0$$

and zero elsewhere.

Type-2 beta density

$$f_{b2}(x) = \frac{\Gamma(\alpha+\beta)}{\Gamma(\alpha)\Gamma(\beta)} x^{\alpha-1}(1+x)^{-(\alpha+\beta)}, \ 0 \le x < \infty, \ \alpha > 0, \ \beta > 0$$

and zero elsewhere.

Logistic density

The basic logistic density is given below, but there are many types of variations and generalizations of this density.

$$f_L(x) = \frac{e^x}{(1+e^x)^2}, \ -\infty < x < \infty.$$

Generalized gamma density

$$f_g(x) = \frac{\delta\, a^{\frac{\gamma}{\delta}}}{\Gamma(\frac{\gamma}{\delta})} x^{\gamma-1} e^{-ax^\delta}, \ 0 \le x < \infty, \ a > 0, \ \delta > 0, \ \gamma > 0$$

and zero elsewhere. This for $\gamma = \delta$ is the Weibull density. Many other densities are also connected to this generalized gamma density such as stretched exponential, gamma, chi-square, Fermi-Dirac, and others.

4.1a. Optimization of Distance Measures

Definition 4.12 A measure of scatter. A measure of scatter in real scalar random variable x from an arbitrary point α is given by $\sqrt{E(x-\alpha)^2}$.

What should be α such that this dispersion is the least? But minimization of $\sqrt{E(x-\alpha)^2}$ is equivalent to minimizing $E(x-\alpha)^2$. But

$$E(x - \alpha)^2 = E(x - E(x) + E(x) - \alpha)^2 \text{ by adding and subtracting } E(x)$$
$$= E(x - E(x))^2 + E(E(x) - \alpha)^2 + 2E(x - E(x))(E(x) - \alpha)$$
$$= E(x - E(x))^2 + (E(x) - \alpha)^2 \tag{4.1}$$

because the cross product term is zero due to the fact that $(E(x) - \alpha)$ is a constant, and therefore, the expected value applies on $(x - E(x))$, that is,

$$E(x - E(x)) = E(x) - E(E(x)) = E(x) - E(x) = 0$$

since $E(x)$ is a constant. In the above computations, we assumed that $E(x)$ is finite or it exists. In (4.1), the only quantity containing α is $[E(x) - \alpha]^2$ where both the quantities $E(x)$ and α are constants, and hence, (4.1) attains its minimum when the nonnegative quantity $[E(x) - \alpha]^2$ attains its minimum which is zero. [We are dealing with real quantities here]. Therefore,

$$[E(x) - \alpha]^2 = 0 \Rightarrow E(x) - \alpha = 0 \Rightarrow E(x) = \alpha.$$

Hence, the minimum is attained when $\alpha = E(x)$. [The maximum value that $E(x-\alpha)^2$ can attain is $+\infty$ because α is arbitrary]. We will state this as a result.

Result 4.1. *For real scalar random variable x, for which $E(x^2)$ exists or a fixed finite quantity, and for an arbitrary real number α*

$$\min_{\alpha}[E(x - \alpha)^2]^{\frac{1}{2}} \Rightarrow \min_{\alpha} E(x - \alpha)^2 \Rightarrow \alpha = E(x). \tag{4.2}$$

In a similar fashion, we can show that the mean deviation is least when the deviation is taken from the median. This can be stated as a result.

Result 4.2. *For a real scalar random variable x, having a finite median, and for an arbitrary real number b,*

$$\min_{b} E|x - b| \Rightarrow b = M$$

where M is the median of x.

The proof is a little lengthy, and hence, it is not given here. Students may try to prove it. *Hint:* Use the properties of the distribution function $F_x(t) = Pr\{x \le t\}$. The median M is the middle value for x in the sense that

$$Pr(x \le M) \ge \frac{1}{2} \text{ and } Pr(x \ge M) \ge \frac{1}{2}$$

where $Pr(\cdot)$ denotes the probability of the event $(\cdot)$. The median M can be unique, or there may be many points qualifying to be the median for a given x depending upon the distribution of x.

4.2.1   Minimum Mean Square Prediction

First we will consider the problem of predicting one real scalar random variable by using one real scalar random variable. Such situations are plenty in nature. Let y be the variable to be predicted, and let x be the variable by using which the variable y is predicted. Sometimes we call the variable y to be predicted as *dependent* variable and the variable x, which is independently preassigned to predict y, is called the *independent variable*. This terminology should not be confused with statistical independence of random variables. Let y be the marks obtained by a student in a class test and x be the amount of study time spent on that subject. The type of question that we would like to ask is the following: Is y a function of x? If so what is the functional relationship so that we can use it to evaluate y at a given value of x. If there is no obvious functional relationship, can we use a preassigned value of x to predict y? Can we answer a question such as if 20 hours of study time is used what will be the grade in the class test? In this case, irrespective of whether there exists a relationship between x and y or not, we would like to use x to predict y.

As another example, let y be the growth of a plant seedling, (measured in terms of height), in one week, against the amount of water x supplied. As another example let y be the amount of evaporation of certain liquid in one hour and x be the total exposed surface area.

If x is a variable that we can preassign or observe what is a prediction function of x in order to predict y? Let $\phi(x)$ be an arbitrary function of x that we want to use as a predictor of y. We may want to answer questions like what is the predicted value of y if the function $\phi(x)$ is used to predict y at $x = x_0$ where x_0 is a given point. Note that infinitely many functions can be used as a predictor for y. Naturally, the predicted value of y at $x = x_0$ will be far off from the true value if $\phi(x)$ is not a good predictor for y. What is the "best" predictor, "best" in some sense? If y is predicted by using $\phi(x)$, then the ideal situation is that $\phi(x)$ at every given value of x coincides with the corresponding observed value of y. This is the situation of a mathematical relationship, which may not be available in a problem in social sciences. For the example of the student studying for the examination if a specific function $\phi(x)$ is there, where x is the number of hours of study, then when $x = 3$ hours $\phi(x)|_{x=3} = \phi(3)$ should produce the actual grade obtained by the student by spending 3 hours of study. Then, $\phi(x)$ should give the correct observation for every given value of x. Naturally, this does not happen. Then, the error in using $\phi(x)$ to predict the value of y at a given x is

$$y - \phi(x) \text{ or we may take as } \phi(x) - y.$$

The aim is to minimize a "distance" between y and $\phi(x)$ and thus construct ϕ. Then, this ϕ will be a "good" ϕ. This is the answer from common sense. We have many mathematical measures of "distance" between y and $\phi(x)$ or measures of scatter in $e = y - \phi(x)$. One such measure is $\sqrt{E[y - \phi(x)]^2}$, and another measure is

$E|y - \phi(x)|$, where y is a real scalar random variable and x is a preassigned value of another random variable or more precisely, for the first measure of scatter the distance is $\sqrt{E[y - \phi(x = x_0)]^2}$, that is, at $x = x_0$. Now, if we take this measure, then the problem is to minimize over all possible functions ϕ.

$$\min_{\phi} \sqrt{E[y - \phi(x = x_0)]^2} \Rightarrow \min_{\phi} E[y - \phi(x = x_0)]^2.$$

But from (2.2), it is clear that the "best" function ϕ, best in the minimum mean square sense, that is, minimizing the expected value or mean value of the squared error, is $\phi = E(y|x = x_0)$ or simply $\phi = E(y|x) = $ conditional expectation of y given x. Hence, this "best predictor", best in the minimum mean square sense, is defined as the regression of y on x. Note that if we had taken any other measure of scatter in the error e, we would have ended up with some other function for the best ϕ. Hereafter, when we say "regression" we will mean the best predictor, best in the minimum mean square sense, or the conditional expectation of y given x. Naturally, for computing $E(y|x)$ we should either have the joint distribution of x and y or at least the conditional distribution of y, given x.

Definition 4.13. Regression of y on x. It is defined as $E(y|x) = $ conditional expectation of y given x whenever it exists.

Example 4.5. If x and y have a joint distribution given by the following density function, both x and y are normalized to the interval $[0, 1]$,

$$f(x, y) = \begin{cases} x + y, & 0 \le x \le 1, \ 0 \le y \le 1 \\ 0, & \text{elsewhere,} \end{cases}$$

what is the "best" predictor of y based on x, best in the minimum mean square sense? Also predict y at (1): $x = \frac{1}{3}$, (2): $x = 1.5$.

Solution 4.5. As per the criterion of "best," we are asked to compute the regression of y on x or $E(y|x)$. From elementary calculus, the marginal density of x, denoted by $f_1(x)$, is given by

$$f_1(x) = \int_y f(x, y) dy = \int_0^1 (x + y) dy = \begin{cases} x + \frac{1}{2}, & 0 \le x \le 1 \\ 0, & \text{elsewhere.} \end{cases}$$

Then, the conditional density of y given x is

$$g(y|x) = \frac{f(x, y)}{f_1(x)} = \frac{x + y}{x + \frac{1}{2}}, \ 0 \le y \le 1$$

and zero elsewhere. Hence, the conditional expectation of y given x is then

$$E(y|x) = \int_{y|x} yg(y|x)dy = \int_{y=0}^{1} y\left[\frac{x+y}{x+\frac{1}{2}}\right]dy$$

$$= \frac{1}{x+\frac{1}{2}} \int_{y=0}^{1} (xy+y^2)dy = \frac{\frac{x}{2}+\frac{1}{3}}{x+\frac{1}{2}}.$$

Hence, the best predictor of y based on x in Example 4.5 is

$$E(y|x) = \frac{\frac{x}{2}+\frac{1}{3}}{x+\frac{1}{2}}$$

for all given admissible values of x. This answers the first question. Now to predict y at a given x, we need to only substitute the value of x. Hence, the predicted value at $x = \frac{1}{3}$ is

$$E(y|x = \frac{1}{3}) = \frac{\frac{1}{6}+\frac{1}{3}}{\frac{1}{3}+\frac{1}{2}} = \frac{3}{5}.$$

The predicted value of y at $x = 1.5$ is not available from the above formula because 1.5 is not an admissible value of x or it is outside $0 \leq x \leq 1$ the probability for which is zero. The question contradicts with what is given as the density in Example 4.5.

Example 4.6. Suppose that it is found that x and y have a joint distribution given by the following: [Again we will use the same notation to avoid introducing too many symbols, even though f in Example 4.5 is different from f in Example 4.6]

$$f(x, y) = \begin{cases} e^{-x-y}, & 0 \leq x < \infty, \ 0 \leq y < \infty \\ 0, & \text{elsewhere.} \end{cases}$$

Evaluate the "best" predictor of y based on x, best in the minimum mean square sense, and predict y at $x = x_0$.

Solution 4.6. We need the conditional expectation of y given x, for which we need the conditional density of y given x. From the above joint density, it is clear that the marginal density of x is

$$f_1(x) = \begin{cases} e^{-x}, & 0 \leq x < \infty \\ 0, & \text{elsewhere,} \end{cases}$$

and hence, the conditional density of y given x is given by

$$g(y|x) = \frac{f(x, y)}{f_1(x)}$$

$$g(y|x) = \begin{cases} e^{-y}, & 0 \leq y < \infty \\ 0, & \text{elsewhere} \end{cases}$$

and the conditional expectation of y given x is

$$E(y|x) = \int_0^\infty ye^{-y}dy = 1$$

[evaluated by using a gamma function as $\Gamma(2) = 1!$ or by integration by parts]. Here, $E(y|x)$ is not a function of x which means that whatever be the preassigned value of x the predicted value of y is simply 1. In other words, there is no meaning in using x to predict the value of y because the conditional distribution is free of the conditioned variable x. This happened because, in this example, x and y are statistically independently distributed. Hence, x cannot be used to predict y and vice versa.

But note that, in Examples 4.5 and 4.6, we have more information about the variables x and y than what is necessary to construct the "best" predictor. The best predictor is $E(y|x)$, and hence, we need only the conditional distribution of y given x to predict y and we do not need the joint distribution. Knowing the joint distribution means knowing the whole surface in a three-dimensional space. Knowing the conditional distribution means knowing only the shape of the curve when this surface $f(x, y)$ is cut by the plane $x = x_0$ for some preassigned x_0. [The reader is asked to look at the geometry of the whole problem in order to understand the meaning of the above statement]. Thus, we can restrict our attention to conditional distributions only for constructing a regression function.

Example 4.7. The strength y of an iron rod is deteriorating depending upon the amount of rust x on the rod. The more rust means less strength, and finally, rust will destroy the iron rod. The conditional distribution of y given x is seen to be of exponential decay model with the density

$$g(y|x) = \begin{cases} \frac{1}{1+x}e^{-\frac{y}{1+x}}, & 0 \le y < \infty, 1 + x > 0 \\ 0, & \text{elsewhere.} \end{cases}$$

Construct the best predictor function for predicting the strength y at the preassigned amount x of rust and predict the strength when $x = 2$ units.

Solution 4.7. From the density given above, it is clear that at every x, the conditional density of y given x is exponential with expected value $1 + x$ (comparing with a negative exponential density). Hence

$$E(y|x) = 1 + x$$

is the best predictor. The predicted value of y at $x = 2$ is $1 + 2 = 3$ units.

Example 4.8. The marks y obtained by the students in an examination is found to be normally distributed with polynomially increasing expected value with respect to the amount x of time spent. The conditional distribution of y, given x, is found to be normal with the density

$$g(y|x) = \frac{1}{4\sqrt{2\pi}} e^{-\frac{1}{2}(y-70-2x-x^2)^2}, \quad -\infty < y < \infty, \ 0 \le x \le 4.$$

Construct the best predictor of y based on x and predict y at $x = 3$ hours.

Solution 4.8. From the density itself, it is clear that the conditional density of y given x is $N(\mu, \sigma^2)$ with $\sigma^2 = 1$ and $\mu = 70 + 2x + x^2$, and therefore,

$$E(y|x) = 70 + 2x + x^2$$

is the best predictor of y, and the predicted marks at $x = 3$ is

$$70 + 2(3) + 3^2 = 85.$$

In Examples 4.5 and 4.8, the regression function, that is, $E(y|x)$, is a nonlinear function of x.

$$E(y|x) = \frac{\frac{x}{2} + \frac{1}{3}}{x + \frac{1}{2}} \text{ in Example 4.5}$$

$$E(y|x) = 70 + 2x + x^2 \text{ in Example 4.8}$$

whereas

$$E(y|x) = 1 + x \text{ in Example 4.7}$$

which is a linear function in x. Thus, the regression of y on x may or may not be linear function in x. In Example 4.7 if x and y had a joint bivariate normal distribution, then we know for sure that $E(y|x)$ is a linear function in x. Thus, one should not conclude that regression being a linear function in x means that the variables are jointly normally distributed because it is already seen that in Example 4.7, the regression is linear in x, but it is not a case of joint normal distribution.

Note 4.2. The word "regression" means to go back, to regress means to go back. But in a regression-type prediction problem, we are not going back to something. We are only computing the conditional expectation. The word "regression" is used for historical reasons. The original problem, when regression was introduced, was to infer something about ancestors by observing offspring and thus going back.

4.3   Regression on Several Variables

Again, let us examine the problem of predicting a real scalar random variable x at preassigned values of many real scalar variables $x_1, x_2, ..., x_k$. As examples, we can cite many situations.

Example 4.9

$$y = \text{the marks obtained by a student in an examination}$$
$$x_1 = \text{the amount of time spent on it}$$
$$x_2 = \text{instructor's ability measured in the scale } 0 \le x_2 \le 10.$$
$$x_3 = \text{instructor's knowledge in the subject matter}$$
$$x_4 = \text{student's own background preparation in the area.}$$

Example 4.10.

$$y = \text{cost of living}$$
$$x_1 = \text{unit price for staple food}$$
$$x_2 = \text{unit price for vegetables}$$
$$x_3 = \text{cost of transportation}$$

and so on. There can be many factors contributing to y in Example 4.9 as well as in Example 4.10. We are not sure in which form these variables $x_1, ..., x_k$ will enter into the picture. If $\psi(x_1, ..., x_k)$ is the prediction function for y, then predicting exactly as in the case of one variable situation the best prediction function, best in the sense of minimum mean square error, is again the conditional expectation of y given $x_1, ..., x_k$. That is, $E(y|x_1, ..., x_k)$. Hence, the regression of y on $x_1, ..., x_k$ is again defined as the conditional expectation.

Definition 4.14. The regression of y on $x_1, ..., x_k$ is defined as the conditional expectation of y given $x_1, ..., x_k$, that is, $E(y|x_1, ..., x_k)$.

As before, depending upon the conditional distribution of y given $x_1, ... , x_k$, the regression function may be a linear function of $x_1, ..., x_k$ or may not be a linear function.

Example 4.11. If y, x_1, x_2 have a joint density

$$f(y, x_1, x_2) = \begin{cases} \frac{2}{3}(y + x_1 + x_2), 0 \le y, x_1, x_2 \le 1 \\ 0, \text{ elsewhere} \end{cases}$$

evaluate the regression of y on x_1 and x_2.

Solution 4.9. The joint marginal density of x_1 and x_2 is given by integrating out y. Denoting it by $f_1(x_1, x_2)$, we have

$$f_1(x_1, x_2) = \int_{y=0}^{1} \frac{2}{3}(y + x_1 + x_2)dy$$

$$= \begin{cases} \frac{2}{3}(\frac{1}{2} + x_1 + x_2), \ 0 \le x_1 \le 1, 0 \le x_2 \le 1 \\ 0 \text{ elsewhere.} \end{cases}$$

Hence, the conditional density of y given x_1 and x_2 is

$$g(y|x_1, x_2) = \begin{cases} \frac{\frac{2}{3}(y+x_1+x_2)}{\frac{2}{3}(\frac{1}{2}+x_1+x_2)} = \frac{y+x_1+x_2}{\frac{1}{2}+x_1+x_2}, & 0 \le y \le 1 \\ 0, & \text{elsewhere.} \end{cases}$$

Therefore, the regression of y on x_1, x_2 is given by

$$E(y|x_1, x_2) = \int_{y=0}^{1} y \left[\frac{y + x_1 + x_2}{\frac{1}{2} + x_1 + x_2} \right] dy = \frac{\frac{1}{3} + \frac{x_1}{2} + \frac{x_2}{2}}{\frac{1}{2} + x_1 + x_2}.$$

This is the best predictor of y. For example, the predicted value of y at $x_1 = 0$, $x_2 = \frac{1}{2}$ is given by

$$\frac{\frac{1}{3} + 0 + \frac{1}{4}}{\frac{1}{2} + 0 + \frac{1}{2}} = \frac{7}{12}.$$

In this example, how can we predict y based on x_2 alone, that is, $E(y|x_2)$? First, integrate out x_1 and obtain the joint marginal density of y and x_2. Then, proceed as in the one variable case. Note that $E(y|x_2)$ is not the same as $E(y|x_2, x_1 = 0)$. These two are two different statements and two different items.

Again, note that for constructing the regression function of y on $x_1, \ldots, x_k$, that is, $E(y|x_1, \ldots, x_k)$, we need only the conditional distribution of y given $x_1, \ldots, x_k$ and we do not require the joint distribution of $y, x_1, \ldots, x_k$.

Example 4.12 If the conditional density of y given $x_1, \ldots, x_k$ is given by

$$g(y|x_1, \ldots, x_k) = \begin{cases} \frac{1}{5+x_1+\ldots+x_k} e^{-\frac{y}{5+x_1+\ldots+x_k}}, & 0 \le y < \infty, \\ \qquad 5 + x_1 + \ldots + x_k > 0 \\ 0, & \text{elsewhere} \end{cases}$$

evaluate the regression function.

Solution 4.10. The conditional density is the exponential density with the mean value

$$E(y|x_1, \ldots, x_k) = 5 + x_1 + \ldots + x_k$$

and this is the regression of y on $x_1, \ldots, x_k$, which here is a linear function in $x_1, \ldots, x_k$ also.

When the variables are all jointly normally distributed also one can obtain the regression function to be linear in the regressed variables. Example 2.8 illustrates that joint normality is not needed for the regression function to be linear. When the regression function is linear, we have some interesting properties.

Exercises 4.3.

4.3.1. Prove that $\min_a E|y - a| \Rightarrow a = $ median of y, where a is an arbitrary constant. Prove the result for continuous, discrete, and mixed cases for y.

4.3.2. Let

$$f(x, y) = \frac{c}{(y + x)^3}, \ 1 \le y < \infty, 0 \le x \le 1,$$

and zero elsewhere. If $f(x, y)$ is a joint density function of x and y, then evaluate (i): the normalizing constant c; (ii): the marginal density of y; (iii): the marginal density of x; (iv): the conditional density of y given x.

4.3.3. By using the joint density in Exercise 4.3.2, evaluate the regression of y on x and then predict y at $x = \frac{1}{2}$.

4.3.4. Consider the function

$$f(x, y) = c \, y^{x^2 - 1}, 0 \le y \le 1, 1 \le x \le 2,$$

and zero elsewhere. If this is a joint density function, then evaluate (i): the normalizing constant c; (ii): the marginal density of x; (iii): the conditional density of y given x; (iv): the regression of y given x; (v): the predicted value of y at $x = \frac{1}{3}$.

4.3.5. Let

$$f(x_1, x_2, x_3) = c(1 + x_1 + x_2 + x_3), 0 \le x_i \le 1, i = 1, 2, 3$$

and zero elsewhere be a density function. Then evaluate the following: (i): the normalizing constant c; (ii): the regression of x_1 on x_2, x_3; (iii): the predicted value of x_1 at $x_2 = \frac{1}{2}, x_3 = \frac{1}{4}$; ($iv$): the predicted value of x_1 at $x_2 = \frac{1}{3}$.

4.4 Linear Regression

Let $y, x_1, ..., x_k$ be real scalar random variables, and let the regression of y on $x_1, ..., x_k$ be a linear function in $x_1, ..., x_k$, that is,

$$E(y|x_1, ..., x_k) = \beta_0 + \beta_1 x_1 + ... + \beta_k x_k \tag{4.3}$$

where $\beta_0, \beta_1, ..., \beta_k$ are constants. If joint moments up to the second order exist, then we can evaluate the constants $\beta_0, \beta_1, ..., \beta_k$ in terms of product moments. In order to achieve this, we will need two results from elementary statistics. These will be listed here as lemmas.

Lemma 4.1.
$$E(u) = E_v[E(u|v)]$$

whenever the expected values exist. Here $E(u|v)$ is in the conditional space of u given v or computed from the conditional distribution of u given v, as a function of v. Then the resulting quantity is treated as a function of the random variable v in the next step of taking the expected value $E_v(\cdot)$.

Lemma 4.2.
$$\text{Var}(u) = \text{Var}[E(u|v)] + E[\text{Var}(u|v)].$$

That is, the sum of the variance of the conditional expectation and the expected value of the conditional variance is the unconditional variance of any random variable u, as long as the variances exist.

All the expected values and variances defined there must exist for the results to hold. The proofs follow from the definitions themselves and are left to the students. Let us look into the implications of these two lemmas with the help of some examples.

Example 4.13. Consider the joint density of x and y, given by

$$f(x, y) = \begin{cases} \frac{1}{x^2} \frac{1}{\sqrt{2\pi}} e^{-\frac{1}{2}(y-2-x)^2}, & -\infty < y < \infty, 1 \le x < \infty \\ 0, & \text{elsewhere.} \end{cases}$$

Evaluate the regression of y on x, and also verify Lemma 4.1.

Solution 4.11. Integrating out y one has the marginal density of x. Integration with respect to y can be effected by looking at a normal density in y with expected value $2 + x$. Then, the marginal density of x is given by

$$f(x) = \begin{cases} \frac{1}{x^2}, & 1 \le x < \infty \\ 0, & \text{elsewhere,} \end{cases}$$

because the joint density is the product of conditional and marginal densities. Therefore, the conditional density of y given x is normal with expected value $2 + x$, and hence, the regression of y on x is given by

$$E(y|x) = 2 + x \tag{4.4}$$

which is a linear function in x and well behaved smooth function of x. Expected value of the right side of (4.4) is then

$$E(2+x) = 2 + E(x)$$
$$= 2 + \int_1^\infty \frac{x}{x^2} dx$$
$$= 2 + [\ln x]_1^\infty = \infty.$$

Thus, the expected value does not exist and Lemma 4.1 is not applicable here.

Note that in Lemma 4.1, the variable v could be a single real scalar variable or a collection of real scalar variables. But since Lemma 4.2 is specific about variance of a single variable, the formula does not work if v contains many variables.

Example 4.14. Verify Lemma 4.1 for the following joint density.

$$f(x, y) = \begin{cases} 2, & 0 \le x \le y \le 1 \\ 0, & \text{elsewhere.} \end{cases}$$

Solution 4.12. Here, the surface $z = f(x, y)$ is a prism sitting on the (x, y)-plane the nonzero part of the density is in the triangle $0 \le x \le y \le 1$. Thus, the region can be defined as either $0 \le x \le y$ and $0 \le y \le 1$ or $x \le y \le 1$ and $0 \le x \le 1$. Marginally, $0 \le x \le 1$ as well as $0 \le y \le 1$. The marginal densities of x and y are respectively

$$f_1(x) = \int_{y=x}^1 2dy = \begin{cases} 2(1-x), & 0 \le x \le 1 \\ 0, & \text{elsewhere.} \end{cases}$$

$$f_2(y) = \int_{x=0}^y 2dx = \begin{cases} 2y, & 0 \le y \le 1 \\ 0, & \text{elsewhere.} \end{cases}$$

Hence

$$E(y) = \int_0^1 y(2y)dy = \frac{2}{3} \text{ and } E(x) = \int_0^1 x[2(1-x)]dx = \frac{1}{3}.$$

The conditional density of y given x is given by

$$g(y|x) = \frac{f(x, y)}{f_1(x)} = \frac{2}{2(1-x)} = \frac{1}{1-x}, x \le y \le 1$$

and zero elsewhere. Note that when x is fixed at some point, then y can only vary from that point to 1. Therefore, the conditional expectation of y, given x,

$$E(y|x) = \int_{y=x}^1 \frac{y}{1-x} dy = \frac{1-x^2}{2(1-x)} = \frac{1+x}{2}.$$

From this, by taking expected value, we have

$$E_x[E(y|x)] = \frac{1}{2}[E(1+x)] = \frac{1}{2}[1 + E(x)]$$
$$= \frac{1}{2}[1 + \frac{1}{3}] = \frac{2}{3} = E(y).$$

Thus, the result is verified. [Note that when we take $E_x(\cdot)$ we replace the preassigned x by the random variable x or we consider all values taken by x and the corresponding density or we switch back to the density of x and we are no longer in the conditional density].

Coming back to the linear regression in (4.3), we have

$$E(y|x_1, ..., x_k) = \beta_0 + \beta_1 x_1 + ... + \beta_k x_k. \tag{4.5}$$

Taking expected value on both sides, which means expected value in the joint marginal density of $x_1, ..., x_k$ and by Lemma 2.1,

$$E[E(y|x_1, ..., x_k)] = E(y)$$
$$E[\beta_0 + \beta_1 x_1 + ... + \beta_k x_k] = \beta_0 + \beta_1 E(x_1) + ... + \beta_k E(x_k). \tag{4.6}$$

Note that taking expected value of x_j in the joint distribution of $x_1, ..., x_k$ is equivalent to taking the expected value of x_j in the marginal distribution of x_j alone because the other variables can be integrated out (or summed up, if discrete) first to obtain the marginal density of x_j alone. [The student may work out an example to grasp this point]. From (4.5) and (4.6), one has

$$E(y|x_1, ..., x_k) - E(y) = \beta_1[x_1 - E(x_1)] + ... + \beta_k[x_k - E(x_k)]. \tag{4.7}$$

Multiply both sides of (4.7) by $x_j - E(x_j)$ for a specific j and then take expected value with respect to $x_1, ..., x_k$. The right side gives the following:

$$\beta_1 \text{Cov}(x_1, x_j) + \beta_2 \text{Cov}(x_2, x_j)$$
$$+ ... + \beta_j \text{Var}(x_j) + ... + \beta_k \text{Cov}(x_k, x_j) \tag{4.8}$$

because

$$E\{[x_j - E(x_j)][x_r - E(x_r)]\} = \begin{cases} \text{Cov}(x_j, x_r), & \text{if } j \neq r \\ \text{Var}(x_j) & \text{if } j = r. \end{cases}$$

The left side of (4.7) leads to the following:

$$E\{[x_j - E(x_j)]E(y)\} = E(y)\{E[x_j - E(x_j)]\} = 0$$

since for any variable x_j, $E[x_j - E(x_j)] = 0$ as long as the expected value exists.

$$E\{[x_j - E(x_j)]E(y|x_1, ..., x_k)\} = E\{E(y(x_j - E(x_j))|x_1, ..., x_k\}$$
$$= E[y(x_j - E(x_j))]$$

since in the conditional expectation $x_1, ..., x_k$ are fixed and therefore, one can take $x_j - E(x_j)$, being constant, inside the conditional expected value and write $y(x_j - E(x_j))$ given $x_1, ..., x_k$. But

$$E[y(x_j - E(x_j))] = \text{Cov}(y, x_j)$$

because for any two real scalar random variables x and y,

$$\text{Cov}(x, y) = E\{(x - E(x))(y - E(y))\}$$
$$= E\{x[y - E(y)]\} = E\{y[x - E(x)]\}$$

because

$$E\{E(x)[y - E(y)]\} = E(x)E\{y - E(y)\} = 0$$

since $E[y - E(y)] = E(y) - E(y) = 0$ and similarly $E\{E(y)[x - E(x)]\} = 0$. Therefore, we have

$$\sigma_{jy} = \beta_1 \sigma_{1j} + \beta_2 \sigma_{2j} + ... + \beta_k \sigma_{kj} \qquad (4.9)$$

where $\sigma_{ij} = \text{Cov}(x_i, x_j)$ and $\sigma_{jy} = \sigma yj = \text{Cov}(x_j, y)$. Writing (4.9) explicitly, one has

$$\sigma_{1y} = \beta_1 \sigma_{11} + \beta_2 \sigma_{12} + ... + \beta_k \sigma_{1k}$$
$$\sigma_{2y} = \beta_1 \sigma_{21} + \beta_2 \sigma_{22} + ... + \beta_k \sigma_{2k}$$
$$\vdots$$
$$\sigma_{ky} = \beta_1 \sigma_{k1} + \beta_2 \sigma_{k2} + ... + \beta_k \sigma_{kk}.$$

Writing in matrix notation, we have

$$\Sigma_y = \Sigma\beta, \quad \Sigma = (\sigma_{ij})$$

$$\beta = \begin{bmatrix} \beta_1 \\ \vdots \\ \beta_k \end{bmatrix}, \quad \Sigma_y = \begin{bmatrix} \sigma_{1y} \\ \vdots \\ \sigma_{ky} \end{bmatrix}, \quad \Sigma = \begin{bmatrix} \sigma_{11} & \sigma_{12} & ... & \sigma_{1k} \\ \vdots & \vdots & ... & \vdots \\ \sigma_{k1} & \sigma_{k2} & ... & \sigma_{kk} \end{bmatrix}.$$

Note that the covariance matrix Σ is symmetric since $\sigma_{ij} = \sigma_{ji}$ for all i and j. If Σ is non-singular then

$$\beta = \Sigma^{-1}\Sigma_y \qquad (4.10)$$

where Σ^{-1} denotes the regular inverse of the covariance matrix or the variance–covariance matrix Σ. This notation Σ is another awkward symbol in statistics. This can be easily confused with the summation symbol $\sum$. But since it is very widely used, we will also use it here. Is Σ likely to be non-singular? If Σ is singular then it means that at least one of the rows (columns) of Σ is a linear function of other rows (columns). This can happen if at least one of the variables $x_1, ..., x_k$ is a linear function of the other variables. Since these variables are preassigned, nobody will preassign one vector $(x_1, ..., x_k)$ and another point as a constant multiple $\alpha(x_1, ..., x_k)$ because the second point does not give any more information. Thus, when the points are preassigned as in a regression problem, one can assume without loss of generality that Σ is non-singular. But when Σ is estimated, since observations are taken on the variables, near singularity may occur. We will look into this aspect later. From (4.10) and (4.6), one has

$$\begin{aligned} \beta_0 &= E(y) - \beta_1 E(x_1) - ... - \beta_k E(x_k) \\ &= E(y) - \beta' E(X) \end{aligned} \tag{4.11}$$

where a prime denotes the transpose

$$\beta = \begin{bmatrix} \beta_1 \\ \vdots \\ \beta_k \end{bmatrix}, \quad E(X) = \begin{bmatrix} E(x_1) \\ \vdots \\ E(x_k) \end{bmatrix}.$$

From (4.10) we have for example, $\beta_1 = \Sigma^{(1)} \Sigma_y$ where $\Sigma^{(1)}$ is the first row of Σ^{-1}, $\beta_j = \Sigma^{(j)} \Sigma_y$ where $\Sigma^{(j)}$ is the j-th row of Σ^{-1} for $j = 1, ..., k$.

Instead of denoting the variables as y and $x_1, ..., x_k$, we may denote the variables simply as $x_1, ..., x_k$ and the problem is to predict x_1 by preassigning $x_2, ..., x_k$. In this notation, we can write the various quantities in terms of the submatrices of Σ. For this purpose, let us write

$$\Sigma = \begin{bmatrix} \sigma_{11} & \sigma_{12} & \cdots & \sigma_{1k} \\ \sigma_{21} & \sigma_{22} & \cdots & \sigma_{2k} \\ \vdots & \vdots & \cdots & \vdots \\ \sigma_{k1} & \sigma_{k2} & \cdots & \sigma_{kk} \end{bmatrix} = \begin{bmatrix} \sigma_{11} & \Sigma_{12} \\ \Sigma_{21} & \Sigma_{22} \end{bmatrix},$$

$$\Sigma_{21} = \Sigma'_{12} = \begin{bmatrix} \sigma_{21} \\ \vdots \\ \sigma_{k1} \end{bmatrix}, \quad \Sigma_{22} = \begin{bmatrix} \sigma_{22} & \cdots & \sigma_{2k} \\ \sigma_{32} & \cdots & \sigma_{3k} \\ \vdots & \cdots & \vdots \\ \sigma_{k2} & \cdots & \sigma_{kk} \end{bmatrix}. \tag{4.12}$$

Then, the best predictor, best in the minimum mean square sense, for predicting x_1 at preassigned values of $x_2, ..., x_k$ is given by $E(x_1|x_2, ..., x_k)$, and if this regression of x_1 on $x_2, ..., x_k$ is linear in $x_2, ..., x_k$, then it is of the form

$$E(x_1|x_2, ..., x_k) = \alpha_0 + \alpha_2 x_2 + ... + \alpha_k x_k \tag{4.13}$$

where $\alpha_0, \alpha_2, ..., \alpha_k$ are constants. Then from (4.10)

$$\alpha = \begin{bmatrix} \alpha_2 \\ \vdots \\ \alpha_k \end{bmatrix} = \Sigma_{22}^{-1}\Sigma_{21} \tag{4.14}$$

and the regression of x_1 on $x_2, ..., x_k$ or the best predictor of x_1 at preassigned values of $x_2, ..., x_k$, when the regression is linear, is given by

$$E(x_1|x_2, ..., x_k) = \alpha_0 + \alpha' X_2, \ \alpha = \begin{bmatrix} \alpha_2 \\ \vdots \\ \alpha_k \end{bmatrix}, \ X_2 = \begin{bmatrix} x_2 \\ \vdots \\ x_k \end{bmatrix}. \tag{4.15}$$

Let us compute the linear correlation between x_1 and its best linear predictor, that is, between x_1 and the predicting function in (4.15).

Example 4.15. If $X' = (x_1, x_2, x_3)$ has the mean value

$$E(X') = (E(x_1), E(x_2), E(x_3)) = (2, 1, -1)$$

and the covariance matrix

$$\text{Cov}(X) = \begin{bmatrix} 2 & -1 & 0 \\ -1 & 2 & 1 \\ 0 & 1 & 1 \end{bmatrix},$$

constructs the regression function for predicting x_1 at given values of x_2 and x_3, if it is known that the regression is linear in x_2, x_3.

Solution 4.13. As per our notation

$$\Sigma_{11} = \sigma_{11} = 2, \ \Sigma_{12} = (-1, 0), \ \Sigma_{21} = \begin{bmatrix} -1 \\ 0 \end{bmatrix}, \ \Sigma_{22} = \begin{bmatrix} 2 & 1 \\ 1 & 1 \end{bmatrix}$$

and hence

$$\Sigma_{22}^{-1} = \begin{bmatrix} 1 & -1 \\ -1 & 2 \end{bmatrix}, \ \Sigma_{12}\Sigma_{22}^{-1} = (-1, 0)\begin{bmatrix} 1 & -1 \\ -1 & 2 \end{bmatrix} = (-1, 1).$$

Hence, the best predictor is

$$E(x_1|x_2, x_3) = E(x_1) + \Sigma_{12}\Sigma_{22}^{-1}(X_2 - E(X_2))$$
$$= 2 + (-1, 1)\begin{bmatrix} x_2 - 1 \\ x_3 + 1 \end{bmatrix}$$
$$= 2 - x_2 + 1 + x_3 + 1 = 4 - x_2 + x_3$$

is the best prediction function for predicting x_1 at preassigned values of x_2 and x_3.

4.4.1 Correlation Between x_1 and Its Best Linear Predictor

In order to compute the correlation between x_1 and $E(x_1|x_2, .., x_k)$, we need the variances of these two quantities and the covariance between them. As per our notation in (2.12), we have

$$\text{Var}(x_1) = \sigma_{11}. \tag{4.16}$$

The variance of $\alpha_0 + \alpha'X$ can be computed by using the result on variance of a linear function of scalar variables. These will be stated as lemmas. These follow directly from the definition itself.

Lemma 4.3. *Consider a linear function of real scalar random variables $y_1, ..., y_n$ with covariances, $\text{Cov}(y_i, y_j) = v_{ij}, i, j = 1, ..., n$ thereby $v_{ii} = \text{Var}(y_i)$ and let the variance-covariance matrix in $(y_1, ..., y_n)$ be denoted by $V = (v_{ij})$. Let*

$$u = a_0 + a_1 y_1 + a_2 y_2 + ... + a_n y_n = a_0 + a'Y$$
$$v = b_0 + b_1 y_1 + b_2 y_2 + ... + b_n y_n = b_0 + b'Y$$

where

$$a = \begin{bmatrix} a_1 \\ \vdots \\ a_n \end{bmatrix}, \ b = \begin{bmatrix} b_1 \\ \vdots \\ b_n \end{bmatrix}, \ Y = \begin{bmatrix} y_1 \\ \vdots \\ y_n \end{bmatrix}$$

and a prime denotes the transpose. Then

$$\text{Var}(u) = a'Va, \ \text{Var}(v) = b'Vb, \ \text{Cov}(u, v) = a'Vb = b'Va.$$

Note that V is a symmetric matrix. Then with the help of Lemma 4.3, we have

$$\text{Var}[E(x_1|x_2, ..., x_k)] = \text{Var}[\alpha_0 + \alpha'X] = \text{Var}[\alpha'X]$$
$$= \alpha'\text{Cov}(X)\alpha = \alpha'\Sigma_{22}\alpha \tag{4.17}$$

where $\text{Cov}(X)$ means the covariance matrix in X. But from (4.14) and (4.17), the variance of the best linear predictor is the following:

$$\alpha' \Sigma_{22} \alpha = [\Sigma_{22}^{-1} \Sigma_{21}]' \Sigma_{22} [\Sigma_{22}^{-1} \Sigma_{21}] = \Sigma_{12} \Sigma_{22}^{-1} \Sigma_{21} \qquad (4.18)$$

because $\Sigma_{22}' = \Sigma_{22}$, $\Sigma_{21}' = \Sigma_{12}$. The covariance between x_1 and its best linear predictor is then,

$$
\begin{aligned}
\text{Cov}[x_1, E(x_1 | x_2, .., x_k)] &= \text{Cov}[x_1, \alpha_0 + \alpha' X] \\
&= \text{Cov}[x_1, \alpha' X] = \alpha' \text{Cov}(x_1, X) = \alpha' \Sigma_{21} \\
&= [\Sigma_{22}^{-1} \Sigma_{21}]' \Sigma_{21} = \Sigma_{12} \Sigma_{22}^{-1} \Sigma_{21}.
\end{aligned}
$$

Strangely enough, the covariance between x_1 and its best linear predictor is the same as the variance of the best linear predictor. Variance being nonnegative, it is clear that the covariance in this case is also nonnegative, and hence, the correlation is also nonnegative. Denoting the correlation by $\rho_{1.(2...k)}$, we have

$$\rho_{1.(2...k)}^2 = \frac{(\Sigma_{12} \Sigma_{22}^{-1} \Sigma_{21})^2}{\sigma_{11}(\Sigma_{12} \Sigma_{22}^{-1} \Sigma_{21})} = \frac{\Sigma_{12} \Sigma_{22}^{-1} \Sigma_{21}}{\sigma_{11}}. \qquad (4.19)$$

But note that the best predictor $E(x_1 | x_2, ..., x_k)$ for predicting x_1 at preassigned values of $x_2, ..., x_k$ need not be linear. We have cited several examples where the regression function is nonlinear. But if the regression is linear, then the correlation between x_1 and its best linear predictor has the nice from given in (4.19).

Exercises 4.4.

4.4.1. Write the following linear functions by using vector, matrix notations. For example, $4 + x_1 - x_2 + 5x_3 = 4 + a'X = b'Y$ where the prime denotes the transpose and

$$
a = \begin{bmatrix} 1 \\ -1 \\ 5 \end{bmatrix}, \; X = \begin{bmatrix} x_1 \\ x_2 \\ x_3 \end{bmatrix}, \; b = \begin{bmatrix} 4 \\ 1 \\ -1 \\ 5 \end{bmatrix}, \; Y = \begin{bmatrix} 1 \\ x_1 \\ x_2 \\ x_3 \end{bmatrix}.
$$

(i) $y = 2 + x_1 + x_2 - x_3$; (ii) $y = 1 + 2x_1 - x_2$; (iii) $y = 5 + x_1 + x_2 - 2x_3 + x_4$.

4.4.2. Write down the following quadratic forms in the form $X'AX$ where $A = A'$.

(i) $x_1^2 + 2x_2^2 - 3x_1 x_2$;

(ii) $2x_1^2 + x_2^2 - x_3^2 + 2x_1 x_2 - x_2 x_3$;

(iii) $x_1^2 + x_2^2 + ... + x_k^2$.

4.4.3. Write the same quantities in Exercise 4.4.2 as $X'AX$ where $A \neq A'$.

4.4.4. Can the following matrices represent covariance matrices, if so prove and if not explain why?

$$A_1 = \begin{bmatrix} 2 & 0 \\ 0 & -3 \end{bmatrix}; \quad A_2 = \begin{bmatrix} 1 & -3 \\ -3 & 2 \end{bmatrix}; \quad A_3 = \begin{bmatrix} 3 & 1 \\ 1 & 2 \end{bmatrix}$$

$$A_4 = \begin{bmatrix} 1 & -1 & 1 \\ -1 & 2 & 0 \\ 1 & 0 & 4 \end{bmatrix}, \quad A_5 = \begin{bmatrix} 3 & 1 & 0 \\ 1 & 2 & 2 \\ 0 & 2 & 2 \end{bmatrix}, \quad A_6 = \begin{bmatrix} 2 & 1 & 0 \\ 1 & -3 & 1 \\ 0 & 1 & 4 \end{bmatrix}.$$

4.4.5. Let $X' = (x_1, x_2, x_3)$ have a joint distribution with the following mean value and covariance matrix:

$$E(X) = \begin{bmatrix} 1 \\ -1 \\ 2 \end{bmatrix}, \quad \text{Cov}(X) = V = \begin{bmatrix} 1 & 1 & -1 \\ 1 & 3 & 0 \\ -1 & 0 & 2 \end{bmatrix}.$$

Let the regression of x_1 on x_2 and x_3 be a linear function of x_2 and x_3. Construct the best linear predictor $E(x_1|x_2, x_3)$ for predicting x_1 at preassigned values of x_2 and x_3; (ii) predict x_1 at $x_2 = 1$, $x_3 = 0$; (iii) compute the variance of the best predictor $E(x_1|x_2, x_3)$; (iv) compute the covariance between x_1 and the best linear predictor; (v) compute the correlation between x_1 and its best linear predictor.

4.5 Multiple Correlation Coefficient $\rho_{1.(2...k)}$

The multiple correlation coefficient is simply defined as

$$\rho_{1.(2...k)} = \sqrt{\frac{\Sigma_{12} \Sigma_{22}^{-1} \Sigma_{21}}{\sigma_{11}}}$$

with the notations as given in (4.12). It does not mean that we are assuming that there is a linear regression. If the regression is linear, then the multiple correlation coefficient is also the correlation between x_1 and its best linear predictor. The expression in (4.19) itself has many interesting properties and the multiple correlation coefficient, as defined in (4.19), has many statistical properties.

Example 4.16. For the same covariance matrix in Example 4.15, compute $\rho_{1.(2.3)}^2$.

Solution 4.14. We need to compute $\frac{\Sigma_{12} \Sigma_{22}^{-1} \Sigma_{21}}{\sigma_{11}}$, out of which $\Sigma_{12} \Sigma_{22}^{-1}$ is already computed in Example 4.15 as $\Sigma_{12} \Sigma_{22}^{-1} = (-1, 1)$, and $\Sigma_{21} = \begin{pmatrix} -1 \\ 0 \end{pmatrix}$. Therefore,

$$\rho_{1.(2.3)}^2 = \frac{\Sigma_{12}\Sigma_{22}^{-1}\Sigma_{21}}{\sigma_{11}} = \frac{1}{2}(-1, 1)\begin{pmatrix} -1 \\ 0 \end{pmatrix} = \frac{1}{2}.$$

4.5.1 Some Properties of the Multiple Correlation Coefficient

Let $b_2x_2 + ... + b_kx_k = b'X_2$ where $b' = (b_2, ..., b_k)$ and $X_2' = (x_2, ..., x_k)$ be an arbitrary linear predictor of x_1. That is, x_1 is predicted by using $b'X_2$. Let us compute the correlation between x_1 and this arbitrary predictor $b'X_2$. Note that, from Lemma 4.3, we have

$$\text{Var}(b'X_2) = b'\Sigma_{22}b \text{ and } \text{Cov}(x_1, b'X_2) = b'\Sigma_{21}.$$

Then, the square of the correlation between x_1 and an arbitrary linear predictor, denoted by η^2, is given by the following:

$$\eta^2 = \frac{(b'\Sigma_{21})^2}{(b'\Sigma_{22}b)\sigma_{11}}.$$

But from Cauchy–Schwartz inequality, we have

$$(b'\Sigma_{21})^2 = [(b'\Sigma_{22}^{\frac{1}{2}})(\Sigma_{22}^{-\frac{1}{2}}\Sigma_{21})]^2 \le [b'\Sigma_{22}b][\Sigma_{12}\Sigma_{22}^{-1}\Sigma_{21}]$$

where $\Sigma_{22}^{\frac{1}{2}}$ is the symmetric positive definite square root of Σ_{22}. Hence,

$$\eta^2 = \frac{(b'\Sigma_{21})^2}{(b'\Sigma_{22}b)\sigma_{11}} \le \frac{[b'\Sigma_{22}b][\Sigma_{12}\Sigma_{22}^{-1}\Sigma_{21}]}{[b'\Sigma_{22}b]\sigma_{11}}$$

$$= \frac{\Sigma_{12}\Sigma_{22}^{-1}\Sigma_{21}}{\sigma_{11}}. \tag{4.20}$$

In other words, the maximum value of η^2 is $\rho_{1.(2...k)}^2$ the square of the multiple correlation coefficient is given in (4.19).

Theorem 4.1. *Multiple correlation coefficient of x_1 on $(x_2, ..., x_k)$, where $x_1, x_2,...,x_k$ are all real scalar random variables, is also the maximum correlation between x_1 and an arbitrary linear predictor of x_1 based on $x_2, ..., x_k$.*

Note 4.3. Cauchy–Schwartz Inequality.

Let $a_1, ..., a_n$ and $b_1, ..., b_n$ be two sequences of real numbers. Then,

$$\sum_{i=1}^{n} a_ib_i \le \left[\sum_{i=1}^{n} a_i^2\right]^{\frac{1}{2}}\left[\sum_{i=1}^{n} b_i^2\right]^{\frac{1}{2}} \tag{N4.1}$$

and the equality holds when $(a_1, ..., a_n)$ and $(b_1, ..., b_n)$ are linearly related. In terms of real scalar random variables x and y this inequality is the following:

$$\text{Cov}(x, y) \le \sqrt{[\text{Var}(x)]}\sqrt{[\text{Var}(y)]} \qquad (N4.2)$$

and the equality holds when x and y are linearly related. Proof is quite simple. (N4.1) is nothing but the statement $|\cos \theta| \le 1$ where θ is the angle between the vectors $\vec{a} = (a_1, ..., a_k)$ and $\vec{b} = (b_1, ..., b_k)$ and (N4.2) is the statement that $|\rho| \le 1$ where ρ is the correlation coefficient between the real scalar variables x and y. There are various variations and extensions of Cauchy–Schwartz inequality. But what we need to use in our discussions are available from (N4.1) and (N4.2).

Note 4.4. Determinants and inverses of partitioned matrices

Consider a matrix A and its regular inverse A^{-1}, when A is non-singular and let A be partitioned as follows:

$$A = \begin{bmatrix} A_{11} & A_{12} \\ A_{21} & A_{22} \end{bmatrix}, \quad A^{-1} = \begin{bmatrix} A^{11} & A^{12} \\ A^{21} & A^{22} \end{bmatrix} \qquad (N4.3)$$

where $A_{11}, A_{12}, A_{21}, A_{22}$ are submatrices in A and $A^{11}, A^{12}, A^{21}, A^{22}$ are submatrices in A^{-1}. For example, let

$$A = \begin{bmatrix} 2 & 0 & 1 \\ 0 & 1 & 1 \\ 2 & 1 & -1 \end{bmatrix} = \begin{bmatrix} A_{11} & A_{12} \\ A_{21} & A_{22} \end{bmatrix}$$

where let

$$A_{11} = [2], A_{12} = [0, 1], A_{21} = \begin{bmatrix} 0 \\ 2 \end{bmatrix}, A_{22} = \begin{bmatrix} 1 & 1 \\ 1 & -1 \end{bmatrix}.$$

Then from elementary theory of matrices and determinants, we have the following, denoting the determinant of A by $|A|$.

$$|A| = |A_{11}| \, |A_{22} - A_{21} A_{11}^{-1} A_{12}| \text{ if } |A_{11}| \neq 0 \qquad (N4.4)$$

$$= |A_{22}| \, |A_{11} - A_{12} A_{22}^{-1} A_{21}| \text{ if } |A_{22}| \neq 0. \qquad (N4.5)$$

For our illustrative example, the determinant of A_{11} is $|A_{11}| = 2$ and

$$|A_{22} - A_{21} A_{11}^{-1} A_{12}| = \left| \begin{bmatrix} 1 & 1 \\ 1 & -1 \end{bmatrix} - \begin{bmatrix} 0 \\ 2 \end{bmatrix} \frac{1}{2}[0, 1] \right|$$

$$= \left| \begin{bmatrix} 1 & 1 \\ 1 & -1 \end{bmatrix} - \begin{bmatrix} 0 & 0 \\ 0 & 1 \end{bmatrix} \right|$$

$$= \begin{vmatrix} 1 & 1 \\ 1 & -2 \end{vmatrix} = -3$$

and therefore

$$|A_{11}|\,|A_{22} - A_{21}A_{11}^{-1}A_{12}| = (2)(-3) = -6.$$

The student may evaluate the determinant of A directly and verify the result and as well as use (N4.5) and verify that result also. The proof for establishing (N4.4) and (N4.5) are quite simple. From the axioms defining a determinant, it follows that if linear functions of one or more rows are added to one or more rows, the value of the determinant remains the same. This operation can be done one at a time or several steps together. Consider (N4.3). What is a suitable linear function of the rows containing A_{11} to be added to the remaining rows containing A_{21} so that a null matrix appears at the position of A_{21}. The appropriate linear combination is obtained by a pre-multiplication by

$$-A_{21}A_{11}^{-1}\left[A_{11}\ A_{12}\right] = \left[-A_{21}\ -A_{21}A_{11}^{-1}A_{12}\right].$$

Hence, the resulting matrix and the corresponding determinant are the following:

$$|A| = \begin{vmatrix} A_{11} & A_{12} \\ A_{21} & A_{22} \end{vmatrix} = \begin{vmatrix} A_{11} & A_{12} \\ O & A_{22} - A_{21}A_{11}^{-1}A_{12} \end{vmatrix}.$$

This is a triangular block matrix, and hence, the determinant is a product of the determinants of the diagonal blocks. That is,

$$|A| = |A_{11}|\,|A_{22} - A_{21}A_{11}^{-1}A_{12}|.$$

If A^{-1} exists, then $AA^{-1} = I = A^{-1}A$. In the partitioned format in (N4.3)

$$AA^{-1} = I \Rightarrow \begin{bmatrix} A_{11} & A_{12} \\ A_{21} & A_{22} \end{bmatrix}\begin{bmatrix} A^{11} & A^{12} \\ A^{21} & A^{22} \end{bmatrix} = \begin{bmatrix} I_r & O \\ O & I_s \end{bmatrix}.$$

Thus, by straight multiplication, the following equations are determined where r and s denote the orders of the identity matrices where we assumed that A_{11} is $r \times r$ and A_{22} is $s \times s$.

$$A_{11}A^{11} + A_{12}A^{21} = I_r$$
$$A_{11}A^{12} + A_{12}A^{22} = O$$
$$A_{21}A^{11} + A_{22}A^{21} = O$$
$$A_{21}A^{12} + A_{22}A^{22} = I_s. \tag{N4.6}$$

Solving the system in (N4.6), we have the following representations, among other results:

$$A^{11} = [A_{11} - A_{12}A_{22}^{-1}A_{21}]^{-1}, A_{11}^{-1} = A^{11} - A^{12}(A^{22})^{-1}A^{21}$$
$$A^{22} = [A_{22} - A_{21}A_{11}^{-1}A_{12}]^{-1}, A_{22}^{-1} = A^{22} - A^{21}(A^{11})^{-1}A^{12}. \qquad (N4.7)$$

Note that the submatrices in the inverse are not the inverses of the corresponding submatrices in the original matrix. That is, $A^{11} \neq A_{11}^{-1}$, $A^{22} \neq A_{22}^{-1}$. From (N4.6), one can also derive formulae for A^{21} and A^{12} in terms of the submatrices in A and vice versa, which are not listed above. [This is left as an exercise to the student]

For our illustrative example, it is easily verified that

$$A^{-1} = \frac{1}{3}\begin{bmatrix} 1 & -\frac{1}{2} & \frac{1}{2} \\ -1 & 2 & 1 \\ 1 & 1 & -1 \end{bmatrix} = \begin{bmatrix} A^{11} & A^{12} \\ A^{21} & A^{22} \end{bmatrix}$$

where

$$A^{11} = [\frac{1}{3}], A^{12} = [-\frac{1}{6}, \frac{1}{6}],$$
$$A^{21} = \begin{bmatrix} -\frac{1}{3} \\ \frac{1}{3} \end{bmatrix}, A^{22} = \begin{bmatrix} \frac{2}{3} & \frac{1}{3} \\ \frac{1}{3} & -\frac{1}{3} \end{bmatrix}.$$

From the computations earlier, we have, for example

$$A_{22} - A_{21}A_{11}^{-1}A_{12} = \begin{bmatrix} 1 & 1 \\ 1 & -2 \end{bmatrix}$$

and hence

$$[A_{22} - A_{21}A_{11}^{-1}A_{12}]^{-1} = \begin{bmatrix} 1 & 1 \\ 1 & -2 \end{bmatrix}^{-1} = \frac{1}{3}\begin{bmatrix} 2 & 1 \\ 1 & -1 \end{bmatrix} = A^{22}.$$

Thus one result is verified. The remaining verifications are left to the students.

Note 4.5. Correlation coefficient

This is very often a misused concept in applied statistics. The phrase "correlation" indicates "relationship," and hence, people misinterpret it as a measure of relationship between two real scalar random variables and the corresponding sample value as measuring the relationship between the pairs of numbers. There is extensive literature trying to evaluate the strength of the relationship, "negative relationship," "positive relationship," "increasing and decreasing nature of the relationship," etc., by studying correlation. But it is very easy to show that correlation does not measure relationship at all. Let ρ denote the correlation between two real scalar random variables x and y. Then, it is easy to show that $-1 \leq \rho \leq 1$. This follows from Cauchy–Schwartz inequality or by using the property $|\cos\theta| \leq 1$ or by considering two random variables:

$$u = \frac{x}{\sigma_1} + \frac{y}{\sigma_2} \text{ and } v = \frac{x}{\sigma_1} - \frac{v}{\sigma_2}$$

where x and y are non-degenerate random variables with $\mathrm{Var}(x) = \sigma_1^2 > 0$ and $\mathrm{Var}(y) = \sigma_2^2 > 0$. Take $\mathrm{Var}(u)$ and $\mathrm{Var}(v)$, and use the fact that they are nonnegative to show that $-1 \leq \rho \leq 1$. In Cauchy–Schwartz inequality, equality is attained or the boundary values $\rho = +1$ and $\rho = -1$ are attained if and only if x and y are linearly related, that is, $y = a + bx$ where a and $b \neq 0$ are constants. This relationship must hold almost surely meaning that there could be nonlinear relationships, but the total probability measure on the nonlinear part must be zero or there must exist an equivalent linear function with probability 1. This aspect will be illustrated with an example later. Coming back to ρ, let us consider a perfect relationship between x and y in the form

$$y = a + bx + cx^2, c \neq 0. \tag{N4.8}$$

Since we are computing correlations, without loss of generality we can omit a. Further, for convenience let us assume that x has a symmetrical distribution so that all odd moments disappear. Then, $E(x) = 0$, $E(x^3) = 0$ and $E(y) = a + cE(x^2)$. Also, we rule out degenerate variables when computing correlation, and hence, it is assumed that $\mathrm{Var}(x) \neq 0$.

$$\begin{aligned}
\mathrm{Cov}(x, y) &= E\{x[y - E(y)]\} = E\{x[bx + c(x^2 - E(x^2))]\} \\
&= 0 + b\mathrm{Var}(x) + 0 = b\mathrm{Var}(x). \\
\mathrm{Var}(y) &= E[bx + c(x^2 - E(x^2))]^2 \\
&= b^2\mathrm{Var}(x) + c^2\{E(x^4) - [E(x^2)]^2\}.
\end{aligned}$$

Then

$$\begin{aligned}
\rho &= \frac{b\mathrm{Var}(x)}{\sqrt{\mathrm{Var}(x)}\sqrt{b^2\mathrm{Var}(x) + c^2\{E(x^4) - (E(x^2))^2\}}} \\
&= \frac{b}{|b|\sqrt{1 + \frac{c^2}{b^2}\left\{\frac{E(x^4) - (E(x^2))^2}{\mathrm{Var}(x)}\right\}}}, \quad \text{for } b \neq 0 \tag{N4.9} \\
&= 0 \text{ if } b = 0 \\
&= \frac{1}{\sqrt{1 + \frac{c^2}{b^2}\left\{\frac{E(x^4) - (E(x^2))^2}{\mathrm{Var}(x)}\right\}}}, \quad \text{if } b > 0 \\
&= -\frac{1}{\sqrt{1 + \frac{c^2}{b^2}\left\{\frac{E(x^4) - (E(x^2))^2}{\mathrm{Var}(x)}\right\}}} \quad \text{if } b < 0.
\end{aligned}$$

Let us take x to be a standard normal variable, that is, $x \sim N(0, 1)$, then we know that $E(x) = 0$, $E(x^2) = 1$, $E(x^4) = 3$. In this case,

$$\rho = \pm \frac{1}{\sqrt{1 + 2\frac{c^2}{b^2}}}, \qquad (N4.10)$$

positive if $b > 0$, negative if $b < 0$, and zero if $b = 0$. Suppose that we would like to have $\rho = 0.01$ and at the same time a perfect mathematical relationship between x and y, such as the one in (N4.8) existing. Then, let $b > 0$ and let

$$\frac{1}{\sqrt{1 + 2\frac{c^2}{b^2}}} = 0.01 \Rightarrow 1 + 2\frac{c^2}{b^2} = \frac{1}{(0.01)^2} = 10000 \Rightarrow$$

$$2\frac{c^2}{b^2} = 9999 \Rightarrow c^2 = \frac{9999b^2}{2}.$$

Take any $b > 0$ such that $c^2 = \frac{9999b^2}{2}$. There are infinitely many choices. For example, $b = 1$ gives $c^2 = \frac{9999}{2}$. Similarly, if we want ρ to be zero, then take any $b = 0$. If we want ρ to be a very high positive number such as 0.999 or a number close to -1 such as -0.99, then also there are infinitely many choices of b and c such that a perfect mathematical relationship existing between x and y, and at the same time, ρ can be anything between -1 and $+1$. Thus, ρ is not an indicator of relationship between x and y. Even when there is a relationship between x and y, other than linear relationship, ρ can be anything between -1 and $+1$, and $\rho = \pm 1$ when and only when there is a linear relationship almost surely. From the quadratic function that we started with, note that increasing values of x can go with increasing as well as decreasing values of y when ρ is positive or negative. Hence, that type of interpretation cannot be given to ρ either.

Example 4.17. Consider the following probability function for x,

$$f(x) = \begin{cases} \frac{1}{2}, & x = \alpha \\ \frac{1}{2}, & x = -\alpha \\ 0, & \text{elsewhere}. \end{cases}$$

Compute ρ and check the quadratic relationship between x and y as given in (N4.8).

Solution 4.15. Here, $E(x) = 0$, $E(x^2) = \alpha^2$, $E(x^4) = \alpha^4$. Then, ρ in (N4.8) becomes

$$\rho = \frac{b}{|b|} = \pm 1$$

but $c \neq 0$ thereby (N4.8) holds. Is there something wrong with Cauchy–Schwartz inequality? This is left to the student to think over.

Then, what is the correlation coefficient ρ? What does it really measure? The numerator of ρ, namely $\text{Cov}(x, y)$, measures the joint variation of x and y or the scatter of the point (x, y), corresponding to the scatter in x, $\text{Var}(x) = \text{Cov}(x, x)$. Then, division by $\sqrt{\text{Var}(x)\text{Var}(y)}$ has the effect of making the covariance scale-free. Hence, ρ really measures the joint variation of x and y or a type of scatter in the point (x, y) in the sense that when $y = x$ it becomes $\text{Var}(x)$ which is the square of a measure of scatter in x. Thus, ρ is more appropriately called a scale-free covariance and this author suggested, through one of the published papers, to call ρ *scovariance* or scale-free covariance. It should never be interpreted as measuring relationship or linearity or near linearity or anything like that. For postulates defining covariance or for an axiomatic definition of covariance, the student may consult the book: Mathai, A.M. and Rathie, P.N. (1975): *Basic Concepts in Information Theory and Statistics: Axiomatic Foundations and Applications*, Wiley, New York, and Wiley, New Delhi.

Exercises 4.5.

4.5.1. Verify equations (N4.4) and (N4.5) for the following partitioned matrices:

$$A = \begin{bmatrix} 1 & -1 & 0 & 2 \\ 3 & 4 & 1 & 1 \\ 2 & 1 & -1 & 1 \\ 1 & 0 & 1 & 1 \end{bmatrix} = \begin{bmatrix} A_{11} & A_{12} \\ A_{21} & A_{22} \end{bmatrix}, \quad A_{11} = \begin{bmatrix} 1 & -1 \\ 3 & 4 \end{bmatrix}$$

$$B = \begin{bmatrix} 1 & 01 & -1 \\ 0 & 2 & -1 & 0 \\ 1 & -1 & 3 & 1 \\ -1 & 0 & 1 & 4 \end{bmatrix} = \begin{bmatrix} B_{11} & B_{12} \\ B_{21} & B_{22} \end{bmatrix}, \quad B_{11} = [1].$$

4.5.2. For real scalar random variable x, let $E(x) = 0$, $E(x^2) = 4$, $E(x^3) = 6$, $E(x^4) = 24$. Let $y = 50 + x + cx^2, c \neq 0$. Compute the correlation between x and y and interpret it for various values of c.

4.5.3. Let x be a standard normal variable, that is, $x \sim N(0, 1)$. Let $y = a + x + 2x^2 + cx^3, c \neq 0$. Compute the correlation between x and y and interpret it for various values of c.

4.5.4. Let x be a type-1 beta random variable with the parameters $(\alpha = 2, \beta = 1)$. Let $y = ax^\delta$ for some parameters a and δ. Compute the correlation between x and y and interpret it for various values of a and δ.

4.5.5. Repeat the Exercise in 4.5.4 if x is type-1 beta with the parameters $(\alpha = 1, \beta = 2)$.

4.6 Regression Analysis Versus Correlation Analysis

As mentioned earlier, for studying regression one needs only the conditional distribution of x_1 given $x_2, \ldots, x_k$ because the regression of x_1 on $x_2, \ldots, x_k$ is the conditional expectation of x_1 given $x_2, \ldots, x_k$, that is, $E(x_1|x_2, \ldots, x_k)$. But for correlation analysis, we need the joint distribution of all the variables involved. For example, in order to compute multiple correlation coefficient $\rho_{1.(2\ldots k)}$, we need the joint moments involving all the variables $x_1, x_2, \ldots, x_k$ up to second-order moments. Hence, the joint distribution, not just the conditional distribution, is needed. Thus, regression analysis and correlation analysis are built up on two different premises and should not be mixed up.

4.6.1 Multiple Correlation Ratio

In many of our examples, it is seen that the regression function is not linear in many situations. Let $E(x_1|x_2, \ldots, x_k) = M(x_2, \ldots, x_k)$, may or may not be a linear function of $x_2, \ldots, x_k$. Consider an arbitrary predictor $g(x_2, \ldots, x_k)$ for predicting x_1. To start with, we are assuming that there is a joint distribution of $x_1, x_2, \ldots, x_k$. Let us compute the correlation between x_1 and an arbitrary predictor $g(x_2, \ldots, x_k)$ for x_1, as a measure of joint scatter in x_1 and g.

$$\text{Cov}(x_1, g) = E\{[x_1 - E(x_1)][g - E(g)]\} = E\{x_1[(g - E(g)]\} \qquad (4.21)$$

as explained earlier since $E\{E(x_1)[g - E(g)]\} = E(x_1)E[g - E(g)]\} = 0$. Let us convert the expected value in (4.21) into an expectation of the conditional expectation through Lemma 4.1. Then,

$$\begin{aligned}
\text{Cov}(x_1, g) &= E\{E[x_1(g - E(g)]|x_2, \ldots, x_k]\} \\
&= E\{(g - E(g))E(x_1|x_2, \ldots, x_k)\}, \text{ since } g \text{ is free of } x_1 \\
&= E\{(g - E(g))M(x_2, \ldots, x_k)\} \\
&= \text{Cov}(g, M) \leq \sqrt{\text{Var}(g)\text{Var}(M)},
\end{aligned}$$

the last inequality follows from the fact that the correlation $\rho \leq 1$. Then, the correlation between x_1 and an arbitrary predictor, which includes linear predictors also, denoted by η is given by the following:

$$\eta = \frac{\text{Cov}(x_1, g)}{\sqrt{\text{Var}(g)}\sqrt{\text{Var}(x_1)}} \leq \frac{\sqrt{\text{Var}(g)}\sqrt{\text{Var}(M)}}{\sqrt{\text{Var}(g)}\sqrt{\text{Var}(x_1)}} = \frac{\sqrt{\text{Var}(M)}}{\sqrt{\text{Var}(x_1)}}. \qquad (4.22)$$

Definition 4.15. **Multiple correlation ratio** The maximum correlation between x_1 and an arbitrary predictor of x_1 by using $x_2, \ldots, x_k$ is given by the following:

$$\max_{g} \eta = \frac{\sqrt{\text{Var}(M)}}{\sqrt{\text{Var}(x_1)}} = \eta_{1.(2\ldots k)}. \tag{4.23}$$

This maximum correlation between x_1 and an arbitrary predictor of x_1 by using $x_2, \ldots, x_k$ is given by $\sqrt{\frac{\text{Var}(M)}{\text{Var}(x_1)}}$, and it is defined as the *multiple correlation ratio* $\eta_{1.(2\ldots k)}$ and the maximum is attained when the arbitrary predictor is the regression of x_1 on $x_2, \ldots, x_k$, namely $E(x_1 | x_2, \ldots, x_k) = M(x_2, \ldots, x_k)$.

Note that when $M(x_2, \ldots, x_k)$ is linear in $x_2, \ldots, x_k$, we have the multiple correlation coefficient given in (4.19). Thus, when g is confined to linear predictors or in the class of linear predictors

$$\eta^2_{1.(2\ldots k)} = \rho^2_{1.(2\ldots k)} = \frac{\Sigma_{12} \Sigma_{22}^{-1} \Sigma_{21}}{\sigma_{11}}. \tag{4.24}$$

Some further properties of $\rho^2_{1.(2\ldots k)}$ can be seen easily. Note that from (N4.7)

$$1 - \rho^2_{1.(2\ldots k)} = \frac{\sigma_{11} - \Sigma_{12} \Sigma_{22}^{-1} \Sigma_{21}}{\sigma_{11}} = \frac{(\sigma^{11})^{-1}}{\sigma_{11}} = \frac{1}{\sigma_{11} \sigma^{11}}. \tag{4.25}$$

Example 4.18. Check whether the following matrix Σ can represent the covariance matrix of $X' = (x_1, x_2, x_3)$ where x_1, x_2, x_3 are real scalar random variables. If so evaluate $\rho_{1.(2.3)}$ and verify (4.25).

$$\Sigma = \begin{bmatrix} 2 & 0 & 1 \\ 0 & 2 & -1 \\ 1 & -1 & 3 \end{bmatrix}.$$

Solution 4.16.

$$2 > 0, \begin{vmatrix} 2 & 0 \\ 0 & 2 \end{vmatrix} = 4 > 0, |\Sigma| = 8 > 0$$

and hence $\Sigma = \Sigma' > 0$ (positive definite) and hence it can represent the covariance matrix of X. For being a covariance matrix, one needs only symmetry plus at least positive semi-definiteness. As per our notation,

$$\sigma_{11} = 2, \ \Sigma_{12} = [0, 1], \ \Sigma_{22} = \begin{bmatrix} 2 & -1 \\ -1 & 3 \end{bmatrix}, \ \Sigma_{21} = \Sigma'_{12}$$

and therefore

$$\Sigma_{22}^{-1} = \frac{1}{5}\begin{bmatrix} 3 & 1 \\ 1 & 2 \end{bmatrix},$$

$$\frac{\Sigma_{12}\Sigma_{22}^{-1}\Sigma_{21}}{\sigma_{11}} = \frac{1}{(5)(2)}[0, 1]\begin{bmatrix} 3 & 1 \\ 1 & 2 \end{bmatrix}\begin{bmatrix} 0 \\ 1 \end{bmatrix}$$

$$= \frac{1}{5} = \rho_{1.(2.3)}^2$$

$$1 - \rho_{1.(2.3)}^2 = 1 - \frac{1}{5} = \frac{4}{5};$$

$$\frac{\sigma_{11} - \Sigma_{12}\Sigma_{22}^{-1}\Sigma_{21}}{\sigma_{11}} = \frac{1}{2}\left[2 - \frac{2}{5}\right] = \frac{4}{5};$$

$$\frac{1}{\sigma_{11}\sigma^{11}} = \frac{1}{2}\left(\frac{8}{5}\right) = \frac{4}{5}.$$

Thus, (4.25) is verified.

4.6.2 Multiple Correlation as a Function of the Number of Regressed Variables

Let $X' = (x_1, ..., x_k)$ and let the variance-covariance matrix in X be denoted by $\Sigma = (\sigma_{ij})$. Our general notation for the multiple correlation coefficient is

$$\rho_{1.(2...k)} = \sqrt{\frac{\Sigma_{12}\Sigma_{22}^{-1}\Sigma_{21}}{\sigma_{11}}}.$$

For $k = 2$,

$$\rho_{1.2} = \sqrt{\frac{\sigma_{12}\sigma_{12}}{\sigma_{22}\sigma_{11}}} = \sqrt{\frac{\sigma_{12}^2}{\sigma_{11}\sigma_{22}}} = \rho_{12}$$

$$= \text{correlation between } x_1 \text{ and } x_2.$$

For $k = 3$,

$$\rho_{1.(2.3)}^2 = \frac{1}{\sigma_{11}}[\sigma_{12}, \sigma_{13}]\begin{bmatrix} \sigma_{22} & \sigma_{23} \\ \sigma_{32} & \sigma_{33} \end{bmatrix}^{-1}\begin{bmatrix} \sigma_{21} \\ \sigma_{31} \end{bmatrix}.$$

Converting everything on the right in terms of the correlations, that is,

$$\sigma_{21} = \sigma_{12} = \rho_{12}\sigma_1\sigma_2, \quad \sigma_{11} = \sigma_1^2, \quad \sigma_{22} = \sigma_2^2,$$

$$\sigma_{31} = \sigma_{13} = \rho_{13}\sigma_1\sigma_3, \quad \sigma_{23} = \rho_{23}\sigma_2\sigma_3,$$

we have the following:

$$\rho_{1.(2.3)}^2 = \frac{1}{\sigma_1^2}[\rho_{12}\sigma_1\sigma_2, \rho_{13}\sigma_1\sigma_3]\begin{bmatrix} \sigma_2^2 & \rho_{23}\sigma_2\sigma_3 \\ \rho_{23}\sigma_2\sigma_3 & \sigma_3^2 \end{bmatrix}^{-1}\begin{bmatrix} \rho_{12}\sigma_1\sigma_2 \\ \rho_{13}\sigma_1\sigma_3 \end{bmatrix}$$

$$= [\rho_{12}\sigma_2, \rho_{13}\sigma_3]\begin{bmatrix} \sigma_2^2 & \rho_{23}\sigma_2\sigma_3 \\ \rho_{23}\sigma_2\sigma_3 & \sigma_3^2 \end{bmatrix}^{-1}\begin{bmatrix} \rho_{12}\sigma_2 \\ \rho_{13}\sigma_3 \end{bmatrix}$$

$$= [\rho_{12}, \rho_{13}]\begin{bmatrix} \sigma_2 & 0 \\ 0 & \sigma_3 \end{bmatrix}\begin{bmatrix} \sigma_2 & 0 \\ 0 & \sigma_3 \end{bmatrix}^{-1}$$

$$\times \begin{bmatrix} 1 & \rho_{23} \\ \rho_{23} & 1 \end{bmatrix}^{-1}\begin{bmatrix} \sigma_2 & 0 \\ 0 & \sigma_3 \end{bmatrix}^{-1}\begin{bmatrix} \sigma_2 & 0 \\ 0 & \sigma_3 \end{bmatrix}\begin{bmatrix} \rho_{12} \\ \rho_{13} \end{bmatrix}$$

$$= [\rho_{12}, \rho_{13}]\begin{bmatrix} 1 & \rho_{23} \\ \rho_{23} & 1 \end{bmatrix}^{-1}\begin{bmatrix} \rho_{12} \\ \rho_{13} \end{bmatrix}$$

$$= \frac{1}{1-\rho_{23}^2}[\rho_{12}, \rho_{13}]\begin{bmatrix} 1 & -\rho_{23} \\ -\rho_{23} & 1 \end{bmatrix}\begin{bmatrix} \rho_{12} \\ \rho_{13} \end{bmatrix}, 1-\rho_{23} > 0,$$

$$= \frac{\rho_{12}^2 + \rho_{13}^2 - 2\rho_{12}\rho_{13}\rho_{23}}{1-\rho_{23}^2}.$$

Then

$$\rho_{1.(2.3)}^2 - \rho_{12}^2 = \frac{\rho_{12}^2 + \rho_{13}^2 - 2\rho_{12}\rho_{13}\rho_{23}}{1-\rho_{23}^2} - \rho_{12}^2$$

$$= \frac{\rho_{13}^2 - 2\rho_{12}\rho_{13}\rho_{23} + \rho_{12}^2\rho_{23}^2}{1-\rho_{23}^2}$$

$$= \frac{(\rho_{13} - \rho_{12}\rho_{23})^2}{1-\rho_{23}^2} \geq 0$$

which is equal to zero only when $\rho_{13} = \rho_{12}\rho_{23}$. Thus, in general,

$$\rho_{1.(2.3)}^2 - \rho_{12}^2 \geq 0 \Rightarrow \rho_{1.(2.3)}^2 \geq \rho_{12}^2.$$

In other words, the multiple correlation coefficient increased when we incorporated one more variable x_3 in the regressed set. It is not difficult to show (left as an exercise to the student) that

$$\rho_{12}^2 \leq \rho_{1.(2.3)}^2 \leq \rho_{1.(2.3.4)}^2 \leq \cdots \tag{4.26}$$

This indicates that $\rho_{1.(2\ldots k)}^2$ is an increasing function of k, the number of variables involved in the regressed set. There is a tendency among applied statisticians to use the sample multiple correlation coefficient as an indicator of how good is a linear regression function by looking at the value of the multiple correlation coefficient, in the sense, bigger the value better the model. From (4.26), it is evident that this is a

fallacious approach. Also, this approach comes from the tendency to look at correlation coefficient as a measure of relationship which again is a fallacious concept.

Exercises 4.6.

4.6.1. Show that $\rho^2_{1.(2.3)} \leq \rho^2_{1.(2.3.4)}$ with the standard notation for the multiple correlation coefficient $\rho_{1.(2....k)}$.

4.6.2. (i): Show that the following matrix V can be a covariance matrix:

$$V = \begin{bmatrix} 1 & 1 & 0 & -1 \\ 1 & 3 & 1 & 0 \\ 0 & 1 & 3 & 1 \\ -1 & 0 & 1 & 2 \end{bmatrix}.$$

(ii): Compute $\rho^2_{1.2}, \rho^2_{1.(2.3)}, \rho^2_{1.(2.3.4)}$. (iii): Verify that $\rho^2_{1.2} \leq \rho^2_{1.(2.3)} \leq \rho^2_{1.(2.3.4)}$.

4.6.3. Let the conditional density of x_1 given x_2 be Gaussian with mean value
$1 + 2x_2 + x_2^2$ and variance 1, and let the marginal distribution of x_2 be uniform over
$[0, 1]$. Compute the square of the correlation ratio of x_1 to x_2, that is

$$\eta^2_{1.2} = \frac{\text{Var}(M)}{\text{Var}(x_1)}, \quad M = E(x_1|x_2)$$

and

$$\text{Var}(x_1) = \text{Var}[E(x_1|x_2)] + E[\text{Var}(x_1|x_2)].$$

4.6.4. Let the conditional density of x_1 given x_2, x_3 be exponential with mean value
$1 + x_2 + x_3 + x_2x_3$, and let the joint density of x_2 and x_3 be

$$f(x_2, x_3) = x_2 + x_3, 0 \leq x_2 \leq 1, 0 \leq x_3 \leq 1$$

and zero elsewhere. Compute the square of the correlation ratio

$$\eta^2_{1.(2.3)} = \frac{\text{Var}(M)}{\text{Var}(x_1)}$$

where

$$M = E(x_1|x_2, x_3)$$

and

$$\text{Var}(x_1) = \text{Var}[E(x_1|x_2, x_3)] + E[\text{Var}(x_1|x_2, x_3)].$$

4.6.5. Let the conditional density of x_1 given x_2, x_3 be Gaussian, $N(x_2 + x_3 + x_2 x_3, 1)$, where let x_2, x_3 have a joint density as in Exercise 4.6.4. Evaluate the square of the correlation ratio $\eta^2_{1.(2.3)}$.

4.7 Residual Effect

Suppose that x_1 is predicted by using $x_2, ..., x_k$. The best predictor function is $E(x_1|x_2, ..., x_k)$. If the regression is linear, then from (4.15)

$$E(x_1|x_2, ..., x_k) = \alpha_0 + \Sigma_{12} \Sigma_{22}^{-1} X_2$$
$$= E(x_1) + \Sigma_{12} \Sigma_{22}^{-1} (X_2 - E(X_2)). \qquad (4.27)$$

Thus, x_1 is predicted by the expression $E(x_1) + \Sigma_{12} \Sigma_{22}^{-1} (X_2 - E(X_2))$ when the prediction function is linear. Then, the residual part, after removing the effect of linear regression from x_1, is given by

$$u_1 = x_1 - E(x_1) - \Sigma_{12} \Sigma_{22}^{-1} (X_2 - E(X_2)). \qquad (4.28)$$

Then,

$$\text{Cov}(x_1, u_1) = \text{Var}(x_1) - \Sigma_{12} \Sigma_{22}^{-1} \Sigma_{21} = \sigma_{11} - \Sigma_{12} \Sigma_{22}^{-1} \Sigma_{21}$$
$$\text{Var}(u_1) = \sigma_{11} + \Sigma_{12} \Sigma_{22}^{-1} \Sigma_{21} - 2\Sigma_{12} \Sigma_{22}^{-1} \Sigma_{21}$$
$$= \sigma_{11} - \Sigma_{12} \Sigma_{22}^{-1} \Sigma_{21} \geq 0. \qquad (4.29)$$

The correlation between x_1 and the residual part, after removing the effect of linear regression from x_1, is then

$$\rho^2_{(x_1, u_1)} = \frac{[\text{Cov}(x_1, u_1)]^2}{\sigma_{11}[\text{Var}(u_1)]} = \frac{\sigma_{11} - \Sigma_{12} \Sigma_{22}^{-1} \Sigma_{21}}{\sigma_{11}} = \frac{1}{\sigma_{11} \sigma^{11}}$$
$$= 1 - \rho^2_{1.(2...k)}. \qquad (4.30)$$

This is another interesting observation that one can make in this regard. Note that if the regression is not linear, then

$$u_1 = x_1 - E(x_1|x_2, ..., x_k) = x_1 - M(x_2, ..., x_k). \qquad (4.31)$$

Then, going through expectation of conditional expectation as before one has the following: (derivation is left as an exercise to the student)

$$\text{Var}(u_1) = \text{Var}(x_1) - \text{Var}(M) = \sigma_{11} - \text{Var}(M) \geq 0$$

$$\text{Cov}(x_1, u_1) = \sigma_{11} - \text{Var}(M)$$

$$\rho^2_{(x_1, u_1)} = \frac{\sigma_{11} - \text{Var}(M)}{\sigma_{11}} = 1 - \frac{\text{Var}(M)}{\sigma_{11}}$$

$$= 1 - \eta^2_{1.(2...k)} \tag{4.32}$$

where $\eta^2_{1.(2...k)}$ is the square of the multiple correlation ratio.

Example 4.19. Consider the following joint density of y, x_1, x_2:

$$f(y, x_1, x_2) = \begin{cases} \frac{(x_1+x_2)}{1+x_1+x_2^2} e^{-\frac{y}{1+x_1+x_2^2}}, & 0 \leq y < \infty, 0 \leq x_1 \leq 1, 0 \leq x_2 \leq 1 \\ 0, & \text{elsewhere.} \end{cases}$$

Evaluate $\sigma^2, \eta^2_{1.(2.3)}$ where $\eta^2_{1.(2.3)}$ is the multiple correlation ratio.

Solution 4.17. Integrating out y, we see that the marginal density of x_1 and x_2 is given by

$$f_1(x_1, x_2) = \begin{cases} x_1 + x_2, & 0 \leq x_1 \leq 1, 0 \leq x_2 \leq 1 \\ 0, & \text{elsewhere} \end{cases}$$

and the conditional density of y given x_1, x_2 is the balance in $f(y, x_1, x_2)$ and the conditional expectation of y given x_1, x_2, denoted by $M(x_1, x_2)$, is the following:

$$M(x_1, x_2) = 1 + x_1 + x_2^2.$$

Therefore,

$$\text{Var}(M) = \text{Var}(1 + x_1 + x_2^2) = \text{Var}(x_1 + x_2^2)$$
$$= \text{Var}(x_1) + \text{Var}(x_2^2) + 2\text{Cov}(x_1, x_2^2).$$

The marginal density of x_2 is $x_2 + \frac{1}{2}$ for $0 \leq x_2 \leq 1$ and zero elsewhere. Hence,

$$E(x_2^2) = \int_0^1 x_2^2 (x_2 + \frac{1}{2}) dx_2 = \frac{5}{12};$$

$$E(x_2^4) = \int_0^1 x_2^4 (x_2 + \frac{1}{2}) dx_2 = \frac{4}{15};$$

$$\text{Var}(x_2^2) = \frac{4}{15} - \left(\frac{5}{12}\right)^2; \quad E(x_1) = \frac{7}{12};$$

$$E(x_1^2) = \frac{5}{12}; \quad \text{Var}(x_1) = \frac{5}{12} - \left(\frac{7}{12}\right)^2;$$

$$E(x_1 x_2^2) = \int_0^1 \int_0^1 x_1 x_2^2 (x_1 + x_2) dx_1 dx_2$$

$$= \int_0^1 x_2^2 \left(\frac{1}{3} + \frac{x_2}{2} \right) dx_2 = \frac{17}{72};$$

$$\text{Cov}(x_1, x_2^2) = \frac{17}{72} - (\frac{7}{12})(\frac{5}{12}).$$

$$\text{Var}(M) = \frac{5}{12} - (\frac{7}{12})^2 + \frac{4}{15} - (\frac{5}{12})^2 + 2 \left[\frac{17}{72} - (\frac{7}{12})(\frac{5}{12}) \right] = \frac{7}{45}.$$

$$= \sigma_y^2 \, \eta_{1.(2.3)}^2$$

Note that the marginal density of y is difficult to evaluate from the joint density. We can evaluate the variance of y by using the formula

$$\text{Var}(y) = \text{Var}[E(y|x_1, x_2)] + E[\text{Var}(y|x_1, x_2)].$$

Observe that the conditional variance of y, given x_1, x_2, is available from the given density, which is $(1 + x_1 + x_2^2)^2$ and $M(x_1, x_2) = (1 + x_1 + x_2^2)$. Thus, from these two items, $\text{Var}(y)$ is available. This is left as an exercise to the student.

Exercises 4.7.

4.7.1. For the covariance matrix in Exercise 4.6.2, compute the residual sum of squares after removing the effect of linear regression.

4.7.2. Compute σ_y^2 in Example 4.19.

4.8 Canonical Correlations

In the regression-type prediction, we were tying to predict one real scalar variable with the help of a set of other scalar variables and we came up with the best predictor as the conditional expectation of x_1 given $x_2, ..., x_k$, that is, $E(x_1|x_2, .., x_k)$, and consequently, we arrived at the maximum correlation between x_1 and its best linear predictor as $\rho_{1.(2...k)}$, where

$$\rho_{1.(2...k)}^2 = \frac{\Sigma_{12} \Sigma_{22}^{-1} \Sigma_{21}}{\sigma_{11}} = \sigma_{11}^{-\frac{1}{2}} \Sigma_{12} \Sigma_{22}^{-1} \Sigma_{21} \sigma_{11}^{-\frac{1}{2}}. \tag{4.33}$$

We will derive an expression for another concept known as the canonical correlation matrix where the expression looks very similar to the right side in (4.33). Consider the problem of predicting one set of real scalar random variables by using another set of real scalar random variables. Let

$$X = \begin{bmatrix} x_1 \\ \vdots \\ x_r \\ x_{r+1} \\ \vdots \\ x_k \end{bmatrix} = \begin{bmatrix} X_1 \\ X_2 \end{bmatrix}, \; X_1 = \begin{bmatrix} x_1 \\ \vdots \\ x_r \end{bmatrix}, \; X_2 = \begin{bmatrix} x_{r+1} \\ \vdots \\ x_k \end{bmatrix}$$

where $x_1, .., x_k$ are real scalar random variables. We would like to predict X_1 with the help of X_2 and vice versa. Observe that linear functions also include individual variables. For example if we take the linear function

$$u = a'X_1 = a_1 x_1 + \dots + a_r x_r, \; a = \begin{bmatrix} a_1 \\ \vdots \\ a_r \end{bmatrix}, \; X_1 = \begin{bmatrix} x_1 \\ \vdots \\ x_r \end{bmatrix}$$

then for $a_1 = 1, a_2 = 0, \dots, a_r = 0$, we have $u = x_1$, and so on. Hence, for convenience, we will take linear functions of X_1 and X_2 and then use these functions to construct prediction functions. To this extent, let

$$u = a'X_1 \text{ and } v = b'X_2, b = \begin{bmatrix} b_1 \\ \vdots \\ b_s \end{bmatrix}, \; X_2 = \begin{bmatrix} x_{r+1} \\ \vdots \\ x_{r+s} \end{bmatrix}, k = r + s$$

so that, for convenience we have assumed that there are r components in X_1 and s components in X_2 so that $r + s = k$. Let the covariance matrix of X be denoted by V, and let it be partitioned according to the subvectors in X. Let

$$V = \begin{bmatrix} V_{11} & V_{12} \\ V_{21} & V_{22} \end{bmatrix}, \; V_{11} \text{ is } r \times r, \; V_{22} \text{ is } s \times s, \; V_{12} = V_{21}'$$

$$V_{11} = \mathrm{Cov}(X_1), V_{22} = \mathrm{Cov}(X_2), V_{12} = \mathrm{Cov}(X_1, X_2)$$

where $\mathrm{Cov}(\cdot)$ means the covariance matrix in $(\cdot)$. As explained earlier, we will consider all possible linear functions of X_1 and X_2. Let

$$\alpha = \begin{bmatrix} \alpha_1 \\ \vdots \\ \alpha_r \end{bmatrix}, \; \beta = \begin{bmatrix} \beta_1 \\ \vdots \\ \beta_s \end{bmatrix}, \; u = \alpha'X_1, \; w = \beta'X_2,$$

$$u = \alpha_1 x_1 + \dots + \alpha_r x_r, \; w = \beta_1 x_{r+1} + \dots + \beta_s x_k, r + s = k$$

where α and β are constant vectors. For convenience, we will also take $\mathrm{Var}(u) = 1$ and $\mathrm{Var}(w) = 1$. This can always be done as long as the variances are finite. But, being linear functions, variances and covariances of linear functions are available from Module 1. For the sake of convenience, these are repeated as notes below.

Note 4.6. Variances and covariances of linear functions.

Let $x_1, ..., x_p$ be real scalar variables with $E(x_j) = \mu_j$, $\text{Var}(x_j) = \sigma_{jj}$, $\text{Cov}(x_i, x_j) = \sigma_{ij}, i, j = 1, ..., p$. Let

$$X = \begin{bmatrix} x_1 \\ \vdots \\ x_p \end{bmatrix}, a = \begin{bmatrix} a_1 \\ \vdots \\ a_p \end{bmatrix}, b = \begin{bmatrix} b_1 \\ \vdots \\ b_p \end{bmatrix}, \Sigma = (\sigma_{ij}) = \begin{bmatrix} \sigma_{11} & \cdots & \sigma_{1p} \\ \sigma_{21} & \cdots & \sigma_{2p} \\ \vdots & \cdots & \vdots \\ \sigma_{p1} & \cdots & \sigma_{pp} \end{bmatrix}$$

where a and b are constant vectors, a prime denotes the transpose, E denotes the expected value, and let $\mu' = (\mu_1, ..., \mu_p)$, $\mu_j = E(x_j), j = 1, ..., p$. As per the definition,

$$\text{Var}(x_j) = E[(x_j - E(x_j))^2],$$
$$\text{Cov}(x_i, x_j) = E[(x_i - E(x_i))(x_j - E(x_j))]; \text{Cov}(x_j, x_j) = \text{Var}(x_j).$$

Then,

$$u = a_1 x_1 + ... + a_p x_p \Rightarrow u = a'X = X'a$$
$$v = b_1 x_1 + ... + b_p x_p = b'X = X'b; E(a'X) = a'E(X) = a'\mu$$
$$E(b'X) = b'E(X) = b'\mu; \text{Var}(a'X) = E[a'X - a'\mu]^2 = E[a'(X - \mu)]^2.$$

From elementary theory of matrices, it follows that if we have a 1×1 matrix c then it is a scalar and its transpose is itself, that is, $c' = c$. Being a linear function $a'(X - \mu)$ is a 1×1 matrix and hence it is equal to its transpose, which is $(X - \mu)'a$. Hence, we may write

$$E[a'(X - \mu)]^2 = E[a'(x - \mu)(x - \mu)'a] = a'E[(X - \mu)(X - \mu)']a$$

since a is a constant the expected value can be taken inside. But

$$E[(X - \mu)(X - \mu)']$$
$$= E \begin{bmatrix} (x_1 - \mu_1)^2 & (x_1 - \mu_1)(x_2 - \mu_2) & \cdots & (x_1 - \mu_1)(x_p - \mu_p) \\ (x_2 - \mu_2)(x_1 - \mu_1) & (x_2 - \mu_2)^2 & \cdots & (x_2 - \mu_2)(x_p - \mu_p) \\ \vdots & \vdots & \cdots & \vdots \\ (x_p - \mu_p)(x_1 - \mu_1) & (x_p - \mu_p)(x_2 - \mu_2) & \cdots & (x_p - \mu_p)^2 \end{bmatrix}.$$

Taking expectations inside the matrix, we have

$$E[(X - \mu)(X - \mu)'] = \begin{bmatrix} \sigma_{11} & \sigma_{12} & \cdots & \sigma_{1p} \\ \sigma_{21} & \sigma_{22} & \cdots & \sigma_{2p} \\ \vdots & \vdots & \cdots & \vdots \\ \sigma_{p1} & \sigma_{p2} & \cdots & \sigma_{pp} \end{bmatrix}$$

$$= \Sigma = \text{covariance matrix in } X.$$

Therefore,

$$\text{Var}(a'X) = a'\Sigma a, \ \text{Var}(b'X) = b'\Sigma b, \text{Cov}(a'X, b'X) = a'\Sigma b = b'\Sigma a$$

since $\Sigma = \Sigma'$.

Coming back to the discussion of canonical correlation, we have the following:

$$\text{Var}(\alpha'X_1) = \alpha'V_{11}\alpha, \ \text{Var}(\beta'X_2) = \beta'V_{22}\beta,$$
$$\text{Cov}(u, w) = \alpha'V_{12}\beta = \beta'V_{21}\alpha.$$

Since our aim is to predict u by using w and vice versa, we must base the procedure on some reasonable principle. One procedure that we can adopt is to maximize a scale-free covariance between u and w so that the joint variation or joint dispersion is a maximum. This is equivalent to saying to maximize the correlation between u and w and construct the pair of variables u and w and use them as the "best" predictors to predict each other, "best" in the sense of having maximum joint variation.

$$\rho_{(u,w)} = \frac{\text{Cov}(u, w)}{\sqrt{\text{Var}(u)\text{Var}(w)}} = \frac{\alpha'V_{12}\beta}{\sqrt{(\alpha'V_{11}\alpha)(\beta'V_{22}\beta)}} = \alpha'V_{12}\beta$$

since we assumed that $\text{Var}(u) = 1 = \text{Var}(w)$. We may do this maximization with the help of Lagrangian multipliers. Taking $-\frac{1}{2}\lambda^*$ and $-\frac{1}{2}\tilde{\lambda}$ as Lagrangian multipliers, consider the function

$$\phi = \alpha'V_{12}\beta - \frac{1}{2}\lambda^*(\alpha'V_{11}\alpha - 1) - \frac{1}{2}\tilde{\lambda}(\beta'V_{22}\beta - 1).$$

Differentiating ϕ with respect to α and β and equating to null matrices, we have the following: [Instead of differentiating with respect to each and every component in α and β, we may use vector derivatives]

$$\frac{\partial \phi}{\partial \alpha} = 0, \quad \frac{\partial \phi}{\beta} = 0 \Rightarrow$$

$$V_{12}\beta - \lambda^* V_{11}\alpha = 0 \tag{4.34}$$

$$-\tilde{\lambda} V_{22}\beta + V_{21}\alpha = 0. \tag{4.35}$$

Premultiplying (4.34) by α' and (4.35) by β' and then using the fact that as per our assumption $\alpha' V_{11}\alpha = 1$ and $\beta' V_{22}\beta = 1$, we see that $\lambda^* = \tilde{\lambda} = \lambda$ (say). Thus, (4.34) and (4.35) can be rewritten as a matrix equation:

$$\begin{bmatrix} -\lambda V_{11} & V_{12} \\ V_{21} & -\lambda V_{22} \end{bmatrix} \begin{bmatrix} \alpha \\ \beta \end{bmatrix} = 0. \tag{4.36}$$

Before solving this equation, we will give below some notes about vector and matrix derivatives. Those who are familiar with these items may skip the notes and continue with the main text.

Note 4.7. Vector and matrix derivatives.

Consider the following vector of partial differential operators. Let

$$Y = \begin{bmatrix} y_1 \\ \vdots \\ y_n \end{bmatrix}, \quad \frac{\partial}{\partial Y} = \begin{bmatrix} \frac{\partial}{\partial y_1} \\ \vdots \\ \frac{\partial}{\partial y_n} \end{bmatrix}, \quad \frac{\partial}{\partial Y}[f] = \begin{bmatrix} \frac{\partial f}{\partial y_1} \\ \vdots \\ \frac{\partial f}{\partial y_n} \end{bmatrix}$$

where f is a real-valued scalar function of Y. For example,

(i) $$f_1(Y) = a_1 y_1 + \ldots + a_n y_n = a'Y, a' = (a_1, \ldots, a_n)$$

is such a function, where a is a constant vector. Here, f_1 is a linear function of Y, something like

$$2y_1 - y_2 + y_3; \quad y_1 + y_2 + \ldots + y_n; \quad y_1 + 3y_2 - y_3 + 2y_4$$

etc.

(ii) $$f_2(y_1, \ldots, y_n) = y_1^2 + y_2^2 + \ldots + y_n^2$$

which is the sum of squares or a simple quadratic form or a general quadratic form in its canonical form.

(iii) $$f_3(y_1, ..., y_n) = Y'AY, A = (a_{ij}) = A'$$

is a general quadratic form where A is a known constant matrix, which can be taken to be symmetric without loss of generality. A few basic properties that we are going to use will be listed here as lemmas.

Lemma N4.1

$$f_1 = a'Y \Rightarrow \frac{\partial f_1}{\partial Y} = a.$$

Note that the partial derivative of the linear function $a'Y = a_1 y_1 + ... + a_n y_n$, with respect to y_j gives a_j for $j = 1, ..., n$ and hence the column vector

$$\frac{\partial f_1}{\partial Y} = \frac{\partial (a'Y)}{\partial Y} = \begin{bmatrix} a_1 \\ \vdots \\ a_n \end{bmatrix} = a.$$

For example, if $a'Y = y_1 - y_2 + 2y_3$ then

$$\frac{\partial}{\partial Y}(a'Y) = \begin{bmatrix} 1 \\ -1 \\ 2 \end{bmatrix}.$$

Lemma N4.2

$$f_2 = y_1^2 + ... + y_n^2 = Y'Y \Rightarrow \frac{\partial f_2}{\partial Y} = 2Y = 2 \begin{bmatrix} y_1 \\ \vdots \\ y_n \end{bmatrix}.$$

Note that $Y'Y$ is a scalar function of Y, whereas YY' is a $n \times n$ matrix, and hence, it is a matrix function of Y. Note also that when $Y'Y$ is differentiated with respect to Y, the Y' disappears and a 2 comes in. We get a column vector because our differential operator is a column vector.

Lemma N4.3

$$f_3 = Y'AY, A = A' \Rightarrow \frac{\partial f_3}{\partial Y} = 2AY.$$

Here, it can be seen that if A is not taken as symmetric then instead of $2AY$ we will end up with $(A + A')Y$. As an illustration of f_3, we can consider

$$f_3 = 2y_1^2 + y_2^2 + y_3^2 - 2y_1y_2 + 5y_2y_3$$

$$= [y_1, y_2, y_3] \begin{bmatrix} 2 & -1 & 0 \\ -1 & 1 & \frac{5}{2} \\ 0 & \frac{5}{2} & 1 \end{bmatrix} \begin{bmatrix} y_1 \\ y_2 \\ y_3 \end{bmatrix} = Y'AY, \ A = A'$$

$$= [y_1, y_2, y_3] \begin{bmatrix} 2 & -2 & 0 \\ 0 & 1 & 5 \\ 0 & 0 & 1 \end{bmatrix} \begin{bmatrix} y_1 \\ y_2 \\ y_3 \end{bmatrix} = Y'BY, \ B \neq B'.$$

In the first representation, the matrix A is symmetric, whereas in the second representation of the same quadratic form, the matrix B is not symmetric. By straight differentiation

$$\frac{\partial f_3}{\partial y_1} = 4y_1 - 2y_2, \quad \frac{\partial f_3}{\partial y_2} = 2y_2 - 2y_1 + 5y_3, \quad \frac{\partial f_3}{\partial y_3} = 2y_3 + 5y_2$$

Therefore,

$$\frac{\partial f_3}{\partial Y} = \begin{bmatrix} \frac{\partial f_3}{\partial y_1} \\ \frac{\partial f_3}{\partial y_2} \\ \frac{\partial f_3}{\partial y_3} \end{bmatrix} = \begin{bmatrix} 4y_1 - 2y_2 \\ 2y_2 - 2y_1 + 5y_3 \\ 2y_3 + 5y_2 \end{bmatrix}$$

$$= 2 \begin{bmatrix} 2 & -1 & 0 \\ -1 & 1 & \frac{5}{2} \\ 0 & \frac{5}{2} & 1 \end{bmatrix} \begin{bmatrix} y_1 \\ y_2 \\ y_3 \end{bmatrix} = 2AY.$$

But

$$B + B' = \begin{bmatrix} 2 & -2 & 0 \\ 0 & 1 & 5 \\ 0 & 0 & 1 \end{bmatrix} + \begin{bmatrix} 2 & 0 & 0 \\ -2 & 1 & 0 \\ 0 & 5 & 1 \end{bmatrix}$$

$$= \begin{bmatrix} 4 & -2 & 0 \\ -2 & 2 & 5 \\ 0 & 5 & 2 \end{bmatrix} = 2 \begin{bmatrix} 2 & -1 & 0 \\ -1 & 1 & \frac{5}{2} \\ 0 & \frac{5}{2} & 1 \end{bmatrix} = 2A.$$

Note that when applying Lemma N4.3, write the matrix in the quadratic form as a symmetric matrix. This can be done without any loss of generality since for any square matrix B, $\frac{1}{2}(B + B')$ is a symmetric matrix. Then when operating with the partial differential operator $\frac{\partial}{\partial Y}$ on $Y'AY$, $A = A'$, the net result is to delete Y' (not Y) and premultiply by 2 or write $2AY$.

4.8.1   First Pair of Canonical Variables

Equation (4.36) has a non-null solution for the vector $\begin{pmatrix} \alpha \\ \beta \end{pmatrix}$ only if the system of linear equations is singular or only if the coefficient matrix has the determinant equal to zero. That is,

$$\begin{vmatrix} -\lambda V_{11} & V_{12} \\ V_{21} & -\lambda V_{22} \end{vmatrix} = 0 \Rightarrow \begin{matrix} |V_{21} V_{11}^{-1} V_{12} - \lambda^2 V_{22}| = 0 \\ |V_{12} V_{22}^{-1} V_{21} - \lambda^2 V_{11}| = 0. \end{matrix} \tag{4.37}$$

The determinant on the left is a polynomial in λ of degree $r + s = k$. Let $\lambda_1 \geq \lambda_2 \geq \ldots \geq \lambda_{r+s}$ be the roots of the determinantal equation is (4.37). Since $\lambda = \alpha' V_{12} \beta$, the largest value of $\alpha' V_{12} \beta$ is attained at the largest root λ_1. With this λ_1, we can compute the corresponding α and β by using the conditions $\alpha' V_{11} \alpha = 1$ and $\beta' V_{22} \beta = 1$. Let these solutions be denoted by $\alpha_{(1)}$ and $\beta_{(1)}$, respectively. Then, $u_1 = \alpha'_{(1)} X_1$ and $w_1 = \beta'_{(1)} X_2$ are the first pair of canonical variables and λ_1 is the first canonical correlation coefficient, in the sense the best predictor of $\alpha' X_1$ is $\beta'_{(1)} X_2$ and the best predictor of $\beta' X_2$ is $\alpha'_{(1)} X_1$ and vice versa in the sense that this pair $(\alpha'_{(1)} X_1, \beta'_{(1)} X_2)$ has the maximum joint dispersion in them or the maximum correlation between them.

Before proceeding with further discussion of the next canonical pair of variables and other properties, let us do one example so that the normalization through $\alpha' V_{11} \alpha = 1$ and $\beta' V_{22} \beta = 1$ will be clear.

Example 4.20. Let $X' = (x_1, x_2, x_3, x_4)$ where x_1, x_2, x_3, x_4 be real scalar random variables. Let X be partitioned into subvectors

$$X = \begin{bmatrix} x_1 \\ x_2 \\ x_3 \\ x_4 \end{bmatrix} = \begin{bmatrix} X_1 \\ X_2 \end{bmatrix}, \quad X_1 = \begin{bmatrix} x_1 \\ x_2 \end{bmatrix}, \quad X_2 = \begin{bmatrix} x_3 \\ x_4 \end{bmatrix}.$$

Suppose that we wish to predict linear functions of X_1 by using linear functions of X_2. Show that the following matrix V can represent the covariance matrix of X, and then by using this V, construct the first pair of canonical variables and the corresponding canonical correlation coefficient.

$$V = \begin{bmatrix} 1 & 1 & 1 & 1 \\ 1 & 2 & -1 & 0 \\ 1 & -1 & 6 & 4 \\ 1 & 0 & 4 & 4 \end{bmatrix}.$$

Solution 4.18. Note that V is already symmetric. Now, let us consider the leading minors of V.

$$1 > 0, \begin{vmatrix} 1 & 1 \\ 1 & 2 \end{vmatrix} = 1 > 0, \begin{vmatrix} 1 & 1 & 1 \\ 1 & 2 & -1 \\ 1 & -1 & 6 \end{vmatrix} = 1 > 0, |V| > 0.$$

Thus, V is symmetric positive definite, and hence, V can represent a non-singular covariance matrix for a 4-vector such as X. Partitioning of V, according to the partitioning of X, into submatrices leads to the following, as per our notations:

$$V = \begin{bmatrix} V_{11} & V_{12} \\ V_{21} & V_{22} \end{bmatrix}, \; V_{11} = \begin{bmatrix} 1 & 1 \\ 1 & 2 \end{bmatrix}, \; V_{12} = V_{21}' = \begin{bmatrix} 1 & 1 \\ -1 & 0 \end{bmatrix},$$

$$V_{22} = \begin{bmatrix} 6 & 4 \\ 4 & 4 \end{bmatrix}, \; V_{11}^{-1} = \begin{bmatrix} 2 & -1 \\ -1 & 1 \end{bmatrix}, \; V_{22}^{-1} = \begin{bmatrix} \frac{1}{2} & -\frac{1}{2} \\ -\frac{1}{2} & \frac{3}{4} \end{bmatrix}.$$

Solving the equation

$$|V_{21} V_{11}^{-1} V_{12} - \nu V_{22}| = 0, \; \nu = \lambda^2,$$

we have

$$\left| \begin{bmatrix} 1 & -1 \\ 1 & 0 \end{bmatrix} \begin{bmatrix} 2 & -1 \\ -1 & 1 \end{bmatrix} \begin{bmatrix} 1 & 1 \\ -1 & 0 \end{bmatrix} - \nu \begin{bmatrix} 6 & 4 \\ 4 & 4 \end{bmatrix} \right| = 0 \Rightarrow$$

$$\left| \begin{bmatrix} 5 & 3 \\ 3 & 2 \end{bmatrix} - \nu \begin{bmatrix} 6 & 4 \\ 4 & 4 \end{bmatrix} \right| = 0 \Rightarrow$$

$$8\nu^2 - 8\nu + 1 = 0 \Rightarrow$$

$$\nu = \frac{1}{2} \pm \frac{1}{4}\sqrt{2}.$$

Let

$$\nu_1 = \frac{1}{2} + \frac{\sqrt{2}}{4} \text{ and } \nu_2 = \frac{1}{2} - \frac{\sqrt{2}}{4}.$$

For ν_1, consider the equation

$$[V_{21} V_{11}^{-1} V_{12} - \nu_1 V_{22}]\beta = O \Rightarrow$$

$$\left[\begin{bmatrix} 5 & 3 \\ 3 & 2 \end{bmatrix} - \begin{bmatrix} \frac{1}{2} + \frac{\sqrt{2}}{4} \end{bmatrix} \begin{bmatrix} 6 & 4 \\ 4 & 4 \end{bmatrix} \right] \begin{bmatrix} \beta_1 \\ \beta_2 \end{bmatrix} = \begin{bmatrix} 0 \\ 0 \end{bmatrix} \Rightarrow$$

$$\begin{bmatrix} 2 - \frac{3}{2}\sqrt{2} & 1 - \sqrt{2} \\ 1 - \sqrt{2} & -\sqrt{2} \end{bmatrix} \begin{bmatrix} \beta_1 \\ \beta_2 \end{bmatrix} = \begin{bmatrix} 0 \\ 0 \end{bmatrix} \Rightarrow$$

$$b_1 = \begin{bmatrix} -2 - \sqrt{2} \\ 1 \end{bmatrix}$$

is one solution. Let us normalize through the relation $1 = b_1' V_{22} b_1$, that is,

$$b_1' V_{22} b_1 = [-2 - \sqrt{2}, 1] \begin{bmatrix} 6 & 4 \\ 4 & 4 \end{bmatrix} \begin{bmatrix} -2 - \sqrt{2} \\ 1 \end{bmatrix}$$

$$= 4(2 + \sqrt{2})^2 = c, \text{ say.}$$

Then,

$$\beta_{(1)} = \frac{b_1}{+\sqrt{c}} = \frac{1}{2(2 + \sqrt{2})} \begin{bmatrix} -(2 + \sqrt{2}) \\ 1 \end{bmatrix} = \begin{bmatrix} -\frac{1}{2} \\ \frac{1}{2(2+\sqrt{2})} \end{bmatrix}.$$

Therefore,

$$w_1 = \beta_{(1)}' X_2 = -\frac{1}{2} x_3 + \frac{x_4}{2(2 + \sqrt{2})}.$$

Now, consider

$$(V_{12} V_{22}^{-1} V_{21} - \nu_1 V_{11}) \alpha = O \Rightarrow$$

$$\begin{bmatrix} 1 & 1 \\ -1 & 0 \end{bmatrix} \begin{bmatrix} \frac{1}{2} & -\frac{1}{2} \\ -\frac{1}{2} & quadfrac34 \end{bmatrix} \begin{bmatrix} 1 & -1 \\ 1 & 0 \end{bmatrix} - \begin{bmatrix} \frac{1}{2} + \frac{\sqrt{2}}{4} \end{bmatrix} \begin{bmatrix} 1 & 1 \\ 1 & 2 \end{bmatrix} \begin{bmatrix} \alpha_1 \\ \alpha_2 \end{bmatrix} = \begin{bmatrix} 0 \\ 0 \end{bmatrix} \Rightarrow$$

$$\begin{bmatrix} 1 + \sqrt{2} & 2 + \sqrt{2} \\ 2 + \sqrt{2} & 2(1 + \sqrt{2}) \end{bmatrix} \begin{bmatrix} \alpha_1 \\ \alpha_2 \end{bmatrix} = \begin{bmatrix} 0 \\ 0 \end{bmatrix} \Rightarrow$$

$$a_1 = \begin{bmatrix} -\sqrt{2} \\ 1 \end{bmatrix}$$

is one solution. Consider the normalization through $a_1' V_{11} a_1 = 1$. Consider

$$a_1' V_{11} a_1 = [-\sqrt{2}, 1] \begin{bmatrix} 1 & 1 \\ 1 & 2 \end{bmatrix} \begin{bmatrix} -\sqrt{2} \\ 1 \end{bmatrix} = 4 - 2\sqrt{2}.$$

Then

$$\alpha_{(1)} = \frac{1}{\sqrt{(4 - 2\sqrt{2})}} \begin{bmatrix} -\sqrt{2} \\ 1 \end{bmatrix}$$

and then

$$u_1 = \alpha_{(1)}' X_1 = \frac{1}{\sqrt{(4 - 2\sqrt{2})}} [-\sqrt{2} x_1 + x_2].$$

Thus, (u_1, w_1) is the first pair of canonical variables and the corresponding canonical correlation is $\lambda_1 = \frac{1}{2} + \frac{\sqrt{2}}{4}$.

From (4.37), we can also have

$$|V_{21} V_{11}^{-1} V_{12} - \lambda^2 V_{22}| = 0 \Rightarrow |V_{22}^{-\frac{1}{2}} V_{21} V_{11}^{-1} V_{12} V_{22}^{-\frac{1}{2}} - \lambda^2 I| = 0$$

and

$$|V_{12} V_{22}^{-1} V_{21} - \lambda^2 V_{11}| = 0 \Rightarrow |V_{11}^{-\frac{1}{2}} V_{12} V_{22}^{-1} V_{21} V_{11}^{-\frac{1}{2}} - \lambda^2 I| = 0.$$

Thus, $\nu = \lambda^2$ is an eigenvalue of the matrix

$$V_{11}^{-\frac{1}{2}} V_{12} V_{22}^{-1} V_{21} V_{11}^{-\frac{1}{2}} \text{ as well as the matrix } V_{22}^{-\frac{1}{2}} V_{21} V_{11}^{-1} V_{12} V_{22}^{-\frac{1}{2}}. \tag{4.38}$$

The matrices in (4.38) are called *canonical correlation matrices* corresponding to the square of the multiple correlation coefficient in equation (4.19) of Section 4.3. The coefficients $\alpha_{(1)}, \beta_{(1)}$ in the first pair of canonical variables also satisfy the relations

$$\alpha'_{(1)} V_{11} \alpha_{(1)} = 1, \ \beta'_{(1)} V_{22} \beta_{(1)} = 1.$$

4.8.2 Second an Subsequent Pair of Canonical Variables

When looking for a second pair of canonical variables, we can require the second pair (u, w) should be such that u is non-correlated with u_1 and w_1, and w is non-correlated with u_1 and w_1, and at the same time, the second pair has maximum correlation between them, subject to the normalization, as before, $u' V_{11} u = 1$ and $w' V_{22} w = 1$. At each stage, we can continue requiring the pair to be non-correlated with all previous pairs as well as having maximum correlation within the pair, and normalized. Using more Lagrangian multipliers, the function to be maximized at the $(r + 1)$th stage is:

$$\phi_{r+1} = \alpha' V_{12} \beta - \frac{1}{2} \lambda^* (\alpha' V_{11} \alpha - 1) - \frac{1}{2} \tilde{\lambda} (\beta' V_{22} \beta - 1)$$
$$+ \sum_{i=1}^{r} \rho_i \alpha' V_{11} \alpha_{(i)} + \sum_{i=1}^{r} s_i \beta' V_{22} \beta_{(i)},$$

where $-\frac{1}{2}\lambda^*, -\frac{1}{2}\tilde{\lambda}, \rho_i, s_i, i = 1, ..., r$ are Lagrangian multipliers. Then, differentiating partially with respect to $\alpha, \beta, \lambda^*, \tilde{\lambda}, \rho_i, s_i, i = 1, ..., r$ and equating to null vectors, we have

$$\frac{\partial \phi_{r+1}}{\partial \alpha} = O, \quad \frac{\partial \phi_{r+1}}{\partial \beta} = O \Rightarrow$$

$$V_{12}\beta - \lambda^* V_{11}\alpha + \sum_{i=1}^{r} \rho_i V_{11}\alpha_{(i)} = O \tag{4.39}$$

$$V_{21}\alpha - \tilde{\lambda} V_{22}\beta + \sum_{i=1}^{r} s_i V_{22}\beta_{(i)} = O. \tag{4.40}$$

Premultiply (4.39) by $\alpha'_{(j)}$ and (4.40) by $\beta'_{(j)}$, then we have

$$0 = \rho_j \alpha'_{(j)} V_{11}\alpha_{(j)} = \rho_j$$

and

$$0 = s_j \beta'_{(j)} V_{22}\beta_{(j)} = s_j.$$

Thus, the Equations (4.39) and (4.40) go back to the original Equations (4.34) and (4.35) and thereby to (4.36) and (4.37). Thus, the second largest root in (4.35), namely λ_2, leads to the second pair of canonical variables

$$(u_2, w_2), u_2 = \alpha'_{(2)} X_1, w_2 = \beta'_{(2)} X_2,$$

and the next largest root λ_3 leads to the third pair (u_3, w_3), and so on. There will be r such pairs if $r \le s$ or s such pairs if $s \le r$.

Example 4.21. Construct the other pairs of canonical variables in Example 4.20.

Solution 4.19. Consider the second largest eigenvalue $\nu_2 = \frac{1}{2} - \frac{\sqrt{2}}{4}$. Consider the equation

$$[V_{21} V_{11}^{-1} V_{12} - \nu_2 V_{22}]\beta = O \Rightarrow$$

$$\left[\begin{bmatrix} 5 & 3 \\ 3 & 2 \end{bmatrix} - \begin{bmatrix} \frac{1}{2} - \frac{\sqrt{2}}{4} \end{bmatrix} \begin{bmatrix} 6 & 4 \\ 4 & 4 \end{bmatrix} \right] \begin{bmatrix} \beta_1 \\ \beta_2 \end{bmatrix} = \begin{bmatrix} 0 \\ 0 \end{bmatrix} \Rightarrow$$

$$b_2 = \begin{bmatrix} \sqrt{2} - 2 \\ 1 \end{bmatrix}$$

is a solution. Let us normalize through $\beta' V_{22}\beta = 1$. Consider

$$b_2' V_{22} b_2 = [\sqrt{2} - 2] \begin{bmatrix} 6 & 4 \\ 4 & 4 \end{bmatrix} \begin{bmatrix} \sqrt{2} - 2 \\ 1 \end{bmatrix} = [2(2 - \sqrt{2})]^2 \Rightarrow$$

$$\beta_{(2)} = \frac{1}{2(2 - \sqrt{2})} \begin{bmatrix} \sqrt{2} - 2 \\ 1 \end{bmatrix} = \begin{bmatrix} -\frac{1}{2} \\ \frac{1}{2(2-\sqrt{2})} \end{bmatrix}.$$

Now, look for $\alpha_{(2)}$. Take the equation

$$[V_{12}V_{22}^{-1}V_{21} - \nu_2 V_{11}]\alpha = O \Rightarrow$$

$$\begin{bmatrix} 1 & 1 \\ -1 & 0 \end{bmatrix}\begin{bmatrix} \frac{1}{2} & -\frac{1}{2} \\ -\frac{1}{2} & \frac{3}{4} \end{bmatrix}\begin{bmatrix} 1 & -1 \\ 1 & 0 \end{bmatrix}$$

$$-\begin{bmatrix} \frac{1}{2} - \frac{\sqrt{2}}{4} \end{bmatrix}\begin{bmatrix} 1 & 1 \\ 1 & 2 \end{bmatrix}\begin{bmatrix} \alpha_1 \\ \alpha_2 \end{bmatrix} = O \Rightarrow$$

$$\begin{bmatrix} -1+\sqrt{2} & -2+\sqrt{2} \\ -2+\sqrt{2} & -2+2\sqrt{2} \end{bmatrix}\begin{bmatrix} \alpha_1 \\ \alpha_2 \end{bmatrix} = \begin{bmatrix} 0 \\ 0 \end{bmatrix} \Rightarrow$$

$$a_2 = \begin{bmatrix} \sqrt{2} \\ 1 \end{bmatrix}$$

is one such solution. Let us normalize it. For this, let us consider

$$a_2' V_{11} a_2 = [\sqrt{2}, 1]\begin{bmatrix} 1 & 1 \\ 1 & 2 \end{bmatrix}\begin{bmatrix} \sqrt{2} \\ 1 \end{bmatrix} = 4 + 2\sqrt{2}.$$

Therefore,

$$\alpha_{(2)} = \frac{1}{\sqrt{(4+2\sqrt{2})}}\begin{bmatrix} \sqrt{2} \\ 1 \end{bmatrix}.$$

Then, the second pair of canonical variables

$$(u_2, w_2) = (\alpha_{(2)}' X_1, \beta_{(2)}' X_2),$$

where

$$u_2 = \frac{1}{\sqrt{(4+2\sqrt{2})}}(\sqrt{2}x_1 + x_2), \quad w_2 = -\frac{1}{2}x_3 + \frac{1}{2(2-\sqrt{2})}x_4.$$

One can easily verify the following facts:

$$\mathrm{Cov}(u_1, u_2) = 0, \mathrm{Cov}(u_1, w_2) = 0,$$
$$\mathrm{Cov}(w_1, w_2) = 0, \ \mathrm{Cov}(w_1, u_2) = 0$$

(This is left to the student). Here, w_1 is the best predictor of u_1 and w_2 is the best predictor of u_2. There is no more canonical pair of variables in our example since $r = 2 = s$.

4.8.1. For the covariance matrix in Exercise 4.6.2, compute the first pair of canonical variables (u_2, w_1) and the first canonical correlation coefficient ρ_1.

4.8.2. For the Exercise 4.8.1, compute the second pair of canonical variables (u_2, w_2) and the second canonical correlation coefficient ρ_2.

4.8.3. For the Exercises in 4.8.1 and 4.8.2, show that the following pairs of variables are non-correlated: $(u_1, u_2), (w_1, w_2), (u_1, w_2), (u_2, w_1)$ and that the correlations between u_1 and w_1 give ρ_1 and that between U_2 and w_2 give ρ_2.

4.9 Estimation of the Regression Function

In the earlier sections, we looked at prediction functions and "best predictors," best in the minimum mean square sense. We found that in this case, the "best" predictor of a dependent variable y at preassigned values of the variables $x_1, ..., x_k$ would be the conditional expectation of y given $x_1, ..., x_k$. For computing this conditional expectation, so that we have a good predictor function, we need at least the conditional distribution of y given $x_1, ..., x_k$. If the joint distribution of y and $x_1, ..., x_k$ is available that is also fine, but in a joint distribution, there is more information than what we need. In most of the practical situations, we may have some idea about the conditional expectation, but we may not know the conditional distribution. In this case, we cannot explicitly evaluate the regression function analytically. Hence, we will consider various scenarios in this section.

The problem that we will consider in this section is the situation that it is known that there exists the conditional expectation, but we do not know the conditional distribution but a general idea is available about the nature of the conditional expectation or the regression function such as that the regression function is linear in the regressed variables or a polynomial type or some such known functions. Then, the procedure is to collect observations on the variables, estimate the regression function, and then use this estimated regression function to estimate the value of the dependent variable y at preassigned values of $x_1, ..., x_k$, the regressed variables. We will start with $k = 1$, namely one variable to be predicted by preassigning one independent variable. Let us start with the linear regression function: Here, "linear" means linear in the regressed variable or the so-called independent variable.

4.9.1 Linear Regression of y on x

This means that the regression of the real scalar random variable y on the real scalar random variable x is believed to be of the form:

$$E(y|x) = \beta_0 + \beta_1 x \qquad (4.41)$$

where β_0 and β_1 are unknown because the conditional distribution is not available and the only information available is that the conditional expectation, or the regression of y on x, is linear of the type (4.41). In order to estimate β_0 and β_1, we will start with the model

$$y = a + bx \qquad (4.42)$$

and try to take observations on the pair (y, x). Let there be n data points $(y_1, x_1),...,$ (y_n, x_n). Then as per the model in (4.42) when $x = x_j$, the estimated value, as per the model (4.42), is $a + bx_j$ but this estimated value need not be equal to the observed y_j of y. Hence, the error in estimating y by using $a + bx_j$ is the following, denoting it by e_j:

$$e_j = y_j - (a + bx_j). \qquad (4.43)$$

When the model is written, the following conventions are used. We write the model as $y = a + bx$ or $y_j = a + bx_j + e_j, j = 1, ..., n$. The error e_j can be positive for some j, negative for some other j, and zero for some other j. Then, trying to minimize the errors by minimizing the sum of the errors is not a proper procedure to be used because the sum of e_j's may be zero, but this does not mean that there is no error. Here, the negative and positive values may sum up to zero. Hence, a proper quantity to be used is a measure of mathematical "distance" between y_j and $a + bx_j$ or a norm in e_j's. The sum of squares of the errors, namely $\sum_{j=1}^{n} e_j^2$, is a squared norm or the square of the Euclidean distance between y_j's and $a + bx_j$'s. For a real quantity, if the square is zero, then the quantity itself is zero; if the square attains a minimum, then we can say that the distance between the observed y and the estimated y, estimated by the model $y = a + bx$, is minimized. For the model in (4.42),

$$\sum_{j=1}^{n} e_j^2 = \sum_{j=1}^{n} (y_j - a - bx_j)^2. \qquad (4.44)$$

If the model corresponding to (4.43) is a general function $g(a_1, .., a_r, x_1, .., x_k)$, for some g where $a_1, ..., a_r$ are unknown constants in the model, $x_1, .., x_k$ are the regressed variables, then the jth observation on $(x_1, ..., x_k)$ is denoted by $(x_{1j}, x_{2j}, ..., x_{kj}), j = 1, ..., n$ and then the error sum of squares can be written as

$$\sum_{j=1}^{n} e_j^2 = \sum_{j=1}^{n} [y_j - g(a_1, ..., a_r, x_{1j}, ..., x_{kj})]^2. \qquad (4.45)$$

The unknown quantities in (4.45) are $a_1, ..., a_r$. If the unknown quantities $a_1, ..., a_r$, which are also called the *parameters in the model*, are estimated by minimizing the error sum of squares then the method is known as the *method of least squares*, introduced originally by Gauss. For our simple model in (4.44), there are two parameters a, b and the minimization is to be done with respect to a and b. Observe that in (4.45) the functional form of g on $x_1, .., x_k$ is unimportant because some observations on these variables only appear in (4.45), but the nature of the parameters in (4.45) is important or (4.45) is a function of the unknown quantities $a_1, ..., a_r$. Thus, when we say that a model is linear, it means linear in the unknowns, namely linear in the parameters. If we say that the model is a quadratic model, then it is a quadratic function in the unknown parameters. Make a note of the subtle difference. When we say that we have a linear regression, then we are talking about the linearity in the regressed variables where the coefficients are known quantities, available from the conditional distribution. But when we set up a model to estimate a regression function, then the unknown quantities in the model are the parameters to be estimated, and hence, the degrees go with the degrees of the parameters.

Let us look at the minimization of the sum of squares of the errors in (4.44). This can be done either by using purely algebraic procedures or by using calculus. If we use calculus, then we differentiate partially with respect to the parameters a and b and equate to zero and solve the resulting equations:

$$\frac{\partial}{\partial a}[\sum_{j=1}^{n} e_j^2] = 0, \quad \frac{\partial}{\partial b}[\sum_{j=1}^{n} e_j^2] = 0 \Rightarrow$$

$$-2\sum_{j=1}^{n}(y_j - a - bx_j) = 0, \Rightarrow \sum_{j=1}^{n}(y_j - \hat{a} - \hat{b}x_j) = 0. \tag{4.46}$$

$$-2\sum_{j=1}^{n}x_j(y_j - a - bx_j) = 0 \Rightarrow \sum_{j=1}^{n}x_j(y_j - \hat{a} - \hat{b}x_j) = 0. \tag{4.47}$$

Equations (4.46) and (4.47) do not hold universally for all values of the parameters a and b. They hold only at the critical points. The critical points are denoted by $\hat{a}$ and $\hat{b}$, respectively. Taking the sum over all terms and over j, one has the following:

$$\sum_{j=1}^{n} y_j - n\hat{a} - \hat{b}\sum_{j=1}^{n} x_j = 0 \text{ and } \sum_{j=1}^{n} x_j y_j - \hat{a}\sum_{j=1}^{n} x_j - \hat{b}\sum_{j=1}^{n} x_j^2 = 0. \tag{4.48}$$

In order to simplify the equations in (4.48), we will use the following convenient notations. [These are also standard notations].

$$\bar{y} = \sum_{j=1}^{n} \frac{y_j}{n}, \quad \bar{x} = \sum_{j=1}^{n} \frac{x_j}{n}, \quad s_x^2 = \sum_{j=1}^{n} \frac{(x_j - \bar{x})^2}{n}$$

$$s_y^2 = \sum_{j=1}^{n} \frac{(y_j - \bar{y})^2}{n}, \quad s_{xy} = \sum_{j=1}^{n} \frac{(x_j - \bar{x})(y_j - \bar{y})}{n} = s_{yx}.$$

These are the sample means, sample variances, and the sample covariance. Under these notations, the first equation in (4.48) reduces to the following, by dividing by n.

$$\bar{y} - \hat{a} - \hat{b}\bar{x} = 0.$$

Substituting for $\hat{a}$ in the second equation in (2.57), and dividing by n, we have

$$\sum_{j=1}^{n} \frac{x_j y_j}{n} - [\bar{y} - \hat{b}\bar{x}]\bar{x} - \hat{b} \sum_{j=1}^{n} \frac{x_j^2}{n} = 0.$$

Therefore,

$$\hat{b} = \frac{\sum_{j=1}^{n} \frac{x_j y_j}{n} - (\bar{x})(\bar{y})}{\sum_{j=1}^{n} \frac{x_j^2}{n} - (\bar{x})^2}$$

$$= \frac{s_{xy}}{s_x^2} = \frac{\sum_{j=1}^{n} (x_j - \bar{x})(y_j - \bar{y})}{\sum_{j=1}^{n} (x_j - \bar{x})^2} \quad \text{and} \quad \hat{a} = \bar{y} - \hat{b}\bar{x}. \tag{4.49}$$

The simplifications are done by using the following formulae. For any set of real numbers $(x_1, y_1), \ldots, (x_n, y_n)$

$$\sum_{j=1}^{n} (x_j - \bar{x}) = 0, \quad \sum_{j=1}^{n} (y_j - \bar{y}) = 0, \quad \sum_{j=1}^{n} (x_j - \bar{x})^2 = \sum_{j=1}^{n} x_j^2 - n(\bar{x})^2$$

$$\sum_{j=1}^{n} (y_j - \bar{y})^2 = \sum_{j=1}^{n} y_j^2 - n(\bar{y})^2, \quad \sum_{j=1}^{n} (x_j - \bar{x})(y_j - \bar{y}) = \sum_{j=1}^{n} (x_j y_j) - n(\bar{x}\bar{y}).$$

When we used calculus to obtain (4.48), we have noted that there is only one critical point $(\hat{a}, \hat{b})$ for our problem under consideration. Does this point $(\hat{a}, \hat{b})$ in the parameter space $\Omega = \{(a, b) | -\infty < a < \infty, -\infty < b < \infty\}$ correspond to a maximum or minimum? Note that since (4.44) is the sum of squares of real numbers, the maximum for $\sum_{j=1}^{n} e_j^2$ for all a and b is at $+\infty$. Hence, the only critical point $(\hat{a}, \hat{b})$ in fact corresponds to a minimum. Thus, our estimated regression function, under the assumption that the regression was of the form $E(y|x) = \beta_0 + \beta_1 x$, and then estimating it by using the method of least squares, is:

$$y = \hat{a} + \hat{b}x, \hat{a} = \bar{y} - \hat{b}\bar{x}, \hat{b} = \frac{s_{xy}}{s_x^2}. \qquad (4.50)$$

Hence, (4.50) is to be used to estimate the values of y at preassigned values of x.

Example 4.22. The growth of a certain plant y, growth measured in terms of its height in centimeters, is guessed to have a linear regression on x the time measured in the units of weeks. Here, $x = 0$ means the starting of the observations, $x = 1$ means at the end of the first week, $x = 2$ means at the end of the second week, and so on. The following observations are made.

x	0	1	2	3	4
y	2.0	4.5	5.5	7.5	10.5

Estimate the regression function and then estimate y at $x = 3.5$, $x = 7$.

Solution 4.20. As per our notation, $n = 5$,

$$\bar{x} = \frac{0+1+2+3+4}{5} = 2, \bar{y} = \frac{2.0+4.5+5.5+7.5+10.5}{5} = 6.$$

If you are using a computer with a built-in or loaded program for "regression," then by feeding the observations $(x, y) = (0, 2), (1, 4.5), (2, 5.5), (3, 7.5), (4, 10.5)$ the estimated linear function is readily printed. The same thing is achieved if you have a programmable calculator. If nothing is available to you readily and if you have to do the problem by hand, then for doing the computations fast, form the following table:

y	x	$y - \bar{y}$	$x - \bar{x}$	$(x - \bar{x})^2$	$(y - \bar{y})(x - \bar{x})$	$\hat{y}$	$y - \hat{y}$	$(y - \hat{y})^2$
2	0	-4.0	-2	4	8.0	2	0.0	0.00
4.5	1	-1.5	-1	1	1.5	4	0.5	0.25
5.5	2	-0.5	0	0	0	6	-0.5	0.25
7.5	3	1.5	1	1	1.5	7	0.5	0.25
10.5	4	4.5	2	4	9.0	10	0.5	0.25
				10	20.0			1.00

Therefore

$$\hat{b} = \frac{\sum_{j=1}^{n}(x_j - \bar{x})(y_j - \bar{y})}{\sum_{j=1}^{n}(x_j - \bar{x})^2} = \frac{20}{10} = 2$$

and

$$\hat{a} = \bar{y} - \hat{b}\bar{x} = 6 - (2)(2) = 2.$$

Note 4.8. Do not round up the estimated values. If an estimated value is 2.1, leave it as 2.1 and do not round it up to 2. Similarly, when averages are taken, then also do not round up the values of $\bar{x}$ and $\bar{y}$. If you are filling up sacks with coconuts and if 4020 coconuts are filled in 100 sacks, then the average number in each sack is $\frac{4020}{100} = 40.2$ and it is not 40 because $40 \times 100 \neq 4020$.

Hence, the estimated regression function is

$$y = 2 + 2x.$$

Then, the estimated value $\hat{y}$ of y at $x = 3.5$ is given by $\hat{y} = 2 + 2(3.5) = 9$.

Note 4.9. The point $x = 7$ is far outside the range of the data points. In the observations, the range of x is only $0 \leq x \leq 4$, whereas we are asked to estimate y at $x = 7$ which is far out from 4. The estimated function $y = 2 + 2x$ can be used for this purpose; if we are 100% sure that the underlying regression is the same function $2 + 2x$ for all values of x, then we can use $x = 7$ and obtain the estimated y as $\hat{y} = 2 + 2(7) = 16$. If there is any doubt as to the nature of the function at $x = 7$, then y should not be estimated at a point for x which is far out of the observational range for x.

Note 4.10. In the above table for carrying out computations, the last 3 columns, namely $\hat{y}$, $y - \hat{y}$, $(y - \hat{y})^2$, are constructed for making some other calculations later on. The least square minimum is given by the last column sum, and in this example, it is 1.

Before proceeding further, let us introduce some more technical terms. If we apply calculus on the error sum of squares under the general model in (4.54), we obtain the following equations for evaluating the critical points.

$$\frac{\partial}{\partial a_1}[\sum_{j=1}^{n} e_j^2] = 0, \dots, \frac{\partial}{\partial a_r}[\sum_{j=1}^{n} e_j^2] = 0. \tag{4.51}$$

These minimizing equations in (4.51) under the least square analysis are often called *normal equations*. This is another awkward technical term in statistics, and it has nothing to do with normality or Gaussian distribution or it does not mean that other equations have some abnormalities. The nature of the equations in (4.51) will depend upon the nature of the involvement of the parameters $a_1, \dots, a_r$ with the regressed variables $x_1, \dots, x_k$.

Note 4.11. What should be the size of n or how many observations are needed to carry out the estimation process? If $g(a_1, \dots, a_r, x_1, \dots, x_k)$ is a linear function of the form

$$a_0 + a_1 x_1 + \dots + a_k x_k,$$

then there are $k + 1$ parameters $a_0, \dots, a_k$ and (4.51) leads to $k + 1$ linear equations in $k + 1$ parameters. This means, in order to estimate $a_0, \dots, a_k$, we need at least $k + 1$ observation points if the system of linear equations is consistent or has at least one solution. Hence, in this case, $n \geq k + 1$. In a nonlinear situation, the number of observations needed may be plenty more in order to estimate all parameters successfully. Hence, the minimum condition needed on n is $n \geq k + 1$ where $k + 1$ is the total number of parameters in a model and the model is linear in these $k + 1$ parameters. Since it is not a mathematical problem of solving a system of linear equations, the practical advice is to take n as large as feasible under the given situation so that a wide range of observational points will be involved in the model.

Note 4.12. As a reasonable criterion for estimating y based on $g(a_1, ..., a_r, x_1, ..., x_k)$, we used the error sum of squares, namely

$$\sum_{j=1}^{n} e_j^2 = \sum_{j=1}^{n} [y_j - g(a_1, ..., a_r, x_{1j}, ..., x_{kj})]^2. \tag{4.52}$$

This is the square of a mathematical distance between y_j and g. We could have used other measures of distance between y_j and g, for example,

$$\sum_{j=1}^{n} |y_j - g(a_1, ..., a_r, x_{1j}, ..., x_{kj})|. \tag{4.53}$$

Then, minimization of this distance and estimation of the parameters $a_1, ..., a_r$ thereby estimating the function g is a valid and reasonable procedure. Then, why did we choose the squared distance as in (4.52) rather than any other distance such as the one in (4.53)? This is done only for mathematical convenience. For example, if we try to use calculus, then differentiation of (4.53) will be rather difficult compared to (4.52).

Exercises 4.9.

4.9.1. The weight gain y in grams of an experimental cow under a certain diet x in kilograms is the following:

x	0	1	2	3	4	5
y	2	6	10	18	30	40

(i): Fit the model $y = a + bx$ to these data; (ii): compute the least square minimum; (iii): estimate the weight gain at $x = 3.5$, $x = 2.6$.

4.9.2. For the same data in Exercise 4.9.1, fit the model $y = a + bx + cx^2$, $c \neq 0$. (i): Compute the least square minimum; (ii): by comparing the least square minima in Exercises 4.9.1 and 4.9.2, check to see which model can be taken as a better fit to the data.

4.9.2 *Linear Regression of y on $x_1, ..., x_k$*

Suppose that the regression of y on $x_1, ..., x_k$ is suspected to be linear in $x_1, ..., x_k$, that is of the form

$$E(y|x_1, ..., x_k) = \beta_0 + \beta_1 x_1 + ... + \beta_k x_k.$$

Suppose that n data points $(y_j, x_{1j}, ..., x_{kj}), j = 1, ..., n$ are available. Since the regression is suspected to be linear and if we want to estimate the regression function, then we will start with the model

$$y = a_0 + a_1 x_1 + ... + a_k x_k.$$

Hence, at the jth data point if the error in estimating y is denoted by e_j, then

$$e_j = y_j - [a_0 + a_1 x_{1j} + ... + a_k x_{kj}], j = 1, ..., n$$

and the error sum of squares is then

$$\sum_{j=1}^{n} e_j^2 = \sum_{j=1}^{n} [y_j - a_0 - a_1 x_{1j} - ... - a_k x_{kj}]^2. \tag{4.54}$$

We obtain the normal equations by differentiating partially with respect to $a_0, a_1, ..., a_k$ and equating to zeros. That is,

$$\frac{\partial}{\partial a_0} [\sum_{j=1}^{n} e_j^2] = 0 \Rightarrow -2[\sum_{j=1}^{n} y_j - n\hat{a}_0 - \hat{a}_1 \sum_{j=1}^{n} x_{1j} - ... - \hat{a}_k \sum_{j=1}^{n} a_{kj}] = 0.$$

We can delete -2 and divide by n. Then,

$$\bar{y} = \hat{a}_0 + \hat{a}_1 \bar{x}_1 + ... + \hat{a}_k \bar{x}_k \text{ or } \hat{a}_0 = \bar{y} - \hat{a}_1 \bar{x}_1 - ... - \hat{a}_k \bar{x}_k \tag{4.55}$$

where

$$\bar{y} = \sum_{j=1}^{n} \frac{y_j}{n}, \bar{x}_i = \sum_{j=1}^{n} \frac{x_{ij}}{n}, i = 1, ..., k$$

and $\hat{a}_0, \hat{a}_i, i = 1, ..., k$ indicate the critical point $(\hat{a}_0, ..., \hat{a}_k)$ or the point at which the equations hold. Differentiating with respect to $a_i, i = 1, ..., k$, we have

$$\frac{\partial}{\partial a_i}[\sum_{j=1}^{n} e_j^2] = 0 \Rightarrow -2\sum_{j=1}^{n} x_{ij}[y_j - \hat{a}_0 - \hat{a}_1 x_{1j} - \dots - \hat{a}_k x_{kj}] = 0$$

$$\Rightarrow \sum_{j=1}^{n} x_{ij} y_j = \hat{a}_0 \sum_{j=1}^{n} x_{ij} + \hat{a}_1 \sum_{j=1}^{n} x_{ij} x_{1j} + \dots + \hat{a}_k \sum_{j=1}^{n} x_{ij} x_{kj}. \qquad (4.56)$$

Substituting the value of $\hat{a}_0$ from (4.55) into (4.56) and rearranging and then dividing by n, we have the following:

$$s_{iy} = \hat{a}_1 s_{1i} + \hat{a}_2 s_{2i} + \dots + \hat{a}_k s_{ki}, i = 1, \dots, k \qquad (4.57)$$

where

$$s_{ij} = \sum_{k=1}^{n} \frac{(x_{ik} - \bar{x}_i)(x_{jk} - \bar{x}_j)}{n} = s_{ji},$$

$$s_{iy} = s_{yi} = \sum_{k=1}^{n} \frac{(y_k - \bar{y})(x_{ik} - \bar{x}_i)}{n}$$

or the corresponding sample variances and covariances. If we do not wish to substitute for $\hat{a}_0$ from (4.55) into (4.56), then we may solve (4.55) and (4.56) together to obtain a solution for $(\hat{a}_0, \hat{a}_1, \dots, \hat{a}_k)$. But from (4.57), we get only $(\hat{a}_1, \dots, \hat{a}_k)$, and then, this has to be used in (4.55) to obtain $\hat{a}_0$. From (4.57), we have the following matrix equation:

$$s_{1y} = \hat{a}_1 s_{11} + \hat{a}_2 s_{12} + \dots + \hat{a}_k s_{1k}$$
$$s_{2y} = \hat{a}_1 s_{21} + \hat{a}_2 s_{22} + \dots + \hat{a}_k s_{2k}$$

$$\vdots$$

$$s_{ky} = \hat{a}_1 s_{k1} + \hat{a}_2 s_{k2} + \dots + \hat{a}_k s_{kk} \text{ or}$$
$$S_y = S\hat{a} \qquad (4.58)$$

where

$$S_y = \begin{bmatrix} s_{1y} \\ \vdots \\ s_{ky} \end{bmatrix}, \hat{a} = \begin{bmatrix} \hat{a}_1 \\ \vdots \\ \hat{a}_k \end{bmatrix}, S = (s_{ij}),$$

$$s_{ij} = \sum_{k=1}^{n} \frac{(x_{ik} - \bar{x}_i)(x_{jk} - \bar{x}_j)}{n} = s_{ji}.$$

From (4.58),

$$\hat{a} = \begin{bmatrix} \hat{a}_1 \\ \vdots \\ \hat{a}_k \end{bmatrix} = S^{-1} S_y, \text{ for } |S| \neq 0. \tag{4.59}$$

From (4.55)

$$\hat{a}_0 = \bar{y} - \hat{a}'\bar{x} = \bar{y} - S'_y S^{-1} \bar{x}, \bar{x} = \begin{bmatrix} \bar{x}_1 \\ \vdots \\ \bar{x}_k \end{bmatrix}. \tag{4.60}$$

Note 4.13. When observations on $(x_1, ..., x_k)$ are involved, even if we take extreme care sometimes near singularity may occur in S. In general, one has to solve the system of linear equations in (4.58) for which many standard methods are available whether the coefficient matrix, in our case S, is non-singular or not. In a regression-type model, as in our case above, the points $(x_{1j}, x_{2j}, ..., x_{kj})$, $j = 1, ..., n$ are preassigned and hence, while preassigning, make sure that data points for $(x_1, ..., x_k)$, which are linear functions of other points which are already included, are not taken as a new data point. If linear functions are taken, then this will result in S being singular.

Example 4.23. In a feeding experiment on cows, it is suspected that the increase in weight y has a linear regression on the amount of green fodder x_1 and the amount of marketed cattle feed x_2 consumed. The following observations are available, and all observations on x_1 and x_2 are in kilograms and the observations on y are in grams.

x_1	1	1.5	2	1	2.5	1
x_2	2	1.5	1	1.5	2	4
y	5	5	6	4.5	7.5	8

Construct the estimating function and then estimate y at the points $(i): (x_1, x_2) = (1, 0), (1, 3), (5, 8)$.

Solution 4.21. As per our notation $n = 6$,

$$\bar{x}_1 = \frac{(1.0 + 1.5 + 2.0 + 1.0 + 2.5 + 1.0)}{6} = 1.5,$$

$$\bar{x}_2 = \frac{(2.0 + 1.5 + 1.0 + 1.5 + 2.0 + 4.0)}{6} = 2,$$

$$\bar{y} = \frac{(5.0 + 5.0 + 6.0 + 4.5 + 7.5 + 8.0)}{6} = 6.$$

Again, if we are using a computer or programmable calculator, then regression problems are there in the computer and the results are instantly available by feeding in the data. For the calculations by hand, the following table will be handy:

y	x_1	x_2	$y - \bar{y}$	$x_1 - \bar{x}_1$	$x_2 - \bar{x}_2$	$(y - \bar{y})(x_1 - \bar{x}_1)$
5.0	1.0	2.0	-1	-0.5	0	0.5
5.0	1.5	1.5	-1	0	-0.5	0
6.0	2.0	1.0	0	0.5	-1	0
4.5	1.0	1.5	-1.5	-0.5	-0.5	0.75
7.5	2.5	2.0	1.5	1.0	0	1.5
8.0	1.0	4.0	2.0	-0.5	2.0	-1.0
						1.75

$(y - \bar{y})(x_2 - \bar{x}_2)$	$(x_1 - \bar{x}_1)^2$	$(x_2 - \bar{x}_2)^2$	$(x_1 - \bar{x}_1)(x_2 - \bar{x}_2)$
0	0.25	0	0
0.5	0	0.25	0
0	0.25	1.0	-0.5
0.75	0.25	0.25	0.25
0	1.0	0	0
4.0	0.25	4.0	-1
5.25	2.0	5.5	-1.25

The equations corresponding to (4.58), without the dividing factor $n = 6$, are the following:

$$1.75 = 2\hat{a}_1 - 1.25\hat{a}_2$$
$$5.25 = -1.25\hat{a}_1 + 5.5\hat{a}_2 \Rightarrow \hat{a}_1 \approx 1.72,\ \hat{a}_2 \approx 1.34.$$

Then,

$$\hat{a}_0 = \bar{y} - \hat{a}_1\bar{x}_1 - \hat{a}_2\bar{x}_2$$
$$\approx 6 - (1.72)(1.5) - (1.34)(2) = 0.74.$$

Hence, the estimated function is given by

$$y = 0.74 + 1.725x_1 + 1.34x_2.$$

The estimated value of y at $(x_1, x_2) = (1, 3)$ is $\hat{y} = 6.48$. The point $(x_1, x_2) = (5, 8)$ is too far out of the observational range, and hence, we may estimate y only if we are sure that the conditional expectation is linear for all possible (x_1, x_2). If the regression is sure to hold for all (x_1, x_2), then the estimated y at $(x_1, x_2) = (5, 8)$ is

$$\hat{y} = 0.74 + (1.72)(5) + (1.34)(8) = 20.06.$$

For example at

$$(x_1, x_2) = (1, 2), \ \hat{y} = 5.14, \ \text{at } (x_1, x_2) = (1.5, 1.5), \ \hat{y} = 5.33;$$
$$\text{at } (x_1, x_2) = (2, 1), \ \hat{y} = 5.52; \ \text{at } (x_1, x_2) = (1, 4), \ \hat{y} = 7.82.$$

Hence, we can construct the following table:

y	$\hat{y}$	$y - \hat{y}$	$(y - \bar{y})^2$
5	5.14	−0.14	0.096
5	5.33	−0.33	0.1089
6	5.52	0.48	0.2304
4.5	4.4	−0.03	0.0009
7.5	7.72	−0.22	0.0484
8	7.82	0.18	0.0324
			0.4406

An estimate of the error sum of squares as well as the least square minimum is 0.4406 in this model.

Exercises 4.9.

4.9.3. If the yield y of corn in a test plot is expected to be a linear function of $x_1 =$ amount of water supplied, in addition to the normal rain, and $x_2 =$ amount of organic fertilizer (cow dung), in addition to the fertility of the soil, the following is the data available:

x_1	0	0	1	2	1.5	2.5	3
x_2	0	1	1	1.5	2	2	3
ry	2	2	5	8	7	9	10

(*i*): Fit a linear model $y = a_0 + a_1 x_1 + a_2 x_2$ by the method of least squares.

(*ii*): Estimate y at the points $(x_1, x_2) = (90.5, 1.5), (3.5, 2.5)$.

(*iii*): Compute the least square minimum.

Chapter 5
Optimal Regression Designs

5.1 Introduction

The general linear model, discussed in Chapter 4, will be re-examined here in order to bring out some more interesting points and to talk about estimability of linear functions of the parameters, Gauss–Markov setup, and related matters.

5.2 The General Linear Model

In linear regression, we have the model,

$$y_i = \theta_0 + \theta_1 x_{i1} + \ldots + \theta_{p-1} x_{i,p-1} + e_i, \quad i = 1, \ldots, n \tag{5.1}$$

where $y_1, \ldots, y_n$ are the response or dependent variables, $\theta_0, \ldots, \theta_{p-1}$ are unknown parameters, x_{ij} are the control or independent or regression variables, $e_1, \ldots, e_n$ are random errors. It is assumed that $E(e_i) = 0$, $E(e_i e_j) = 0$ $i \neq j$, $E(e_i)^2 = \sigma^2$, that is,

$$E(e) = O, \quad \text{Cov}(e) = \sigma^2 I_n \tag{5.2}$$

where Cov(.) is the covariance matrix and I_n is the unit or identity matrix of order n. In vector, matrix notation, we have

$$Y = X\theta + e \tag{5.3}$$

This chapter is based on the lectures of Professor Stratis Kounias of the University of Athens, Greece, and the University of Cyprus.

where $Y = (y_1, \ldots, y_n)'$, $X = (x_{ij}), i = 1, \ldots n$, $j = 0, 1, \ldots, p-1$, usually $x_{i0} = 1$, $\theta = (\theta_0, \theta_1, \ldots, \theta_{p-1})'$, $e = (e_1, \ldots, e_n)'$, then $E(Y) = X\theta$. The $n \times p$ matrix X is called the model or design matrix.

If the distribution of the errors is not known, then the method of least squares is employed for the estimation of the parameters, that is,

$$\min_\theta S(\theta) = \min_\theta (Y - X\theta)'(Y - X\theta). \tag{5.4}$$

Differentiating $S(\theta)$ with respect to the vector θ, we obtain the normal equations,

$$X'X\hat{\theta} = X'Y. \tag{5.5}$$

If rank of X is p, that is, the model matrix X is of full rank, then (5.5) gives

$$\hat{\theta} = (X'X)^{-1}X'Y. \tag{5.6}$$

The expected value and the covariance matrix of the estimator $\hat{\theta}$, from (5.2) and (5.6), are,

$$E(\hat{\theta}) = O, \quad \text{Cov}(\hat{\theta}) = \sigma^2 (X'X)^{-1}. \tag{5.7}$$

Definition 5.1. The $p \times p$ matrix $\frac{X'X}{\sigma^2}$ is called *precision matrix*, its inverse is the covariance matrix of the estimator of the unknown parameters.

5.3   Estimable Linear Functions

Definition 5.2. If $c_0, c_1, \ldots, c_{p-1}$ are constants, then the linear function $\sum_{i=0}^{p-1} c_i \theta_i = c'\theta$ is estimable if there exists a linear function $\sum_{i=1}^{n} a_i y_i = a'Y$ of the observations, so that

$$E(a'Y) = c'\theta \tag{5.8}$$

for all values of $\theta \in R^p$.

Proposition 5.1. *If in the model (5.1), $\theta \in R^p$, then the linear function $c'\theta$ is estimable iff c is a linear function of the rows of the model matrix X, that is, $c = X'a$ for some $a \in R^n$.*

Proof From (5.3) and (5.8), we have

$$E(a'Y) = a'X\theta = c'\theta \tag{5.9}$$

for all values of $\theta \in R^p$, hence $a'\theta = c'$. QED

Remark 5.1. If the model (5.3) is of full rank, that is, rank of X is p, then all linear functions of the parameters are estimable.

Remark 5.2. If rank of X is less than p, then only the linear functions $a'X\theta$ are estimable, for all $a \in R^n$, in this case

$$E(X\hat{\theta}) = X\theta, \text{Cov}(X\hat{\theta}) = \sigma^2 X(X'X)^- X$$

where A^- denotes the generalized inverse of the $m \times n$ matrix A.

From (5.3) and (5.6), the fitted values of Y are,

$$\hat{Y} = X\hat{\theta} = X(X'X)^- X'Y = H(X)Y \tag{5.10}$$

where $H(X) = X(X'X)^- X'$ is the hat matrix of X, that is, $H(X)$ is idempotent and is the $n \times n$ orthogonal projection onto the linear space of the columns of X. The diagonal elements h_{ii} of the hat matrix are called *leverages*.

The residuals are defined as the difference of the observed minus the fitted values, that is,

$$\hat{e}_i = y_i - \hat{y}_i, \quad i = 1, 2, \ldots, n$$

or

$$\hat{e} = Y - \hat{Y} = (I_n - H(X))Y. \tag{5.11}$$

Then, from (5.11) and (5.2),

$$E(\hat{e}) = 0, \quad \text{Cov}(\hat{e}) = \sigma^2(I_n - H(X)).$$

The predicted value of $y(x)$ at a point $x = (1, x_1, \ldots, x_{p-1})'$, where $x' = X'a$, is

$$\hat{y}(x) = \hat{\theta}_0 + \hat{\theta}_1 x_1 + \ldots + \hat{\theta}_{p-1} x_{p-1} = x'\hat{\theta}.$$

Then

$$E(\hat{y}(x)) = x'\theta, \text{Var}(\hat{Y}(x)) = \sigma^2 x'(X'X)^- x.$$

The residual sum of squares, denoted as *RSS*, is the minimum value of $S(\theta)$ given in (5.4), that is,

$$RSS = \sum_{i=1}^n \hat{e}_i^2 = \hat{e}'\hat{e}$$

or

$$RSS = \min_\theta S(\theta) = S(\hat{\theta}) = Y'(I_n - H(X))Y. \tag{5.12}$$

5.3.1   The Gauss–Markov Theorem

The conditions $E(e_i) = 0$, $E(e_i e_j) = 0$, $i \neq j$, $E(e_i)^2 = \sigma^2$, given in (5.2), are called G–M conditions. In most applications, we are interested in estimates of some linear function $c'\theta$ of the parameters. We shall confine ourselves to linear functions $a'Y$ of the observations which also are unbiased estimators of $c'\theta$.

The Gauss–Markov theorem states that among all these linear unbiased estimators of $c'\theta$, the least squares estimator $c'\hat{\theta} = c'(X'X)^{-1}X'Y$ has the smallest variance. Such an estimator is called a *best linear unbiased estimator (BLUE)*

Theorem 5.1. (Gauss–Markov) *If $Y = X\theta + e$ and $(X'X)\hat{\theta} = X'Y$ then, under G–M conditions, $c'\hat{\theta}$ is BLUE of $c'\theta$.*

Proof. Since $E(a'Y) = c'\theta$ for all $\theta \in R^p$, then $a'X = c'$. Now

$$\mathrm{Var}(c'Y) = \sigma^2 c'c$$

and

$$\mathrm{Var}(a'\hat{\theta}) = \sigma^2 a'X(X'X)^- X'a.$$

Therefore

$$\mathrm{Var}(c'Y) - \mathrm{Var}(a'\hat{\theta}) = \sigma^2 c'(I_n - X(X'X)^- X')c. \tag{5.13}$$

But the $n \times n$ matrix $(I_n - X(X'X)^- X')$ is idempotent, which means that it is non-negative definite and the result follows.   QED

If we want to estimate $k > 1$ linear functions of the unknown parameters, say $C'\theta$, and $A'Y$ is an unbiased estimator of $C'\theta$, then $AX = C$, where C is $k \times p$ matrix and A is $k \times n$ matrix. Therefore, taking the difference of the covariance matrices, we have

$$\mathrm{Cov}(C'Y) - \mathrm{Cov}(A'\hat{\theta})) = \sigma^2 C'(I_n - X(X'X)^- X')C$$

which is nonnegative definite.

In the model (5.3) under the G–M conditions, then the fitted values and the residuals are also uncorrelated since from (5.10), (5.11) we have,

$$\mathrm{cov}(\hat{e}, \hat{Y}) = \sigma^2 H(X)(I_n - H(X)) = O. \tag{5.14}$$

5.4   Correlated Errors

If in the model (5.1) the errors do not have equal variances or if they are correlated, that is, their covariance matrix $\mathrm{Cov}(e)$ is

$$\mathrm{Cov}(e) = E(ee') = \sigma^2 V, \tag{5.15}$$

where V is a known symmetric, positive definite matrix or order n, then the method of ordinary least squares (OLS) defined in (5.4) does not give the best estimates of the unknown parameters θ. Instead the method of weighted (generalized) least squares (GLS) is applied, that is,

$$\min_{\theta} S(\theta) = \min_{\theta} (Y - X\theta)' V^{-1} (Y - X\theta). \tag{5.16}$$

Then the estimated parameters, denoted by $\hat{\theta}_\omega$ are,

$$X'V^{-1}X\hat{\theta}_\omega = X'V^{-1}Y. \tag{5.17}$$

The expected value and the covariance matrix of the estimator $\hat{\theta}_\omega$ are

$$E(\hat{\theta}_\omega) = \theta, \text{Cov}(\hat{\theta}_\omega) = \sigma^2 (X'V^{-1}X)^{-1}. \tag{5.18}$$

Actually to go from OLS to GLS, in the case of correlated errors, it is enough to set $X'V^{-1}X$ instead of $X'X$, $X'V^{-1}Y$ instead of $X'Y$ and $Y'V^{-1}Y$ instead of $Y'Y$.

5.4.1 Estimation of Some of the Parameters

If we are interested only in some of the parameters, say $\theta_0, \theta_1, \ldots, \theta_{r-1}$, then the design matrix is written $X = (X_1, X_2)$ and $\theta' = (b_1', b_2')$, with $b_1 = (\theta_0, \theta_1, \ldots, \theta_{r-1})'$, $b_2 = (\theta_r, \ldots, \theta_{p-1})'$. The columns of X_1 correspond to the parameters $(\theta_0, \theta_1, \ldots, \theta_{r-1})$, and the columns of X_2 correspond to the parameters $(\theta_r, \ldots, \theta_{p-1})$.

Proposition 5.2. *The parameter estimation of $(\theta_0, \theta_1, \ldots, \theta_{r-1})$ and their covariance matrix are given by the following relations:*

$$X_1'(I_n - H(X_2))X_1\hat{b}_1 = X_1'(I_n - H(X_2))Y \tag{5.19}$$

$$\text{Cov}(\hat{b}_1) = \sigma^2 Q^{-1} \tag{5.20}$$

where $Q = X_1'(I_n - H(X_2))X_1$ and $H(X) = X(X'X)^{-1}X'$ is the $n \times n$ orthogonal projection matrix onto the linear space of the columns of X.

Proof. The method of least squares gives

$$X'X\hat{\theta} = X'Y \Rightarrow \begin{bmatrix} X_1'X_1 & X_1'X_2 \\ X_2'X_1 & X_2'X_2 \end{bmatrix} \begin{bmatrix} \hat{b}_1 \\ \hat{b}_2 \end{bmatrix} = \begin{bmatrix} X_1' \\ X_2' \end{bmatrix} Y \Rightarrow$$
$$X_1'X_1\hat{b}_1 + X_1X_2\hat{b}_2 = X_1'Y$$
$$X_2'X_1\hat{b}_1 + X_2'X_2\hat{b}_2 = X_2'Y.$$

Multiplying on the left both sides of the last equation by $X_1' X_2 (X_2' X_2)^-$, we get

$$X_1' X_1 \hat{b}_1 + X_1' X_2 \hat{b}_2 = X_1' Y$$
$$X_1' H(X_2) X_1 \hat{b}_1 + X_1' X_2 \hat{b}_2 = X_1' H(X_2) Y.$$

By subtraction we get relation (5.19). To find the covariance matrix of $\hat{b}_1$, go to (5.20) and then

$$Q \text{Cov}(\hat{b}_1) Q = \sigma^2 Q \Rightarrow \text{Cov}(\hat{b}_1) = \sigma^2 Q^{-1}$$

where $Q = X_1' (I_n - H(X_2)) X_1$.   QED

5.5 Other Distance Measures

The method of least squares was first applied in 1805 by Andrien Marie Legendre (1752–1833) and in (1809) by Carl Friedrich Gauss. Other distance measures are:

$$L_1 - \text{norm } S_1(y, \hat{y}) = \min\left(\sum_{i=1}^{n} |y_i - E(y_i)|\right)$$

$$L_\infty - \text{norm } S_\infty(y, \hat{y}) = \min\left(\max_{i=1,\dots,n} |y_i - E(y_i)|\right).$$

$$L_2 - \text{norm } S_2(y, \hat{y}) = \min\left(\sum_{i=1}^{n} (y_i - E(y_i))^2\right)$$

$$S_p - \text{norm } S_p(y, \hat{y}) = \min\left(\sum_{i=1}^{n} |y_i - E(y_i)|^p\right)$$

where $y_i - E(y_i) = y_i - \sum_{j=0}^{p-1} \theta_i x_{ij}$.

In the above criteria, all observations have equal weight, because the errors are uncorrelated with mean zero and constant variance.

5.6 Means and Covariances of Random Vectors

If $Y = (y_1, \dots, y_n)'$ and y_i, $i = 1, \dots, n$ are real random variables, then Y is a random vector.

Definition 5.3.

(i) The mean value of Y is $E(Y) = (E(y_1), \dots, E(y_n))'$.

(ii) The covariance matrix of Y is $\mathrm{Cov}(Y) = \mathrm{cov}(y_i, y_j)$, $i, j = 1, \ldots, n$ is an $n \times n$ matrix, that is,

$$\mathrm{Cov}(Y) = E(Y - E(Y))(Y - E(Y))'.$$

(iii) If Y and Z are random vectors of order n and m, their covariance matrix is $n \times m$,

$$\mathrm{Cov}(Y, Z) = E(Y - E(Y))(Z - E(Z))'.$$

5.6.1 Quadratic Forms and Generalized Inverses

Definition 5.4. If $A = (a_{ij})$ is a $n \times n$ symmetric matrix with real elements $a_{ij} \in R$ and $X = (x_1, \ldots, x_n)'$ is a real vector, then the function,

$$X'AX = \sum_{i=1}^{n} a_{ij} x_i x_j$$

is called a quadratic form with respect to X.

Definition 5.5. The $n \times n$ matrix A is positive definite (PD(n)), denoted $A > 0$, if $A = A'$ and $X'AX > 0$ for all $X \neq O$. If $A = A'$ and $X'AX \geq 0$ for all $X \neq O$ and $X'AX = 0$ for some $X \neq O$, then A is called nonnegative definite (NND(n)), denoted $A \geq 0$.

Definition 5.6. The $n \times n$ matrix A is idempotent if $A^2 = A$.

Definition 5.7. If A is $n \times m$ real matrix, then the $m \times n$ matrix A^- is called generalized inverse of A if the following relation is satisfied,

$$AA^-A = A.$$

If rank $A = n$ its inverse A^{-1} is unique and $A^- = A^{-1}$. If rank $A < n$ or if A is $m \times n$ with $m \neq n$, then there exist more than one generalized inverse of A.

Definition 5.8. If A, B are NND(n) and $A - B > 0$, then we say that $A > B$, similarly $A \geq B$ is defined.

5.6.2 Distribution of Quadratic Forms

Proposition 5.3. *If $Y \sim N(\mu, \sigma^2 I_n)$, B is $n \times n$ real symmetric matrix, then*

$$Y'BY \sim \sigma^2 \chi^2_{r,\delta} \tag{5.21}$$

where $r = rank\ B, \delta = \mu' B \mu / \sigma^2$ if and only if B is idempotent of rank r, then rank B = trace B

δ is called the non-centrality parameter of the chi-square distribution, and r denotes the degrees of freedom. If $\delta = 0$, then we write χ_r^2.

Proposition 5.4. *If $Y \sim N(\mu, V)$ and (i) V is $PD(n)$, then*

$$(Y - \mu)'V^{-1}(Y - \mu) \sim \chi_n^2$$
$$Y'V^{-1}Y \sim \chi_{n,\delta}^2, \quad \delta = \mu'V^{-1}\mu.$$

(ii) If V is $NND(n)$ with rank $V = r < n$, then

$$(Y - \mu)'V^-(Y - \mu) \sim \chi_r^2.$$

Theorem 5.2. *(Cochran) If*

(i) $Y \sim N(\mu, \sigma^2 I_n)$
(ii) $A_1, A_2, ..., A_k$ are $n \times n$ matrices, then
(a) If $A_1 + A_2 + ... + A_k = I_n$ and one of the following equivalent relations holds

> *(1) $A_1, ..., A_k$ idempotent*
> *(2) $A_i A_j = O, i \neq j$*
> *(3) rank $A_1 + ... + $ rank $A_k = n$*

then the random variables $(Y'AY)/\sigma^2$ are independently distributed as χ_{r_i,δ_i}^2 with $r_i = rank\ A_i$, $\delta_i = (\mu' A_i \mu)/\sigma^2, i = 1, ..., k$.
(b) If $(Y'A_iY)/\sigma^2, i = 1, ..., k$ are independently distributed as χ_{r_i,δ_i}^2 and $r_1 + ... + r_k = n$ then $A_1, ..., A_k$ are idempotent with $r_i = rank\ A_i$ and $\delta_i = (\mu' A_i \mu)/\sigma^2$, $i = 1, ..., k$.

As an application consider the sum of squares due to the model, denoted by $RegSS$ (repression sum of squares), defined by

$$RegSS = \sum_{i=1}^{n}(\hat{Y} - \bar{Y})^2 = \hat{Y}'\bar{Y} - n\bar{Y}^2 = Y'(H(X) - \frac{1}{n}I_n)Y.$$

Observe that the $n \times n$ symmetric matrices

$$\frac{1}{n}J_n, \quad H(X) - \frac{1}{n}J_n, \quad (I_n - H(X))$$

are idempotent and the sum is I_n, hence $n\bar{Y}^2$, RSS, $RegSS$ are independently distributed and

$$n\bar{Y}^2 \sim \sigma^2 \chi^2_{1,\delta_1}, \quad RegSS \sim \sigma^2 \chi^2_{p-1,\delta_2}, \quad RSS \sim \sigma^2 \chi^2_{n-p}.$$

If $TSS = \sum_{i=1}^{n}(Y_i - \bar{Y})^2$ is the total (corrected) sum of squares, then

$$TSS = RegSS + RSS.$$

5.6.3 Normal Errors

If the model (5.3) is correct and the errors have a normal distribution, that is, $e \sim N(O, \sigma^2 I_n)$, then

$$X\hat{\theta} \sim N(X\theta, \sigma^2 H(X)) \tag{5.22}$$

independently distributed of RSS which follows a chi-square distribution with $n - p$ degrees of freedom, that is,

$$RSS \sim \sigma^2 \chi^2_{n-p}.$$

A $100(1 - \alpha)\%$ confidence region for θ is

$$(\theta - \hat{\theta})'X'X(\theta - \hat{\theta}) < ps^2 F_{p,n-p,1-\alpha}. \tag{5.23}$$

As a function of θ this is an ellipsoid, and the content of this ellipsoid is proportional to $|X'X|^{-\frac{1}{2}}$.

5.6.4 Rank and Nullity of a Matrix

If $B = (b_{ij})$ is an $n \times m$ matrix, then range B is the linear space generated by the columns of B, that is,

$$\text{range} \quad B = \{BX : X \in R^m\}$$

the dimension of this linear space is the rank B. If C is $m \times q$, then range $BC \subseteq B$ hence rank $BC \leq$ rank B.

The null space of B is the linear space orthogonal to the rows of B, that is,

$$\text{null space} \quad B = \{X \in R^n : BX = O\}.$$

The dimension of this space is called the nullity of B. From the definition, it follows that

(i) $\{\text{range } B\}^{\perp} = \text{null space } B'$

(ii) $\text{range } B + \text{null space } B' = R^n$

(iii) $\text{range } BB' + \text{null space } BB' = R^n$

(iv) $\text{null space } B' = \text{null space } BB' \Rightarrow$

(v) $\text{range } B = \text{range } BB'.$

Hence and the rank of $B =$ the rank of BB' and equals the nullity of B' and it is $= n$-rank of B.

Exercises 5.

5.1. Let $y_1, \ldots, y_m$ and $y_{m+1}, \ldots, y_{m+n}$ be two sets of observations and we fit the model $y_i = \alpha_1 + \beta x_i + e_i$ to the first set and the model $y_i = \alpha_2 + \beta x_i + e_i$ to the second set of observations. Assume that the errors e_i, $i = 1, \ldots, m+n$ are independently distributed with zero mean values and constant variance σ^2. Find the least squares estimator of $\alpha_1, \alpha_2, \beta$, and σ^2. How do you proceed if the variances in the first and second set are σ_1^2 and σ_2^2 and $\sigma_2^2 = c\sigma_1^2$ with c known?

5.2. Show that in the simple linear model $y_i = \theta_0 + x_i\theta_1 + e_i, i = 1, \ldots, n$, where $x_i \neq x_j$ for some $i \neq j, x = (x_1, \ldots, x_n)'$, we have

(i) $\text{Var}(\hat{\theta}_1) = \sigma^2 / \Sigma_{i=1}^n (x_i - \bar{x})^2,$

(ii) $\text{Var}(\hat{\theta}_0) = \sigma^2 \sum_{i=1}^n x_i^2 / (n \sum_{i=1}^n (x_i - \bar{x})^2),$

(iii) $\text{Cov}(\hat{\theta}_0, \hat{\theta}_1) = -\sigma^2 \bar{x} / \sum_{i=1}^n (x_i - \bar{x})^2,$

(iv) $\hat{y}(z) = \hat{\theta}_0 + \hat{\theta}_1 z = \bar{y} + \hat{\theta}_1 (z - \bar{x})$ for any z and

(v) $E(\hat{y}(z)) = \theta_0 + \theta_1 z, \text{Var}(\hat{y}(z)) = \sigma^2 \left(\frac{1}{n} + \frac{(z - \bar{x})^2}{\sum_{i=1}^n (x_i - \bar{x})^2} \right).$

5.3. (i) If a column X_j of the model matrix X is linearly dependent on the other columns of X, show that the corresponding parameter θ_{X_j} is not estimable. Show also that if X_j is not linearly dependent on the other columns of X, then θ_{X_j} is estimable. (ii) In the simple linear model, if $x_1 = x_2 = \ldots = x_{p-1} = x$, then are θ_0, θ_1 estimable?. What type of function of the parameters is estimable?

5.4. Let the model be $y_i = \theta_0 + \theta_1 x_{i1} + \theta_2 x_{i2} + e_i$, $i = 1, 2, 3, 4$ and the four rows of the model matrix X are $(3, 4, 2), (2, 7, -3), (-4, 1, -9), (6, 5, 7)$. (i) Are the 3 parameters of the model estimable? (ii) Is $\theta_2 + \theta_3$ estimable? (iii) Give a linear function of the parameters which is estimable.

5.5. If A is idempotent of order n, then show that (i) the eigenvalues of A are 0 or 1 and (ii) Rank A = trace A.

5.6. If A is $m \times n$ then show that: (i) (AA^-) and (A^-A) are idempotent; (i) rank (A) = rank (AA^-) = rank (A^-A); (iii) rank $(A) \leq$ rank (A^-); (iv) If A is $n \times n$ symmetric matrix and a, b are $n \times 1$ vectors in the range of A, that is, $a = Ax, b = Ay$ for some $n \times 1$ vectors x, y then $a'A^-b$ remains invariant for all A^-; (v) The hat (projection) matrix $H(X) = X(X'X)^{-1}X'$ remains invariant for all generalized inverses $(X'X)^-$.

5.7. If V is $NND(n)$ and X is $n \times k$, then show that *range* $X'VX$ = *range* $X'V$.

5.8. If $H(X) = (h_{ij})$ is the hat matrix of X, show that $H(X)$ and $I_n - H(X)$ are idempotent and hence $NND(n)$ and

(i)
$$h_{ij}^2 \leq h_{ii}h_{jj}$$

(ii)
$$|h_{ii}| \leq 1$$

(iii)
$$h_{ii} = \sum_{j=1}^n h_{ij}^2,$$

(iv)
$$h_{ij}^2 \leq h_{ii}(1 - h_{ii}) \leq \tfrac{1}{4} \text{ for all } i \neq j$$

(v) If X is $n \times p$, $p < n$ and rank $X = p$, then $\sum_{i=1}^n h_{ii} = p$.

5.9. Show that a real symmetric A is $PD(n)$ if and only if all its eigenvalues are real and positive.

5.10. If L, M, G are $m \times n, r \times n, n \times n$ matrices and Y is an $n \times 1$ random vector, then show that

(i) $E(LY) = LE(Y)$,
(ii) The covariance matrix $\text{Cov}(LY) = L\text{Cov}(Y)L'$,
(iii) The covariance matrix $\text{Cov}(LY, MY) = L\text{Cov}(Y)M'$,
(iv) $E(Y'GY) = \text{trace } (G\Sigma) + \mu'G\mu$, where $\mu = E(Y), \Sigma = \text{Cov}(Y)$.

5.11. If $A = RLR'$, where A, R are $n \times n$ and L is diagonal $(l_1, ...l_n)$, then show that a generalized inverse of A is $A^- = RL^-R'$ where L^- has diagonal elements $1/l_i$ when $l_i \neq 0$ and 0 when $l_i = 0$.

5.12. Let A be a symmetric $k \times k$ matrix, then show that

$$A \in NND(n) \Rightarrow \text{trace}(AB) \geq 0 \text{ for all } B \in NND(k) \text{ and conversely}$$
$$A \in PD(k) \Rightarrow \text{trace}(AB) > 0 \text{ for all } 0 \neq B \in NND(k) \text{ and conversely.}$$

5.13. If Y is a $n \times 1$ random vector with mean μ and covariance matrix V, then show that $(Y - \mu) \in range\ V$ with probability 1.

5.14. If A and B are $PD(n)$, then show that

$$aA^{-1} + (1 - a)B^{-1} \geq [aA + (1 - a)B^{-1}, 0 < a < 1,$$

equality holds only if $A = B$. This means that the inverses of $PD(n)$ are convex functions.

5.15. Establish the following results for block matrices:

(i)
$$M = \begin{bmatrix} A & B \\ C & D \end{bmatrix} \Rightarrow M^{-1} = \begin{bmatrix} A^{-1} + A^{-1}BH^{-1}CA^{-1} & -A^{-1}BH^{-1} \\ -H^{-1}CA^{-1} & H^{-1} \end{bmatrix}$$

where $H = D - CA^{-1}B$. For $K = A - BD^{-12}C$,

(ii)
$$M^{-1} = \begin{bmatrix} K^{-1} & -K^{-1}BD^{-1} \\ -D^{-1}CK^{-1} & D^{-1} + D^{-1}CK^{-1}BD^{-1} \end{bmatrix}.$$

(iii)
$$|M| = |A|\,|D - CA^{-1}B| = |A - BD^{-1}C|\,|D|,$$

where $|M|$ denotes the determinant of M.

5.16. Show that in $PD(n)$, all the diagonal square block matrices have positive determinants.

5.17. Establish the Cauchy–Swartz inequality,

$$(a'b)^2 \leq (a'M^{-1}a)(b'Mb)$$

with equality only when $a = cMb$ where c is a constant. Hence show that for the diagonal elements m^{ii}, m_{ii} of M^{-1}, M we have $1 \leq m^{ii}m_{ii}$.

5.18. If A, B are $PD(n)$ and $A \geq B > 0$ then show that $0 < A^{-1} \leq B^{-1}$ and conversely.

5.19. If A and B are $PD(n)$ and $0 < a < 1$, then show that

$$|aA + (1 - a)B| \geq |A|^a|B|^{1-a},$$

where $|A|$ denotes the determinant of A. That is, $\ln|A|$ is concave.

5.20. (Hadamard's inequality) (i) If the positive definite matrix M is such that

$$M = \begin{bmatrix} A & B \\ B' & D \end{bmatrix} \text{ then show that } |M| \leq |A|\,|D| \leq \{\prod_{i=1}^{m} a_{ii}\}\{\prod_{j=1}^{n} b_{jj}\}$$

where A is $m \times m$ and B is $n \times n$. (ii) If $A = (a_{ij})$ is $m \times m$ real matrix, then show that

$$|AA'| = |A|^2 \leq \prod_{i=1}^{m} \sum_{j=1}^{m} a_{ij}^2.$$

5.21. If we know the distribution of the errors, then the method of maximum likelihood gives estimates of the parameters which are asymptotically unbiased with the minimum variance. If the model is $y_i = \alpha + \beta x_i + e_i$ and the G–M conditions hold, find the maximum likelihood estimators of α, β and σ^2 when the errors have density:

(i) $$f(e_i) = \frac{1}{\sigma\sqrt{2\pi}} \exp(-\frac{(e_i-\mu_i)^2}{2\sigma^2}), \quad -\infty < e_i < \infty, \quad (N(\mu, \sigma^2))$$

(ii) $$f(e_i) = \frac{1}{2\theta} \exp(-|e_i - \mu_i|/\theta), \quad -\infty < e_i < \infty \text{ (Laplace)}$$

(iii) $$f(e_i) = \frac{1}{2\alpha}, \quad -\alpha < e_i - \mu_i < \alpha, \quad \text{(uniform), with } \mu_i = \alpha + \beta x_i.$$

Compare these results with the least squares estimates. Give the parameter estimates when the data are: $(x_i, y_i) = (0.7, -1.5), (1.4, 0.8), (2.1, -3.2), (2.8, -0.5), (3.5, 2.4)$.

5.22. If $A_1, ..., A_k$ are $n \times n$ matrices and $A_1 + ... + A_k = I_n$, then show that the following relations are equivalent:

(i) $A_1, ..., A_k$ are idempotent;
(ii) $A_i A_j = O, i \neq j$;
(iii) rank $A_1 + ... + $ rank $A_k = n$.

5.7 Exact or Discrete Optimal Designs

The linear model (5.1) or (5.3) can be written

$$E(Y_i) = f'(x_i)\,\theta, \quad i = 1, 2, \ldots, n \tag{5.24}$$

where $f'(x_i) = (x_{i0}, x_{i1}, \ldots, x_{i(p-1)})$, usually $x_{i0} = 1$. The design space $\mathcal{X}$ is a subset of the p-Euclidean space $\Re^p$, that is $x_i \in \Re^p$.

If the point $x_i = (x_{i0}, x_{i1}, \ldots x_{i(p-1)})'$, $i = 1, 2, \ldots, k$ is taken n_i times, with

$$n_1 + n_2 + \cdots + n_k = n$$

then the design ξ is defined if the points x_i, $i = 1, \ldots, k$ and their multiplicities are specified, that is,

$$\xi = \begin{pmatrix} x_1 & x_2 & \ldots & x_k \\ n_1 & n_2 & \ldots & n_k \end{pmatrix}$$

Example 5.1. Consider the simple linear model

$$y_i = \theta_0 + x_i\theta_1 + e_i$$

with $-1 \le x_i \le 1$, hence the design region is $\mathcal{X} = [-1, 1]$.

Problem 5.1. Choose the design ξ so that

(i) $\mathrm{Var}(\hat{\theta}_1)$ is minimized.
(ii) The variance of the predicted response $\hat{y}(z)$, at a point z, is minimized.
(iii) The covariance matrix of $(\hat{\theta}_0, \hat{\theta}_1)$ is minimized.

If a design ξ is given, it is often helpful to consider the standardized variance of a random variable, or the standardized covariance matrix of a random vector $y(x)$, that is,

$$d(x, \xi) = (n/\sigma^2)\mathrm{Var}(y(x)).$$

Let us consider first two designs, each with 3 observations, design 5.1:$(-1, 0, 1)$ and design 5.2:$(-1, 1, 1)$, that is.

$$\xi_{5.1} = \begin{pmatrix} -1 & 0 & 1 \\ -1 & 1 & 1 \end{pmatrix}, \quad \xi_{5.2} = \begin{pmatrix} -1 & 1 \\ 1 & 2 \end{pmatrix}.$$

The value of the standardized variance in (i), (ii), (iii), from Exercise 5.2 is,
(i)

$$\xi_{5.1} : \bar{x} = 0, \ d(x, \xi_{5.1}) = 3/2 = 1.5,$$
$$\xi_{5.2} : \bar{x} = 1/3, d(x, \xi_{5.2}) = 3/(24/9) = 9/8.$$

So the design $\xi_{5.1}$ has smaller standardized variance and hence smaller variance.
(ii)

$$d(z, \xi_{5.1}) = (1 + \frac{3z^2}{2}), \ d(z, \xi_{5.2}) = (1 + \frac{(3z-1)^2}{8}).$$

Hence

$$d(z, \xi_{5.1}) < d(z, \xi_{5.2}) \text{ if } z \in [-1, 0, 155)$$

and

$$d(z, \xi_{5.1}) > d(z, \xi_{5.2}) \text{ if } z \in [0, 155, 1].$$

(iii) The standardized covariance matrix of a design ξ, from Exercise 5.2, is $n(M(x, \xi))^{-1}$, where

$$M(x, \xi) = X'X = \begin{pmatrix} n & n\bar{x} \\ n\bar{x} & \sum_{i=1}^{n} x_i^2 \end{pmatrix}.$$

We would like to see when $n(M(x, \xi_{5.1}))^{-1} < n(M(x, \xi_{5.2}))^{-1}$. From Exercise 5.18, it is enough to see if $(M(x, \xi_{5.1})) > (M(x, \xi_{5.2}))$, that is,

$$M(x, \xi_{5.1}) = \begin{pmatrix} 3 & 0 \\ 0 & 2 \end{pmatrix} > M(x, \xi_{5.2}) = \begin{pmatrix} 3 & 1 \\ 1 & 3 \end{pmatrix}.$$

Observe that none of the following relations holds

$$M(x, \xi_{5.1}) \leq M(x, \xi_{5.1}), \quad M(x, \xi_{5.1}) \geq M(x, \xi_{5.1}).$$

Now let us go back to Problem 5.1.

Proposition 5.5. *(i) The optimal design ξ^*, which minimizes $\mathrm{Var}(\hat{\theta}_1)$, when $-1 \leq x_i \leq 1$, is the following: If n is even, half of the observations are at $x_i = 1$ and half are at $x_i = -1$ and the minimum standardized variance is $d(x, \xi^*) = 1$. If n is odd, then $(n \pm 1)/2$ observations are at $x_i = 1$ and $(n \mp 1)/2$ are at $x_i = -1$ and then $d(x, \xi^*) = (1 - (1/n^2))^{-1}$.*

Proof. If n is even, then $d(x, \xi) = n/\sum_{i=1}^{n}(x_i - \bar{x})^2 \geq n/\sum_{i=1}^{n} x_i^2 \geq 1$ with equality only if $\bar{x} = 0$ and $x_i = \pm 1$. For odd n, the optimal design ξ^* has all its values $x_i = \pm 1$, otherwise there exists at least one point $-1 < x_j < 1$. If we take a design which has $-1 \leq x_j + \varepsilon \leq 1$ for some ε, then $\bar{x}$ becomes $\bar{x} + \varepsilon/n$ and $\sum_{i=1}^{n} x_i - n\bar{x}^2$ becomes,

$$\sum_{i=1}^{n} x_i^2 - n\bar{x}^2 + \varepsilon^2(1 - (1/n)) + 2(x_j - \bar{x})\varepsilon \leq \sum_{i=1}^{n} x_i^2 - n\bar{x}^2 + \varepsilon^2(1 - (1/n))$$

for a proper choice of ε, hence $d(x, \xi^*)$ has all its values $x_i = \pm 1$. If $x_i = 1$ for r values of x_i and $x_i = -1$ for $n - r$ values of x_i, then

$$\sum_{i=1}^{n}(x_i - \bar{x})^2 = 4r(n - r)/n \leq n - (1/n)$$

with equality only if $r = (n \pm 1)/2$. QED

Proposition 5.6. *(ii) The optimal design ξ^*, which minimizes the variance of the predicted response $\hat{y}(z)$, at a point $z \in [-1, +1]$, is to take the points x_i so that $\bar{x} = z$ and then $d(z, \xi^*) = 1$.*

Proof. $d(z, \xi^*) = 1 + (n(z - \bar{x})^2)/\sum_{i=1}^{n}(x_i - \bar{x})^2 \geq 1$, with equality only if $\bar{x} = z$. If $z = \pm 1$, take all the observations at the same point 1 or -1 and then $\hat{y}(z) = \bar{y}$ and $d(z, \xi^*) = 1$. QED

Proposition 5.7. *(iii) There does not exist an optimal design that minimizes the covariance matrix of $(\hat{\theta}_0, \hat{\theta}_1)$.*

Proof. If ξ^* is the optimal design, then $d(x, \xi^*) = n(M(x, \xi^*))^{-1}$, where

$$M(x, \xi) = X'X = \begin{pmatrix} n & n\bar{x} \\ n\bar{x} & \sum_{i=1}^{n} x_i^2 \end{pmatrix}.$$

If ξ is any other design, then $d(x, \xi^*) = n(M(x, \xi^*))^{-1} \le d(x, \xi)$ or, from Exercise 5.18,

$$M(x, \xi^*) = X'X = \begin{pmatrix} n & n\bar{x} \\ n\bar{x} & \sum_{i=1}^{n} x_i^2 \end{pmatrix} \ge M(x', \xi) = \begin{pmatrix} n & n\bar{x}' \\ n\bar{x}' & \sum_{i=1}^{n} (x')_i^2 \end{pmatrix}.$$

This means that $x_i = \pm 1$ and $\bar{x} = \bar{x}'$ which is not valid for all ξ.   QED.

5.7.1 Optimality Criteria

The optimal design for one parameter is the design ξ that minimizes the variance or equivalently the standardized variance of the parameter estimate. However, if we are interested in two or more parameters, then we have the covariance matrix of the parameter estimate. We refer either to the model given in (5.3) or to the one given in (5.24). The information matrix is,

$$M(\xi) = X'X = \sum_{i=1}^{k} n_i f(x_i) f'(x_i).$$

Definition 5.9. (Optimality in the Loewner sense): In linear models, the design ξ^*, for estimating two or more parameters, is optimal if $M(\xi^*) \ge M(\xi) \ge 0$ for any other competing design ξ.

Since this rarely happens, we are looking for other optimality criteria, which have good properties. The most used criteria of optimality are the following:

D-optimality: A design ξ is D-optimal, or D-optimum, if the determinant of the standardized covariance matrix of the parameter estimates $d(\xi) = n(M(\xi))^{-1}$ is minimized, this determinant is called generalized variance. This is equivalent to maximizing the determinant $|X'X|$ or to minimizing $-\ln(|X'X|)$, that is,

$$\max_{\xi} |X'X|.$$

A-optimality: An A-optimal design minimizes the total variance or the trace of the covariance matrix of the parameter estimates, that is the trace $M^{-1}(\xi)$, equivalent to minimizing the trace of $d(\xi)$ over all competing designs ξ, that is:

$$\min_{\xi} (\text{trace } M^{-1}).$$

G-optimality: A G-optimal design minimizes the maximum over the design space $\mathcal{X}$ of the standardized variance $d(x, \xi)$, or minimize the maximum variance of the estimated response surface, that is,

$$\min_{\xi} \max_{x \in \mathcal{X}} \text{Var}(f'(x)\hat{\theta})/\sigma^2 = \min_{\xi} \max_{x \in \mathcal{X}} f'(x)M^{-1}f(x)$$

where $x = (1, x_1, \ldots, x_{p-1})'$, $x \in \mathcal{X}$, $\mathcal{X}$ is a subset of $\Re^p$.

E-optimality: An E-optimal design minimizes the maximum variance of $c'\hat{\theta}$, over c, subject to the constraint $c'c = 1$, where c is $p \times 1$ vector, that is,

$$\min_{\xi} \max_{c} c'M^{-1}c, \ c'c = 1.$$

We have shown in Proposition 5.7 that there does not exist optimal design in the Loewner sense, that is, minimizing the covariance matrix of the two parameters $(\hat{\theta}_0, \hat{\theta}_1)$. The D-optimal design, for estimating these two parameters, maximizes the determinant of $X'X$, that is, for $-1 \le x_i \le 1$ it maximizes $n \sum_{i=1}^{n} (x_i - \bar{x})^2$.

This D-optimal design ξ^* exists and is the same as the one given in Proposition 4.5.

The relation (5.23), in Section 5.6.3, gives a $100(1 - \alpha)\%$ confidence region for θ in the linear model. As a function of θ this is an ellipsoid, and the content of this ellipsoid is proportional to $|X'X|^{-1/2}$. Hence the D-optimal design gives confidence regions with maximum content.

Another property of D-optimality is that the D-optimal design is invariant under a linear transformation of the parameters. If $Y = X\theta + e$ and takes the parameter transformation $b = A\theta$, where A is $(n + 1) \times (k + 1)$ non-singular matrix, then the model becomes $Y = Fb + e$, with $F = XA^{-1}$ and then

$$F'F = A'^{-1}X'XA^{-1} \Rightarrow |F'F| = |X'X|(|A|^{-2}).$$

5.7.2 Two-Level Factorial Designs of Resolution III

If we have q factors, $F_1, F_2, \ldots, F_q$, each at two levels, and the interactions of two or more factors are negligible, then we have a design of resolution III or of first order. For convenience, call the two levels as high and low. Now if the effect of the high level of factor F_i is a_i and the effect of the low level is b_i, with $\theta_i = a_i - b_i$, then the model is:

$$y_i = \theta_0 + \sum_{j=1}^{q} x_{ij}\theta_j + e_i, \quad i = 1, \ldots, n \tag{5.25}$$

where $x_{ij} = 1$ if the factor F_j enters the experiment at high level and $x_{ij} = -1$ if the factor F_j enters at low level, θ_0 is the general mean and $\theta_1, \ldots, \theta_q$ are the main effects.

The design matrix X is $n \times (q + 1)$ with $n \geq q + 1$, if $n = q + 1$ the design is called saturated. The first column of X is the $n \times 1$ vector of 1, and the other elements are ± 1.

Problem 5.2. Find the optimal designs for estimating the parameters $\theta_0, \theta_1, \ldots, \theta_q$ Here $M = X'X$ is $(q + 1) \times (q + 1)$ with all the diagonal elements equal to n and the non-diagonal elements $m_{ij} = x_i' x_j$, where $x_i, i = 0, \ldots, q$ is the i-th column of the design matrix X. Also $M = \sum_{j=1}^{n} R_j R_j'$, where R_j' is the j-th row of X, remains invariant if any row of X is multiplied by -1, so we can take all elements of the first column of X to be equal to 1. If a column of X, is multiplied by -1, then the corresponding row and column of M is multiplied by -1. Since $m_{ij} = n \bmod 4$ or $m_{ij} = (n + 2) \bmod 4$, then by multiplying certain columns of X by -1 we can make all elements $m_{ij} = n \bmod 4$.

Proposition 5.8. *(i) Optimal design in the Loewner sense does not exist. (ii) If $n = 0$ mod 4 and the design matrix X has orthogonal columns, that is, $X'X = nI_{q+1}$, the design is D, A, G, E-optimal. (iii) If $n = 1 \bmod 4$ and $M = X'X = (n - 1)I_{(q+1)} + J_{(q+1)}$, that is, all non-diagonal elements are equal to 1, the design is D, A, G-optimal. (iv) If $n = 2 \bmod 4$ and $M = X'X$ is a block matrix with two diagonal blocks, M_1, M_2 of sizes $s_1, s_2, |s_1 - s_2| \leq 1, s_1 + s_2 = q + 1$,*

$$M = \begin{pmatrix} M_1 & M_{12} \\ M_{12}' & M_2 \end{pmatrix}$$

M_1 and M_2 have all off-diagonal elements equal to 2, and the $s_1 \times s_2$ matrix M_{12} has all elements equal to -2, the design is D, A, G optimal. If $n = 3 \bmod 4$ and M is a block matrix with s block of sizes differing at most by 1, each off-diagonal element in every block is equal to 3 and outside the diagonal blocks each element is equal to -1, the design is D, A, G optimal.

For the construction of the designs with the above properties, work has been done by Atkinson et al. [1], Kounias et al. [5], Sathe and Shenoy [6], Shah and Sinha [7] and others.

Definition 5.10. A Hadamard matrix H_n of order n is an $n \times n$ matrix with elements ± 1, with the property $H_n' H_n = H_n H_n' = I_n$. Hadamard matrices exist for $n = 1, 2$ and $n = 0 \bmod 4$.

If $n = 0 \bmod 4$, then the D, A, G, E-optimal design of resolution III is constructed by taking $(k + 1)$ columns of an H_n, $n \geq q + 1$. For D and A-optimal designs of resolution III, with q factors, each at three levels.

ϕ-**optimality**: Let C be $NND(s)$ and a function ϕ from the nonnegative definite $s \times s$ matrices into the real line, that is,

$$\phi(C) \to \mathfrak{R}.$$

We say that C is at least as good as the $NND(s)$ D if $\phi(C) \geq \phi(D)$.

Definition 5.11. A function ϕ defined on the $NND(s)$, with $\phi(C) \to \mathfrak{R}$ is called information function if, (i) $C \geq D$, then $\phi(C) \geq \phi(D)$; (ii) ϕ is concave, that is, $\phi((1-a)C + aD) \geq (1-a)\phi(C) + a\phi(D)$ for all $a \in (0,1)$, C, $DNND(s)$; (iii) $\phi(C) = \delta\phi(C)$ for all $\delta > 0$, $C \geq 0$; (iv) $\phi(C) = \phi(H'CH)$ for any permutation matrix H of order s and

$$D = \sum_i a_i H_i' C H_i \Rightarrow$$

$$\phi(D) = \phi(\sum_i a_i H_i' C H_i) \geq \sum_i a_i \phi(H_i' C H_i) = \phi(C)$$

for every convex combination of the permutation matrices H_i of order s.

If H is a permutation matrix of order s, that is, a matrix obtained from the identity matrix I_s by permuting its columns, then $H'H = HH' = I_s$. So if $b = H\theta$ is a linear transformation of the parameters in a linear model, the new design matrix is XH' and the new information matrix is $H'X'XH = H'CH$. Since optimality in the Loewner sense rarely exists, we introduce the notion of universal optimality introduced by Kiefer [4].

Definition 5.12. (Kiefer optimality) A design ξ with information matrix M is universally optimal if $\phi(M) \geq \phi(C)$ for all information matrices C of competing designs and all information functions ϕ.

If C is $PD(s)$ (positive definite) then the function,

$$\phi_p(C) = \begin{cases} \lambda_{max}(C) \text{ for } p = \infty \\ (\frac{1}{s}\text{trace } C^p)^{1/p} \text{ for } p \neq 0, \pm\infty \\ (|C|)^{1/s} \text{ for } p = 0 \\ \lambda_{min}(C) \text{ for } p = -\infty. \end{cases}$$

If $\lambda_1, \ldots, \lambda_s$ are the eigenvalues of C and $z_1, \ldots, z_s$ the corresponding eigenvectors, then $C^p = \sum_{i=1}^s \lambda_i^p z_i z_i'$. For $p \leq 1$ this function is concave and for $p \geq 1$ is convex. Hence $\max_\xi \phi_p(C)$ gives for $p = 0$ the D-optimal design, for $p = -1$ the A-optimal and for $p = -\infty$ the E-optimal design.

5.7.3 Optimal Block Designs

We have k treatments $T_1, T_2, \ldots, T_k$ applied to a homogeneous population and n experimental units. If treatment T_i is applied to n_i experimental units, with

$n_1 + \cdots + n_k = n$, then the model is

$$y_{ij} = \mu_{A_i} + e_i, \ j = 1, \ldots, n_i, \ i = 1, \ldots, k,$$

If $n = km$, the ϕ-optimal design is to allocate $m = [n/k]$ units to each treatment. If we have $k - 1$ treatments and treatment T_k is the control, the problem is to find the optimal design for estimating $\mu_1 - \mu_k, \ldots, \mu_{k-1} - \mu_k$, see Hedayat, Jacroux, and Majundar [3].

In the cases where the population is not homogeneous in one or two directions, we have the row and row and column designs, see Shah and Sinha [7].

5.7.4 Approximate or Continuous Optimal Designs

It is difficult to find the exact optimal design in many problems, a way to overcome it is to consider the corresponding approximate design problem, where differential calculus can be applied. From Section 5.7, the design ξ is defined if the points $x_1, \ldots, x_k$, $x_i \in \Re^p$, at which the observations are taken, and their multiplicities $n_1, \ldots, n_k$ are given, with $n_1 + \cdots + n_k = n$. Then the information matrix is,

$$M = X'X = \sum_{i=1}^{k} n_i x_i x_i' = n \sum_{i=1}^{k} w_i x_i x_i', \ w_i = n_i/n.$$

Since $w_1 + \cdots + w_k = 1$, this is a probability distribution defined on the design space $\mathcal{X}$ and then $M = nE(xx')$, $x \in \Re^p$. Extending this idea let $\mathcal{X}$ be a compact subset of the Euclidean p-space, to be our design space. Let ξ be a probability measure on the Borel sets of $\mathcal{X}$. For the approximate design ξ, the information matrix is

$$M(\xi) = E_\xi(f(x)f'(x)) = \int_{\mathcal{X}} f(x)f'(x)\xi(dx).$$

Our problem is to determine ξ^* to maximize $\phi(M(\xi))$ over all competing designs ξ, $M(\xi)$ is a symmetric $NND(p)$ and the set of all information matrices

$$\mathcal{M} = \{M(\xi)\}$$

is convex, indeed is the closed convex hull of $\{f(x)f'(x) : x \in \mathcal{X}\}$.

Definition 5.13. (Silvey [8] p. 18) The Fréchet derivative of ϕ at M_1 in the direction of M_2 is:

$$F_\phi(M_1, M_2) = \frac{1}{\epsilon} \lim_{\epsilon \to 0^+} [\phi((1 - \epsilon)M_1 + \epsilon M_2) - \phi(M_1)].$$

Since $\mathcal{M}$ is convex, then $((1 - \epsilon)M_1 + \epsilon M_2) \in \mathcal{M}$ and $F_\phi \leq 0$ if and only if $M_1 = M(\xi^*)$.

Theorem 5.3. (Silvey [8] p. 19) *If $\phi(M)$ is concave on $\mathcal{M}$ and differentiable at $M(\xi^*)$, then ξ^* is ϕ-optimal if and only if*

$$F_\phi(M(\xi^*), f(x)f'(x)) \leq 0 \ \text{for all} \ x \in \mathcal{X}. \tag{5.26}$$

The maximum is attained at the points of support of ξ^*, that is,

$$\max_{x \in \mathcal{X}} F_\phi(M(\xi^*), f(x)f'(x)) = 0$$

$$\min_{\xi} \max_{x \in \mathcal{X}} F_\phi(M(\xi), f(x)f'(x)) = 0.$$

In the case of D-optimality F_ϕ takes the form

$$F_\phi(M_1, M_2) = \text{tr}(M_2 M_1^{-1}) - p \leq 0$$

the maximum value 0 is attained if $M_2 = M_1$.

Proposition 5.9. *The design ξ^* is D-optimal if and only if*

$$f'(x)M(\xi^*)^{-1}f(x) \leq p \ \text{for all} \ x \in \mathcal{X}$$

where p is the number of parameters of interest.

This is because the function $\phi(M) = \ln|M|$ is concave (see Exercise 5.19)

$$F_\phi(M_1, M_2) = \text{tr}(M_2 M_1^{-1}) - p$$

and

$$F_\phi(M_1, f(x)f'(x)) = f'(x)M_1^{-1}f(x) - p \ \text{for all} \ x \in \mathcal{X}.$$

Proposition 5.10. *If ξ^* is D-optimal, then*

$$\max_{x \in \mathcal{X}} f'(x)M(\xi^*)^{-1}f(x) - p = 0$$

and the maximum is attained at the points x_i of support of ξ^*

$$M(\xi^*) = \sum_i^s a_i f(x_i)f(x_i)', \ a_1 + \cdots + a_s = 1, \ a_i \geq 0.$$

Such convex representation always exists by Carathéodory's Theorem.
   Now since $f'(x)M(\xi)^{-1}f(x)$ is a convex function of x, the maximum is attained at the extreme points of $\mathcal{X}$.

Proposition 5.11. *If a design ξ is not D-optimal, then from (5.26),*

$$\max_{x \in \mathcal{X}} f'(x)M(\xi)^{-1}f(x) - p \geq 0$$

and also

$$\min_{\xi} \max_{x \in \mathcal{X}} f'(x)M(\xi)^{-1}f(x) = p.$$

Hence ξ^* is also G-optimal.

This is known as the *General Equivalence Theorem.*

Example 5.2. (Quadratic regression) Suppose we have the quadratic regression,

$$E(y) = \theta_0 + \theta_1 x + \theta_2 x^2 = f'(x)\theta$$

where $f(x) = (1, x, x^2)'$ and $1 \leq x \leq 1$, here $p = 3$. If we have $n = 3m$ observations and take the design ξ^* which assigns probability $1/3$ at each one of the points $f(-1) = (-1, 1, 1)'$, $f(0) = (1, 0, 0)'$, $f(1) = (1, 1, 1)'$, then

$$M_1 = (1/3)[f(-1)f'(-1) + f(0)f'(0) + f(1)f'(1)]$$

$$= \frac{1}{3}\begin{pmatrix} 1 & 0 & 2 \\ 0 & 2 & 0 \\ 2 & 0 & 2 \end{pmatrix}$$

and

$$M_1^{-1} = (3/4)\begin{pmatrix} 4 & 0 & -4 \\ 0 & 2 & 0 \\ -4 & 0 & 6 \end{pmatrix}.$$

With $f(x) = (1, x, x^2)'$ we have

$$\max_{-1 \leq x \leq 1} f'(x)M_1^{-1}f(x) = \max_{-1 \leq x \leq 1} (3/4)(4 - 6x^2 + 6x^4) = 3.$$

The maximum is equal to 3 and is attained at $x = (-1, 0, 1)$. Hence the design ξ^* is D-optimal and G-optimal, that is, in quadratic regression take $(n/3)$ observations at each one of the points $x = -1$, $x = 0$, $x = 1$. In n is not a multiple of 3, the case needs further study.

5.7.5   Algorithms

The above method verifies if a given design ξ is D-optimal or ϕ-optimal, otherwise we find the D-optimal design using algorithms.

Algorithm (Fedorov [2]) Step(0):

We start with an arbitrary design measure ξ_0, for example take s vectors in $\mathcal{X}$ each with probability $1/s$ with $s \geq p$ or take

$$\xi_0 = \begin{cases} x_1 \ x_2 \ \ldots \ x_s \\ p_1 \ p_2 \ \ldots \ p_s \end{cases} \quad p_1 + \cdots + p_s = 1, \ s \geq p$$

so that $M_0(\xi) = \sum_{i=1}^{s} p_i f(x_i) f'(x_i)$ is not singular and write

$$d(x, \xi_0) = f'(x) M_0^{-1} f(x).$$

Step(1): Find the vector $x_1 \in \mathcal{X}$ such that

$$d(x_1, \xi_0) = \max_{x \in \mathcal{X}} d(x, \xi_0) = f'(x_1) M_0^{-1} f(x_1).$$

Create the new design $\xi_1 = (1 - a_1)\xi_0 + a_1 x_1$, that is the new design ξ_1 has $s + 1$ vectors, the first s vectors have their probabilities multiplied by $(1 - a_1)$ and the new vector x_1 has probability a_1, then $M_1 = (1 - a_1)M_0 + a_1 f(x_1) f'(x_1)$ and write $d(x, \xi_1) = f'(x) M_1^{-1} f(x)$.

Step (2): If we are at step k proceed as in step (1) and find the vector $x_k \in \mathcal{X}$ such that

$$d(x_k, \xi_{k-1}) = \max_{x \in \mathcal{X}} d(x, \xi_{k-1}) = f'(x_k) M_{k-1}^{-1} f(x_k).$$

Construct the new design $\xi_k = (1 - a_k)\xi_{k-1} + a_k x_k$, then

$$M_k = (1 - a_k)M_{k-1} + a_k f(x_k) f'(x_k)$$

and write $d(x, \xi_k) = f'(x) M_k^{-1} f(x)$.

Step(3): Terminate the process when

$$\frac{|M_{k+1}| - |M_k|}{|M_{k+1}|} < \gamma$$

for some prespecified value γ.

The problem is how to choose the step length a_k and the difficulty is in finding the inverse M_k^{-1}. Note that

$$M_k^{-1} = \frac{M_{k-1}^{-1}}{(1 - a_k)} - \frac{a_k M_{k-1}^{-1} f(x_k) f'(x_k) M_{k-1}^{-1}}{(1 - a_k)^2 (1 + a_k d(x_k, \xi_{k-1})/(1 - a_k))}.$$

The step length a_k is chosen to maximize $f'(x_k)M_k^{-1}f(x_k) - p$, then

$$a_k = \frac{d(x_k, \xi_{k-1}) - p}{p(d(x_k, \xi_{k-1}) - 1)}.$$

The difficulty then is to compute M_0^{-1} and to find in every step the

$$\max_{x \in \mathcal{X}} f'(x)M_k^{-1}f(x),$$

for more details see Section 2.5 of Fedorov [2].

Algorithm (Wynn): Choose as the step length a_k the members of any sequence $a_1, a_2 \ldots$, such that $\lim_{n \to \infty} a_n = 0$ and $\sum_{i=1}^{\infty} a_i = \infty$, for example, $a_k = 1/(k+1)$.

The drawback of these algorithms is that if we start with a bad design measure, it takes many iterations to reach convergence and also we do not get rid of the initial points.

Remarks. Let $\mathcal{M}$ be the set of the symmetric, $NND(p)$ matrices M, every $M \in \mathcal{M}$ is specified by its $p(p+1)/2$ elements, that is, p diagonal and $p(p-1)/2$ non-diagonal elements. Each element of $\mathcal{M}$, by Carathéodory's Theorem, can be expressed as a convex combination

$$M(\xi) = \sum_{i=1}^{m} c_i f(x_i) f'(x_i), \quad x_i \in \mathcal{X}$$

with $m \leq 1 + p(p+1)/2$ and $M(\xi)$ is a boundary point of $\mathcal{M}$.

This means that every design measure on $\mathcal{X}$ gives the same information matrix M as a discrete design measure with m points of support, with $m \leq 1 + p(p+1)/2$.

Exercises 5.

5.23. Let $E(y) = \theta_0 + \theta_1 x$, $-1 \leq x \leq 1$ and consider the design ξ_1. Show that if ξ_2 is the mirror image of ξ_1 with respect to zero, then the design $\xi = (1/2)\xi_1 + (1/2)\xi_2$ is better D-optimal design. It is also better ϕ-optimal design, when ϕ is strictly concave. Hence the ϕ-optimal design has its points of support symmetrically around zero. Is this true for any polynomial regression?

5.24. Let $E(y) = \theta_0 + \theta_1 x$, $\mathrm{Var}(y) = \sigma^2$, $0 \leq x \leq 1$. (a) Find the D-optimal approximate design and the D-optimal exact n observation design. (b) Do the same if $var(y) = \sigma^2/x$.

5.25. Let

$$E(y) = \theta_0 + \theta_1 x_1 + \theta_2 x_2, \quad -1 \leq x_1 \leq 1, \quad -1 \leq x_2 \leq 1, \quad \mathrm{Var}(y) = \sigma^2.$$

Here the design space $\mathcal{X}$ is a square. Compare, with respect to D-optimality, the designs which are concentrated: (a) At 3 vertices of the square with probabilities $p_i = 1/3$, $i = 1, 2, 3$. (b) At the 4 vertices of the square with probabilities $p_i = 1/4$, $i = 1, 2, 3, 4$. Is any of these designs D-optimal?

5.26. In the quadratic regression $y_i = \theta_0 + \theta_1 x_i + \theta_2 x_i^2 + e_i$, $|x| \leq 1$ consider the 3 discrete designs

$$\xi_1 = \begin{pmatrix} -1 & 0 & 1 \\ 1 & 2 & 1 \end{pmatrix}, \quad \xi_2 = \begin{pmatrix} -1 & -1/3 & 1/3 & 1 \\ 1 & 1 & 1 & 1 \end{pmatrix}, \quad \xi_3 = \begin{pmatrix} -1 & 0 & 1 \\ 2 & 1 & 1 \end{pmatrix}.$$

Compare them with respect to the criteria D, A, G, E-optimality, when we are interested in the following: (a) for all the parameters; (b) for the parameters $\theta_1, \theta_2, \theta_3$; (c) for the parameter θ_3.

5.27. For the quadratic regression given above, find the exact D, G, A optimal designs.

5.28. For the quadratic regression given above, find the value of c so that the following approximate design is A-optimal.

$$\xi_1 = \begin{pmatrix} -1 & 0 & 1 \\ c & 1 - 2c & c \end{pmatrix}, \quad 0 \leq c \leq 0.5.$$

5.29. For the quadratic regression through the origin $y_i = \theta_1 x_i + \theta_2 x_i^2 + e_i$, $|x| \leq 1$, find the D, G-optimal approximate and exact designs.

5.30. Apply Fedorov's and Wynn's algorithms for the quadratic regression.

5.31. Construct the exact D-optimal saturated factorial design of resolution III when $n = 7$ and $n = 9$.

5.32. Express the relations given in (5.26) in terms of eigenvalues.

References

1. Atkinson, A. C., Donev, A. N., & Tobias, R. D. (2007). *Optimum experimental designs.* Oxford: Oxford University Press.
2. Fedorov, V. V. (1972). *Theory of optimal experiments.* New York: Academic Press.
3. Hedayat, A. S., Jacroux, M., & Majundar, D. (1988). Optimal designs for comparing test treatments with controls. *Statistical Science, 3*(4), 462–491.
4. Kiefer, J. C. (1975). Construction and optimality of generalized Youden designs. In Srivastava (Ed.), *A survey of statistical design and linear models* (pp. 333–353). North Holland: Amsterdam.
5. Kounias, S., Lefkopoulou, M., & Bagiatis, C. (1983). G-optimal N-observation first order 2^k designs. *Discrete Mathematics, 43*, 21–31.

6. Sathe, Y. S., & Shenoy, R. G. (1989). A-optimal weighing designs when $N = 3$ mod 4. *The Annals of Statistics, 17*(4), 1906–1915.
7. Shah, K. R., & Sinha, B. K. (1989). *Theory of optimal designs* (Vol. 54), Lecture Notes in Statistics. New York: Springer.
8. Silvey, S. D. (1980). *Optimal design*. London: Chapman and Hall.

Chapter 6
Pathway Models

6.1 Introduction

Here, we look at model building in general. First, we will deal with real scalar variable cases. Then generalized models, matrix-variate cases, etc., will be examined. First, let us look at the effect of power transformations and exponentiations on a given model. This will give some ideas about the changes required in a given situation of building models over a given data at hand. When building models for the data in hand, one usually takes a member from a parametric family of functions. It is often found that the data require a model with a thicker or thinner tail, or different shapes near zero, or in the middle or of a slightly different shape in general. The effects of power transformations, exponentiations, limiting processes, and pathway idea will be explored to study the changes that take place in a given model. We start with type 1 and type 2 beta models. These models are used in practice to model different situations in various fields.

6.2 Power Transformation and Exponentiation of a Type 1 Beta Density

Let us see the effect of power transformations and exponentiations on standard density functions in current use in statistical sciences. This will give an indication of the types of changes that are required when constructing a model for a real-life situation. [The following material is based on Mathai's paper on power transformation and exponentiation.] Let us start with a type 1 beta model

---

This chapter is based on the lectures of Professor A.M. Mathai of CMS India and McGill University, Canada.

© Springer International Publishing AG 2017
A.M. Mathai and H.J. Haubold, *Fractional and Multivariable Calculus*,
Springer Optimization and Its Applications 122, DOI 10.1007/978-3-319-59993-9_6

$$f(x) = a^\alpha \frac{\Gamma(\alpha + \beta)}{\Gamma(\alpha)\Gamma(\beta)} x^{\alpha-1}(1 - ax)^{\beta-1}, \ 0 \le x \le \frac{1}{a}, \ a > 0 \qquad (6.2.1)$$

and 0 elsewhere, $\Re(\alpha) > 0$, $\Re(\beta) > 0$, and the extended model

$$f^*(x) = a^\alpha \frac{\Gamma(\alpha + \beta)}{2\Gamma(\alpha)\Gamma(\beta)} |x|^{\alpha-1}(1 - a|x|)^{\beta-1}, \ -\frac{1}{a} \le x \le \frac{1}{a}, \ a > 0 \qquad (6.2.2)$$

and 0 elsewhere, for $\Re(\alpha) > 0$, $\Re(\beta) > 0$. In statistical model building situations, all parameters are real, but the results will hold for complex parameters as well, and hence the conditions are stated in terms of the real values of complex parameters. Let us consider a power transformation. Put $x = y^\delta$, $\delta > 0 \Rightarrow dx = \delta y^{\delta-1} dy$ for $y > 0$. For $0 < \delta \le 1$, the rate of increase is retarded whereas when $\delta > 1$ there is acceleration in the rate of increase of x. In a practical situation if the shape of the curve is to be regulated, then this type of a transformation is useful. Under a power transformation, the model becomes

$$f_1(x) = \frac{\delta a^\alpha \Gamma(\alpha + \beta)}{\Gamma(\alpha)\Gamma(\beta)} y^{\alpha\delta-1}(1 - ay^\delta)^{\beta-1}, \ 0 \le y \le a^{-1/\delta}, \ \delta > 0, \ a > 0 \quad (6.2.3)$$

and zero elsewhere, for $\Re(\alpha) > 0$, $\Re(\beta) > 0$. The model in (6.2.2) is Mathai's pathway model for $a = b(1 - q)$, $b > 0$, $\beta - 1 = \frac{\eta}{1-q}$ and for the cases $q < 1, q > 1$, $q \to 1$, producing a large number of densities and their transitional forms, see Mathai [6] and Mathai and Haubold [7]. We can examine this model when the scale parameter a is becoming smaller and smaller, and when $\beta = 1/a$. This will produce an extreme-value type density. If we replace $a > 0$ by ab, with $b > 0$ and put $\beta = \frac{1}{a}$ and take the limit when $a \to 0$, then observing that

$$\lim_{a \to 0} a^\alpha \frac{\Gamma(\alpha + \frac{1}{a})}{\Gamma(\frac{1}{a})} (1 - aby^\delta)^{\frac{1}{a}-1} = e^{-by^\delta} \qquad (6.2.4)$$

we have the limiting density in the following form:

$$f_2(y) = \frac{\delta b^\alpha}{\Gamma(\alpha)} y^{\alpha\delta-1} e^{-by^\delta}, \ 0 \le y < \infty, \ b > 0, \ \Re(\alpha) > 0 \qquad (6.2.5)$$

and zero elsewhere. For taking limit on gamma function in (6.2.4), we use the asymptotic expansion of gamma functions. For the first approximation, there is Stirling's formula

$$\Gamma(z + \alpha) \approx \sqrt{2\pi} z^{z+\alpha-\frac{1}{2}} e^{-z}$$

for $|z| \to \infty$ and α is a bounded quantity. This density in (6.2.5) is a type of extreme-value density and can be called Gamma-Weibull density because it is a slight generalization of the Weibull density or a special case of the generalized gamma density. When the parameter a moves toward 0, a large number of models in the transitional stages are available. In a physical situation, if (6.2.5) is the stable situation, then the unstable neighborhoods are described by the transitional stages when the parameter a moves to 0.

If the shape of the density in (6.2.1) is to be further modified in a practical situation, where the growth or decay of x is more or less of exponential type, then we can consider exponentiation. Let us consider the transformation $x = e^{-\gamma y}$, $\gamma > 0$ in (6.2.1). Then when $x \to 0$, $y \to \infty$ and when x goes to $a^{-1/\delta}$, $y \to \frac{\ln a}{\gamma}$ and zero for $a = 1$. Then, the density in (6.2.1) becomes

$$f_3(y) = \frac{\gamma \, a^\alpha \Gamma(\alpha + \beta)}{\Gamma(\alpha)\Gamma(\beta)} e^{-\gamma\alpha y}(1 - ae^{-\gamma y})^{\beta-1}, \quad \frac{\ln a}{\gamma} \le y < \infty, \tag{6.2.6}$$

for $\gamma > 0$, $\Re(\alpha) > 0$, $\Re(\beta) > 0$, $a > 0$, and zero elsewhere. Note that for both $\gamma > 0$ and $\gamma < 0$ it can produce a density, one will have the support $\frac{\ln a}{\gamma} \le y < \infty$ and the other will have the support $-\infty < y \le -\frac{\ln a}{\gamma}$. For $\beta = \frac{1}{a}$, $a \to 0$, we have the pathway form of the density

$$f_4(y) = \frac{\gamma}{\Gamma(\alpha)} e^{-\gamma\alpha y} e^{-e^{-\gamma y}}, \quad -\infty < y < \infty \tag{6.2.7}$$

which is also an extreme-value density. This is also available from (6.2.5) by exponentiation. On the other hand, if $a > 0$ is replaced by ab with $b > 0$, put $\beta = \frac{1}{a}$ and then take the limit $a \to 0$, then we have the density

$$f_5(y) = \frac{\gamma b^\alpha}{\Gamma(\alpha)} e^{-\gamma\alpha y} e^{-be^{-\gamma y}}, \quad -\infty < y < \infty, \ b > 0, \ \gamma > 0, \ \alpha > 0. \tag{6.2.8}$$

Note that for $a = e^\delta$ in $f_3(y)$, one has a generalized form of Bose–Einstein density in physics and engineering sciences, see Mathai [3] for a form of Bose–Einstein density. For each of the above situations, if we replace y by $|y|$ and multiply the normalizing constant with $\frac{1}{2}$, then we have the extended symmetric forms of the corresponding densities.

$$f_1^*(y) = \frac{\delta \, a^\alpha \Gamma(\alpha + \beta)}{2\Gamma(\alpha)\Gamma(\beta)} |y|^{\alpha\delta-1}(1 - a|y|^\delta)^{\beta-1}, \quad -a^{-1/\delta} \le y \le a^{-1/\delta} \tag{6.2.9}$$

for $\delta > 0$, $\alpha > 0$, $\beta > 0$, $a > 0$, and zero elsewhere.

$$f_2^*(y) = \frac{\delta b^\alpha}{2\Gamma(\alpha)} |y|^{\alpha\delta-1} e^{-b|y|^\delta}, \quad -\infty < y < \infty, \ b > 0, \ \alpha > 0. \tag{6.2.10}$$

$$f_3^*(y) = \frac{\gamma a^\alpha \Gamma(\alpha + \beta)}{2\Gamma(\alpha)\Gamma(\beta)} e^{-\gamma\alpha|y|}(1 - ae^{-\gamma|y|})^{\beta-1}, \quad \frac{\ln a}{\gamma} \le |y| < \infty, \ \gamma > 0. \tag{6.2.11}$$

$$f_5^*(y) = \frac{\gamma b^\alpha}{2\Gamma(\alpha)} e^{-\gamma\alpha|y|} e^{-be^{-\gamma|y|}}, \quad -\infty < y < \infty, \ b > 0, \ \gamma > 0, \ \alpha > 0. \tag{6.2.12}$$

In $f_2^*(y)$ if $\alpha = 1$, $\delta = 1$, then we have the Laplace or double exponential density. When $\alpha = 1$ in (6.2.10), the model is a power transformed Laplace density.

6.2.1   Asymmetric Models

Our extended models so far are all symmetric models in the sense that we were simply replacing x by $|x|$. In many practical situations, asymmetric models are relevant. We will introduce asymmetry through scaling parameters. There are many ways of introducing asymmetry. The usual procedure is to put different weights for $x < 0$ and for $x > 0$. By taking different scaling parameters, we can create asymmetric models. We will consider a procedure of changing the shape of the curve itself and obtain an asymmetric version, which may be more relevant in practical situations. For all the models listed below the conditions $\Re(\alpha) > 0$, $\Re(\beta) > 0$ hold. For example, an asymmetric model corresponding to $f^*(x)$ can be formulated as follows, denoting the asymmetric versions by a double star notation, by $f^{**}(x)$:

$$f^{**}(x) = \frac{(a_1 a_2)^\alpha \Gamma(\alpha + \beta)}{(a_1^\alpha + a_2^\alpha)\Gamma(\alpha)\Gamma(\beta)} |x|^{\alpha-1}$$
$$\times \begin{cases} [1 - a_1|x|]^{\beta-1}, & -\frac{1}{a_1} \le x \le 0, \ a_1 > 0 \\ [1 - a_2|x|]^{\beta-1}, & 0 \le x \le \frac{1}{a_2}, \ a_2 > 0 \end{cases} \quad (6.2.13)$$

Asymmetric versions for other densities are the following:

$$f_1^{**}(y) = \frac{\delta(a_1 a_2)^\alpha \Gamma(\alpha + \beta)}{(a_1^\alpha + a_2^\alpha)\Gamma(\alpha)\Gamma(\beta)} |y|^{\delta\alpha-1}$$
$$\times \begin{cases} [1 - a_1|y|^\delta]^{\beta-1}, & -a_1^{-1/\delta} \le y \le 0, \ a_1 > 0, \ \delta > 0 \\ [1 - a_2|y|^\delta]^{\beta-1}, & 0 \le y \le a_2^{-1/\delta}, \ \delta > 0, \ a_2 > 0 \end{cases} \quad (6.2.14)$$

$$f_2^{**}(y) = \frac{(b_1 b_2)^\alpha}{(b_1^\alpha + b_2^\alpha)\Gamma(\alpha)} |y|^{\alpha\delta-1}$$
$$\times \begin{cases} e^{-b_1|y|^\delta}, & -\infty < y \le 0, \ b_1 > 0, \ \delta > 0 \\ e^{-b_2 y^\delta}, & 0 \le y < \infty, \ b_2 > 0, \ \delta > 0 \end{cases} \quad (6.2.15)$$

$$f_3^{**}(y) = \frac{\gamma(a_1 a_2)^\alpha \Gamma(\alpha + \beta)}{(a_1^\alpha + a_2^\alpha)\Gamma(\alpha)\Gamma(\beta)} e^{-\gamma\alpha|y|}$$
$$\times \begin{cases} [1 - a_1 e^{-\gamma|y|}]^{\beta-1}, & -\infty < y \le 0, \ a_1 > 0, \ \gamma > 0 \\ [1 - a_2 e^{-\gamma y}]^{\beta-1}, & 0 \le y < \infty, \ a_2 > 0, \ \gamma > 0 \end{cases} \quad (6.2.16)$$

Note 6.2: In all the models that we have considered so far, as well as in all the models that we will consider later on, if a location parameter is desirable, then replace the variable x or y by $x - \mu$ or $y - \mu$, where μ will act as a relocation parameter.

Exercises 6.2.

By using MAPLE or MATHEMATICA or any other mathematical word processor with graphic facilities, draw the graphs of the following functions and examine the changes:

6.2.1. $f(x)$ of (6.2.1) for $(i): \alpha = 2, \beta = 3, a = 1$; $(ii): \alpha = 2, \beta = 3, a = \frac{1}{2}$; $(iii)\ \alpha = 2, \beta = 3, a = \frac{1}{3}$.

6.2.2. $f_1(x)$ of (6.2.3) for $(i): \alpha = 2, \beta = 3, a = \frac{1}{2}, \delta = 1$; $(ii)\ \alpha = 2, \beta = 3, a = \frac{1}{2}, \delta = 2$; $(iii)\ \alpha = 2, \beta = 3, a = \frac{1}{2}, \delta = \frac{1}{2}$.

6.2.3. $f_2(x)$ of (6.2.5) for $(i): \alpha = 3, b = 1, \delta = 3, 2, 1$; $(ii)\ \alpha = 2, \delta = 1, b = 1, 2, 3$.

6.2.4. $f_3(x)$ of (6.2.5) for $(i): \alpha = 2, \beta = 3, a = \frac{1}{2}, \gamma = 1, 2\frac{1}{2}$; $(ii): \alpha = 2, \beta = 3, a = \frac{1}{2}, \gamma = 1, 2, \frac{1}{2}$.

6.2.5. $f_5(x)$ of (6.2.8) for $(i): \alpha = 2, b = 2, \gamma = 1, 2, \frac{1}{2}$; $(ii): \alpha = 2, \gamma = 2, b = 1, 2, \frac{1}{2}$.

6.3 Power Transformation and Exponentiation on a Type 2 Beta Model

Let us start with a standard type 2 beta model and examine the effect of power transformation and exponentiation on this model. In all the discussions to follow in this section the basic conditions $\Re(\alpha) > 0$, $\Re(\beta) > 0$ hold, if the parameters are real, then $\alpha > 0$, $\beta > 0$ hold. Also, only the nonzero part of the density is given along with its support. The basic type 2 beta model, with a scaling factor $a > 0$, and its extended version are the following:

$$g(x) = \frac{a^\alpha \Gamma(\alpha + \beta)}{\Gamma(\alpha)\Gamma(\beta)} x^{\alpha-1}(1 + ax)^{-(\alpha+\beta)},\ 0 \le x < \infty \qquad (6.3.1)$$

for $\Re(\alpha) > 0$, $\Re(\beta) > 0$, $a > 0$.

$$g^*(x) = \frac{a^\alpha \Gamma(\alpha + \beta)}{2\Gamma(\alpha)\Gamma(\beta)} |x|^{\alpha-1}(1 + a|x|)^{-(\alpha+\beta)},\ -\infty < x < \infty,\ a > 0 \qquad (6.3.2)$$

In (6.3.1), if $\alpha = 1$, $\beta + \alpha = \frac{1}{q-1}$, and $a = q - 1$, then (6.3.1) reduces to Tsallis statistics in non-extensive statistical mechanics. It is said that over 5000 researchers are working in this area, and they have produced over 3000 papers from 1990 to 2010 period. Both for $q < 1$ and $q > 1$, it will produce Tsallis statistics. For the original starting paper see Tsallis [11]. Also in this case, the functional part follows a power law. That is, consider (6.3.1) for the case $\alpha = 1$, $\beta + \alpha = (q - 1)^{-1}$, $a = (q - 1)$

and if the resulting $g(x)$ is $g_{(1)}(x) = c(1 + (q-1)x)^{-\frac{1}{q-1}}$, then $h(x) = \frac{g_{(1)}(x)}{c}$ has the property that

$$\frac{dh(x)}{dx} = -[h(x)]^q$$

which is the power law. The same (6.3.1) for $a = (q-1)$, $\alpha + \beta = \frac{1}{q-1}$, $q > 1$ is the current hot topic of superstatistics in statistical mechanics. Dozens of papers are available on this topic also, see for the original papers Beck and Cohen [2] and Beck [1]. The power transformation $x = y^\delta$, $\delta > 0$ produces the following density, along with its extended version, denoted by $g_1(x)$ and $g_1^*(x)$, respectively.

$$g_1(y) = \frac{\delta\, a^\alpha \Gamma(\alpha + \beta)}{\Gamma(\alpha)\Gamma(\beta)} y^{\alpha\delta-1}[1 + ay^\delta]^{-(\alpha+\beta)}, \ 0 \le y < \infty \qquad (6.3.3)$$

$$g_1^*(y) = \frac{\delta\, a^\alpha \Gamma(\alpha + \beta)}{2\Gamma(\alpha)\Gamma(\beta)} |y|^{\alpha\delta-1}[1 + a|y|^\delta]^{-(\alpha+\beta)}, \ -\infty < y < \infty. \qquad (6.3.4)$$

This (6.3.4) with $\alpha\delta$ in $y^{\alpha\delta-1}$ is replaced by a general parameter $\gamma > 0$, and if $\alpha + \beta$ is replaced by $\frac{\eta}{q-1}$, $\eta > 0$ and a is replaced by $b(q-1)$, $b > 0$, then we have the pathway density of Mathai [6] in the scalar case. A limiting form or a pathway form of the density in (6.3.3) is available by replacing $a > 0$ by ab, $b > 0$, taking $\beta = \frac{1}{a}$ and then taking the limit when $a \to 0$. Then, we get an extreme-value form of a density. This and its extended versions are given below.

$$g_2(y) = \frac{\delta\, b^\alpha}{\Gamma(\alpha)} y^{\alpha\delta-1} e^{-by^\delta}, \ 0 \le y < \infty \qquad (6.3.5)$$

$$g_2^*(y) = \frac{\delta\, b^\alpha}{2\Gamma(\alpha)} |y|^{\alpha\delta-1} e^{-b|y|^\delta}, \ -\infty < y < \infty. \qquad (6.3.6)$$

Note 6.3. Model in (6.3.6), for $\delta = 2$, $\alpha = \delta^{-1}$, $b = (2\sigma^2)^{-1}$, $\sigma > 0$, and y replaced by $y - \mu$, is the lognormal density. Note that our original transformation was $x = e^y$ in (6.3.1), which is nothing but $y = \ln x$ for $x > 0$. Then, the model in (6.3.6) also produces a pathway for going from the type 2 beta model to a lognormal model.

Note that in models (6.3.5) and (6.3.6) one can take $\delta < 0$ or $\delta > 0$, and both cases will produce densities. With the appropriate normalizing constant, if $\alpha\delta$ in $y^{\alpha\delta-1}$ is replaced by a general parameter $\gamma > 0$, then (6.3.5) produces the generalized gamma density. One can also consider exponentiation in the basic model in (6.3.1). Let $x = e^{-\gamma y}$, $\gamma > 0$. Then when $x \to 0$, $y \to \infty$ and when $x \to \infty$, $y \to -\infty$. This will produce the model given as $g_3(y)$ below. When $x = e^{\gamma y}$, $\gamma > 0$ this will also produce a model, denoted by $g_3^*(y)$ below. Thus, the models become

$$g_3(y) = \frac{\gamma \, a^\alpha \Gamma(\alpha + \beta)}{\Gamma(\alpha)\Gamma(\beta)} e^{-\alpha\gamma y}[1 + ae^{-\gamma y}]^{-(\alpha+\beta)}, \quad -\infty < y < \infty. \tag{6.3.7}$$

$$g_3^*(y) = \frac{\gamma \, a^\alpha \Gamma(\alpha + \beta)}{\Gamma(\alpha)\Gamma(\beta)} e^{\alpha\gamma y}[1 + ae^{\gamma y}]^{-(\alpha+\beta)}, \quad -\infty < y < \infty. \tag{6.3.8}$$

The transformation to obtain (6.3.7) and (6.3.8) can also be looked upon as logarithmic transformations $y = \frac{1}{\gamma} \ln x$ for $y = -\frac{1}{\gamma} \ln x$. Thus, the logarithmic versions, such as lognormal distribution, are also available from the above and the following models. We may replace y by $\frac{1}{\gamma} \ln x$ or $-\frac{1}{\gamma} \ln x$ and dy by $\frac{1}{\gamma x} dx$. If we put $\alpha = 1, \beta = 1, a = 1, \delta = 1$ in (6.3.7), then we get the standard logistic density, which is applicable in many areas of model building situations. If this particular case is denoted by $g_{(3)}(y)$, then it has many interesting mathematical properties. If the distribution function is

$$F(y) = \int_{-\infty}^{y} g_{(3)}(x)dx = \frac{e^x}{1 + e^x}$$

$$F(x)[1 - F(x)] = f(x) = \frac{d}{dx}F(x), \quad \frac{d}{dx}\frac{f(x)}{1 - F(x)} = f(x).$$

Note that (6.3.7) is an aspect of the generalized logistic density studied in Mathai and Provost [8, 9] and Mathai [5]. Let us examine some special forms of (6.3.7) and (6.3.8). If one of α or β is absent, can a function such as $c\,(1 + e^x)^{-\delta}$, $\delta > 0$ create a density where $c > 0$ is a constant? We can see that $\int_{-\infty}^{\infty}(1 + e^x)^{-\delta}dx$ does not converge for all δ. If the support is only $0 \le x < \infty$, can the function $c(1 + ae^x)^{-1}$ create a density, for $c > 0, a > 0$?

$$c\int_0^\infty \frac{1}{(1 + ae^x)}dx = c\int_1^\infty \frac{1}{u(1 + au)}du < \infty$$

and, in fact, it can produce a density. If $x = \beta y$ and $a = e^\alpha$ with $\beta > 0$ and if the support is $0 \le y < \infty$, then it can produce a density and, this density is the Fermi–Dirac density.

In (6.3.8) if a is replaced by e^δ, $\delta > 0$, then we have a generalized form of Fermi–Dirac density, see Mathai [3] for one form of Fermi–Dirac density. The pathway version of $g_3(y)$ is available by putting $\beta = \frac{1}{a}$, replacing $a > 0$ by $ab, b > 0$, and then taking the limit when $a \to 0$. Then, we have

$$g_4(y) = \frac{\gamma \, b^\alpha}{\Gamma(\alpha)} e^{-\alpha\gamma y} e^{-be^{-\gamma y}}, \quad -\infty < y < \infty, \ b > 0, \ \alpha > 0, \ \gamma > 0. \tag{6.3.9}$$

Note that $g_4(y)$ remains as a density for the transformation $x = e^{\gamma y}$ and for the transformation $x = e^{-\gamma y}$. Hence, we have a companion density

$$g_4^*(y) = \frac{\gamma \, b^\alpha}{\Gamma(\alpha)} e^{\alpha\gamma y} e^{-be^{\gamma y}}, \quad b > 0, \ \alpha > 0, \ \gamma > 0, \ -\infty < y < \infty. \tag{6.3.10}$$

In order to see the movements with reference to various parameter values, some of the above densities are plotted with one parameter varying and others fixed for each case.

6.3.1   Asymmetric Version of the Models

The asymmetric versions can be worked out parallel to those in Section 2. They are listed below:

$$g^{**}(x) = \frac{(a_1 a_2)^\alpha \Gamma(\alpha + \beta)}{(a_1^\alpha + a_2^\alpha)\Gamma(\alpha)\Gamma(\beta)} |x|^{\alpha-1}$$

$$\times \begin{cases} [1 + a_1|x|]^{-(\alpha+\beta)}, & x < 0, \ a_1 > 0 \\ [1 + ax]^{-(\alpha+\beta)}, & x \geq 0, \ a_1 > 0, \ a_2 > 0 \end{cases} \qquad (6.3.11)$$

$$g_1^{**}(y) = \frac{\delta \, (a_1 a_2)^\alpha \Gamma(\alpha + \beta)}{(a_1^\alpha + a_2^\alpha)\Gamma(\alpha)\Gamma(\beta)} |y|^{\alpha\delta-1}$$

$$\times \begin{cases} [1 + a_1|y|^\delta]^{-(\alpha+\beta)}, & y < 0, \ a_1 > 0, \ \alpha > 0, \ \delta > 0 \\ [1 + a_2 y^\delta]^{-(\alpha+\beta)}, & y \geq 0, \ a_2 > 0 \end{cases} \qquad (6.3.12)$$

$$g_2^{**}(y) = \frac{\delta (b_1 b_2)^\alpha}{(b_1^\alpha + b_2^\alpha)\Gamma(\alpha)} |y|^{\alpha\delta-1}$$

$$\times \begin{cases} e^{-b_1|y|^\delta}, & b_1 > 0, \ \delta > 0, \ y < 0 \\ e^{-b_2|y|^\delta}, & b_2 > 0, \ y \geq 0 \end{cases} \qquad (6.3.13)$$

$$g_3^{**}(y) = \frac{\gamma(a_1 a_2)^\alpha \Gamma(\alpha + \beta)}{(a_1^\alpha + a_2^\alpha)\Gamma(\alpha)\Gamma(\beta)} e^{-\alpha\gamma|y|}$$

$$\times \begin{cases} [1 + a_1 e^{-\gamma|y|}]^{-(\alpha+\beta)}, & a_1 > 0, \ \gamma > 0, \ y < 0 \\ [1 + a_2 e^{-\gamma y}]^{-(\alpha+\beta)}, & a_2 > 0, \ y \geq 0 \end{cases} \qquad (6.3.14)$$

$$g_4^{**}(y) = \frac{\gamma(b_1 b_2)^\alpha}{(b_1^\alpha + b_2^\alpha)\Gamma(\alpha)} e^{-\alpha\gamma|y|}$$

$$\times \begin{cases} e^{-b_1 e^{-|y|}}, & y < 0, \ b_1 > 0, \ \gamma > 0, \ \alpha > 0 \\ e^{-b_2 e^{-y}}, & y \geq 0, \ b_2 > 0 \end{cases} \qquad (6.3.15)$$

Similar procedure can be adopted to construct the asymmetrical versions of other densities.

Exercises 6.3.

By using MAPLE or MATHEMATICA or any other mathematical word processor with graphic facilities, draw the graphs of the following functions and examine the changes:

6.3.1. $g(x)$ of (6.3.1) for (i) : $\alpha = 2, \beta = 3, a = 1$;   ($ii$) : $\alpha = 2, \beta = 3, a = 2$; ($iii$) : $\alpha = 2, \beta = 3, a = 3$.

6.3.2. $g_1(x)$ of (6.3.3) for (i) : $\alpha = 2, \beta = 2, a = 2, \delta = 1$; (ii) : $\alpha = 2, \beta = 3, a = 2, \delta = 2$; (iii) : $\alpha = 2, \beta = 3, a = 2, \delta = \frac{1}{2}$.

6.3.3. $g_2(x)$ of (6.3.5) for (i) : $\alpha = 2, b = 2, \delta = 1$; (ii) : $\alpha = 2, b = 2, \delta = 2$; (iii) : $\alpha = 2, b = 2, \delta = \frac{1}{2}$.

6.3.4. $g_3(x)$ of (6.3.7) for (i) : $\alpha = 2, \beta = 3, a = 1, \gamma = 1$; (ii) : $\alpha = 2, \beta = 3, a = 2, \gamma = 2$; (iii) : $\alpha = 2, \beta = 3, a = 2, \gamma = \frac{1}{2}$.

6.3.5. $g_3^*(x)$ of (6.3.8) for (i) : $\alpha = 2, \beta = 3, a = 2, \gamma = 1$; (ii) : $a = 2, \beta = 3, \alpha = 2, \gamma = 2$; (iii) : $\alpha = 2, \beta = 3, a = 2, \gamma = \frac{1}{2}$.

6.3.6. $g_4(x)$ of (6.3.9) for (i) : $\alpha = 2, b = 2, \gamma = 1$; (ii) : $\alpha = 2, b = 2, \gamma = 2$; (iii) : $\alpha = 2, b = 2, \gamma = \frac{1}{2}$.

6.3.7. $g_4^*(x)$ of (6.3.10) for (i) : $\alpha = 2, b = 2, \gamma = 1$; (ii) : $\alpha = 2, b = 2, \gamma = 2$; (iii) : $\alpha = 2, b = 2, \gamma = \frac{1}{2}$.

6.4 Laplace Transforms and Moment Generating Functions

The Laplace transforms of the densities in Sections 6.2 and 6.3 can be derived with the help of Mellin convolution procedures. A few of them are derived here and others can be done in a similar way. To start with, consider $g_1(y)$ with Laplace parameter t. This integral will be evaluated by using the Mellin convolution technique.

$$L_{g_1}(t) = \frac{\delta \, a^\alpha \Gamma(\alpha + \beta)}{\Gamma(\alpha)\Gamma(\beta)} \int_0^\infty e^{-ty} y^{\alpha\delta - 1} [1 + ay^\delta]^{-(\alpha+\beta)} dy. \qquad (6.4.1)$$

Consider the Mellin convolution, interpreted in terms of real scalar positive random variables, will be equivalent to evaluating the density of u of the form $u = \frac{x_1}{x_2}$ where x_1 and x_2 are statistically independently distributed real scalar random variables with x_1 having the density $f_1(x_1)$ and x_2 having the density $f_2(x_2)$ with $f_1(x_1) = c_1 e^{-x_1}$ and $f_2(x_2) = c_2 x_2^{\alpha\delta - 1}[1 + ay^\delta]^{-(\alpha+\beta)}$ where c_1 and c_2 are normalizing constants. Then, the density of u is of the form $\int_0^\infty v f_1(uv) f_2(v) dv, u = t$, where the Mellin transforms, denoted by $M_{(\cdot)}(s)$, are the following:

$$M_{f_1}(s) = \Gamma(s), \quad \text{for } \Re(s) > 0$$

$$M_{f_2}(s) = \frac{\Gamma(\alpha - \frac{1}{\delta} + \frac{s}{\delta})\Gamma(\beta - \frac{s}{\delta} + \frac{1}{\delta})}{\Gamma(\alpha)\Gamma(\beta)} \quad \text{for } -\alpha\delta + 1 < \Re(s) < \beta\delta + 1$$

$$M_{f_2}(2 - s) = \frac{\Gamma(\alpha + \frac{1}{\delta} - \frac{s}{\delta})\Gamma(\beta - \frac{1}{\delta} + \frac{s}{\delta})}{\Gamma(\alpha)\Gamma(\beta)} \Rightarrow$$

$$M(L_{g_1}(t) : s) = \frac{\Gamma(s)\Gamma(\alpha + \frac{1}{\delta} - \frac{s}{\delta})\Gamma(\beta - \frac{1}{\delta} + \frac{s}{\delta})}{\Gamma(\alpha)\Gamma(\beta)}.$$

Denoting statistical expectation by $E(\cdot)$, observe that

$$E(u^{s-1}) = E(x_1^{s-1})E(x_2^{(2-s)-1}) = M_{f_1}(s)M_{f_2}(2-s).$$

Taking the inverse Mellin transform, we have

$$L_{g_1}(t) = \frac{\delta\, a^\alpha}{\Gamma(\alpha)\Gamma(\beta)} H_{1,2}^{2,1}\left[t \Big|_{(0,1),(\beta-\frac{1}{\delta},\frac{1}{\delta})}^{(1-\alpha-\frac{1}{\delta},\frac{1}{\delta})}\right] \tag{6.4.2}$$

$$= \frac{a^\alpha}{\Gamma(\alpha)\Gamma(\beta)} H_{1,2}^{2,1}\left[t^\delta \Big|_{(0,\delta),(\beta-\frac{1}{\delta},1)}^{(1-\alpha-\frac{1}{\delta},1)}\right], \tag{6.4.3}$$

where $H(\cdot)$ is the H-function. For the theory and applications of the H-function, see Mathai et al. [10]. The particular cases of (6.4.1) also provide the Laplace transforms (moment generating functions) of a type 2 beta density, F-density, and folded Student-t density. As another example, let us look at the moment generating function of g_3, with parameter t, denoted by $M_{g_3}(t)$. That is,

$$M_{g_3}(t) = \frac{\gamma\, a^\alpha \Gamma(\alpha+\beta)}{\Gamma(\alpha)\Gamma(\beta)} \int_{-\infty}^{\infty} e^{tx-\alpha\gamma x}[1+ae^{-\gamma x}]^{-(\alpha+\beta)}dx.$$

Put $y = e^{-x}$, then $z = y^\gamma$ to obtain the following:

$$M_{g_3}(t) = \frac{a^\alpha \Gamma(\alpha+\beta)}{\Gamma(\alpha)\Gamma(\beta)} \int_0^\infty z^{\alpha-\frac{t}{\gamma}-1}[1+az]^{-(\alpha+\beta)}dz$$

$$= \frac{a^{\frac{t}{\gamma}}}{\Gamma(\alpha)\Gamma(\beta)} \Gamma(\alpha-\frac{t}{\gamma})\Gamma(\beta+\frac{t}{\gamma}) \tag{6.4.4}$$

for $-\Re(\beta\gamma) < \Re(t) < \Re(\alpha\gamma)$. The moments are available by differentiation and evaluating at $t = 0$. For example, in $g_3(x)$, the mean value $E(x)$ is given by

$$E(x) = \frac{1}{\gamma}[1 + \psi(\beta) - \psi(\alpha)] \tag{6.4.5}$$

where $\psi(\cdot)$ is the psi function. As another example, the moment generating function of g_4^* is the following:

$$M_{g_4^*}(t) = \frac{b^\alpha \delta}{\Gamma(\alpha)} \int_{-\infty}^{\infty} e^{tx+\alpha\delta x} e^{-be^{\delta x}} dx \tag{6.4.6}$$

$$= \frac{b^\alpha \delta}{\Gamma(\alpha)} \int_0^\infty y^{t+\alpha\delta-1} e^{-by^\delta} dy$$

$$= b^{-t/\delta} \frac{\Gamma(\alpha+\frac{t}{\delta})}{\Gamma(\alpha)}, \quad \Re(t) > -\Re(\alpha\delta). \tag{6.4.7}$$

The moments are available in terms of psi and zeta functions. For example, the first moment is given by

$$E(x) = \frac{1}{\delta}[-\ln b + \psi(\alpha)]. \tag{6.4.8}$$

Proceeding in the same fashion, the moment generating function of $f_5(x)$ is given by the following:

$$M_{f_5}(t) = \frac{\gamma b^\alpha}{\Gamma(\alpha)} \int_{-\infty}^{\infty} e^{tx - \gamma \alpha x} e^{-be^{-\gamma x}} dx \tag{6.4.9}$$

$$= b^{\frac{t}{\gamma}} \frac{\Gamma(\alpha - \frac{t}{\gamma})}{\Gamma(\alpha)}, \quad \Re(t) < \Re(\alpha\gamma). \tag{6.4.10}$$

And the first moment is $E(x) = \frac{1}{\gamma}[\ln b - \psi(\alpha)]$. As another example, consider the moment generating function of $f_1(x)$.

$$M_{f_1}(t) = \frac{\delta \, a^\alpha \Gamma(\alpha + \beta)}{\Gamma(\alpha)\Gamma(\beta)} \int_0^{a^{-1/\delta}} e^{tx} x^{\alpha\delta - 1}[1 - ax^\delta]^{\beta - 1} dx$$

$$= \sum_{k=0}^{\infty} \frac{t^k}{k!} a^\alpha \frac{\Gamma(\alpha + \beta)}{\Gamma(\alpha)\Gamma(\beta)} \int_0^{1/a} y^{\alpha + \frac{(k+t)}{\delta} - 1}(1 - ay)^{\beta - 1} dy$$

$$= a^{-\frac{t}{\delta}} \sum_{k=0}^{\infty} \frac{\Gamma(\alpha + \frac{(t+k)}{\delta})}{\Gamma(\alpha)} \frac{\Gamma(\alpha + \beta)}{\Gamma(\alpha + \beta + \frac{(t+k)}{\delta})} \frac{(ta^{-1/\delta})^k}{k!} \tag{6.4.11}$$

for $\Re(\alpha + \frac{t}{\delta}) > 0$. This can be written as a Wright's function.

$$M_{f_1}(t) = \frac{\Gamma(\alpha + \beta)}{\Gamma(\alpha)} a^{-\frac{t}{\delta}} {}_1\psi_1 \left[\frac{t}{a^{1/\delta}} \, \middle| \, \begin{matrix} (\alpha + \frac{t}{\delta}, \frac{1}{\delta}) \\ (\alpha + \beta + \frac{t}{\delta}, \frac{1}{\delta}) \end{matrix} \right], \tag{6.4.12}$$

which can be written in terms of a confluent hypergeometric series ${}_1F_1$ for $\delta = 1$. Moments are available by differentiation and can be written in terms of psi and zeta functions.

6.5 Models with Appended Bessel Series

Another way of controlling the tail behavior of a given model is to append a Bessel series to it so that one can create models with thicker or thinner tails compared to the original model without the appended Bessel series. We will illustrate this procedure on $f_2(y)$ of (6.2.5). Consider the appended model, denoted by $\tilde{f}_2(y)$ with the new parameter c, where c could be positive or negative or zero. When $c = 0$, the Bessel series will be taken as 1.

$$\tilde{f}_2(y) = \sum_{k=0}^{\infty} \frac{c^k \delta}{k!} \frac{y^{\delta k}}{(\alpha)_k} \frac{b^\alpha e^{-c/b} y^{\alpha\delta-1} e^{-by^\delta}}{\Gamma(\alpha)}, \quad 0 \le y < \infty, \delta > 0, \alpha > 0. \quad (6.5.1)$$

The net effect of appending such a Bessel series can be seen from the following graphs, where $c = 0$ corresponds the graph without the appended Bessel series.

Exercises 6.5.

By using MAPLE or MATHEMATICA or any other mathematical word processor with graphic facilities, draw the graphs of the following functions and examine the changes:

6.5.1. $\tilde{f}_2(y)$ of (6.5.1) for (i) : $\delta = 1, b = 1, \alpha = 2, c = 0$; (ii) : $\delta = 2, b = 1, \alpha = 2, c = 1$; (iii) : $\delta = 2, b = 1, \alpha = 2, c = 2$.

6.5.2. $\tilde{f}_2(y)$ of (6.5.1) for negative c: (i) : $\delta = 2, b = 1, \alpha = 2, c = 0$; (ii) : $\delta = 2, b = 1, \alpha = 2, c = -1$; (iii) : $\delta = 2, b = 1, \alpha = 2, c = -2$.

6.6 Models with Appended Mittag-Leffler Series

Another technique of controlling the shape of a curve is to append it with a Mittag-Leffler series. We will illustrate this on the same model $f_2(y)$ of (6.2.5). Consider the following density:

$$f_{21}(y) = \sum_{k=0}^{\infty} \frac{c^k y^{\delta\gamma k}}{\Gamma(\alpha + \gamma k)} \delta b^\alpha (1 - cb^{-\gamma}) y^{\alpha\delta-1} e^{-by^\delta}, \quad 0 \le y < \infty, b > 0, \delta > 0, \alpha > 0,$$

$$(6.6.1)$$

for $|cb^{-\gamma}| < 1$, where c could be positive or negative. We can also consider a more general Mittag-Leffler series of the following form:

$$f_{22}(y) = \sum_{k=0}^{\infty} \frac{c^k (\eta)_k y^{\delta\gamma k}}{k! \Gamma(\alpha + \gamma k)} (1 - cb^{-\gamma})^\eta \delta b^\alpha y^{\alpha\delta-1} e^{-by^\delta}, \quad 0 \le y < \infty, b > 0, \alpha > 0, \delta > 0, \gamma > 0$$

$$(6.6.2)$$

and for $|cb^{-\gamma}| < 1$ and η free. Note that the case $c = 0$ is the one without the appended form. The behavior may be noted from the following graphs:

Exercises 6.6.

By using MAPLE or MATHEMATICA or any other mathematical word processor with graphic facilities, draw the graphs of the following functions and examine the changes:

6.6.1. $f_{21}(y)$ of (6.6.1) for (i): $\delta = 2, b = 1, \alpha = 2, c = 0$; (ii): $\delta = 2, b = 1, \alpha = 2, c = \frac{1}{3}$; (iii): $\delta = 2, b = 1, \alpha = 2, c = \frac{1}{2}$.

6.6.2. $f_{21}(y)$ of (6.6.1) for negative x: (i) $\delta = 2, b = 1, \alpha = 2, c = 0$; (ii): $\delta = 2, b = 1, \alpha = 2, c = -1$; (iii): $\delta = 2, b = 1, \alpha = 2, c = -\frac{1}{2}$.

6.6.3. $f_{22}(y)$ of (6.6.2) for (i): $\delta = 2, b = 1, \alpha = 2, \eta = 2, c = 0$; (ii): $\delta = 2, b = 1, \alpha = 2, \eta = 2, c = \frac{1}{3}$; (iii): $\alpha = 2, \delta = 2, b = 1, \eta = 2, c = \frac{1}{2}$.

6.6.4. $f_{22}(y)$ of (6.6.2) for negative c: (i): $\alpha = 2, b = 1, \delta = 2, \eta = 2, c = 0$; (i): $\alpha = 2, b = 1, \delta = 2, \eta = 2, c = -\frac{1}{3}$; (iv): $\alpha = 2, b = 1, \delta = 2, \eta = 2, c = -\frac{1}{2}$.

6.7 Bayesian Outlook

Let us start with $g_4^*(y|b)$ with $x = \gamma y$, b replaced by $b\delta$, α replaced by γ. Let the parameter b in $g_4^*(y|b)$ has a prior exponential density e^{-b}, $0 \le b < \infty$ and zero elsewhere, then the unconditional density is given by

$$\frac{\delta^\gamma (e^x)^\gamma}{\Gamma(\gamma)} \int_0^\infty b^\gamma e^{-b(1+\delta e^x)} db$$

$$= \frac{\gamma \delta^\gamma (e^x)^\gamma}{[1 + \delta(e^x)]^{\gamma+1}}, \quad -\infty < x < \infty.$$

This is a particular case of the generalized logistic model introduced by Mathai and Provost [9]. If b has a prior density of the form

$$f_b(b) = \frac{t^\eta b^{\eta-1}}{\Gamma(\eta)} e^{-tb}, \quad t > 0, b \ge 0,$$

then the unconditional density of x is given by

$$f_x(x) = \frac{\delta^\gamma (e^x)^\gamma t^\eta}{\Gamma(\gamma)\Gamma(\eta)} \int_0^\infty b^{\gamma+\eta-1} e^{-b(t+\delta e^x)} db$$

$$= \frac{\delta^\gamma \Gamma(\gamma+\eta)}{t^\eta \Gamma(\gamma)\Gamma(\eta)} (e^x)^\gamma [1 + \frac{\delta}{t} e^x]^{-(\gamma+\eta)}, \quad -\infty < x < \infty$$

for $\gamma > 0, \eta > 0, \delta > 0, t > 0$. A more general form is available by taking $x = \beta y$ for some β. Then, this is a generalized form of $g_3^*(y)$ with the scaling factor $\frac{\delta}{t}$ for e^x. Note that if $u = e^x$, then the density of u is a generalized type 2 beta density, and for $\gamma = \frac{m}{2}, \eta = \frac{n}{2}$, $m, n = 1, 2, \ldots$, then the generalized type 2 beta is a variance-ratio density or F-density. If $y = e^x$ in $f_x(x)$, then the density is that of Beck-Cohen superstatistics in physics.

6.8   Pathway Model in the Scalar Case

A model which can switch around from one functional form to another is the pathway idea so that one can capture the transitional stages or in between two functional forms. If one functional form is the stable case in a physical situation, then the unstable neighborhoods are also captured by the pathway idea, a pathway leading to the stable physical situation. Consider the following function $\phi_1(x)$ of a real scalar variable x

$$\phi_1(x) = c_1 \, |x|^{\gamma-1}[1 - a(1-\alpha)|x|^{\delta}]^{\frac{\eta}{1-\alpha}}, \quad -\infty < x < \infty \qquad (6.8.1a)$$

where $a > 0, \delta > 0, \eta > 0$, and c_1 can act as a normalizing constant if we wish to create a statistical density out of $\phi_1(x)$. Here, α is the pathway parameter. For $\alpha < 1$, (6.8.1a) is a generalized type 1 beta form and the normalizing constant c_1 can be evaluated to be the following, if we wish to make a statistical density out of the form in (6.8.1a):

$$c_1^{-1} = \frac{1}{2} \frac{\Gamma(\frac{\gamma}{\delta})\Gamma(\frac{\eta}{1-\alpha}+1)}{\delta \, [a(1-\alpha)]^{\frac{\gamma}{\delta}}\Gamma(1+\frac{\eta}{1-\alpha}+\frac{\gamma}{\delta})}, \quad \alpha < 1 \qquad (6.8.1b)$$

For $\alpha > 1$, the model in (6.8.1a) goes to the form

$$\phi_2(x) = c_2|x|^{\gamma-1}[1 + a(\alpha-1)|x|^{\delta}]^{-\frac{\eta}{\alpha-1}}, \quad -\infty < x < \infty \qquad (6.8.2a)$$

where c_2 can act as a normalizing constant. Note that (6.8.2a) is an extended form of type 2 beta model where a number of standard densities such as the F-density, Student-t density, and Fermi–Dirac density are special cases. The normalizing constant can be seen to be the following:

$$c_2^{-1} = \frac{1}{2} \frac{\Gamma(\frac{\gamma}{\delta})\Gamma(\frac{\eta}{\alpha-1}-\frac{\gamma}{\delta})}{\delta \, [a(\alpha-1)]^{\frac{\gamma}{\delta}}\Gamma(\frac{\eta}{\alpha-1})}, \quad \alpha > 1 \qquad (6.8.2b)$$

for $\Re(\frac{\eta}{\alpha-1} - \frac{\gamma}{\delta}) > 0$ and other conditions on the parameters remain the same. Note that when $\alpha \to 1$, from the left or from the right, $\phi_1(x)$ and $\phi_2(x)$ go to the limiting form

$$\phi_3(x) = c_3|x|^{\gamma-1}e^{-a\eta|x|^{\delta}}, \quad -\infty < x < \infty \qquad (6.8.3a)$$

where the conditions on the parameters remain the same as before, and if we wish to have a statistical density out of $\phi_3(x)$, then the normalizing constant c_3 will be of the following form:

$$c_3^{-1} = \frac{1}{2} \frac{\Gamma(\frac{\gamma}{\delta})}{\delta \, [a\eta]^{\frac{\gamma}{\delta}}} \qquad (6.8.3b)$$

where the conditions on the parameters remain the same.

Note that in a physical situation if the exponential form or the model in (6.8.3a) is the limiting form or the stable situation, then all neighborhoods leading to this stable form or the paths leading to the stable situation will be described by (6.8.1a) and (6.8.2a). Here, the parameter α acts as a pathway parameter providing all paths leading to the limiting situation in (6.8.3a). The models in (6.8.1a), (6.8.2a) and (6.8.3a) cover a wide range of statistical densities in different fields, in fact, almost all univariate statistical densities in current use, as models for different physical situations, are special cases of the model in (6.8.1a) for the three cases, $\alpha < 1, \alpha > 1$, and $\alpha \to 1$. Some of the special cases are the following:

$\alpha \to 1, \gamma = 1, a = 1, \delta = 2$	Gaussian or normal density for $-\infty < x < \infty$
$\alpha \to 1, \gamma = 2, a = 1, \delta = 2, x > 0$	Maxwell-Bolzmann density in physics
$\alpha \to 1, \gamma = 1, a = 1, \delta = 2, x > 0$	Rayleigh density
$\alpha \to 1, \gamma = n - 3, \eta = \frac{n}{2\sigma^2}, \sigma > 0,$ $\delta = 2, x > 0$	—— Helmert density
$\alpha = 2, a = \frac{1}{\nu}, \eta = \frac{\nu+1}{2}, \gamma = 1, \delta = 2$	Student-t density, $-\infty < x < \infty$
$\alpha = 2, \gamma = 1, \eta = 1, \delta = 2$	Cauchy density, $-\infty < x < \infty$
$\alpha = 0, a = 1, \gamma > 0, \delta = 1, \eta = \beta - 1,$ $\beta > 0, x > 0$	Standard type 1 beta density ——
$\alpha = 2, a = 1, \delta = 1, \gamma > 0, \eta = \gamma + \epsilon,$ $\epsilon > 0, x > 0$	—— Standard type 2 beta density
$\gamma = 1, a = 1, \eta = 1, \delta = 1, x > 0$	Tsallis statistic for $\alpha < 1, \alpha > 1,$ $\alpha \to 1$
$- - - - - - - - - - - - - - - -$	Power law, q-binomial density
$\alpha = 0, \gamma = 1, \delta = 1$	Triangular density
$\alpha = 2, \gamma = \frac{m}{2}, a = \frac{m}{n}, \eta = \frac{m+n}{2}, \delta = 1, x > 0$	F-density
$\alpha \to 1, \gamma = 1, a = 1, \delta = 1, \eta = \frac{mg}{KT}, x > 0$	Helley's density in physics
$\alpha \to 1, a = 1, \delta = 1, x > 0$	Gamma density
$\alpha \to 1, \gamma = \frac{\nu}{2}, a = 1, \delta = 1, \eta = \frac{1}{2}, x > 0$	Chisquare density with ν degrees of freedom
$\alpha \to 1, a = 1, \gamma = 1, \delta = 1$	Double exponential of Laplace density
$\alpha \to 1, a = 1, \gamma = 1, x > 0$	Generalized gamma density
$\alpha \to 1, a = 1, \delta = \gamma$	Weibull density
$\alpha = 2, a = 1, \gamma = 1, x = e^y, \delta = 1, \eta = 2,$ $x > 0, -\infty < y < \infty$	—— Logistic density
$\alpha = 2, \gamma = 1, a = 1, x = e^{a+by}, \eta = 1, y > 0$	Has Fermi–Dirac desnity

$q < 1, a(1-q) = 1$	Extended type 1 beta density
$q < 1, a(1-q) = 1, Y = XBX'$	Nonstandard type 1 beta density
$q < 1, a(1-q) = 1, Y = A^{\frac{1}{2}}XBX'A^{\frac{1}{2}}$	Standard type 1 beta density
$q < 1, a(1-q) = 1, \alpha = 0, \beta = 0$	Extended uniform density
$q < 1, a(1-q) = 1, \alpha + \frac{r}{2} = \frac{p+1}{2}, \beta = 0,$	———
$\quad Y = A^{\frac{1}{2}}XBX'A^{\frac{1}{2}}$	Standard uniform density
$q < 1, a(1-q) = 1, \alpha = 0, \beta = 1$	A q-binomial density
$q < 1, a(1-q) = 1, \alpha = 0,$	
$\quad \frac{\beta}{1-q} = \frac{1}{2}(m-p-r-1)$	Inverted T density of Dickey
$q > 1, a(q-1) = 1$	Extended type 2 beta density
$q > 1, a(q-1) = 1, Y = XBX'$	Nonstandard type 2 beta density
$q > 1, a(q-1) = 1, \frac{\beta}{q-1} = \frac{m}{2}, \alpha = 0$	T density of Dickey
$q > 1, a(q-1) = 1, \alpha + \frac{r}{2} = \frac{p+1}{2}, \frac{\beta}{q-1} = 1,$	———
$\quad Y = A^{\frac{1}{2}}XBX'A^{\frac{1}{2}}$	Standard T density
$q > 1, a(q-1) = 1, \alpha + \frac{r}{2} = \frac{p+1}{2}, \frac{\beta}{q-1} = 1,$	———
$\quad Y = A^{\frac{1}{2}}XBX'A^{\frac{1}{2}}$	Standard Cauchy density
$q > 1, q(q-1) = \frac{m}{n}, \alpha + \frac{r}{2} = \frac{m}{2}, \frac{\beta}{q-1} = \frac{m+n}{2},$	———
$\quad Y = A^{\frac{1}{2}}XBX'A^{\frac{1}{2}}$	Standard F-density
$q = 1, a = 1, \beta = 1, Y = A^{\frac{1}{2}}XBX'A^{\frac{1}{2}}$	Standard gamma density
$q = 1, a = 1, \beta = 1, \alpha = 0$	Gaussian density
$q = 1, a = 1, \beta = 1, \alpha + \frac{r}{2} = \frac{n}{2},$	———
$\quad Y = XBX', A = \frac{1}{2}V^{-1}$	Wishart density

6.9   Pathway Model in the Matrix-Variate Case

Here, we consider the original pathway model of Mathai [6]. Only the situation of real case is considered here but the corresponding case, where the elements are in the complex domain, is also available. Consider a $p \times r$ matrix $X = (x_{ij}), i = 1, ..., p, j = 1, ..., r$ of full rank p with $r \geq p$. Let x_{ij}'s be distinct real variables. Let us consider a real-valued scalar function of X,

$$f(X) = c|A^{\frac{1}{2}}XBX'A^{\frac{1}{2}}|^{\alpha}|I - a(1-q)A^{\frac{1}{2}}XBX'A^{\frac{1}{2}}|^{\frac{\beta}{1-q}} \qquad (6.9.1)$$

where A and B are constant $p \times p$ and $r \times r$ real positive definite matrices, respectively, that is $A = A' > O, B = B' > O$. Also, $|(\cdot)|$ denotes the determinant of $(\cdot)$, and $\text{tr}(\cdot)$ denotes the trace of the matrix $(\cdot)$. The positive definite square roots of A

and B are denoted by $A^{\frac{1}{2}}$ and $B^{\frac{1}{2}}$, respectively. A prime will denote the transpose, and I denotes the identity matrix. If (6.9.1) is to be treated as a statistical density, then c can act as a normalizing constant and it can be evaluated by using the following procedure. Let $Y = A^{\frac{1}{2}}XB^{\frac{1}{2}} \Rightarrow dY = |A|^{\frac{r}{2}}|B|^{\frac{p}{2}}dX$ where dY stands for the wedge product of the pr differentials in Y and the corresponding wedge product in X is denoted by dX. For example,

$$dX = \wedge_{i=1}^{p} \wedge_{j=1}^{r} dx_{ij}. \tag{6.9.2}$$

For the Jacobians of matrix transformations, see Mathai [4]. Under the transformation $YY' = U$ and integrating over the Stiefel manifold, we have [see also Mathai [4]]

$$U = YY' \Rightarrow dY = \frac{\pi^{\frac{rp}{2}}}{\Gamma(\frac{r}{2})}|U|^{\frac{r}{2}-\frac{p+1}{2}} \tag{6.9.3}$$

where, for example,

$$\Gamma_p(\alpha) = \pi^{\frac{p(p-1)}{4}}\Gamma(\alpha)\Gamma(\alpha - \frac{1}{2})...\Gamma(\alpha - \frac{p-1}{2}), \Re(\alpha) > \frac{p-1}{2} \tag{6.9.4}$$

is the matrix-variate gamma in the real case. Now, let

$$V = a(1-q)U \Rightarrow dV = [a(1-q)]^{\frac{p(p+1)}{2}}dU.$$

If $f(X)$ is a density, then the total integral over X must be unity and therefore

$$1 = \int_X f(X)dX = \frac{c}{|A|^{\frac{r}{2}}|B|^{\frac{p}{2}}} \int_Y |YY'|^{\alpha}|I - a(1-q)YY'|^{\frac{\beta}{1-q}}dY$$

$$= \frac{\pi^{\frac{rp}{2}}}{\Gamma_p(\frac{r}{2})|A|^{\frac{r}{2}}|B|^{\frac{p}{2}}} \int_U |U|^{\alpha+\frac{r}{2}-\frac{p+1}{2}}|I - a(1-q)U|^{\frac{\beta}{1-q}}dU. \tag{6.9.5}$$

From here, we consider three cases separately, namely $q < 1, q > 1, q \to 1$. Let the normalizing constants be denoted by $c_{*1}, c_{*2}, and\ c_{*3}$, respectively. For $q < 1$, $a(1-q) > 0$ and $V = a(1-q)U$ give

$$c_{*1}^{-1} = \frac{\pi^{\frac{rp}{2}}}{\Gamma_p(\frac{r}{2})|A|^{\frac{r}{2}}|B|^{\frac{p}{2}}[a(1-q)]^{p(\alpha+\frac{r}{2})}}$$

$$\times \int_V |V|^{\alpha+\frac{r}{2}-\frac{p+1}{2}}|I - V|^{\frac{\beta}{1-q}}dV.$$

This integral can be evaluated by using a matrix-variate real type 1 beta integral [see Mathai [4]] and then for $q < 1$,

$$c_{*1}^{-1} = \frac{\pi^{\frac{rp}{2}}}{\Gamma_p(\frac{r}{2})|A|^{\frac{r}{2}}|B|^{\frac{p}{2}}[a(1-q)]^{p(\alpha+\frac{r}{2})}}$$
$$\times \frac{\Gamma_p(\alpha+\frac{r}{2})\Gamma_p(\frac{\beta}{1-q}+\frac{p+1}{2})}{\Gamma_p(\alpha+\frac{r}{2}+\frac{\beta}{1-q}+\frac{p+1}{2})} \tag{6.9.5a}$$

for $\alpha + \frac{r}{2} > \frac{p-1}{2}$. When $q > 1$, the transformation is $V = a(q-1)U$ and proceeding as before and evaluating the integral by using a matrix-variate real type 2 integral [see Mathai [4]] we have, for $q > 1$,

$$c_{*2}^{-1} = \frac{\pi^{\frac{rp}{2}}}{\Gamma_p(\frac{r}{2})|A|^{\frac{r}{2}}|B|^{\frac{p}{2}}[a(q-1)]^{p(\alpha+\frac{r}{2})}}$$
$$\times \frac{\Gamma_p(\alpha+\frac{r}{2})\Gamma_p(\frac{\beta}{q-1}-\alpha-\frac{r}{2})}{\Gamma_p(\frac{\beta}{q-1})} \tag{6.9.5b}$$

for $\alpha + \frac{r}{2} > \frac{p-1}{2}$, $\frac{\beta}{q-1} - \alpha - \frac{r}{2} > \frac{p-1}{2}$. When $q \to 1$, then both the forms in (6.9.5a) and (6.9.5b) will go to the following form when $q \to 1$:

$$c_{*3}^{-1} = \frac{\pi^{\frac{rp}{2}}\Gamma_p(\alpha+\frac{r}{2})}{\Gamma_p(\frac{r}{2})|A|^{\frac{r}{2}}|B|^{\frac{p}{2}}a^{p(\alpha+\frac{r}{2})}} \tag{6.9.5c}$$

for $\alpha + \frac{r}{2} > \frac{p-1}{2}$. The functional part in the type 1 and type 2 types of beta integrals can be seen to go to the exponential type integral in the light of the following lemma.

Lemma 1. *For $a > 0$, U any $p \times p$ matrix [in our case, U is positive definite also]*

$$\lim_{q \to 1} |I - a(1-q)U|^{\frac{\beta}{1-q}} = e^{-a\beta \text{tr}(U)}. \tag{6.9.6}$$

Note that if the eigenvalues of U are $\lambda_j, j = 1, ..., p$, determinant being the product of the eigenvalues and the eigenvalues of $I - a(1-q)U$ being $1 - a(1-q)\lambda_j$, we have

$$|I - a(1-q)U| = \prod_{j=1}^{p}(1 - a(1-q)\lambda_j) \Rightarrow$$

$$\lim_{q \to 1} |I - a(1-q)U|^{\frac{\beta}{1-q}} = \prod_{j=1}^{p} \lim_{q \to 1}[1 - a(1-q)\lambda_j]^{\frac{\beta}{1-q}}$$

$$= \prod_{j=1}^{p} e^{-a\beta\lambda_j} = e^{-a\beta \sum_{j=1}^{p}\lambda_j}$$

$$= e^{-a\beta \text{tr}(U)}.$$

Then, from Stirling's formula or from the first approximation of the asymptotic expansion of gamma functions

$$\Gamma(z+a) \approx \sqrt{2\pi} z^{z+a-\frac{1}{2}} e^{-z} \text{ for } |z| \to \infty,$$

and a bounded, we have the following result:

Lemma 2. *For $q \to 1$*

$$\lim_{q \to 1} c_{*1} = \lim_{q \to 1} c_{*2} = c_{*3}.$$

Thus, the densities corresponding to $q < 1$ and $q > 1$ will go to the density for the case of $q \to 1$. Thus, we have a matrix-variate model which switches around among three different functional forms, namely, the generalized matrix-variate type 1 beta model, generalized matrix-variate type 2 beta model, and the generalized matrix-variate gamma model. Thus, q here is the pathway parameter covering all families within the cases $q < 1, q > 1$, and $q \to 1$ as well as all transitional stages for $-\infty < q < \infty$.

6.9.1 Arbitrary Moments

Arbitrary h-th moments are available for the three cases $q < 1, q > 1$, and $q \to 1$ of the pathway density from the normalizing constant itself. This will be stated in the form of a theorem.

Theorem 6.1.

$$E|A^{\frac{1}{2}}XBX'A^{\frac{1}{2}}|^h = \frac{1}{[a(1-q)]^{ph}} \frac{\Gamma_p(\alpha+h+\frac{r}{2})}{\Gamma_p(\alpha+\frac{r}{2})} \frac{\Gamma_p(\alpha+\frac{r}{2}+\frac{\beta}{1-q}+\frac{p-1}{2})}{\Gamma_p(\alpha+h+\frac{r}{2}+\frac{\beta}{1-q}+\frac{p-1}{2})}$$

$$\text{for } q < 1, \alpha+h+\frac{r}{2} > \frac{p-1}{2} \tag{6.9.7a}$$

$$= \frac{1}{[a(q-1)]^{ph}} \frac{\Gamma_p(\alpha+h+\frac{r}{2})}{\Gamma_p(\alpha+\frac{r}{2})} \frac{\Gamma_p(\frac{\beta}{q-1}-\alpha-h-\frac{r}{2})}{\Gamma_p(\frac{\beta}{q-1}-\alpha-\frac{r}{2})}$$

$$\text{for } q > 1, \frac{\beta}{q-1}-\alpha-h-\frac{r}{2} > \frac{p-1}{2}, \alpha+h+\frac{r}{2} > \frac{p-1}{2} \tag{6.9.7b}$$

$$= \frac{1}{(a\beta)^{ph}} \frac{\Gamma_p(\alpha+h+\frac{r}{2})}{\Gamma_p(\alpha+\frac{r}{2})}$$

$$\text{for } q \to 1, \alpha+h+\frac{r}{2} > \frac{p-1}{2}. \tag{6.9.7c}$$

6.9.2   Some Special Cases

Some special cases of the matrix-variate pathway model are given below. It may be noted that almost all real matrix-variate densities in current use are available from the pathway model described above.

$q < 1, a(1 - q) = 1$	Extended type 1 beta density
$q < 1, a(1 - q) = 1, Y = XBX'$	Nonstandard type 1 beta density
$q < 1, a(1 - q) = 1, Y = A^{\frac{1}{2}}XBX'A^{\frac{1}{2}}$	Standard type 1 beta density
$q < 1, a(1 - q) = 1, \alpha = 0, \beta = 0$	Extended uniform density
$q < 1, a(1 - q) = 1, \alpha + \frac{r}{2} = \frac{p+1}{2}, \beta = 0,$	———
$Y = A^{\frac{1}{2}}XBX'A^{\frac{1}{2}}$	Standard uniform density
$q < 1, a(1 - q) = 1, \alpha = 0, \beta = 1$	A q-binomial density
$q < 1, a(1 - q) = 1, \alpha = 0,$	
$\frac{\beta}{1-q} = \frac{1}{2}(m - p - r - 1)$	Inverted T density of Dickey
$q > 1, a(q - 1) = 1$	Extended type 2 beta density
$q > 1, a(q - 1) = 1, Y = XBX'$	Nonstandard type 2 beta density
$q > 1, a(q - 1) = 1, \frac{\beta}{q-1} = \frac{m}{2}, \alpha = 0$	T density of Dickey
$q > 1, a(q - 1) = 1, \alpha + \frac{r}{2} = \frac{p+1}{2}, \frac{\beta}{q-1} = 1,$	———
$Y = A^{\frac{1}{2}}XBX'A^{\frac{1}{2}}$	Standard T dendity
$q > 1, a(q - 1) = 1, \alpha + \frac{r}{2} = \frac{p+1}{2}, \frac{\beta}{q-1} = 1,$	———
$Y = A^{\frac{1}{2}}XBX'A^{\frac{1}{2}}$	Standard Cauchy density
$q > 1, q(q - 1) = \frac{m}{n}, \alpha + \frac{r}{2} = \frac{m}{2}, \frac{\beta}{q-1} = \frac{m+n}{2},$	———
$Y = A^{\frac{1}{2}}XBX'A^{\frac{1}{2}}$	Standard F-density
$q = 1, a = 1, \beta = 1, Y = A^{\frac{1}{2}}XBX'A^{\frac{1}{2}}$	Standard gamma density
$q = 1, a = 1, \beta = 1, \alpha = 0$	Gaussian density
$q = 1, a = 1, \beta = 1, \alpha + \frac{r}{2} = \frac{n}{2},$	———
$Y = XBX', A = \frac{1}{2}V^{-1}$	Wishart density

References

1. Beck, C. (2006). Stretched exponentials from superstatistics. *Physica A, 365,* 96–101.
2. Beck, C., & Cohen, E. G. D. (2003). Superstatistics. *Physica A, 322,* 267–275.
3. Mathai, A. M. (1993). *A handbook of generalized special functions for statistical and physical sciences.* Oxford: Oxford University Press.

4. Mathai, A. M. (1997). *Jacobians of matrix transformations and functions of matrix argument.* New York: World Scientific Publishing.
5. Mathai, A. M. (2003). Order statistics from a logistic distribution and applications to survival & reliability analysis. *IEEE Transactions on Reliability, 52*(2), 200–206.
6. Mathai, A. M. (2005). A pathway to matrix-variate gamma and normal densities. *Linear Algebra and Its Applications, 396,* 317–328.
7. Mathai, A. M., & Haubold, H. J. (2007). Pathway model, superstatistics, Tsallis statistics and a generalized measure of entropy. *Physica A, 375,* 110–123.
8. Mathai, A. M., & Provost, S. B. (2004). On the distribution of order statistics from generalized logistic samples. *Metron, LXII*(1), 63–71.
9. Mathai, A. M., & Provost, S. B. (2006). On q-logistic and related distributions. *IEEE Transactions on Reliability, 55*(2), 237–244.
10. Mathai, A. M., Saxena, R. K., & Haubold, H. J. (2010). *The H-function: theory and applications.* New York: Springer.
11. Tsallis, C. (1988). Possible generalization of Boltzmann-Gibbs statistics. *Journal of Statistical Physics, 52,* 479–487.

Index

A
Asymmetric models, 212

B
Bayesian outlook, 219
Bessel series, 219
Birth and death process, 109
Branching process, 109
Buffon's clean tile problem, 114

C
Canonical variables, 164
Cauchy problem, 25
Cauchy–Swartz inequality, 42
Conditional expectation, 127
Correlation, 146
Correlation analysis, 150
Covariance, 137

D
Differentiability, 53
Dot product, 42

E
Entropy, 96
Eulerian functions, 3
Expected value, 118

F
Fractional calculus, 1
Fractional derivative, 16
Fractional differential equation, 23

Fractional diffusion, 23
Fractional integral, 15
Fractional oscillation, 19
Fractional relaxation, 19

G
Gamma function, 3
Gauss–Markov theorem, 186
Generalized inverse, 185

I
Input–output models, 94
Input–output process, 73

M
Method of least squares, 172
Mittag-Leffler function, 5
Multiple correlation coefficient, 142

O
Optimal designs, 183
Optimal recovery, 74
Optimality criterion, 198
Optimization, 64

P
Partitioned matrices, 144
Pathway model, 95, 210
Power transformation, 209
Probability, 112
Pythagorean identity, 43

© Springer International Publishing AG 2017
A.M. Mathai and H.J. Haubold, *Fractional and Multivariable Calculus*,
Springer Optimization and Its Applications 122, DOI 10.1007/978-3-319-59993-9

R
Regression, 127

S
Sandwich theorem, 51
Signaling problem, 25

T
Thermal field, 46

V
Variance, 140
Vector, 40

W
Wave equation, 26
Wright's function, 8

Author Index

A
Abramowitz, M., 13
Atkinson, A. C., 200

B
Bagian, 14
Bagiatis, 198
Beck, C., 214
Bender, C. M., 12
Buchen, P. W., 15

C
Caputo, M., 24
Cohen, E. G. D., 214

D
Davis, T. A., 98
Doetsch, G., 13–15
Dzherbashian, M. M., 14

F
Fedorov, V. V., 205, 206

G
Golomb, M., 87
Gorenflo, R., 9, 17, 23, 28

H
Haubold, H. J., 93–95, 103, 104, 210
Hedayat, A. S., 202
Holmes, R. B., 87

Humbert, P., 15

J
Jacroux, M., 202

K
Kounias, S., 200

L
Lichko, Yu., 9, 28

M
Mainardi, F., 1, 9, 11–15, 17, 23, 24, 28, 29
Majumdar, D., 202
Marichev, O. I., 2, 3
Mathai, A. M., 93–96, 98, 103, 104, 149,
    210, 211, 214, 215, 218, 224–226
Mhaskar, H. N., 76, 79, 84
Micchelli, C. A., 83, 87, 88
Mikusiʹnki, J., 15
Mittag-Leffler, M. G., 1, 5–8, 13, 14, 19, 20,
    103, 218
Mura, A., 24, 28

O
Orszag, S. A., 12

P
Pagnini, G., 28
Pai, D. V., 76, 79, 84
Pollard, H., 15
Provost, S. B., 215

© Springer International Publishing AG 2017
A.M. Mathai and H.J. Haubold, *Fractional and Multivariable Calculus*,
Springer Optimization and Its Applications 122, DOI 10.1007/978-3-319-59993-9

R
Rathie, P. N., 149
Rivlin, T. J., 76, 83, 87, 88

S
Sathe, Y. S., 200
Saxena, R. K., 103, 104
Schneider, W. R., 24
Shah, K. R., 200, 202
Shenoy, R. G., 200
Silvey, S. D., 202, 203
Sinha, B. K., 200, 202
Stankoviè, B., 13, 15
Stegun, I. A., 13

T
Tomirotti, M., 11, 12, 24, 29
Tricomi, F. G., 9
Tsallis, C., 94, 98, 213

W
Weinberger, H., 87
Winograd, S., 83
Wong, A., 9
Wright, E. M., 8

Z
Zhao, Y.-Q., 9

Printed in the United States
By Bookmasters

Lecture Notes in Computer Science   13386

More information about this series at https://link.springer.com/bookseries/558

Alessa Hering · Julia Schnabel ·
Miaomiao Zhang · Enzo Ferrante ·
Mattias Heinrich · Daniel Rueckert (Eds.)

Biomedical Image Registration

10th International Workshop, WBIR 2022
Munich, Germany, July 10–12, 2022
Proceedings

 Springer

Editors
Alessa Hering
Radboud University Nijmegen Medical
Center
Nijmegen, The Netherlands

Miaomiao Zhang ⓘ
University of Virginia
Charlottesville, VA, USA

Mattias Heinrich ⓘ
Universität zu Lübeck
Lübeck, Germany

Julia Schnabel ⓘ
Helmholtz Center Munich
Neuherberg, Germany

Enzo Ferrante ⓘ
CONICET, Universidad Nacional del Litoral
Santa Fe, Argentina

Daniel Rueckert ⓘ
GALILEO
Technical University of Munich
Munich, Germany

ISSN 0302-9743 ISSN 1611-3349 (electronic)
Lecture Notes in Computer Science
ISBN 978-3-031-11202-7 ISBN 978-3-031-11203-4 (eBook)
https://doi.org/10.1007/978-3-031-11203-4

Preface

The 10th International Workshop on Biomedical Image Registration (WBIR 2022, https://2022.wbir.info) was held in Munich, Germany, during July 10–12, 2022. After many missed conferences due to the COVID-19 pandemic, we sincerely hoped that 2022 would be a fresh restart for in-person and hybrid meetings that support exchange and collaboration within the WBIR community.

The WBIR 2022 meeting was a two-and-a-half-day workshop endorsed by the MICCAI society and supported by its new "Special Interest Group on Biomedical Image Registration" (SIG-BIR). It was organized in close spatial and temporal proximity to the Medical Imaging with Deep Learning (MIDL 2022) conference to enable interested researchers to minimize travel. Preceding editions of WBIR have run mostly as standalone two-day workshops at various locations: Bled, Slovenia (1999); Philadelphia, USA (2003); Utrecht, The Netherlands (2006); Lübeck, Germany (2010); Nashville, USA (2012); London, UK (2014); Las Vegas, USA (2016); Leiden, The Netherlands (2018), and Portorož, Slovenia (2020 / virtually). As with previous editions, the major appeal of WBIR 2022 was bringing together researchers from different backgrounds and countries, and at different points in their academic careers, who all share a great interest in image registration. Based on our relaxed two-and-a-half-day format with tutorials, three keynotes, and a main scientific program with short and long oral presentations, as well as in-person poster presentations, WBIR 2022 enabled space for lots of interaction and ample discussion among peers. As everyone's mindset is on image registration, it makes it easier for students to approach and meet their distinguished colleagues.

The WBIR 2022 proceedings, published in the Lecture Notes in Computer Science series, were established through two cycles of peer-review using OpenReview (for the first time). Full papers were reviewed in a double-blind fashion, with each submission evaluated by at least three members of the Program Committee. The Program Committee consisted of 25 experienced scientists in the field of medical image registration. All papers and reviews were afterwards discussed in an online meeting by the Paper Selection Committee to reach decisions. Short papers were categorized as either exciting early-work or abstracts of recently published/submitted long articles. Those submissions went through a lighter peer-review process, each being assigned to two members of the Paper Selection Committee. From a total of 34 submissions, 30 were selected for oral and poster presentation and 26 original works are included in these proceedings. Prominent topics include optimization, deep learning architectures, neuroimaging, diffeomorphisms, uncertainty, topology, and metrics. The presenting authors at WBIR 2022 represented a delightful diverse community with approximately 45% female speakers, nine papers from groups outside of Europe (primarily the USA), and academic levels ranging from Master's and PhD students to lecturers. To further stimulate participation from Asia, Africa, and South America we established a scholarship program that received 25 applications.

We were grateful to have three excellent keynote speakers at WBIR 2022. With rich experience in conducting numerous medical image computing projects from early feasibility to product implementation, Wolfgang Wein from ImFusion (Germany) spoke about combining visual computing with machine learning for improved registration in image-guided interventions. Maria Vakalopoulou, who is an expert on deep learning for biomedical image analysis from Paris-Saclay University (France), discussed classical and deep learning-based registration methods and their impacts on clinical diagnosis. Finally, Josien Pluim, head of the Medical Image Analysis group at Eindhoven University of Technology (The Netherlands), provided a historical overview of trends in image registration, going back to the first papers on the topic and taking us through some of the most important advances until today.

Many people contributed to the organization and success of WBIR 2022. In particular, we would like to thank the members of the Program Committee and the additional Paper Selection Committee members (Stefan Klein and Žiga Špiclin) for their work that assured the high quality of the workshop. We thank the MICCAI SIG-BIR group for their financial support and the MICCAI Society for their endorsement. Finally, we would like to thank all authors and participants of WBIR 2022 for their contributions.

June 2022

Mattias Heinrich
Alessa Hering
Julia Schnabel
Daniel Rückert
Enzo Ferrante
Miaomiao Zhang

Organization

General Chairs

Mattias Heinrich   University of Lübeck, Germany
Alessa Hering   Fraunhofer MEVIS, Germany, and Radboudumc,
     The Netherlands
Julia Schnabel   Helmholtz Zentrum Munich and Technical
     University of Munich, Germany
Daniel Rückert   Technical University of Munich, Germany

Program Committee Chairs

Enzo Ferrante   CONICET, Universidad Nacional del Litoral,
     Argentina
Miaomiao Zhang   University of Virginia, USA

Paper Selection Committee

Stefan Klein   Erasmus MC, The Netherlands
Žiga Špiclin   University of Ljubljana, Slovenia

Program Committee

Annkristin Lange   Fraunhofer MEVIS, Germany
Bartlomiej Papiez   University of Oxford, UK
Bernhard Kainz   Friedrich-Alexander-Universität
     Erlangen-Nürnberg, Germany
Tony C. W. Mok   Hong Kong University of Science and
     Technology, Hong Kong
Deepa Krishnaswamy   Brigham and Women's Hospital, USA
Demian Wassermann   Inria, France
Gary E. Christensen   University of Iowa, USA
Hanna Siebert   Universität zu Lübeck, Germany
Hari Om Aggrawal   Technical University of Denmark, Denmark
Ivor J. A. Simpson   University of Sussex, UK
Josien P. W. Pluim   Eindhoven University of Technology,
     The Netherlands
Lilla Zollei   MGH and Harvard Medical School, USA

Contents

Atlases/Topology

Unsupervised Non-correspondence Detection in Medical Images Using
an Image Registration Convolutional Neural Network 3
  *Julia Andresen, Timo Kepp, Jan Ehrhardt, Claus von der Burchard,
  Johann Roider, and Heinz Handels*

Weighted Metamorphosis for Registration of Images with Different
Topologies 8
  *Anton François, Matthis Maillard, Catherine Oppenheim, Johan Pallud,
  Isabelle Bloch, Pietro Gori, and Joan Glaunès*

LDDMM Meets GANs: Generative Adversarial Networks
for Diffeomorphic Registration 18
  Ubaldo Ramon, Monica Hernandez, and Elvira Mayordomo

Towards a 4D Spatio-Temporal Atlas of the Embryonic and Fetal Brain
Using a Deep Learning Approach for Groupwise Image Registration 29
  *Wietske A. P. Bastiaansen, Melek Rousian,
  Régine P. M. Steegers-Theunissen, Wiro J. Niessen, Anton H. J. Koning,
  and Stefan Klein*

Uncertainty

DeepSTAPLE: Learning to Predict Multimodal Registration Quality
for Unsupervised Domain Adaptation 37
  *Christian Weihsbach, Alexander Bigalke, Christian N. Kruse,
  Hellena Hempe, and Mattias P. Heinrich*

A Method for Image Registration via Broken Geodesics 47
  *Alphin J. Thottupattu, Jayanthi Sivaswamy,
  and Venkateswaran P. Krishnan*

Deformable Image Registration Uncertainty Quantification Using Deep
Learning for Dose Accumulation in Adaptive Proton Therapy 57
  A. Smolders, T. Lomax, D. C. Weber, and F. Albertini

Distinct Structural Patterns of the Human Brain: A Caveat for Registration 67
  Frithjof Kruggel

Architectures

A Multi-organ Point Cloud Registration Algorithm for Abdominal CT
Registration 75
*Samuel Joutard, Thomas Pheiffer, Chloe Audigier, Patrick Wohlfahrt,
Reuben Dorent, Sebastien Piat, Tom Vercauteren, Marc Modat,
and Tommaso Mansi*

Voxelmorph++: Going Beyond the Cranial Vault with Keypoint
Supervision and Multi-channel Instance Optimisation 85
Mattias P. Heinrich and Lasse Hansen

Unsupervised Learning of Diffeomorphic Image Registration
via TransMorph 96
Junyu Chen, Eric C. Frey, and Yong Du

SuperWarp: Supervised Learning and Warping on U-Net for Invariant
Subvoxel-Precise Registration 103
*Sean I. Young, Yaël Balbastre, Adrian V. Dalca, William M. Wells,
Juan Eugenio Iglesias, and Bruce Fischl*

Optimisation

Learn to Fuse Input Features for Large-Deformation Registration
with Differentiable Convex-Discrete Optimisation 119
Hanna Siebert and Mattias P. Heinrich

Multi-magnification Networks for Deformable Image Registration
on Histopathology Images 124
*Oezdemir Cetin, Yiran Shu, Nadine Flinner, Paul Ziegler, Peter Wild,
and Heinz Koeppl*

Realtime Optical Flow Estimation on Vein and Artery Ultrasound
Sequences Based on Knowledge-Distillation 134
*Till Nicke, Laura Graf, Mikko Lauri, Sven Mischkewitz,
Simone Frintrop, and Mattias P. Heinrich*

Metrics/Losses

Motion Correction in Low SNR MRI Using an Approximate Rician
Log-Likelihood 147
*Ivor J. A. Simpson, Balázs Örzsik, Neil Harrison, Iris Asllani,
and Mara Cercignani*

Cross-Sim-NGF: FFT-Based Global Rigid Multimodal Alignment
of Image Volumes Using Normalized Gradient Fields 156
  Johan Öfverstedt, Joakim Lindblad, and Nataša Sladoje

Identifying Partial Mouse Brain Microscopy Images from the Allen
Reference Atlas Using a Contrastively Learned Semantic Space 166
  Justinas Antanavicius, Roberto Leiras, and Raghavendra Selvan

Transformed Grid Distance Loss for Supervised Image Registration 177
  *Xinrui Song, Hanqing Chao, Sheng Xu, Baris Turkbey,
  Bradford J. Wood, Ge Wang, and Pingkun Yan*

Efficiency

Deformable Lung CT Registration by Decomposing Large Deformation 185
  Jing Zou, Lihao Liu, Youyi Song, Kup-Sze Choi, and Jing Qin

You only Look at Patches: A Patch-wise Framework for 3D Unsupervised
Medical Image Registration 190
  *Lihao Liu, Zhening Huang, Pietro Liò, Carola-Bibiane Schönlieb,
  and Angelica I. Aviles-Rivero*

Recent Developments of an Optimal Control Approach to Nonrigid Image
Registration 194
  Zicong Zhou and Guojun Liao

2D/3D Quasi-Intramodal Registration of Quantitative Magnetic Resonance
Images 198
  Batool Abbas, Riccardo Lattanzi, Catherine Petchprapa, and Guido Gerig

Deep Learning-Based Longitudinal Intra-subject Registration of Pediatric
Brain MR Images 206
  Andjela Dimitrijevic, Vincent Noblet, and Benjamin De Leener

Real-Time Alignment for Connectomics 211
  Neha Goyal, Yahiya Hussain, Gianna G. Yang, and Daniel Haehn

Weak Bounding Box Supervision for Image Registration Networks 215
  *Mona Schumacher, Hanna Siebert, Ragnar Bade, Andreas Genz,
  and Mattias Heinrich*

Author Index 221

Atlases/Topology

Unsupervised Non-correspondence Detection in Medical Images Using an Image Registration Convolutional Neural Network

Julia Andresen[1]([✉])(iD), Timo Kepp[1](iD), Jan Ehrhardt[1,2],
Claus von der Burchard[3](iD), Johann Roider[3], and Heinz Handels[1,2](iD)

[1] Institute of Medical Informatics, University of Lübeck, Ratzeburger Allee 160, 23562 Lübeck, Germany
j.andresen@uni-luebeck.de
[2] German Research Center for Artificial Intelligence, Lübeck, Germany
[3] Department of Ophthalmology, Christian-Albrechts-University of Kiel, Kiel, Germany

1 Introduction

Medical image registration allows comparing images from different patients, modalities or time-points, but often suffers from missing correspondences due to pathologies and inter-patient variations. The handling of non-corresponding regions has been tackled with several approaches in the literature. For evolving processes, metamorphoses models have been used that model both spatial and appearance offsets to align images from different time-points [1,2]. Other approaches mask out [3] or weight down [4,5] the image distance measure in non-corresponding regions based on outlier detection [3], estimation of matching uniqueness [4] or correspondence probabilities [5].

Our recently published paper "Deep learning-based simultaneous registration and unsupervised non-correspondence segmentation of medical images with pathologies" [6] proposes a convolutional neural network (CNN) for joint image registration and detection of non-corresponding regions. As in previous iterative approaches [3], non-correspondences are considered as outliers in the image distance measure and are masked out. The conversion to a deep learning-based approach allows a two-step training procedure that results in better separation of spatial displacement and non-correspondence segmentation. Network training does not require manual segmentations of non-correspondences that are found in a single run, overcoming limitations of other CNN-based approaches [7–10].

A. Hering et al. (Eds.): WBIR 2022, LNCS 13386, pp. 3–7, 2022.
https://doi.org/10.1007/978-3-031-11203-4_1

2   Materials and Methods

The joint non-correspondence detection and image registration network (NCR-Net) is inspired by the U-Net [11] but follows a Y-shaped architecture with one encoder and two separate decoders. The decoders output a diffeomorphic deformation field ϕ and a non-correspondence segmentation S, respectively. Both decoders are connected to the encoder with skip connections. Moving image M and fixed image F serve as network input and outputs are generated on three resolution levels. At each resolution level, the loss function is computed to enable in-depth supervision of the network, with finer resolution levels being given more weight.

Segmentation and registration performances of NCR-Net are extensively evaluated on two datasets. The first dataset consists of longitudinal OCT images from 40 patients suffering from age-related macular degeneration. Three boundary segmentations, but no pathological labels are given for these data. The second dataset is the LPBA40 dataset, containing 40 whole-head MRI volumes from healthy probands and manual segmentations of 56 anatomical regions. To introduce known non-correspondences into the images, we simulate four different stroke lesions, two of which are quite large and the other two are smaller.

The network training takes place in two phases. First, the encoding part of the network as well as the deformation decoder are pre-trained with the "standard" objective function for image registration

$$\mathcal{L}_{\text{Reg}}(\theta; \text{M}, \text{F}) = \sum_{\mathbf{x} \in \Omega} \mathcal{D}[\text{F}, \phi \circ \text{M}] + \alpha \mathcal{R}_\phi + \lambda \mathcal{L}_{\text{opt}} \tag{1}$$

consisting of image distance measure $\mathcal{D}$ and regularization of the deformation $\mathcal{R}_\phi$. The last term $\mathcal{L}_{\text{opt}}$ is optional and may be used to provide supervision to the registration or segmentation task. In this work, we use the Dice loss comparing brain masks for MRI and retinal masks for OCT data in moving and fixed images to support the registration task. In the second training phase, the entire CNN is updated using

$$\mathcal{L}(\theta; \text{M}, \text{F}) = \sum_{\mathbf{x} \in \Omega} (1 - S) \cdot \mathcal{D}[\text{F}, \phi \circ \text{M}] + \alpha \mathcal{R}_\phi + \beta \mathcal{R}_S + \lambda \mathcal{L}_{\text{opt}} \tag{2}$$

as loss function. Here, the image distance is evaluated in corresponding regions only and the segmentation S is regularized with $\mathcal{R}_S$ consisting of segmentation volume and perimeter.

3   Results

In a first experiment, ablation studies are performed on the OCT data, comparing supervised and unsupervised versions of NCR-Net, i.e. versions trained with and without $\mathcal{L}_{\text{opt}}$, as well as versions trained with the proposed two-phase training or with loss function (2) from scratch. Two main results arise from this experiment. First, unsupervised and supervised NCR-Net perform comparably, allowing its use even for datasets without any given annotations. Second, the two-phase training scheme significantly improves Hausdorff and average surface distance of all three segmented retinal boundaries, indicating better disentanglement of spatial deformation and non-correspondence segmentation.

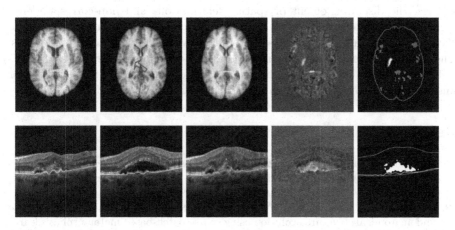

Fig. 1. Exemplary results for MRI (top row) and OCT (bottom row) data. Shown are moving, fixed, warped moving and the difference image after registration as well as the generated non-correspondence maps. Manually segmented retinal borders and automatically generated brain masks are given in blue. For the MRI data, segmentation results before and after region-growing are displayed in gray and white, respectively. The ground truth lesion is outlined in red. (Color figure online)

The registration performance of NCR-Net is further evaluated on the LPBA40 data by calculating average Jaccard indices of the given anatomical labels and comparing NCR-Net to state-of-the-art registration algorithms in 2D and 3D. NCR-Net significantly outperforms the competitive methods in the presence of large pathologies and performs comparable for images with small or no lesion. Network training with small and large simulated lesions leads to improved robustness against non-correspondences.

Finally, we evaluate the non-correspondence detection and segmentation performance of NCR-Net using the MRI data. The generated segmentations are compared to the ground truth lesion masks in two ways, first directly and second after applying region growing inside the lesions. In 2D, mean Dice scores of 0.871, 0.870, 0.630 and 0.880 are achieved for the four lesion types considered. Even though the segmentation performance in 3D is inferior, lesion detection rates are still high with 83.7 % for the worst performing lesion type.

4   Discussion

Our NCR-Net closes the gap between deep learning and iterative approaches for joint image registration and non-correspondence detection. The proposed network achieves state-of-the-art and robust registration of pathological images while additionally segmenting non-correspondent areas. With a two-step training scheme, the disentanglement of spatial deformations and non-correspondence segmentation is improved. Manual annotations may provide more supervision to the registration task, but can also be omitted without much performance loss. The simulated stroke lesions are detected as non-correspondent regions by NCR-Net very reliably and the generated segmentations are shown to be usable for unsupervised lesion segmentation and for the monitoring of evolving diseases.

References

1. Niethammer, M., et al.: Geometric metamorphosis. In: Fichtinger, G., Martel, A., Peters, T. (eds.) MICCAI 2011. LNCS, vol. 6892, pp. 639–646. Springer, Heidelberg (2011). https://doi.org/10.1007/978-3-642-23629-7_78
2. Rekik, I., Li, G., Wu, G., Lin, W., Shen, D.: Prediction of infant MRI appearance and anatomical structure evolution using sparse patch-based metamorphosis learning framework. In: Wu, G., Coupé, P., Zhan, Y., Munsell, B., Rueckert, D. (eds.) Patch-MI 2015. LNCS, vol. 9467, pp. 197–204. Springer, Cham (2015). https://doi.org/10.1007/978-3-319-28194-0_24
3. Chen, K., Derksen, A., Heldmann, S., Hallmann, M., Berkels, B.: Deformable image registration with automatic non-correspondence detection. In: Aujol, J.-F., Nikolova, M., Papadakis, N. (eds.) SSVM 2015. LNCS, vol. 9087, pp. 360–371. Springer, Cham (2015). https://doi.org/10.1007/978-3-319-18461-6_29
4. Ou, Y., Sotiras, A., Paragios, N., Davatzikos, C.: DRAMMS: deformable registration via attribute matching and mutual-saliency weighting. Med. Image Anal. **15**(4), 622–639 (2011) https://doi.org/10.1016/j.media.2010.07.002
5. Krüger, J., Schultz, S., Handels, H., Ehrhardt, J.: Registration with probabilistic correspondences-accurate and robust registration for pathological and inhomogeneous medical data. Comput. Vis. Image Underst. **190** (2020). https://doi.org/10.1016/j.cviu.2019.102839
6. Andresen, J., Kepp, T., Ehrhardt, J., von der Burchard, C., Roider, J., Handels, H.: Deep learning-based simultaneous registration and unsupervised non-correspondence segmentation of medical images with pathologies. Int. J. CARS **17**, 699–710 (2022). https://doi.org/10.1007/s11548-022-02577-4
7. Sedghi, A., Kapur, T., Luo, J., Mousavi, P., Wells, W.M.: Probabilistic image registration via deep multi-class classification: characterizing uncertainty. In: Greenspan, H., et al. (eds.) CLIP/UNSURE -2019. LNCS, vol. 11840, pp. 12–22. Springer, Cham (2019). https://doi.org/10.1007/978-3-030-32689-0_2
8. Yang, X., Kwitt, R., Styner, M., Niethammer, M.: Quicksilver: fast predictive image registration - a deep learning approach. NeuroImage **158**, 378–396 (2017)
9. Sentker, T., Madesta, F., Werner, R.: GDL-FIRE^{4D}: deep learning-based fast 4D CT image registration. In: Frangi, A.F., Schnabel, J.A., Davatzikos, C., Alberola-López, C., Fichtinger, G. (eds.) MICCAI 2018. LNCS, vol. 11070, pp. 765–773. Springer, Cham (2018). https://doi.org/10.1007/978-3-030-00928-1_86

10. Zhou, T., Krähenbühl, P., Aubry, M., Huang, Q., Efro, A.A.: Learning dense correspondence via 3D-guided cycle consistency. In: 2016 IEEE Conference on Computer Vision and Pattern Recognition (CVPR), pp. 117–126 (2016)
11. Ronneberger, O., Fischer, P., Brox, T.: U-net: convolutional networks for biomedical image segmentation. In: Navab, N., Hornegger, J., Wells, W.M., Frangi, A.F. (eds.) MICCAI 2015. LNCS, vol. 9351, pp. 234–241. Springer, Cham (2015). https://doi.org/10.1007/978-3-319-24574-4_28

Weighted Metamorphosis for Registration of Images with Different Topologies

Anton François[1,2(✉)], Matthis Maillard[2], Catherine Oppenheim[3],
Johan Pallud[3], Isabelle Bloch[2,4], Pietro Gori[2(✉)], and Joan Glaunès[1(✉)]

[1] Université de Paris-Cité, Paris, France
{anton.francois,alexis.glaunes}@parisdescartes.fr
[2] LTCI Télécom Paris, Institut Polytechnique de Paris, Paris, France
pietro.gori@parisdescartes.fr
[3] UMR 1266 INSERM, IMA-BRAIN, IPNP, Paris, France
[4] Sorbonne Université, CNRS, LIP6, Paris, France

Abstract. We present an extension of the Metamorphosis algorithm to align images with different topologies and/or appearances. We propose to restrict/limit the metamorphic intensity additions using a time-varying spatial weight function. It can be used to model prior knowledge about the topological/appearance changes (e.g., tumour/oedema). We show that our method improves the disentanglement between anatomical (i.e., shape) and topological (i.e., appearance) changes, thus improving the registration interpretability and its clinical usefulness. As clinical application, we validated our method using MR brain tumour images from the BraTS 2021 dataset. We showed that our method can better align healthy brain templates to images with brain tumours than existing state-of-the-art methods. Our PyTorch code is freely available here: https://github.com/antonfrancois/Demeter_metamorphosis.

Keywords: Image registration · Metamorphosis · Topology variation · Brain tumour

1 Introduction

When comparing medical images, for diagnosis or research purposes, physicians need accurate anatomical registrations. In practice, this is achieved by mapping images voxel wise with a plausible anatomical transformation. Possible applications are: computer assisted diagnosis or therapy, multi-modal fusion or surgical planning. These mappings are usually modelled as diffeomorphisms, as they allow for the creation of a realistic one to one deformation without modifying the topology of the source image. There exists a vast literature dealing with this subject. Some authors proposed to use stationary vectors fields, using the Lie algebra vector field exponential [1,2,14], or, more recently, Deep-Learning based methods [5,16,19,21,22,29]. Other authors used the Large Diffeomorphic

A. Hering et al. (Eds.): WBIR 2022, LNCS 13386, pp. 8–17, 2022.
https://doi.org/10.1007/978-3-031-11203-4_2

Deformation Metric Mapping (LDDMM) that uses time varying vector fields to define a right-invariant Riemannian metric on the group of diffeomorphisms. One advantage of this metric is that it can be used to build a shape space, providing useful notions of geodesics, shortest paths and distances between images [3,6,30,31]. A shortest path represents the registration between two images.

However, clinical or morphometric studies often include an alignment step between a healthy template (or atlas) and images with lesions, alterations or pathologies, like white matter multiple sclerosis or tumour. In such applications, source and target images show a different topology, thus preventing the use of diffeomorphisms, which are by definitions one-to-one mappings. Several solutions have been proposed in order to take into account such topological variations. One of the first methods was the Cost-Function Masking [7], where authors simply excluded the lesions from the image similarity cost. It is versatile and easy to implement, but it does not give good results when working with big lesions. Sdika et al. [24] proposed an inpainting method which only works on small lesions. Niethammer et al. proposed Geometric Metamorphosis [20], that combines two deformations to align pathological images which need to have the same topology. Another strategy, when working with brain images with tumours, is to use bio-physical models [10,23] to mimic the growth of a tumour into an healthy image and then perform the registration (see for instance GLISTR [11]). However, this solution is slow, computationally heavy, specific to a particular kind of tumour and needs many different imaging modalities. Other works proposed to solve this problem using Deep-Learning techniques [8,12,15,25]. However, these methods strongly depend on the data-set and on the modality they have been trained on, and might not correctly disentangle shape and appearance changes.

The Metamorphic framework [13,27,30] can be seen as a relaxed version of LDDMM in which residual time-varying intensity variations are added to the diffeomorphic flow, therefore allowing for topological changes. Nevertheless, even if metamorphosis leads to very good registrations, the disentanglement between geometric and intensity changes is not unique and it highly depends on user-defined hyper-parameters. This makes interpretation of the results hard, thus hampering its clinical usage. For instance, in order to align a healthy template to an image with a tumour, one would expect that the method adds intensities only to create new structures (i.e., tumours) or to compensate for intensity changes due to the pathology (i.e. oedema). All other structures should be correctly aligned solely by the deformations. However, depending on the hyper-parameters, the algorithm might decide to account for morphological differences (i.e. mass effect of tumours) by changing the appearance rather than applying deformations. This limitation mainly comes from the fact that the additive intensity changes can theoretically be applied all over the image domain. However, in many clinical applications, one usually has prior knowledge about the position of the topological variations between an healthy image and a pathological one (e.g., tumour and oedema position).

To this end, we propose an extension of the Metamorphosis (M) model [13,27], called Weighted Metamorphosis (WM), where we introduce a time-varying spatial weight function that restricts, or limits, the intensity addition only to some specified areas. Our main contributions are: 1./ A novel time-varying spatial weight function that restricts, or limits, the metamorphic intensity additions [13,27] only to some specified areas. 2./ A new cost function that results in a set of geodesic equations similar to the ones in [13,27]. Metamorphosis can thus be seen as a specific case of our method. 3./ Evaluation on a synthetic shape dataset and on the BraTS 2021 dataset [17], proposing a simple and effective weight function (i.e., segmentation mask) when working with tumour images. 4./ An efficient PyTorch implementation of our method, available at https://github.com/antonfrancois/Demeter_metamorphosis.

2 Methods

Weighted Metamorphosis. Our model can be seen as an extension of the model introduced by Trouvé and Younès [27,30]. We will use the same notations as in [9]. Let $S, T : \Omega \rightarrow [0,1]$ be grey-scale images, where Ω is the image domain. To register S on T, we define, similarly to [27,30], the evolution of an image I_t ($t \in [0,1]$) using the action of a vector field v_t, defined as $v \cdot I_t = -\langle \nabla I_t, v_t \rangle$, and additive intensity changes, given by the residuals z_t, as:

$$\dot{I}_t = -\langle \nabla I_t, v_t \rangle + \mu M_t z_t, \quad \text{s.t. } I_0 = S, \ I_1 = T, \ \mu \in \mathbb{R}^+. \tag{1}$$

where we introduce the weight function $M_t : x \in \Omega \rightarrow [0,1]$ (at each time $t \in [0,1]$) that multiplies the residuals z_t at each time step t and at every location x. We assume that M_t is smooth with compact support and that it can be fully computed before the optimisation. Furthermore, we also define a new pseudo-norm $\| \bullet \|_{M_t}$ for z. Since we want to consider the magnitude of z only at the voxels where the intensity is added, or in other terms, where $M_t(x)$ is not zero, we propose the following pseudo-norm:

$$\|z_t\|_{M_t}^2 = \left\| \sqrt{M_t} z_t \right\|_{L^2}^2 = \langle z_t, M_t z_t \rangle_{L^2} \tag{2}$$

This metric will sum up the square values of z inside the support of M_t. As usual in LDDMM, we assume that each $v_t \in V$, where V is a Hilbert space with a reproducing kernel K_σ, which is chosen here as a Gaussian kernel parametrized by σ [18,28]. Similarly to [27,30], we use the sum of the norm of z and the one of v (i.e., the total kinetic energy), balanced by ρ, as cost function:

$$E_{\text{WM}}(v, I) = \int_0^1 \|v_t\|_V^2 + \rho \|z_t\|_{M_t}^2 dt, \quad \text{s.t. } I_0 = S, \ I_1 = T, \ \rho \in \mathbb{R}^+ \tag{3}$$

where z depends on I through Eq. 1. By minimising Eq. 3, we obtain an exact matching.

Theorem 1. *The geodesics associated to Eq. 3 are:*

$$
\begin{cases}
v_t = -\frac{\rho}{\mu} K_\sigma \star (z_t \nabla I_t) \\
\dot{z}_t = - \quad \nabla \cdot (z_t v_t) \\
\dot{I}_t = -\langle \nabla I_t, v_t \rangle + \mu M_t z_t
\end{cases}
\tag{4}
$$

where $\nabla \cdot (v)$ is the divergence of the field v and $\star$ represents the convolution.

Proof. This proof is similar to the one in [30], Chap. 12, but needs to be treated carefully due to the pseudo-norm $\|z_t\|_{M_t}^2 = \left\langle z_t, \frac{1}{\mu}(\dot{I}_t + v_t \cdot \nabla I_t) \right\rangle_{L^2}$. We aim at computing the variations of Eq. 3 with respect to I and v and compute the Euler-Lagrange equations. To this end, we define two Lagrangians: $L_I(t, I, \dot{I}) = E_{\mathrm{WM}}(\bullet, v)$ and $L_v(t, v, \dot{v}) = E_{\mathrm{WM}}(I, \bullet)$ and start by computing the variations h with respect to v:

$$
D_v L_v \cdot h = \int_0^1 \langle 2(K^{-1} v_t + \frac{\rho}{\mu} z_t \nabla I_t), h_t \rangle_{L^2} dt
\tag{5}
$$

Then, noting that $\nabla_v L_v = 2(K^{-1} v_t + \frac{\rho}{\mu} z_t \nabla I_t)$ and since $\nabla_{\dot{v}} L_v = 0$, the Euler-Lagrange equation is:

$$
\nabla_v L_v - \dot{\nabla}_{\dot{v}} L_v = 0 \Leftrightarrow v_t = -\frac{\rho}{\mu} K \star (z_t \nabla I_t)
\tag{6}
$$

as in the classical Metamorphosis framework [30]. Considering the variation of I, we have $D_I \left\| \sqrt{M_t} z \right\|_{L_2}^2 = \left\langle z_t, \frac{1}{\mu} v_t \cdot \nabla h_t \right\rangle_{L^2}$, thus obtaining:

$$
D_I L_I \cdot h = 2 \int_0^1 \left\langle z_t, \frac{1}{\mu} \nabla h_t \cdot v_t \right\rangle_{L^2} dt = \int_0^1 \left\langle -\frac{2}{\mu} \nabla \cdot (z_t v_t), h_t \right\rangle_{L^2} dt
\tag{7}
$$

and $D_{\dot{I}} L_I \cdot h = \int_0^1 \langle \frac{2}{\mu} z_t, h_t \rangle_{L^2} dt$. We deduce that $\nabla_I L_I = \frac{2}{\mu} \nabla \cdot (z_t v_t)$ and as $\nabla_{\dot{v}} L_v = \frac{2}{\mu} z_t$, its Euler-Lagrange equation is:

$$
\nabla_I L_I - \dot{\nabla}_{\dot{I}} L_I = 0 \Leftrightarrow \dot{z}_t = -\nabla \cdot (z_t v_t)
\tag{8}
$$

We can first notice that, by following the geodesic paths, the squared norms over time are conserved ($\forall t \in [0, 1], \|v_0\|_V^2 = \|v_t\|_V^2$) and thus one can actually optimise using only the initial norms. Furthermore, since v_0 can be computed from z_0 and I_0, the only parameters of the system are z_0 and I_0. As it is often the case in the image registration literature, we propose to convert Eq. 3 into an unconstrained inexact matching problem, thus minimising:

$$
J_{\mathrm{WM}}(z_0) = \|I_1 - T\|_{L_2}^2 + \lambda \left[\|v_0\|_V^2 + \rho \|z_0\|_M^2 \right], \quad \lambda \in \mathbb{R}^+, I_0 = S
\tag{9}
$$

where I_1 is integrated with Eq. 4, $\|v_0\|_V^2 = \langle z_0 \nabla S, K_\sigma \star (z_0 \nabla S) \rangle$ and λ is the trade-off between the data term (based here on a L2-norm, but any metric could be used as well) and the total regularisation.

Weighted Function Construction. The definition of the weight function M_t is quite generic and could be used to register any kind of topological/appearance differences. Here, we restrict to brain tumour images and propose to use an evolving segmentation mask as weight function. We assume that we already have the binary segmentation mask B of the tumour (comprising both oedema and necrosis) in the pathological image and that healthy and pathological images are rigidly registered, so that B can be rigidly moved onto the healthy image. Our goal is to obtain an evolving mask $M_t : [0,1] \times \Omega \to [0,1]$ that somehow mimics the tumour growth in the healthy image starting from a smoothed small ball in the centre of the tumour (M_0) and smoothly expanding it towards B. We generate M_t by computing the LDDMM registration between M_0 and B. Please note that here one could use an actual biophysical model [10,23] instead of the proposed simplistic approximation based on LDDMM. However, it would require prior knowledge, correct initialisation and more than one imaging modality. The main idea is to smoothly and slowly regularise the transformation so that the algorithm first modifies the appearance only in a small portion of the image, trying to align the surrounding structure only with deformations. In this way, the algorithm tries to align all structures with shape changes adding/removing intensity only when necessary. This should prevent the algorithm from changing the appearance instead of applying deformations (i.e. better disentanglement) and avoid wrong overlapping between new structures (e.g. tumour) and healthy ones. Please refer to Fig. 1 for a visual explanation.

3   Results and Perspectives

Implementation Details. Our Python implementation is based on PyTorch for automatic differentiation and GPU support, and it uses the semi-Lagrangian formulation for geodesic shooting presented in [9]. For optimisation we use the PyTorch's Adadelta method.

Synthetic Data. Here, we illustrate our method on a 300×300 grey-scale image registration toy-example (Fig. 1). We can observe the differences in the geodesic image evolution for LDDMM, Metamorphosis (M) and Weighted Metamorphosis (WM) with a constant and evolving mask. First, LDDMM cannot correctly align all grey ovals and Metamorphosis results in an image very similar to the target. However, most of the differences are accounted for with intensity changes rather than deformations. By contrast, when using the proposed evolving mask (fourth row), the algorithm initially adds a small quantity of intensity in the middle of the image and then produces a deformation that enlarges it and correctly pushes away the four grey ovals. In the third row, a constant mask ($M_t = M_1, \forall t \in [0,1]$) is applied. One can observe that, in this case, the bottom and left ovals overlap with the created central triangle and therefore pure deformations cannot correctly match both triangle and ovals. In all methods, the registration was done with the same field smoothness regularisation σ and integration steps. Please note that the four grey ovals at the border are not correctly matched with LDDMM and, to a lesser extent, also with our method. This is due to the

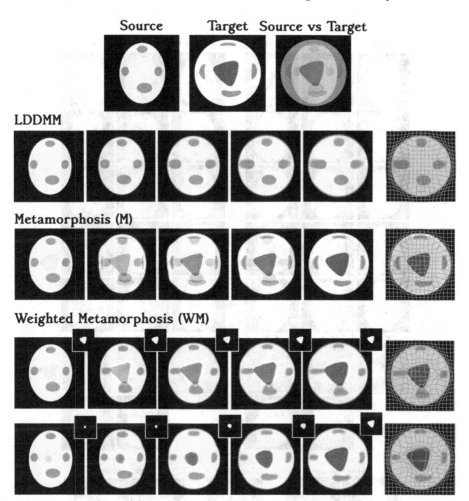

Fig. 1. Comparison between LDDMM, Metamorphosis and our method.
Image registration toy example. Differently from the Source image (S), the Target
image (T) has a big central triangle that has grown "pushing" the surroundings ovals.
Note that the bottom and left ovals in S overlap with the triangle in T. The two last
rows show our method using a constant and time evolving mask (see Sect. 2). The used
mask is displayed on the top right corner of each image. see animations in GitHub
in notebook : toyExample_weightedMetamorphosis.ipynb

L2-norm data term since these shapes do not overlap between the initial source
and target images and therefore the optimiser cannot match them.

Validation on 2D Real Data. For evaluation, we used T1-w MR images from
the BraTS 2021 dataset [4,17]. For each patient, a tumour segmentation is provided. We selected the same slice for 50 patients resizing them to 240×240
and making sure that a tumour was present. We then proceeded to register the

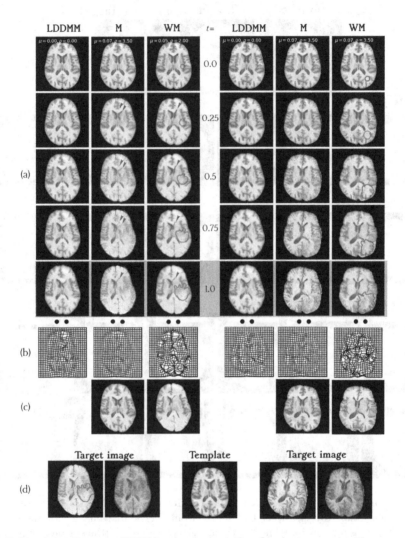

Fig. 2. Registrations on MRI brains presenting brain tumours. Two examples from BraTS database [4,17]. Comparison of geodesic shooting for LDDMM, Metamorphosis (M) and Weighted Metamorphosis (WM). (a&d) On the target images and the geodesic integration, the temporal mask is indicated by the red outline. The final result of each integration can be seen in the green outlined row. (b) The deformation grids retrieved from each method and (c) the template image deformed without intensity additions for each concerned method. Purple arrows in columns 2 and 3 in the top right part of each image show the evolution of one ventricle through registration: while M makes the ventricle disappear and reappear, WM coherently displaces the structure. (d) Target images with the segmentation outlined in red; the colored image is its superposition with the source. see animations in GitHub in notebook : brains_weightedMetamorphosis.ipynb

healthy brain template SRI24 [26] to each of the selected slices (see Fig. 2 for two examples). To evaluate the quality of the alignment we used three different measures in Table 1: 1./ the Sum of Squared Differences (SSD) (i.e. L2-norm) between the target (T) and the transformed source (S) images. This is a natural choice as it is used in the cost function. 2./ the SSD between T and the deformed S without considering intensity changes. This is necessary since Metamorphoses could do a perfect matching without using deformations but only intensity changes. 3./ A Dice score between the segmentations of the ventricles in the deformed S and T. The ventricles were manually segmented. All methods should correctly align the ventricles using solely pure deformations since theses regions are (theoretically) not infiltrated by the tumour (*i.e.*, no intensity modifications) and they can only be displaced by the tumor mass effect.

Table 1. Quantitative evaluation for different registration methods. Results were computed on a test set of 50 2D 240 × 240 images from BraTS 2021 dataset. - (∗) SSD for CFM is computed over the domain outside the mask.

Method	LDDMM [18]	Meta. [9]	WM (ours)	MAE [8]	Voxelm. [5]	CFM [7]
SSD (final)	223 ± 51	**36 ± 9**	65 ± 71	497 ± 108	166.71 ± 37	49∗ ± 28
SSD (def.)	–	112 ± 21	**102 ± 76**	865 ± 172	–	–
Dice score	68.6 ± 11.9	74.1 ± 9.3	**77.2 ± 10.1**	60.6 ± 8.79	66.8 ± 10	45.0 ± 13.5

We compared our method with LDDMM [6], Metamorphosis [27], using the implementation of [9], Metamorphic Auto-Encoder (MAE) [8], Voxelmorph [5] and Cost Function Masking (CFM) [7] (see Table 1). Please note that we did not include other deep-learning methods, such as [12,15], since they only work the other way around, namely they can only register images with brain tumours to healty templates. As expected, Metamorphosis got the best score for SSD (final) as it is the closest to an exact matching method. However, WM outperformed all methods in terms of Dice score obtaining a very low SSD (both final and deformation-only). This means that our method correctly aligned the ventricles, using only the deformation, and at the same time it added intensity only where needed to globally match the two images (i.e., good disentanglement between shape and appearance).

Perspectives and Conclusion. In this work, we introduced a new image registration method, Weighted Metamorphosis, and showed that it successfully disentangles deformation from intensity addition in metamorphic registration, by using prior information. Furthermore, the use of a spatial mask makes our method less sensitive to hyper-parameter choice than Metamorphosis, since it spatially constrains the intensity changes. We also showed that WM improves the accuracy of registration of MR images with brain tumours from the BRATS 2021 dataset. We are confident that this method could be applied to any kind of medical images showing exogenous tissue growth with mass-effect. A future research direction will be the integration of methods from topological data analysis, such as persistent homology, to improve even more the disentanglement between geometric and appearance changes. We also plan to adapt our method to 3D data.

Acknowledgement. M. Maillard was supported by a grant of IMT, Fondation Mines-Télécom and Institut Carnot TSN, through the "Futur & Ruptures" program.

References

1. Arsigny, V., Commowick, O., Pennec, X., Ayache, N.: A log-euclidean framework for statistics on diffeomorphisms. In: Larsen, R., Nielsen, M., Sporring, J. (eds.) MICCAI 2006. LNCS, vol. 4190, pp. 924–931. Springer, Heidelberg (2006). https://doi.org/10.1007/11866565_113
2. Ashburner, J.: A fast diffeomorphic image registration algorithm. NeuroImage **38**(1), 95–113 (2007)
3. Avants, B.B., Epstein, C.L., Grossman, M., Gee, J.C.: Symmetric diffeomorphic image registration with cross-correlation: evaluating automated labeling of elderly and neurodegenerative brain. Med. Image Anal. **12**(1), 26–41 (2008)
4. Baid, U., et al.: The RSNA-ASNR-MICCAI BraTS 2021 benchmark on brain tumor segmentation and radiogenomic classification. arXiv:2107.02314 (2021)
5. Balakrishnan, G., Zhao, A., Sabuncu, M.R., Guttag, J., Dalca, A.V.: VoxelMorph: a learning framework for deformable medical image registration. IEEE Trans. Med. Imaging **38**(8), 1788–1800 (2019)
6. Beg, M.F., Miller, M.I., Trouvé, A., Younes, L.: Computing large deformation metric mappings via geodesic flows of diffeomorphisms. Int. J. Comput. Vis. **61**(2), 139–157 (2005)
7. Brett, M., Leff, A., Rorden, C., Ashburner, J.: Spatial normalization of brain images with focal lesions using cost function masking. NeuroImage **14**(2), 486–500 (2001)
8. Bône, A., Vernhet, P., Colliot, O., Durrleman, S.: Learning joint shape and appearance representations with metamorphic auto-encoders. In: Martel, A.L., et al. (eds.) MICCAI 2020. LNCS, vol. 12261, pp. 202–211. Springer, Cham (2020). https://doi.org/10.1007/978-3-030-59710-8_20
9. François, A., Gori, P., Glaunès, J.: Metamorphic image registration using a semi-lagrangian scheme. In: SEE GSI (2021)
10. Gooya, A., Biros, G., Davatzikos, C.: Deformable registration of glioma images using EM algorithm and diffusion reaction modeling. IEEE Trans. Med. Imaging **30**(2), 375–390 (2011)
11. Gooya, A., Pohl, K., Bilello, M., Cirillo, L., Biros, G., Melhem, E., Davatzikos, C.: GLISTR: glioma image segmentation and registration. IEEE Trans. Med. Imaging **31**, 1941–54 (2012)
12. Han, X., et al.: A deep network for joint registration and reconstruction of images with pathologies. In: Liu, M., Yan, P., Lian, C., Cao, X. (eds.) MLMI 2020. LNCS, vol. 12436, pp. 342–352. Springer, Cham (2020). https://doi.org/10.1007/978-3-030-59861-7_35
13. Holm, D.D., Trouvé, A., Younes, L.: The euler-poincaré theory of metamorphosis. Q. Appl. Math. **67**(4), 661–685 (2009)
14. Lorenzi, M., Ayache, N., Frisoni, G.B., Pennec, X.: LCC-Demons: a robust and accurate symmetric diffeomorphic registration algorithm. NeuroImage **81**, 470–483 (2013)

15. Maillard, M., François, A., Glaunès, J., Bloch, I., Gori, P.: A deep residual learning implementation of metamorphosis. In: IEEE ISBI (2022)
16. Mansilla, L., Milone, D.H., Ferrante, E.: Learning deformable registration of medical images with anatomical constraints. Neural Netw. **124**, 269–279 (2020)
17. Menze, B.H., et al.: The multimodal brain tumor image segmentation benchmark (brats). IEEE Trans. Med. Imaging **34**(10), 1993–2024 (2015)
18. Miller, M.I., Trouvé, A., Younes, L.: Geodesic shooting for computational anatomy. J. Math. Imaging Vis. **24**(2), 209–228 (2006)
19. Mok, T.C.W., Chung, A.C.S.: large deformation diffeomorphic image registration with laplacian pyramid networks. In: MICCAI (2020)
20. Niethammer, M., et al.: Geometric Metamorphosis. In: Fichtinger, G., Martel, A., Peters, T. (eds.) MICCAI 2011. LNCS, vol. 6892, pp. 639–646. Springer, Heidelberg (2011). https://doi.org/10.1007/978-3-642-23629-7_78
21. Niethammer, M., Kwitt, R., Vialard, F.X.: Metric learning for image registration. In: CVPR, pp. 8455–8464 (2019)
22. Rohé, M.M., Datar, M., Heimann, T., Sermesant, M., Pennec, X.: SVF-Net: learning deformable image registration using shape matching. In: MICCAI, p. 266 (2017)
23. Scheufele, K., et al.: Coupling brain-tumor biophysical models and diffeomorphic image registration. Comput. Methods Appl. Mech Eng. **347**, 533–567 (2019)
24. Sdika, M., Pelletier, D.: Nonrigid registration of multiple sclerosis brain images using lesion inpainting for morphometry or lesion mapping. Hum. Brain Mapp. **30**(4), 1060–1067 (2009)
25. Shu, Z., et al.: Deforming autoencoders: unsupervised disentangling of shape and appearance. In: ECCV (2018)
26. Torsten, R., Zahr, N.M., Sullivan, E.V., Pfefferbaum, A.: The SRI24 multichannel atlas of normal adult human brain structure. Hum. Brain Mapp. **31**(5), 798–819 (2010)
27. Trouvé, A., Younes, L.: Local geometry of deformable templates. SIAM J. Math. Anal. **37**(1), 17–59 (2005)
28. Vialard, F.X., Risser, L., Rueckert, D., Cotter, C.J.: Diffeomorphic 3D image registration via geodesic shooting using an efficient adjoint calculation. Int. J. Comput. Vis. **97**(2), 229–241 (2011)
29. Yang, X., Kwitt, R., Styner, M., Niethammer, M.: Quicksilver: fast predictive image registration - a deep learning approach. NeuroImage **158**, 378–396 (2017)
30. Younes, L.: Deformable objects and matching functionals. In: Shapes and Diffeomorphisms. AMS, vol. 171, pp. 243–289. Springer, Heidelberg (2019). https://doi.org/10.1007/978-3-662-58496-5_9
31. Zhang, M., Fletcher, P.T.: Fast diffeomorphic image registration via fourier-approximated lie algebras. Int. J. Comput. Vis. **127**(1), 61–73 (2018)

LDDMM Meets GANs: Generative Adversarial Networks for Diffeomorphic Registration

Ubaldo Ramon[✉], Monica Hernandez, and Elvira Mayordomo

Department of Computer Science and Systems Engineering, School of Engineering and Architecture, University of Zaragoza, Zaragoza, Spain
{uramon,mhg,elvira}@unizar.es

Abstract. The purpose of this work is to contribute to the state of the art of deep-learning methods for diffeomorphic registration. We propose an adversarial learning LDDMM method for pairs of 3D monomodal images based on Generative Adversarial Networks. The method is inspired by the recent literature on deformable image registration with adversarial learning. We combine the best performing generative, discriminative, and adversarial ingredients from the state of the art within the LDDMM paradigm. We have successfully implemented two models with the stationary and the EPDiff-constrained non-stationary parameterizations of diffeomorphisms. Our unsupervised learning approach has shown competitive performance with respect to benchmark supervised learning and model-based methods.

Keywords: Large deformation diffeomorphic metric mapping · Generative Adversarial Networks · Geodesic shooting · Stationary velocity fields

1 Introduction

Since the 80s, deformable image registration has become a fundamental problem in medical image analysis [1]. A vast literature on deformable image registration methods exists, providing solutions to important clinical problems and applications. Up to the ubiquitous success of methods based on Convolutional Neural Networks (CNNs) in computer vision and medical image analysis, the great majority of deformable image registration methods were based on energy minimization models [2]. This traditional approach is model-based or optimization-based, in contrast with recent deep-learning approaches that are

---

U. Ramon, M. Hernandez and E. Mayordomo—With the ADNI Consortium.

Supplementary Information The online version contains supplementary material available at https://doi.org/10.1007/978-3-031-11203-4_3.

known as learning-based or data-based. Diffeomorphic registration constitutes the inception point in Computational Anatomy studies for modeling and understanding population trends and longitudinal variations, and for establishing relationships between imaging phenotypes and genotypes in Imaging Genetics [3,4]. Model-based diffeomorphic image registration is computationally costly. In fact, the huge computational complexity of large deformation diffeomorphic metric mapping (LDDMM) [5] is considered the curse of diffeomorphic registration, where very original solutions such as the stationary parameterization [6–8], the EPDiff constraint on the initial velocity field [9], or the band-limited parameterization [10] have been proposed to alleviate the problem. Since the advances that made it possible to learn the optical flow using CNNs (FlowNet [11]), dozens of deep-learning data-based methods have been proposed to approach the problem of deformable image registration in different clinical applications [12], some specifically for diffeomorphic registration [13–22]. Overall, all data-based methods yield fast inference algorithms for diffeomorphism computation once the difficulties with training have been overcome. Generative Adversarial Networks (GANs) is an interesting unsupervised approach where some interesting proposals for non-diffeomorphic deformable registration have been made [23] (2D) and [24,25] (3D). GANs have also been used for diffeomorphic deformable template generation [26], where the registration sub-network is based on an established U-net architecture [22,27], or for finding deformations for other purposes like interpretation of disease evidence [28]. A GAN combines the interaction of two different networks during training: a generative network and a discrimination network. The generative network itself can be regarded as an unsupervised method that, once included in the GAN system, is trained with the feedback of the discrimination network. The discriminator helps further update the generator during training with information regarding how the appearance of plausible warped source images. The main contribution of this work is the proposal of a GAN-based unsupervised learning LDDMM method for pairs of 3D mono-modal images, the first to use GANs for diffeomorphic registration. The method is inspired by the recent literature for deformable image registration with adversarial learning [24,25] and combines the best performing components within the LDDMM paradigm. We have successfully implemented two models for the stationary and the EPDiff-constrained non-stationary parameterizations and demonstrate the effectiveness of our models in both 2D simulated and 3D real brain MRI data.

2   Background on LDDMM

Let $\Omega \subseteq \mathbb{R}^d$ be the image domain. Let $Diff(\Omega)$ be the LDDMM Riemannian manifold of diffeomorphisms and V the tangent space at the identity element. $Diff(\Omega)$ is a Lie group, and V is the corresponding Lie algebra [5]. The Riemannian metric of $Diff(\Omega)$ is defined from the scalar product in V, $\langle v, w \rangle_V = \langle Lv, w \rangle_{L^2}$, where L is the invertible self-adjoint differential operator associated with the differential structure of $Diff(\Omega)$. In traditional LDDMM methods, $L = (Id - \alpha\Delta)^s, \alpha > 0, s \in \mathbb{R}$ [5]. We will denote with K the inverse of operator L. Let I_0 and I_1 be the source and the target images. LDDMM is formulated from the minimization of the variational problem

$$E(v) = \frac{1}{2}\int_0^1 \langle Lv_t, v_t \rangle_{L^2} dt + \frac{1}{\sigma^2} \| I_0 \circ (\phi_1^v)^{-1} - I_1 \|_{L^2}^2. \tag{1}$$

The LDDMM variational problem was originally posed in the space of time-varying smooth flows of velocity fields, $v \in L^2([0,1], V)$. Given the smooth flow $v : [0,1] \to V$, $v_t : \Omega \to \mathbb{R}^d$, the solution at time $t = 1$ to the evolution equation

$$\partial_t(\phi_t^v)^{-1} = -v_t \circ (\phi_t^v)^{-1} \tag{2}$$

with initial condition $(\phi_0^v)^{-1} = id$ is a diffeomorphism, $(\phi_1^v)^{-1} \in Diff(\Omega)$. The transformation $(\phi_1^v)^{-1}$, computed from the minimum of $E(v)$, is the diffeomorphism that solves the LDDMM registration problem between I_0 and I_1. The most significant limitation of LDDMM is its large computational complexity. In order to circumvent this problem, the original LDDMM variational problem is parameterized on the space of initial velocity fields

$$E(v_0) = \frac{1}{2}\langle Lv_0, v_0 \rangle_{L^2} + \frac{1}{\sigma^2} \| I_0 \circ (\phi_1^v)^{-1} - I_1 \|_{L^2}^2. \tag{3}$$

where the time-varying flow of velocity fields v is obtained from the EPDiff equation

$$\partial_t v_t + K[(Dv_t)^T \cdot Lv_t + DLv_t \cdot v_t + Lv_t \cdot \nabla \cdot v_t] = 0 \tag{4}$$

with initial condition v_0 (geodesic shooting). The diffeomorphism $(\phi_1^v)^{-1}$, computed from the minimum of $E(v_0)$ via Eqs. 4 and 2, verifies the momentum conservation constraint (MCC) [29], and, therefore, it belongs to a geodesic path on $Diff(\Omega)$. Simultaneously to the MCC parameterization, a family of methods was proposed to further circumvent the large computational complexity of the original LDDMM [6–8]. In all these methods, the time-varying flow of velocity fields v is restricted to be steady or stationary [30]. In this case, the solution does not belong to a geodesic.

3   Generative Adversarial Networks for LDDMM

Similarly to model-driven approaches for estimating LDDMM diffeomorphic registration, data-driven approaches for learning LDDMM diffeomorphic registration aim at the inference of a diffeomorphism $(\phi_1^v)^{-1}$ such that the LDDMM energy is minimized for a given (I_0, I_1) pair. In particular, data-driven approaches compute an approximation of the functional

$$\mathcal{S}(\arg\min_{v \in V} E(v, I_0, I_1)) \tag{5}$$

where $\mathcal{S}$ represents the operations needed to compute $(\phi_1^v)^{-1}$ from v, and the energy E is either given by Eqs. 1 or 3. The functional approximation is obtained via a neural network representation with parameters learned from a representative sample of image pairs. Unsupervised approaches assume that the LDDMM parameterization in combination with the minimization of the energy E considered as a loss function are enough for the inference of suitable diffeomorphic transformations after training. Therefore, there is no need for ground truth deformations. GAN-based approaches depart from unsupervised approaches by the definition of two different networks: the generative network (G) and the discrimination network (D), and are trained in an adversarial fashion as follows. The discrimination network D learns to distinguish between a warped source image $I_0 \circ (\phi_1^v)^{-1}$ generated by G and a plausible warped source image. It is trained using the loss function

$$L_D = \begin{cases} -\log(p) & c \in P^+ \\ -\log(1-p) & c \in P^- \end{cases} \tag{6}$$

where c indicates the input case, P^+ and P^- indicate positive or negative cases for the GAN, and p is the probability computed by D for the input case. In the first place, D is trained on a positive case $c \in P^+$ representing a target image I_1 and a warped source image I_0^w plausibly registered to I_1 with a diffeomorphic transformation. The warped source image is modeled from I_0 and I_1 with a strictly convex linear combination: $I_0^w = \beta I_0 + (1-\beta)I_1$. It should be noticed that, although the warped source image would ideally be I_1, the selection of $I_0^w = I_1$ (e.g. $\beta = 0$) empirically leads to the discriminator rapidly outperforming the generator. This approach to discriminators has been successfully used in adversarial learning methods for deformable registration [25]. Next, D is trained on a negative case $c \in P^-$ representing a target image I_1 and a warped source image I_0^w obtained from the generator network G. The generative network in this context is the diffeomorphic registration network. G is aimed at the approximation of the functional given in Eq. 5 similarly to unsupervised approaches for the inference of $(\phi_1^v)^{-1}$. It is trained using the combined loss function

$$L_G = L_{\text{adv}} + \lambda E(v, I_0, I_1). \tag{7}$$

where L_{adv} is the adversarial loss function, defined from $L_{\text{adv}} = -\log(p)$ where p is computed from D; E is the LDDMM energy given by Eqs. 1 or 3; and λ is the weight for balancing the adversarial and the generative losses. For each sample pair (I_0^w, I_1), G is fed with the pair of images and updates the network parameters from the back-propagation of the information of the loss function values coming from the LDDMM energy and the discriminator probability of being a pair generated by G.

3.1 Proposed GAN Architecture

Generator Network. In this work, the diffeomorphic registration network G is intended to learn LDDMM diffeomorphic registration parameterized on the space of steady velocity fields or the space of initial velocity fields subject to the EPDiff equation (Eq. 4). The diffeomorphic transformation $(\phi_1^v)^{-1}$ is obtained from these velocity fields either from scaling and squaring [7,8] or the solution of the deformation state equation [5]. Euler integration is used as PDE solver for all the involved differential equations. A number of different generator network architectures have been proposed in the recent literature, with predominance of simple fully convolutional (FC) [23] or U-Net like architectures [24,25]. In this work, we propose to use the architecture by Duan et al. [24] adapted to fit our purposes. The network follows the general U-net design of utilizing an encoder-decoder structure with skip connections. However, during the encoding phase, the source and target images are fed to two encoding streams with different resolution levels. The combination of the two encoding streams allows a larger receptive field suitable to learn large deformations. The upsampling is performed with a deconvolutional operation based on transposed convolutional layers [31]. We have empirically noticed that the learnable parameters of these layers help reduce typical checkerboard GAN artifacts in the decoding [32].

Discriminator Network. The discriminator network D follows a traditional CNN architecture. The two input images are concatenated and passed through five convolutional blocks. Each block includes a convolutional layer, a RELU activation function, and a size-two max-pooling layer. After the convolutions, the 4D volume is flattened and passed through three fully connected layers. The output of the last layer is the probability of the input images to come from a registered pair not generated by G.

Generative-Discriminative Integration Layer. The generator and the discriminator networks G and D are connected through an integration layer. This integration layer allows calculating the diffeomorphism $(\phi_1^v)^{-1}$ that warps the source image I_0. The selected integration layer depends on the velocity parameterization: stationary (SVF-GAN) or EPDiff-constrained time-dependent (EPDiff-GAN). The computed diffeomorphisms are applied to the source image via a second 3D spatial transformation layer [33] with no learnable parameters.

Parameter Selection and Implementation Details We selected the parameters $\lambda = 1000$, $\sigma^2 = 1.0$, $\alpha = 0.0025$, and $s = 4$ and a unit-domain discretization of the image domain Ω [5]. Scaling and squaring and Euler integration were performed in 8 and 10 time samples respectively. The parameter β for the convex linear modeling of warped images was selected equal to 0.2. Both the generator network and the discriminator network were trained with Adam's optimizer with default parameters and learning rates of $5e^{-5}$ for G and $1e^{-6}$ for D, respectively. The experiments were run on a machine equipped with one NVidia Titan RTX with 24 GBS of video memory and an Intel Core i7 with 64 GBS of DDR3 RAM, and developed in Python with Keras and a TensorFlow backend.

Source Target| DD SVF |St. LDDMM SVF |Flash V_0 |SVF-GAN SVF |EPDiff-GAN V_0

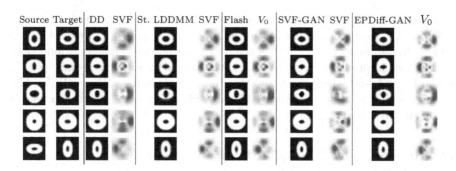

Fig. 1. Example of simulated 2D registration results. Up: source and target images of five selected experiments. Down, left to right: deformed images and velocity fields computed from diffeomorphic Demons (DD), stationary LDDMM (St. LDDMM), Flash, and our proposed SVF-GAN and EPDiff-GAN. SVF stands for a stationary velocity field and V_0 for the initial velocity field of a geodesic shooting approach, respectively.

4   Experiments and Results

2D Simulated Dataset. We simulated a total of 2560 torus images by varying the parameters of two ellipse equations, similarly to [19]. The parameters were drawn from two Gaussian distributions: $\mathcal{N}(4, 2)$ for the inner ellipse and $\mathcal{N}(12, 4)$ for the outer ellipse. The simulated images were of size 64×64. The networks were trained during 1000 epochs with a batch size of 64 samples.

3D Brain MRI Datasets. We used a total of 2113 T1-weighted brain MRI images from the Alzheimer's Disease Neuroimaging Initiative (ADNI). The images were acquired at the baseline visit and belong to all the available ADNI projects (1, 2, Go, and 3). The images were preprocessed with N3 bias field correction, affinely registered to the MNI152 atlas, skull-stripped, and affinely registered to the skull-stripped MNI152 atlas. The evaluation of our generated GAN models in the task of diffeomorphic registration was performed in NIREP dataset [34], where one image was chosen as reference and pair-wise registration was performed with the remaining 15. All images were scaled to size $176 \times 224 \times 176$, and in this case trained for 50 epochs with a batch size of 1 sample. Inference of either a stationary or a time dependent velocity field takes 1.3 s.

Results in the 2D Simulated Dataset. Figure 1 show the deformed images and the velocity fields obtained in the 2D simulated dataset by diffeomorphic Demons [7], a stationary version of LDDMM (St. LDDMM) [8], the spatial version of Flash [10], and our proposed SVF and EPDiff GANs. Apart from diffeomorphic Demons that uses Gaussian smoothing for regularization, all the considered methods use the same parameters for operator L. Therefore, St. LDDMM and SVF-GAN can be seen as a model-based and a data-based approach for the minimization of the same variational problem. The same happens with Flash and EPDiff-GAN. From the figure, it can be appreciated that our proposed GANs are able to obtain accurate warps of the source to the target images, similarly to model-based approaches. For SVF-GAN, the inferred velocity fields are visually similar to model-based approaches in three of five experiments. For EPDiff-GAN, the inferred initial velocity fields are visually similar to model-based approaches in four of five experiments.

4.1   Results in the 3D NIREP Dataset

Quantitative Assessment. Figure 2 shows the Dice similarity coefficients obtained with diffeomorphic Demons [7], St. LDDMM [8], Voxelmorph II [16], the spatial version of Flash [10], Quicksilver [14] and our proposed SVF and EPDiff GANs. SVF-GAN shows an accuracy similar to St. LDDMM and competitive with diffeomorphic Demons. Our proposed method tends to overpass Voxelmorph II in the great majority of the structures. On the other hand, EPDiff-GAN shows an accuracy similar to Flash and Quicksilver in the great majority of regions, with the exception of the temporal pole (TP) and the orbital frontal gyrus (OFG), two small localized and difficult to register regions. Furthermore, the two-stream architecture greatly improves the accuracy obtained by a simple U-Net. SVF-GAN outperforms the ablation study model in which no discriminator was used, though EPDiff-GAN only shows clear performance improvements in some structures. It drives our attention that Flash underperformed in the superior frontal gyrus (SFG). All tested methods generate smooth deformations with almost no foldings, as can be seen in table 1 from the supplementary material.

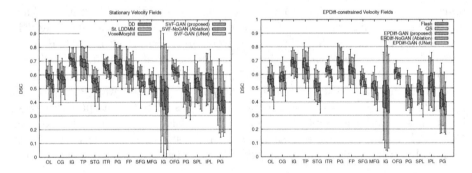

Fig. 2. Evaluation in NIREP. Dice scores obtained by propagating the diffeomorphisms to the segmentation labels on the 16 NIREP brain structures. Left, methods parameterized with stationary velocity fields: diffeomorphic Demons (DD), stationary LDDMM (St. LDDMM), Voxelmorph II, our proposed SVF-GAN with the two-stream architecture, SVF-GAN without discriminator and SVF-GAN with a U-net. Right, geodesic shooting methods: Flash, Quicksilver (QS), our proposed EPDiff-GAN, EPDiff-GAN without discriminator, and EPDiff-GAN with a U-net.

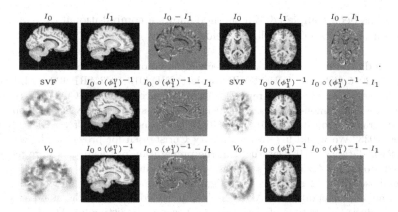

Fig. 3. Example of 3D registration results. First row, sagittal and axial views of the source and the target images and the differences before registration. Second row, inferred stationary velocity field, warped image, and differences after registration for SVF-GAN. Third row, inferred initial velocity field, warped image, and differences after registration for EPDiff-GAN.

Qualitative Assessment. For a qualitative assessment of the quality of the registration results, Fig. 3 shows the sagittal and axial views of one selected NIREP registration result. In the figure, it can be appreciated a high matching between the target and the warped ventricles, and more difficult to register regions like the cingulate gyrus (observable in the sagittal view) or the insular cortex (observable in the axial view).

5   Conclusions

We have proposed an adversarial learning LDDMM method for the registration of 3D mono-modal images. We have successfully implemented two models: one for the stationary parameterization and the other for the EPDiff-constrained non-stationary parameterization (geodesic shooting). The performed ablation study shows how GANs improve the results of the proposed registration networks. Furthermore, our experiments have shown that the inferred velocity fields are comparable to the solutions of model-based approaches. In addition, the evaluation study has shown the competitiveness of our approach with state of the art model- and data- based methods. It should be remarked that our methods perform similarly to Quicksilver, a supervised method that uses patches for training, and therefore, it learns in a rich-data environment. In contrast, our method is unsupervised and uses the whole image for training in a data-hungry environment. Indeed, our proposed methods outperform Voxelmorph II, an unsupervised method for diffeomorphic registration usually selected as benchmark in the state of the art. Finally, our proposal may constitute a good candidate for the massive computation of diffeomorphisms in Computational Anatomy studies, since once training has been completed, our method shows a computational time of over a second for the inference of velocity fields.

Acknowledgement. This work was partially supported by the national research grant TIN2016-80347-R (DIAMOND project),PID2019-104358RB-I00 (DL-Ageing project), and Government of Aragon Group Reference $T64\_20R$ (COSMOS research group). Ubaldo Ramon-Julvez's work was partially supported by an Aragon Government grant. Project PID2019-104358RB-I00 granted by MCIN/AEI/10.13039/501100011033. We would like to thank Gary Christensen for providing the access to the NIREP database [34]. Data used in the preparation of this article were partially obtained from the Alzheimer' s Disease Neuroimaging Initiative (ADNI) database (adni.loni.usc.edu). As such, the investigators within the ADNI contributed to the design and implementation of ADNI and/or provided data but did not participate in the analysis or writing of this report. A complete listing of ADNI investigators can be found at: https://adni.loni.usc.edu/wp-content/uploads/how_to_apply/ADNI_Acknowledgement_List.pdf.

References

1. Sotiras, A., Davatzikos, C., Paragios, N.: Deformable medical image registration: a survey. IEEE Trans. Med. Imaging **32**(7), 1153–1190 (2013)
2. Modersitzki, J.: FAIR: Flexible Algorithms for Image Registration. SIAM, New Delhi (2009)
3. Hua, X.: ADNI: tensor-based morphometry as a neuroimaging biomarker for Alzheimer's disease: an MRI study of 676 AD, MCI, and normal subjects. Neuroimage **43**(3), 458–469 (2008)
4. Liu, Y., Li, Z., Ge, Q., Lin, N., Xiong, M.: Deep feature selection and causal analysis of Alzheimer's disease. Front. Neurosci. **13**, 1198 (2019)
5. Beg, M.F., Miller, M.I., Trouve, A., Younes, L.: Computing large deformation metric mappings via geodesic flows of diffeomorphisms. Int. J. Comput. Vision **61**(2), 139–157 (2005)
6. Ashburner, J.: A fast diffeomorphic image registration algorithm. Neuroimage **38**(1), 95–113 (2007)
7. Vercauteren, T., Pennec, X., Perchant, A., Ayache, N.: Diffeomorphic demons: efficient non-parametric image registration. Neuroimage **45**(1), S61–S72 (2009)
8. Hernandez, M.: Gauss-Newton inspired preconditioned optimization in large deformation diffeomorphic metric mapping. Phys. Med. Biol. **59**(20), 6805 (2014)
9. Vialard, F.X., Risser, L., Rueckert, D., Cotter, C.J.: Diffeomorphic 3D image registration via geodesic shooting using an efficient adjoint calculation. Int. J. Comput. Vision **97**(2), 229–241 (2011)
10. Zhang, M., Fletcher, T.: Fast diffeomorphic image registration via fourier-approximated lie algebras. Int. J. Comput. Vision **127**, 61–73 (2018)
11. Dosovitskiy, A., Fischere, P., Ilg, E., Hausser, P., Hazirbas, C., Golkov, V.: Flownet: learning optical flow with convolutional networks. In: Proceedings of the 16th IEEE International Conference on Computer Vision (ICCV 2015), pp. 2758–2766 (2015)
12. Boveiri, H., Khayami, R., Javidan, R., Mehdizadeh, A.: Medical image registration using deep neural networks: a comprehensive review. Comput. Electr. Eng. **87**, 106767 (2020)

13. Rohé, M.-M., Datar, M., Heimann, T., Sermesant, M., Pennec, X.: SVF-Net: learning deformable image registration using shape matching. In: Descoteaux, M., Maier-Hein, L., Franz, A., Jannin, P., Collins, D.L., Duchesne, S. (eds.) MICCAI 2017. LNCS, vol. 10433, pp. 266–274. Springer, Cham (2017). https://doi.org/10.1007/978-3-319-66182-7_31

14. Yang, X., Kwitt, R., Styner, M., Niethammer, M.: Quicksilver: fast predictive image registration - a deep learning approach. Neuroimage **158**, 378–396 (2017)

15. Dalca, A.V., Balakrishnan, G., Guttag, J., Sabuncu, M.R.: Unsupervised learning for fast probabilistic diffeomorphic registration. In: Frangi, A.F., Schnabel, J.A., Davatzikos, C., Alberola-López, C., Fichtinger, G. (eds.) MICCAI 2018. LNCS, vol. 11070, pp. 729–738. Springer, Cham (2018). https://doi.org/10.1007/978-3-030-00928-1_82

16. Balakrishnan, G., Zhao, A., Sabuncu, M.R., Guttag, J., Dalca, A.V.: Voxelmorph: a learning framework for deformable medical image registration. IEEE Trans. Med. Imaging **38**(8), 1788–1800 (2019)

17. Krebs, J., Delingetter, H., Mailhe, B., Ayache, N., Mansi, T.: Learning a probabilistic model for diffeomorphic registration. IEEE Trans. Med. Imaging **38**, 2165–2176 (2019)

18. Fan, J., Cao, X., Yap, P., Shen, D.: BIRNet: brain image registration using dual-supervised fully convolutional networks. Med. Image Anal. **54**, 193–206 (2019)

19. Wang, J., Zhang, M.: DeepFLASH: an efficient network for learning-based medical image registration. In: Proceedings of the IEEE Conference on Computer Vision and Pattern Recognition (CVPR 2020) (2020)

20. Mok, T.C.W., Chung, A.C.S.: Fast symmetric diffeomorphic image registration with convolutional neural networks. In: Proceedings of the IEEE Conference on Computer Vision and Pattern Recognition (CVPR 2020) (2020)

21. Hoffmann, M., Billot, B., Greve, D.N., Iglesias, J.E., Fischl, B., Dalca, A.V.: Synthmorph: learning contrast-invariant registration without acquired images. IEEE Trans. Med. Imaging **41**(3), 543–558 (2021)

22. Dalca, A.V., Balakrishnan, G., Guttag, J., Sabuncu, M.R.: Unsupervised learning of probabilistic diffeomorphic registration for images and surfaces. Med. Image Anal. **57**, 226–236 (2019)

23. Mahapatra, D., Antony, B., Sedai, S., Garvani, R.: Deformable medical image registration using generative adversarial networks. In: IEEE International Symposium on Biomedical Imaging (ISBI 2018) (2018)

24. Duan, L., et al.: Adversarial learning for deformable registration of brain MR image using a multi-scale fully convolutional network. Biomed. Signal Process. Control **53**, 101562 (2018)

25. Fan, J., Cao, X., Wang, Q., Yap, P., Shen, D.: Adversarial learning for mono- or multi-modal registration. Med. Image Anal. **58**, 1015–1045 (2019)

26. Dey, N., Ren, M., Dalca, A.V., Gerig, G.: Generative adversarial registration for improved conditional deformable templates. In: Proceedings of the 18th IEEE International Conference on Computer Vision (ICCV 2021) (2021)

27. Dalca, A.V., Rakic, M., Guttag, J.V., Sabuncu, M.R.: Learning conditional deformable templates with convolutional networks. In: NeurIPS (2019)

28. Bigolin Lanfredi, R., Schroeder, J.D., Vachet, C., Tasdizen, T.: Interpretation of disease evidence for medical images using adversarial deformation fields. In: Martel, A.L., et al. (eds.) MICCAI 2020. LNCS, vol. 12262, pp. 738–748. Springer, Cham (2020). https://doi.org/10.1007/978-3-030-59713-9_71

29. Younes, L.: Jacobi fields in groups of diffeomorphisms and applications. Q. Appl. Math. **65**, 113–134 (2007)

30. Arsigny, V., Commonwick, O., Pennec, X., Ayache, N.: Statistics on diffeomorphisms in a Log-Euclidean framework. In: Proceedings of the 9th International Conference on Medical Image Computing and Computer Assisted Intervention (MICCAI 2006), Lecture Notes in Computer Science, vol. 4190, pp. 924–931 (2006)
31. Zeiler, M.D., Taylor, G.W., Fergus, R.: Adaptive deconvolutional networks for mid and high level feature learning. In: ICCV, vol. 2011, pp. 2018–2025 (2011)
32. Odena, A., Dumoulin, V., Olah, C.: Deconvolution and checkerboard artifacts. Distill **1**(10), e3 (2016)
33. Jaderberg, M., Simonyan, K., Zissermann, A., Kavukcuoglu, K.: Spatial transformer networks. In: Proceedings of Conference on Neural Information Processing Systems (NeurIPS 2015) (2015)
34. Christensen, G.E., et al.: Introduction to the non-rigid image registration evaluation project (NIREP). In: Proceedings of 3rd International Workshop on Biomedical Image Registration (WBIR 2006), vol. 4057, pp. 128–135 (2006)

Towards a 4D Spatio-Temporal Atlas of the Embryonic and Fetal Brain Using a Deep Learning Approach for Groupwise Image Registration

Wietske A. P. Bastiaansen[1,2]([✉]), Melek Rousian[2],
Régine P. M. Steegers-Theunissen[2], Wiro J. Niessen[1], Anton H. J. Koning[3],
and Stefan Klein[1]

[1] Department of Radiology and Nuclear Medicine, Biomedical Imaging Group
Rotterdam, Erasmus MC, Rotterdam, Netherlands
w.bastiaansen@erasmusmc.nl
[2] Department of Obstetrics and Gynecology, Erasmus MC,
Rotterdam, Netherlands
[3] Department of Pathology, Erasmus MC, Rotterdam, Netherlands

Abstract. Brain development during the first trimester is of crucial importance for current and future health of the fetus, and therefore the availability of a spatio-temporal atlas would lead to more in-depth insight into the growth and development during this period. Here, we propose a deep learning approach for creation of a 4D spatio-temporal atlas of the embryonic and fetal brain using groupwise image registration. We build on top of the extension of Voxelmorph for the creation of learned conditional atlases, which consists of an atlas generation and registration network. As a preliminary experiment we trained only the registration network and iteratively updated the atlas. Three-dimensional ultrasound data acquired between the 8th and 12th week of pregnancy were used. We found that in the atlas several relevant brain structures were visible. In future work the atlas generation network will be incorporated and we will further explore, using the atlas, correlations between maternal periconceptional health and brain growth and development.

Keywords: Embryonic and fetal brain atlas · Groupwise image registration · First trimester ultrasound · Deep learning

1 Introduction

Normal growth and development of the human embryonic and fetal brain during the first trimester is of crucial importance for current and future health of the fetus [14,17]. Currently, this is monitored by manual measurements, such as the circumference and volume of the brain [13,15]. However, these measurements lack

A. Hering et al. (Eds.): WBIR 2022, LNCS 13386, pp. 29–34, 2022.
https://doi.org/10.1007/978-3-031-11203-4_4

overview: it is unclear how the different measurements relate. The availability of an atlas i.e., a set of brain templates for a range of gestational ages, could overcome these challenges by offering a unified and automatic framework to compare development across subjects.

In literature several atlases are available [5–8, 11, 12, 16, 18, 19]. However, these are based on magnetic resonance imaging and/or acquired during the second and third trimester of pregnancy. Here, we present to the best of our knowledge the first framework for the development of a brain atlas describing growth of the human embryo and fetus between 56 and 90 days gestational age (GA) based on ultrasound imaging.

2   Method

The atlas is generated from three-dimensional (3D) ultrasound images $I_{i,t}$, for subject i imaged at time t, where t is the GA in days. The atlas A_t is obtained by groupwise registration of $I_{i,t}$ for every pregnancy $i = 1, ..., k$ on every time point t, followed by taking the mean over the deformed images: $A_t = \frac{1}{k} \sum_i I_{i,t} \circ \phi_i$. Hereby the constraint $\sum_{i,t} \phi_{i,t} \approx 0$ is applied, as proposed by Balci et al. and Bhatia et al. [1,3]. To ensure invertibility of the deformations we used diffeomorphic non-rigid deformations with the deformation field $\phi_{i,t}$, obtained by integrating the velocity field $\nu_{i,t}$.

The framework is based on the extension of Voxelmorph for learning conditional atlases by Dalca et al. [4]. An overview of the framework can be found in Fig. 1. Here, we only train the registration framework and we initialize the atlas for every time t as the voxelwise median over all images $I_i \forall i$. The median was chosen over the mean, since this resulted in a sharper initial atlas. Next, the atlas is updated for iteration n as the mean of $I_{i,t} \circ \phi_{i,t}^n$ for every time t. Subsequently, the network is trained until $A_t^n \approx A_t^{n-1}$.

The loss function is defined as follows:

$$\mathcal{L}\left(A_t, I_{i,t}, \phi_{i,t}, \phi_{i,t}^{-1}\right) = \lambda_{\mathrm{sim}} \mathcal{L}_{\mathrm{similarity}}\left(A_t \circ \phi_{i,t}^{-1}, I_{i,t}\right) + \lambda_{\mathrm{group}} \mathcal{L}_{\mathrm{groupwise}}\left(\phi_{i,t}\right)$$
$$+ \lambda_{\mathrm{mag}} \mathcal{L}_{\mathrm{magnitude}}\left(\phi_{i,t}^{-1}\right) + \lambda_{\mathrm{dif}} \mathcal{L}_{\mathrm{diffusion}}\left(\phi_{i,t}^{-1}\right)$$

$$(1)$$

The first term computes the similarly between the atlas and image, we used the local squared normalized cross-correlation, which was used before on this dataset [2]. The second term approximates the constraint for groupwise registration by minimizing the running average over the last c deformation fields obtained during training. To balance the influence of this constraint with respect to time, we sorted the data based on day GA within every epoch and took as window c the average number of images per day GA in the dataset. Finally, the deformations are regularized by: $\mathcal{L}_{\mathrm{mag}} = \|\phi_{i,t}^{-1}\|_2^2$ and $\mathcal{L}_{\mathrm{dif}} = \|\nabla \phi_{i,t}^{-1}\|_2^2$.

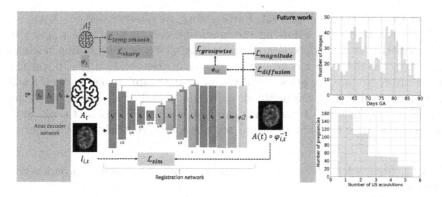

Fig. 1. Overview of the proposed framework and characteristics of the used dataset.

3   Data and Experiments

The Rotterdam Periconceptional Cohort (Predict study) is a large hospital-based cohort study conducted at the Erasmus MC, University Medical Center Rotterdam, the Netherlands. This prospective cohort focuses on the relationships between periconceptional maternal and paternal health and embryonic and fetal growth and development [14,17]. 3D ultrasound scans are acquired at multiple points in time during the first trimester. Here, to model normal development, we included only singleton pregnancies with no adverse outcome and spontaneous conception with a regular menstrual cycle.

We included 871 ultrasound images of 398 pregnancies acquired between 56 and 90 days GA. For each day GA, we have at least 10 ultrasound images, as shown in top-right graph in Fig. 1. The data was split such that for every day GA 80% of the data is in the training set and 20% in the test set. We first spatially aligned and segmented the brain using our previously developed algorithm for multi-atlas segmentation and registration of the embryo [2]. Next, we resized all images to a standard voxelsize per day GA, to ensure that the brain always filled a similar field of view despite the fast growth of the brain. This standard voxelsize per day GA was determined by linear interpolation of the median voxelsize per week GA. We trained the network using the default hyperparameters proposed by Dalca et al. [4] for $\lambda_{\text{group}} \in \{0, 1, 10, 100\}$. We reported the mean percentage of voxels having a non-positive Jacobian determinant $\%|J| \leq 0$, the groupwise loss $\mathcal{L}_{\text{group}}$ and the similarity loss $\mathcal{L}_{\text{sim}}$. Finally, for the best set of hyperparameters the atlas was updated iteratively, and we visually analyzed the result.

4   Results

From the results given in Table 1 for iteration $n = 1$ we concluded that all tested hyperparameters resulted in smooth deformation fields, since the percentage of voxels with a non-positive Jacobian determinant $\%|J| \leq 0$ over the whole dataset was less then one percent. Furthermore, we observe that for $\lambda_{\text{group}} = 1$ $\mathcal{L}_{\text{group}}$ is similar to not enforcing the groupwise constraint. For $\lambda_{\text{group}} = 100$, we observed that $\mathcal{L}_{\text{sim}}$ deteriorated, indicating that the deformation fields are excessively restricted by the groupwise constraint. Hence, $\lambda_{\text{group}} = 10$ was used to iteratively update the atlas. Finally, note that the difference between results for training and testing are minimal: indicating a limited degree of overfitting. In Fig. 2 a visualization of the results can be found for $t = 68$ and $t = 82$. In the showed axial slices the choroid plexus and the fourth ventricle can be observed.

Table 1. Results for different hyperparameters, with the standard deviation given between brackets.

Hyperparameters				Training			Test		
λ_{sim}	λ_{group}	λ_{mag}	λ_{dif}	$\%|J| \leq 0$	$\mathcal{L}_{\text{group}}$	$\mathcal{L}_{\text{sim}}$	$\%|J| \leq 0$	$\mathcal{L}_{\text{group}}$	$\mathcal{L}_{\text{sim}}$
1	0	0.01	0.01	0.26 (0.47)	1.45e$-$3	0.126	0.36 (0.55)	1.90e$-$3	0.130
1	1	0.01	0.01	0.25 (0.38)	1.27e$-$3	0.125	0.32 (0.42)	1.62e$-$3	0.129
1	10	0.01	0.01	0.17 (0.27)	7.80e$-$4	0.118	0.22 (0.28)	9.40e$-$4	0.126
1	100	0.01	0.01	0.04 (0.06)	1.61e$-$4	0.091	0.05 (0.07)	1.85e$-$4	0.101

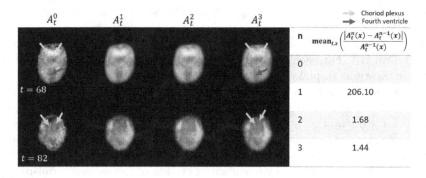

Fig. 2. Axial slice of the atlas for different GA and iterations 0, 1, 2 and 3.

5 Discussion and Conclusion

We propose a deep learning approach for creation of a 4D spatio-temporal atlas of the embryonic and fetal brain using groupwise image registration. Here, we trained the registration network iteratively and visually inspected the resulting atlas. We found that the registration network results in smooth deformation field, and that several relevant brain structures were visible in the atlas.

In this work, the window c of the groupwise loss term was set to the mean number of samples per day GA, in future work this hyperparameter will be varied to study its influence. As shown in Fig. 1, in future work also the atlas generator network will be incorporated, where constraints for temporal smoothness and sharp edges in the atlas can directly be incorporated in the loss. Finally, we will evaluate if the relevant brain measurements of the atlas are close to clinically known values and we will analyze if the morphology of the brain, modelled by the deformations $\phi_{i,t}$, shows the known correlation with maternal periconceptional health factors found in previous research [9,10].

References

1. Balci, S.K., Golland, P., Shenton, M.E., Wells, W.M.: Free-form B-spline deformation model for groupwise registration. Med. Image Comput. Comput. Assist. Interv. **10**, 23–30 (2007)
2. Bastiaansen, W.A., Rousian, M., Steegers-Theunissen, R.P., Niessen, W.J., Koning, A.H., Klein, S.: Multi-atlas segmentation and spatial alignment of the human embryo in first trimester 3D ultrasound. arXiv:2202.06599 (2022)
3. Bhatia, K., Hajnal, J., Puri, B., Edwards, A., Rueckert, D.: Consistent groupwise non-rigid registration for atlas construction. In: 2004 2nd IEEE International Symposium on Biomedical Imaging: Nano to Macro, vol. 1, pp. 908–911 (2004)
4. Dalca, A., Rakic, M., Guttag, J., Sabuncu, M.: Learning conditional deformable templates with convolutional networks. In: Advances in Neural Information Processing Systems, vol. 32 (2019)
5. Dittrich, E., et al.: A spatio-temporal latent atlas for semi-supervised learning of fetal brain segmentations and morphological age estimation. Med. Image Anal. **18**(1), 9–21 (2014)
6. Gholipour, A.: A normative spatiotemporal MRI atlas of the fetal brain for automatic segmentation and analysis of early brain growth. Sci. Rep. **7**(1), 1–13 (2017)
7. Habas, P.A., et al.: A spatiotemporal atlas of MR intensity, tissue probability and shape of the fetal brain with application to segmentation. Neuroimage **53**(2), 460–470 (2010)
8. Khan, S., et al.: Fetal brain growth portrayed by a spatiotemporal diffusion tensor MRI atlas computed from in utero images. Neuroimage **185**, 593–608 (2019)
9. Koning, I., et al.: Growth trajectories of the human embryonic head and periconceptional maternal conditions. Hum. Reprod. **31**(5), 968–976 (2016)
10. Koning, I., Dudink, J., Groenenberg, I., Willemsen, S., Reiss, I., Steegers-Theunissen, R.: Prenatal cerebellar growth trajectories and the impact of periconceptional maternal and fetal factors. Hum. Reprod. **32**(6), 1230–1237 (2017)
11. Kuklisova-Murgasova, M., et al.: A dynamic 4D probabilistic atlas of the developing brain. Neuroimage **54**(4), 2750–2763 (2011)
12. Namburete, A.I.L., van Kampen, R., Papageorghiou, A.T., Papież, B.W.: Multi-channel groupwise registration to construct an ultrasound-specific fetal brain atlas. In: Melbourne, A., et al. (eds.) PIPPI/DATRA -2018. LNCS, vol. 11076, pp. 76–86. Springer, Cham (2018). https://doi.org/10.1007/978-3-030-00807-9_8
13. Paladini, D., Malinger, G., Birnbaum, R., Monteagudo, A., Pilu, G., Salomon, L.: ISUOG practice guidelines (updated): sonographic examination of the fetal central nervous system. Part 1: performance of screening examination and indications for targeted neurosonography. Ultrasound Obstet. Gynecol. **56**, 476–484 (2020)

14. Rousian, M., et al.: Cohort profile update: the Rotterdam Periconceptional Cohort and embryonic and fetal measurements using 3D ultrasound and virtual reality techniques. Int. J. Epidemiol. **50**, 1–14 (2021)
15. Salomon, L.J., et al.: Practice guidelines for performance of the routine mid-trimester fetal ultrasound scan. Ultrasound Obstet. Gynecol. **37**(1), 116–126 (2011)
16. Serag, A., et al.: Construction of a consistent high-definition Spatio-temporal atlas of the developing brain using adaptive kernel regression. Neuroimage **59**(3), 2255–2265 (2012)
17. Steegers-Theunissen, R., et al.: Cohort profile: the Rotterdam Periconceptional cohort (predict study). Int. J. Epidemiol. **45**, 374–381 (2016)
18. Uus, A., et al.: Multi-channel 4D parametrized atlas of macro-and microstructural neonatal brain development. Frontiers Neurosci., 721 (2021)
19. Wu, J., et al.: Age-specific structural fetal brain atlases construction and cortical development quantification for Chinese population. Neuroimage **241**, 118412 (2021)

Uncertainty

DeepSTAPLE: Learning to Predict Multimodal Registration Quality for Unsupervised Domain Adaptation

Christian Weihsbach[(✉)], Alexander Bigalke, Christian N. Kruse,
Hellena Hempe, and Mattias P. Heinrich

Institute of Medical Informatics, Universität zu Lübeck,
Ratzeburger Allee 160, 23538 Lübeck, Germany
christian.weihsbach@uni-luebeck.de
https://www.imi.uni-luebeck.de/en/institute.html

Abstract. While deep neural networks often achieve outstanding results on semantic segmentation tasks within a dataset domain, performance can drop significantly when predicting domain-shifted input data. Multi-atlas segmentation utilizes multiple available sample annotations which are deformed and propagated to the target domain via multimodal image registration and fused to a consensus label afterwards but subsequent network training with the registered data may not yield optimal results due to registration errors. In this work, we propose to extend a curriculum learning approach with additional regularization and fixed weighting to train a semantic segmentation model along with data parameters representing the atlas confidence. Using these adjustments we can show that registration quality information can be extracted out of a semantic segmentation model and further be used to create label consensi when using a straightforward weighting scheme. Comparing our results to the STAPLE method, we find that our consensi are not only a better approximation of the oracle-label regarding Dice score but also improve subsequent network training results.

Keywords: Domain adaptation · Multi-atlas registration · Label noise · Consensus · Curriculum learning

1 Introduction

Deep neural networks dominate the state-of-the-art medical image segmentation [10,14,20], but their high performance is depending on the availability of large-scale labelled datasets. Such labelled data is often not available in the target domain and direct transfer learning leads to performance drops due to domain shift [27]. To overcome these issues transferring existing annotations from a labeled source to the target domain is desirable. Mutli-atlas segmentation is a popular method, which accomplishes such a label transfer in two steps: First,

© The Author(s), under exclusive license to Springer Nature Switzerland AG 2022
A. Hering et al. (Eds.): WBIR 2022, LNCS 13386, pp. 37–46, 2022.
https://doi.org/10.1007/978-3-031-11203-4_5

multiple sample annotations are transferred to target images via image registration [7,18,24] resulting in multiple "optimal" labels [1]. Secondly label fusion can be applied to build the label consensus. Although many methods for finding a consensus label have been developed [1,6,19,25,26], the resulting fused labels are still not perfect and exhibit label noise, which complicates the training of neural networks and degrades performance.

Related Work. In the past, various label fusion methods have been proposed, which use weighted voting on registered label candidates to output a common consensus label [1,6,19,26]. More elaborate fusion methods also use image intensities [25], however when predicting across domains source and target intensities can differ substantially complicating intensity-based fusion and would therefore require handling of the intensity gap i.e. with image-to-image translation techniques [29]. When using the resulting consensus labels from non-optimal registration and fusion for subsequent CNN training, noisy data is introduced to the network [12]. Network training can then be improved with techniques of curriculum learning to estimate label noise (i.e. difficulty) and guide the optimization process accordingly [3,22] but the techniques have not been used in the context of noise introduced through registered pixel-wise labels [2,3,11,22,28] or employ more specialized and complex pipelines [4,5,15]. Other deep learning-based techniques to address ambiguous labels are probabilistic networks [13].

Contributions. We propose to use data parameters [22] to weight noisy atlas samples as a simple but effective extension of semantic segmentation models. During training the data parameters (scalar values assigned to each instance of a registered label) can estimate the label trustworthiness globally across all multi-atlas candidates of all images. We extend the original formulation of data parameters by additional *risk regularization* and *fixed weighting* terms to adapt to the specific characteristics of the segmentation task and show that our adaptation improves network training performance for 2D and 3D tasks in the single-atlas scenario. Furthermore, we apply our method to the multi-atlas 3D image scenario where the network scores do not improve but yield equal performance in comparison to normal cross-entropy loss training when using out-of-line back-propagation. Nonetheless, we still can achieve an improvement by deriving an optimized consensus label from the extracted weights and applying a straight-forward weighted-sum on the registered atlases.

2   Method

In this section, we will describe our data parameter adaption[1] and introduce our proposed extensions when using it in semantic segmentation tasks, namely a special regularization and a fixed weighting scheme. Furthermore, a multi-atlas specific extension will be described, which improves training stability.

---

[1] Our code is openly available on GitHub: https://github.com/multimodallearning/deep_staple.

Data Parameters. Saxena et al. [22] formulate their data parameter and curriculum learning approach as a modification altering the logits input of the loss function. By a learnable logits-weighting improvements could be shown in different scenarios when either noisy training samples and/or classes were weighted during training. Our implementation and experiments focus on per-sample parameters $\mathbf{DP_S}$ of a dataset $S = \{(\mathbf{x_s}, \mathbf{y_s})\}_{s=1}^{n}$ with images x_s and labels y_s containing n training samples. Since weighting schemes for multi-atlas label fusion like STAPLE [26] use a confidence weight of 0 indicating "no confidence" and 1 indicating "maximum confidence we slightly changed the initial formulation of data parameters:

$$\mathbf{DP}_\sigma = sigmoid\,(\mathbf{DP_S}) \tag{1}$$

According to Eq. 1 we limit the data parameters applied to our loss to $DP_\sigma \in (0, 1)$ where a value of 0 indicates "no confidence" and 1 indicates "maximum confidence" such as weighting schemes like STAPLE [26]. The data parameter loss ℓ_{DP} is calculated as

$$\ell_{DP}\,(f_\theta\,(\mathbf{x_B})\,, \mathbf{y_B}) = \sum_{b=1}^{|B|} \ell_{CE,spatial}\,(f_\theta\,(\mathbf{x_b})\,, \mathbf{y_b}) \cdot DP_{\sigma_b}, \quad \text{with} \quad B \subseteq S \tag{2}$$

where B is a training batch, $\ell_{CE,spatial}$ is the cross-entropy loss reduced over spatial dimensions and f_θ the model. As in the original implementation, the parameters require a sparse implementation of the Adam optimizer to avoid diminishing momenta. Note, that the data parameter layer is omitted for inference—inference scores are only affected indirectly by data parameters through optimized model training.

Risk Regularisation. Even when a foreground class is present in the image and a registered target label only contains background voxels, the network can achieve a zero-loss value by overfitting. As a consequence, upweighting the over-fitted samples will be of no harm in terms of loss reduction which leads to the upweighting of maximal noisy (empty) samples. We therefore add a so called *risk regularisation* encouraging the network to take *risk*

$$\ell = \ell_{DP} - \sum_{b=1}^{|B|} \frac{\#\{f_\theta\,(\mathbf{x_b}) = c\}}{\#\{f_\theta\,(\mathbf{x_b}) = c\} + \#\{f_\theta\,(\mathbf{x_b}) = \bar{c}\}} \cdot DP_{\sigma_b} \tag{3}$$

where $\#\{f_\theta\,(\mathbf{x_b}) = c\}$ and $\#\{f_\theta\,(\mathbf{x_b}) = \bar{c}\}$ indicate positive and negative predicted voxel count. According to this regularisation the network can reduce loss when predicting more target voxels under the restriction that the sample has a high data parameter value i.e. is classified as a clean sample. This formulation is balanced because predicting more positive voxels will increase the cross-entropy term if the prediction is inaccurate.

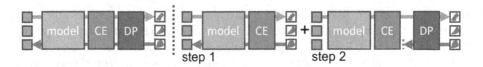

Fig. 1. Left: Inline backpropagation updating (red arrow) model and data parameters together. **Right:** Out-of-line backpropagation first steps on model (gray arrow) using normal cross-entropy loss and then steps on data parameters using the model's weights of the first step. (Color figure online)

Fixed Weighting Scheme. We found that the parameters have a strong correlation with the ground-truth voxels present in their values. Applying a fixed compensation weighting to the data parameters DP_{σ_b} can improve the correlation of the learned parameters and our target scores

$$DP_{\tilde{\sigma}_b} = \frac{DP_{\sigma_b}}{log\left(\#\left\{(\mathbf{y_b} = c\right\} + e\right) + e} \tag{4}$$

where $\#\left\{\mathbf{y_b} = c\right\}$ denotes the count of ground-truth voxels and e Euler's number.

Out-of-Line Backpropagation Process for Improved Stability. The interdependency of data parameters and model parameters can cause convergence issues when training *inline*, especially during earlier epochs when predictions are inaccurate. We found that a two-step forward-backward pass, first through the main model and in the second step through the main model and the data parameters can maintain stability while still estimating label noise (see Fig. 1). First only the main model parameters will be optimized. Secondly only the data parameters will be optimized *out-of-line*. When using the *out-of-line*, two-step approach data parameter optimization becomes a hypothesis of *"what would help the model optimizing right now?"* without intervening. Due to the optimizer momentum the parameter values still become reasonably separated.

Consensus Generation via Weighted Voting. To create a consensus $\mathbf{C_M}$ we use a simple weighted-sum over a set of multi-atlas labels M associated to a fixed image that turned out to be effective

$$\mathbf{C_M} = \left(\sum_{m=1}^{|M|} softmax(\mathbf{DP_M})_m \cdot \mathbf{y_m}\right) > 0.5 \quad \text{with} \quad M \subset S \tag{5}$$

where $\mathbf{DP_M}$ are the parameters associated to the set of multi-atlas labels $\mathbf{y_M}$.

3   Experiments

In this section, we will describe general dataset and model properties as well as our four experiments which increase in complexity up to the successful application of our method in 3D multi-atlas label noise estimation. We will refer to oracle-labels[2] as the real target labels which belong to an image and "registered/training/ground-truth"-labels as image labels that the network used to update its weights. Oracle-Dice refers to the overlapping area of oracle-labels and "registered/training/ground-truth"-labels.

Dataset. For our experiments, we chose a challenging multimodal segmentation task which was part of the CrossMoDa challenge [23]. The data contains contrast-enhanced T1-weighted brain tumour MRI scans and high-resolution T2-weighted images (initial resolution of $384/448 \times 348/448 \times 80$ *vox* @ 0.5 mm $\times$ 0.5 mm $\times$ $1.0-1.5$ mm and $512 \times 512 \times 120$ *vox* @ $0.4 \times 0.4 \times 1.0-1.5$ mm). We used the original TCIA dataset [23] to provide omitted labels of the CrossModa challenge which served as oracle-labels. Prior to training isotropic resampling to 0.5 mm $\times$ 0.5 mm $\times$ 0.5 mm was performed as well as cropping the data to $128 \times 128 \times 128$ *vox* around the tumour. We omitted the provided cochlea labels and train on binary masks of background/tumour. As the tumour is either contained on the right- or left side of the hemisphere, we flipped the right samples to provide pre-oriented training data and omit the data without tumour structures. For the 2D experiments we sliced the last data dimension.

Model and Training Settings. For 2D segmentation, we employ a LR-ASPP MobileNetV3-Large model [9]. For 3D experiments we use a custom 3D-MobileNet backbone similar as proposed in [21] with an adapted 3D-LR-ASPP head [8]. 2D training was performed with an AdamW [17] optimizer with a learning rate of $\lambda_{2D} = 0.0005$, $|B|_{2D} = 32$, cosine annealing [16] as scheduling method with restart after $t_0 = 500$ batch steps and multiplication factor of 2.0. For the data parameters, we used the SparseAdam-optimizer implementation together with the sparse Embedding structure of PyTorch with a learning rate of $\lambda_{DP} = 0.1$, no scheduling, $\beta_1 = 0.9$ and $\beta_2 = 0.999$. 3D training was conducted with learning rate of $\lambda_{3D} = 0.01$, $|B|_{3D} = 8$ due to memory restrictions and exponentially decayed scheduling with factor of $d = 0.99$. As opposed to Saxena et al. [22] during our experiments we did not find weight-clipping, weight decay or ℓ_2-regularisation on data parameters to be necessary. Parameters DP_s were initialized with a value of 0.0. For all experiments, we used spatial affine- and b-spline-augmentation and random-noise-augmentation on image intensities. Prior to augmenting we upscaled the input images and labels to 256×256 *px* in 2D- and $192 \times 192 \times 192$ *vox* in 3D-training. Data was split into 2/3 training and 1/3 validation images during all runs and used global class weights $1/n_{bins}^{0.35}$.

---

[2] "The word oracle [...] properly refers to the priest or priestess uttering the prediction.". "Oracle." Wikipedia, Wikimedia Foundation, 03 Feb 2022, en.wikipedia.org/wiki/Oracle.

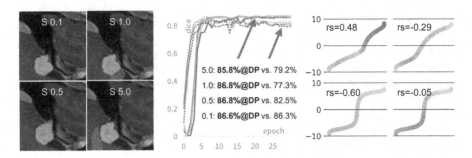

Fig. 2. Left: Sample disturbance ▪ at strengths [0.1, 0.5, 1.0, 5.0]. **Middle:** Validation Dice when training with named disturbance strengths, either with data parameters enabled (—) or disabled (- -). **Right:** Parameter distribution for combinations of risk regularization (RR) and fixed weighting (FW): RR+FW ▪ | RR ▪ | FW ▪ | NONE ▪. Saturated data points indicate higher oracle-Dice. Value of ranked Spearman-correlation r_s between data parameters and oracle-Dice given. (Color figure online)

Experiment I: 2D Model Training, Artificially Disturbed Ground-Truth Labels. This experiment shows the general applicability of data parameters in the semantic segmentation setting when using one parameter per 2D slice. To simulate label-noise, we shifted 30% of the non-empty oracle-slices with different strengths (Fig. 2, left) to see how the network scores behave (Fig. 2, middle) and whether the data parameter distribution captures the artificially disturbed samples (Fig. 2, right). In case of runs with data parameters the optimization was enabled after 10 epochs.

Experiment II: 2D Model Training, Quality-Mixed Registered Single-Atlas Labels. Extending experiment I, in this setting we train on real registration noise with 2D slices on single-atlases. We use 30 T1-weighted images as fixed targets (non-labelled) and T2-weighted images and labels as moving pairs. For registration we use the deep learning-based algorithm Convex Adam [24]. We select two registration qualities to show quality influence during training: *Best*-quality registration means the single best registration with an average of around 80% oracle-Dice across all atlas registrations. *Combined*-quality means a clipped, gaussian-blurred sum of all 30 registered atlas registrations (some sort of consensus). We then input a mix of 50%/50% randomly selected best/combined labels into training. Afterwards we compare the 100% best, 50%/50% mixed and 100% combined selections focusing on the mixed setting where we train with and without data parameters. Validation scores were as follows (descending): best@no-data-parameters 81.1%, mix@data-parameters 74.1%, mix@no-data-parameters 69.6% and combined@no-data-parameters 61.9%.

Experiment III: 3D Model Training, Registered Multi-atlas Labels. Extending experiment II, in this setting we train on real registration noise but with 3D volumes and multiple atlases per image. We follow the CrossMoDa [23] challenge task and use T2-weighted images as fixed targets (non-labelled) and

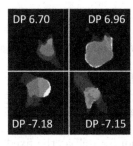

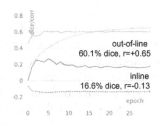

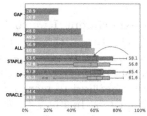

Fig. 3. Selected samples with low- and high parameters: Oracle-label ☐, network prediction ■ and deeds registered label ■ (Color figure online)

Fig. 4. Inline ■ and out-of-line ■ backpropagation. Validation Dice (—) and Spearman-corr. of params. and oracle-Dice (- -) (Color figure online)

Fig. 5. FG: Box plots of STAPLE and DP consensus quality, mean value on the right. **BG:** Bar plot of nnUNet scores; deeds ■, Convex Adam ■ (Color figure online)

T1-weighted images and labels as moving pairs. We conducted registration with two algorithms (iterative deeds [7] and deep learning-based algorithm Convex Adam [24]). For each registration method 10 registered atlases per image are fed to the training routine expanding the T2-weighted training size from 40 to 400 label-image pairs each. Figure 4 shows a run with inline and out-of-line (see Sect. 2) data parameter training on the deeds registrations as an example how training scores behave.

Experiment IV: Consensus Generation and Subsequent Network Training. Using the training output of experiment III, we built 2 × 40 consensi: [10 deeds registered @ 40 fixed] and [10 Convex Adam registered @ 40 fixed]. Consensi were built by applying the STAPLE algorithm as baseline and opposed to that our proposed weighted-sum method on data parameters (DP) (see Sect. 2). On these, we trained several powerful nnUnet-models for segmentation [10]. In Fig. 5 in the foreground four box plots show the quality range of generated consensi regarding the oracle dice: [deeds, Convex Adam registrations]@[STAPLE, DP]. In the background the mean validation Dice of nnUnet-model trainings (150 epochs) is shown. As a reference, we trained directly on the T1-moving data with strong data augmentation (nnUNet "insane" augmentation) trying to overcome the domain gap directly (GAP). Furthermore, we trained on 40 randomly selected atlas labels (RND), all 400 atlas labels (ALL), STAPLE consensi, data parameter consensi (DP) and oracle-labels either on deeds or Convex Adam registered data. Note that the deeds data contained 40 unique moving atlases whereas the Convex Adam data contained 20 unique moving atlases, both warped to 40 fixed images as stated before (Fig. 3).

4   Results and Discussion

In **experiment I** we could show that our usage of data parameters is generally effective in the semantic segmentation scenario under artificial label noise. Figure 2 (middle) shows an increase of validation scores when activating stepping on data parameters after 10 epochs for disturbance strengths >0.1. Stronger disturbances lead to more severe score drops but can be recovered by using data parameters. In Fig. 2 (right) one can see that data parameters and oracle-Dice correlate most, when using the proposed risk regularization as well as the fixed weighting-scheme configuration (see Sect. 2). We did not notice any validation score improvements when switching between configurations and therefore conclude that a sorting of samples can also be learned inherently by the network. However, properly weighted data parameters can extract this information, make it explicitly visible and increase explainability. In **experiment II** we show that our approach works for registration noise during 2D training: When comparing different registration qualities, we observed that training scores drop from 81.1% to 69.6% Dice when lowering registration input quality. By using data parameters we can recover to a score of 74.1% meaning an improvement of +4.5%. **Experiment III** covers our target scenario—3D training with registered multi-atlas labels. With inline training of data parameters (used in the former experiments), validation scores during training drop significantly. Furthermore the data parameters do not separate high- and low quality registered atlases well (see Fig. 4, inline). When using our proposed out-of-line training approach (see Sect. 2) validation Dice and ranked correlation of data parameter values and oracle-Dice improve. **Experiment IV** shows that data parameters can be used to create a weighted-sum consensus as described in Sect. 2: Using data parameters, we can improve mean consensus-Dice for both, deeds and Convex Adam registrations over STAPLE [26] from 58.1% to 64.3% (+6.2%, ours, deeds data) and 56.8% to 61.6% (+4.8%, ours, Convex Adam data). When using the consensi in a subsequent nnUNet training [10], scores behave likewise (see Fig. 5). Regarding training times of over an hour with our LR-ASPP MobileNetV3-Large training, one has to consider that applying the STAPLE algorithm is magnitudes faster.

5   Conclusion and Outlook

Within this work, we showed that using data parameters in a multimodal prediction setting with propagated source labels is a valid approach to improve network training scores, get insight into training data quality and use the extracted info about sample quality in subsequent steps namely to generate consensus segmentations and provide these to further steps of deep learning pipelines. Our improvements over the original data parameter approach for semantic segmentation show strong results in both 2D- and 3D-training settings. Although we could extract sample quality information in the multi-atlas setting successfully, we could not improve network training scores in this setting directly since using the data parameters inline of the training loop resulted in unstable training. Regarding that, we want to continue investigating how an inline training can directly improve training scores in the multi-atlas setting. Furthermore our empirically chosen fixed weighting needs more theoretical foundation. The consensus generation could be further improved by trying more complex weighting schemes or incorporating the network predictions itself. Also we would like to compare our registration-segmentation pipeline against specialized approaches of Ding et al. and Liu et al. [4,5,15] which we consider as very interesting baselines.

References

1. Artaechevarria, X., Munoz-Barrutia, A., Ortiz-de Solorzano, C.: Combination strategies in multi-atlas image segmentation: application to brain MR data. IEEE Trans. Med. Imaging **28**(8), 1266–1277 (2009)
2. Bengio, Y., Louradour, J., Collobert, R., Weston, J.: Curriculum learning. In: Proceedings of the 26th Annual International Conference on Machine Learning, pp. 41–48 (2009)
3. Castells, T., Weinzaepfel, P., Revaud, J.: SuperLoss: a generic loss for robust curriculum learning. Adv. Neural. Inf. Process. Syst. **33**, 4308–4319 (2020)
4. Ding, Z., Han, X., Niethammer, M.: VoteNet: a deep learning label fusion method for multi-atlas segmentation. In: Shen, D., et al. (eds.) MICCAI 2019. LNCS, vol. 11766, pp. 202–210. Springer, Cham (2019). https://doi.org/10.1007/978-3-030-32248-9_23
5. Ding, Z., Han, X., Niethammer, M.: VoteNet+: an improved deep learning label fusion method for multi-atlas segmentation. In: 2020 IEEE 17th International Symposium on Biomedical Imaging (ISBI), pp. 363–367. IEEE (2020)
6. Heckemann, R.A., Hajnal, J.V., Aljabar, P., Rueckert, D., Hammers, A.: Automatic anatomical brain MRI segmentation combining label propagation and decision fusion. Neuroimage **33**(1), 115–126 (2006)
7. Heinrich, M.P., Jenkinson, M., Brady, S.M., Schnabel, J.A.: Globally optimal deformable registration on a minimum spanning tree using dense displacement sampling. In: Ayache, N., Delingette, H., Golland, P., Mori, K. (eds.) MICCAI 2012. LNCS, vol. 7512, pp. 115–122. Springer, Heidelberg (2012). https://doi.org/10.1007/978-3-642-33454-2_15
8. Hempe, H., Yilmaz, E.B., Meyer, C., Heinrich, M.P.: Opportunistic CT screening for degenerative deformities and osteoporotic fractures with 3D DeepLab. In: Medical Imaging 2022: Image Processing. SPIE (2022)
9. Howard, A., et al.: Searching for mobilenetv3. In: Proceedings of the IEEE/CVF International Conference on Computer Vision, pp. 1314–1324 (2019)
10. Isensee, F., Jaeger, P.F., Kohl, S.A., Petersen, J., Maier-Hein, K.H.: NNU-Net: a self-configuring method for deep learning-based biomedical image segmentation. Nat. Methods **18**(2), 203–211 (2021)
11. Jiang, L., Zhou, Z., Leung, T., Li, L.J., Fei-Fei, L.: MentorNet: learning data-driven curriculum for very deep neural networks on corrupted labels. In: International Conference on Machine Learning, pp. 2304–2313. PMLR (2018)
12. Karimi, D., Dou, H., Warfield, S.K., Gholipour, A.: Deep learning with noisy labels: exploring techniques and remedies in medical image analysis. Med. Image Anal. **65**, 101759 (2020)

13. Kohl, S., et al.: A probabilistic U-Net for segmentation of ambiguous images. Adv. Neural Inf. Process. Syst. **31** (2018)
14. Liu, X., Song, L., Liu, S., Zhang, Y.: A review of deep-learning-based medical image segmentation methods. Sustainability **13**(3), 1224 (2021)
15. Liu, Z., et al.: Style curriculum learning for robust medical image segmentation. In: de Bruijne, M., et al. (eds.) MICCAI 2021. LNCS, vol. 12901, pp. 451–460. Springer, Cham (2021). https://doi.org/10.1007/978-3-030-87193-2_43
16. Loshchilov, I., Hutter, F.: SGDR: stochastic gradient descent with warm restarts. arXiv preprint arXiv:1608.03983 (2016)
17. Loshchilov, I., Hutter, F.: Decoupled weight decay regularization. arXiv preprint arXiv:1711.05101 (2017)
18. Marstal, K., Berendsen, F., Staring, M., Klein, S.: SimpleElastix: a user-friendly, multi-lingual library for medical image registration. In: Proceedings of the IEEE Conference on Computer Vision and Pattern Recognition Workshops, pp. 134–142 (2016)
19. Rohlfing, T., Russakoff, D.B., Maurer, C.R.: Performance-based classifier combination in atlas-based image segmentation using expectation-maximization parameter estimation. IEEE Trans. Med. Imaging **23**(8), 983–994 (2004)
20. Ronneberger, O., Fischer, P., Brox, T.: U-Net: convolutional networks for biomedical image segmentation. In: Navab, N., Hornegger, J., Wells, W.M., Frangi, A.F. (eds.) MICCAI 2015. LNCS, vol. 9351, pp. 234–241. Springer, Cham (2015). https://doi.org/10.1007/978-3-319-24574-4_28
21. Sandler, M., Howard, A., Zhu, M., Zhmoginov, A., Chen, L.C.: MobileNetV2: inverted residuals and linear bottlenecks. In: Proceedings of the IEEE Conference on Computer Vision and Pattern Recognition, pp. 4510–4520 (2018)
22. Saxena, S., Tuzel, O., DeCoste, D.: Data parameters: a new family of parameters for learning a differentiable curriculum. Adv. Neural Inf. Process. Syst. **32** (2019)
23. Shapey, J., et al.: Segmentation of vestibular schwannoma from magnetic resonance imaging: an open annotated dataset and baseline algorithm. The Cancer Imaging Archive (2021)
24. Siebert, H., Hansen, L., Heinrich, M.P.: Fast 3D registration with accurate optimisation and little learning for learn2Reg 2021. arXiv preprint arXiv:2112.03053 (2021)
25. Wang, H., Yushkevich, P.: Multi-atlas segmentation with joint label fusion and corrective learning-an open source implementation. Front. Neuroinform. **7**, 27 (2013)
26. Warfield, S.K., Zou, K.H., Wells, W.M.: Simultaneous truth and performance level estimation (staple): an algorithm for the validation of image segmentation. IEEE Trans. Med. Imaging **23**(7), 903–921 (2004)
27. Yan, W., et al.: The domain shift problem of medical image segmentation and vendor-adaptation by Unet-GAN. In: Shen, D., et al. (eds.) MICCAI 2019. LNCS, vol. 11765, pp. 623–631. Springer, Cham (2019). https://doi.org/10.1007/978-3-030-32245-8_69
28. Zhang, Z., Zhang, H., Arik, S.O., Lee, H., Pfister, T.: Distilling effective supervision from severe label noise. In: Proceedings of the IEEE/CVF Conference on Computer Vision and Pattern Recognition, pp. 9294–9303 (2020)
29. Zhu, J.Y., Park, T., Isola, P., Efros, A.A.: Unpaired image-to-image translation using cycle-consistent adversarial networks. In: Proceedings of the IEEE International Conference on Computer Vision, pp. 2223–2232 (2017)

A Method for Image Registration via Broken Geodesics

Alphin J. Thottupattu[1(✉)], Jayanthi Sivaswamy[1],
and Venkateswaran P. Krishnan[2]

[1] International Institute of Information Technology, Hyderabad 500032, India
alphinj.thottupattu@research.iiit.ac.in
[2] TIFR Centre for Applicable Mathematics, Bangalore 560065, India

Abstract. Anatomical variabilities seen in longitudinal data or inter-subject data is usually described by the underlying deformation, captured by non-rigid registration of these images. Stationary Velocity Field (SVF) based non-rigid registration algorithms are widely used for registration. However, these methods cover only a limited degree of deformations. We address this limitation and define an approximate metric space for the manifold of diffeomorphisms $\mathcal{G}$. We propose a method to break down the large deformation into finite set of small sequential deformations. This results in a broken geodesic path on $\mathcal{G}$ and its length now forms an approximate registration metric. We illustrate the method using a simple, intensity-based, log-demon implementation. Validation results of the proposed method show that it can capture large and complex deformations while producing qualitatively better results than state-of-the-art methods. The results also demonstrate that the proposed registration metric is a good indicator of the degree of deformation.

Keywords: Large deformation · Inter-subject registration ·
Approximate registration metric

1 Introduction

Computational anatomy is an area of research focused on developing computational models of biological organs to study the anatomical variabilities in the deformation space. Anatomical variations arise due to structural differences across individuals and changes due to growth or atrophy in an individual. These variations are studied using the deformation between the scans captured by a registration step. The registration algorithms typically optimize an energy functional based on a similarity function computed between the fixed and moving images.

Many initial image registration attempts use energy functionals inspired by physical processes to model the deformation as an elastic deformation [11], or

© The Author(s), under exclusive license to Springer Nature Switzerland AG 2022
A. Hering et al. (Eds.): WBIR 2022, LNCS 13386, pp. 47–56, 2022.
https://doi.org/10.1007/978-3-031-11203-4_6

viscous flow [20] or diffusion [14]. The diffusion-based approaches have been explored for 3D medical images in general [6] and with deformations constrained to be diffeomorphic [23] to ensure preservation of the topology. The two main approaches used to capture diffeomorphisms are parametric and nonparametric methods. The Free Form Deformation (FFD) model [5,12] is a widely used parametric deformation model for medical image registration, where a rectangular grid with control points is used to model the deformation. Large diffeomorphic deformations [12] are handled by concatenating multiple FFDs. Deformable Registration via Attribute Matching and Mutual-Saliency Weighting (DRAMMS) [7] is a popular FFD-based method, which also handles inter-subject registration. DRAMMS matches Gabor features and prioritizes the reliable matching between images while performing registration. The main drawback of the deformations captured by FFD models is that they do not guarantee invertibility. The nonparametric methods represent the deformation with stationary or time varying velocity vector field. The diffeomorphic log-demon [23] is an example of the former while the Large Deformation Diffeomorphic Metric Mapping (LDDMM) [15] inspired from [8] is an example of the latter approach. In LDDMM, deformations are defined as geodesics on a Riemannian manifold, which is attractive; however, the methods based on this framework are computationally complex. The diffeomorphic log-demon framework [23], on the other hand, assigns a Lie group structure and assumes a stationary velocity field (SVF) which leads to computationally efficient methods, which is of interest to the community for practical purposes. This has motivated the exploration of a stationary LDDMM framework [16] that leverages the SVF advantage. The captured deformations are constrained to be symmetric in time-varying LDDMM [1] and log-demon [21] methods. Choosing an efficient optimization scheme such as Gauss-Newton as in [10] reduces the computational complexity of LDDMM framework. However, the log-demon framework is of interest to the community for practical purposes because of its computational efficiency and simplicity.

The Lie group structure gives a locally defined group exponential map to map the SVF to the deformation. Thus log-demon framework is meant to capture only neighboring elements in the manifold, i.e., only a limited degree of deformations can be captured. This will be referred to as the limited coverage issue of the SVF methods in this paper. Notwithstanding the limited coverage, several SVF based methods have been reported for efficient medical image registration with different similarity metrics, sim, such as local correlation between the images [17], spectral features [3], modality independent neighborhood descriptors [19] and wavelet features [9,18].

SVF based algorithms cannot handle complex deformations because the deformations are constrained to be smooth for the entire image and thus constrain the possible degree of deformation to some extent. We address this drawback by splitting the large deformation into finite set of smaller deformations. The key contributions of the paper are: i) an SVF-based registration framework to handle large deformations such as inter-subject variations computationally efficiently ii) an approximate metric to quantify structural variations between two images.

1.1 Background

Let G be a finite-dimensional Lie group with Lie algebra $\mathfrak{g}$. Recall that $\mathfrak{g}$ is the tangent space T_eG at the identity e of G. The exponential map $\exp : \mathfrak{g} \to G$ is defined as follows: Let $v \in \mathfrak{g}$. Then $\exp(v) = \gamma_v(1)$, where γ is the unique one-parameter subgroup of the Lie group G with v being its tangent vector at e. The vector v is called the infinitesimal generator of γ. The exponential map is a diffeomorphism from a small neighborhood containing 0 in the Lie algebra $\mathfrak{g}$ to a small neighborhood containing e of G.

Due to the fact that a bi-invariant metric may not exist for most of the Lie groups considered in medical image registration, the deformations considered here are elements of a Lie group with the Cartan-Schouten Connection [24]. This is the same as the one considered in the log-demon framework [23]. This is a left invariant connection [22] in which geodesics through the identity are one-parameter subgroups. The group geodesics are the geodesics of the connection. Any two neighboring points can be connected with a group geodesic. That is, if the stationary velocity field v connecting two images in the manifold $\mathcal{G}$ is small enough, then its group exponential map forms a geodesic. Similarly every $\mathfrak{g} \in G$ has a geodesically convex open neighbourhood [22].

2 Method

SVF based registration methods capture only a limited degree of deformation because exponential mappings are only locally defined. In order to perform registration of a moving image towards a fixed image, SVF is computed iteratively by updating it with a smoothed velocity field. This update is computed via a similarity metric that measures the correspondence between the moving and fixed images. The spatial smoothing has a detrimental effect as we explain next. A complex deformation typically consists of spatially independent deformations in a local neighbourhood. Depending on the smoothing parameter value, only major SVF updates in each region is considered for registration. Thus, modeling complex deformations with a smooth stationary velocity field is highly dependent on the similarity metric and the smoothing parameter in a registration algorithm. Finding an ideal similarity metric and an appropriate smoothing parameter applicable for any registration problem, irrespective of the complexity of the deformation and the type of data, is difficult.

We propose to address this issue as follows: Deform the moving image toward the fixed image by sequentially applying an SVF based registration. The SVF based algorithm chooses the major or the predominant (correspondence-based) deformation component among the spatially independent deformations in all the neighbourhoods to register along these predominant directions. The subsequent steps in the algorithm captures the next set of predominant directions sequentially. These sequentially captured deformations has a decreasing order of degree of pixel displacement caused by the deformations. Mathematically speaking, the discussion above can be summarised as follows. Consider complex deformations as a set of finite group geodesics and use a registration metric approximation to quantify the deformation between two images in terms of the length of a broken geodesic connecting them; a broken geodesic is a piecewise smooth curve, where each curve segment is a geodesic.

In the proposed method, the similarity-based metric selects the predominant deformation in each sequential step. The deformation that can bring the moving image in a step maximally closer to the target is selected from the one-parameter subgroup of deformations. In the manifold $\mathcal{G}$ every geodesic is contained in a unique maximal geodesic. Hence the maximal group geodesic γ_i computed using log-demon registration framework deforms the sequential image S_{i-1} in the previous step maximally closer to S_N. The maximal group geodesic paths are composed to get the broken geodesic path. As the deformation segments are diffeomorphic, the composed large deformation of the segments also preserves diffeomorphism to some extent.

In the proposed method, the coverage of the SVF method and the degree of deformation determines the number of subgroups N needed to cover the space. The feature based SVF methods in general, give more coverage for a single such subgroup and reduce the value of N.

A broken geodesic $\gamma : [0, T] \rightarrow M$ has finite number of geodesic segments γ_i for partitions of the domain $0 < t_1 < t_2 < \cdots < t_i < \cdots t_N = T$ where $i = 1, \ldots N$. The proposed algorithm to deform S_0 towards S_N is given in Algorithm 1. We have chosen the registration algorithm from [21] to compute SVF, u_i, in Algorithm 1. The Energy term is defined as: Energy$(S_i, S_N) = \text{sim}(S_N, S_i) + \text{Reg}(\gamma_i)$ where the first term is a functional of the similarity measure, which captures the correspondence between images, with $\text{sim}(S_N, S_i) = S_N - S_{i-1} \circ \exp(v_i)$. The second term is a regularization term, with $\text{Reg}(\gamma_i) = \|\nabla \gamma_i\|^2$.

Algorithm 1. Proposed Algorithm

1: Input: S_0 and S_N
2: Result: Transformation $\gamma = \exp(v_1) \circ \exp(v_2) \circ \ldots \exp(v_N)$
3: Initialization: $E_{\min} = \text{Energy}(S_0, S_N)$
4: **repeat**
5:      Register S_{i-1} to $S_N \rightarrow u_i$
6:      Temp $= S_{i-1} \circ \exp(u_i)$
7:      $E_i = \text{Energy}(\text{Temp}, S_N)$
8:      **if** $E_i < E_{\min}$ **then**
9:          $v_i = u_i$
10:          $E_{\min} = E_i$
11:          $S_i = \text{Temp}$
12:      **end if**
13: **until** Convergence

2.1   Registration Metric Approximation

Let γ be a broken geodesic decomposed into N geodesics γ_i with stationary field v_i, i.e. $\dot{\gamma}_i = v_i(\gamma(t)) \in T_{\gamma_i(t)}M$. Each of the constant velocity paths γ_i is parameterized by the time interval $[t_{i-1}, t_i]$, and $N \in \mathbb{N}$ is minimized by requiring each of the geodesics in the broken geodesic to be maximal geodesics. The length of the broken geodesic is defined as,

$$l(\gamma) = \sum_i^N l(\gamma_i) = \sum_i^N d(S_{i-1}, S_i) \tag{1}$$

where, d is a distance metric defined in Eq. 2.

$$d(S_{i-1}, S_i) = \inf\{\|v_i\|_V, S_{i-1} \circ \exp(v_i) = S_i\}. \tag{2}$$

A registration metric needs to be defined to quantify the deformation between two images. The shape metric approximation in [25] can be used for the group geodesics of the Cartan-Schouten connection defined in the finite dimensional case as no bi-invariant metric exists. The length of a broken geodesic $l(\gamma)$ on the manifold $\mathcal{G}$ connecting S_0 and S_N, computed by Eq. 1 is defined as the proposed approximate metric.

3   Results

The proposed method was implemented using a simple intensity based log-demon technique [4] for illustrating the concept which is openly available at: http://dx. doi.org/10.17632/29ssbs4tzf.1. This choice also facilitates understanding the key strengths of the method independently. Two state-of-the-art (SOTA) methods are considered for performance comparison with the proposed method: the symmetric LDDMM implementation in ANTs [1] and DRAMMS which is a feature based, free-form deformation estimation method [7]. These two methods are considered to be good tools for inter-subject registration [26]. Publicly available codes were used for the SOTA methods with parameter settings as suggested in [26] for optimal performance. Both methods were implemented with B-spline interpolation, unless specified. 3D registration was done, and the images used in the experiments are 1.5T T1 MRI scans sourced from [2] and [13] unless specified otherwise. The number of maximum pieces in the broken geodesic path is set as five in all the experiments. The proposed image registration algorithm was used to register MRIs of different individuals.

3.1   Visual Assessment of Registration

To analyse the performance visually, six 3T MRI scans were collected. Three images collected from 20–30 year old male subjects were considered as moving images and three images collected from 40–50 year old female subjects were considered as fixed images. Performing a good registration is challenging with this selection of moving and fixed images. The high resolution MRI scans used for this experiment are openly available at http://dx.doi.org/10.17632/gnhg9n76nn. 1. The registration results for these three different pairs are shown in Fig. 1-A. where only a sample slice is visualized for the 3 cases. The quality of registration can be assessed by observing the degree of match between images in the last two rows of each column. The mean squared error (MSE) was used as a similarity metric along with cubic interpolation. The results indicate that the proposed method is good at capturing complex inter-subject deformations.

The performance of the proposed method on medical images was compared with the state-of-the-art methods in Fig. 1-B. To apply the computed deformation, linear interpolation was used in all the methods. ANTs and the proposed method used MSE as a similarity metric for fair comparison and DRAMMS used its Gabor feature-based metric as it is a feature based method. The results shows that the deformations at the sulcal regions are better captured by the proposed method.

The quality of inverted deformations captured with ANTs and proposed method were also compared as follows. In Fig. 1-C the moving image deformed with moving-fixed deformation and fixed image deformed with inverted moving-fixed deformation are analysed for both the methods. The arrows overlaid on the registered images highlight regions where the proposed method yields error-free results as opposed to the other method. The results with proposed method shows better visual similarity with the target images in each case.

3.2   Quantitative Assessment of Registration

We present a quantitative comparison of the proposed method compared with ANTs and DRAMMS under the same setting. The average MSE for 10 image pair registrations with ANTs was 0.0036 ± 0.0009, with DRAMMS it was 0.0113 ± 0.0068 and with the proposed method it was $0.0012 \pm 7.0552e{-08}$.

The computed deformations in each method were used to transfer region segmentation (labels) from the moving image to the fixed image. The transferred segmentations are assessed using the Dice metric. Figure 2 shows a box plot of the obtained Dice values calculated by registering 10 pairs of brain MRIs with the fixed image, for white matter (WM), grey matter (GM) and 2 structures (L & R-hippocampus). The segmentation results for larger structures (i.e., WM and GM) are better with the proposed method compared to the other methods, whereas the smaller structure segmentation is comparable to DRAMMS.

3.3   Validation of Proposed Registration Metric

Finally, a validation of the proposed registration metric was done using two age-differentiated (20–30 versus 70–90 years) sets of MRIs, of 6 female subjects. Images from these 3D image sets were registered to an (independently drawn) MRI of a 20 year-old subject. The proposed registration metric was computed for the 6 pairs of registrations. A box plot of the registration metric value for each age group is shown in Fig. 3. Since the fixed image is that of a young subject,

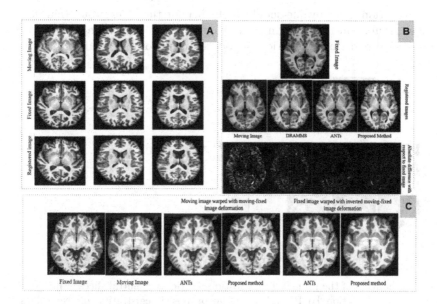

Fig. 1. A) Inter-subject image registration with proposed method for 3 pairs of volumes (in 3 columns) using cubic interpolation. Only sample slices are shown. B) Inter-subject image registration with 3 methods: DRAMMS, ANTs and the proposed method, implemented with linear interpolation. The regions near same colour arrows can be compared to check the registration accuracy. C) Forward and Backward Image Registration. Blue (Red) arrow shows where proposed method yields error-free results in moving (fixed) images, fixed (moving) images and warped moving (fixed) image using moving-fixed (inverted moving-fixed) deformation. Inverted moving-fixed deformation applied on fixed image and proposed method captures finer details compared to ANTs. (Color figure online)

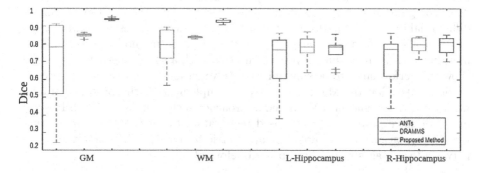

Fig. 2. Assessment of registration via segmentation of different structures using ANTS (magenta), DRAMMS (red), and the proposed method (blue). Box plots for the Dice coefficient are shown for White Matter (WM), Gray Matter (GM) and the Left and Right Hippocampi. (Color figure online)

the registration metric value should be higher for the older group than for the younger group, which is confirmed by the plot. Hence, it can be concluded that the proposed registration metric is a good indicator of natural deformations.

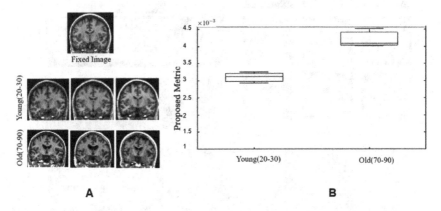

Fig. 3. Validation of the proposed registration metric. A) Central slices of images used to perform registration B) Box plots of the proposed registration metric values for registration of the fixed image with images of young and old subject group.

4   Discussion

Group exponential map based methods, with simple similarity registration metrics, fail to capture large deformations as the map is local in nature. We have addressed this issue in this paper by modelling large deformations with broken geodesic paths with the path length taken to be the associated registration metric. From the experiments it is observed that five pieces in the broken geodesic path is enough to capture very complex deformations. The proposed method does not guarantee diffeomorphism in a strict mathematical sense of infinite differentiability as the paths are modelled as piecewise geodesics. However, the experiments we have done suggest that the proposed method produces diffeomorphic paths. The results of implementation with a simple log-demon method show the performance to be superior to SOTA methods for complex/large deformations. We plan to extend this work by implementing the proposed framework using more efficient SVF based approaches such as in [3,9,17–19]. In summary, we have proposed a SVF-based registration framework that can capture large deformations and an approximate metric to quantify the shape variations between two images using the captured deformations.

References

1. Avants, B.B., et al.: Symmetric diffeomorphic image registration with cross-correlation: evaluating automated labeling of elderly and neurodegenerative brain. Med. Image Anal. **12**, 26–41 (2008)

2. Landman, B., Warfield, S.: MICCAI 2012 Workshop on Multi-atlas Labeling, vol. 2. Create Space Independent Publishing Platform, Nice (2012)
3. Lombaert, H., Grady, L., Pennec, X., Ayache, N., Cheriet, F.: Spectral log-demons: diffeomorphic image registration with very large deformations. Int. J. Comput. Vision **107**(3), 254–271 (2013). https://doi.org/10.1007/s11263-013-0681-5
4. Lombaert, H.: Diffeomorphic Log Demons Image Registration. MATLAB Central File Exchange (2020)
5. Declerck, J., et al.: Automatic registration and alignment on a template of cardiac stress and rest reoriented SPECT images. IEEE Trans. Med. Imaging **16**, 727–37 (1997)
6. Pennec, X., Cachier, P., Ayache, N.: Understanding the "Demon's algorithm": 3D non-rigid registration by gradient descent. In: Taylor, C., Colchester, A. (eds.) MICCAI 1999. LNCS, vol. 1679, pp. 597–605. Springer, Heidelberg (1999). https://doi.org/10.1007/10704282_64
7. Ou, Y., et al.: DRAMMS: deformable registration via attribute matching and mutual-saliency weighting. Med. Image Anal. **15**(4), 622–639 (2011)
8. Trouvé, A.: Diffeomorphisms groups and pattern matching in image analysis. Int. J. Comput. Vision **28**, 213–221 (1998)
9. A, B: Good Afternoon. Conference, pp. 4–6 (2018)
10. Ashburner, J., Friston, K.J.: Diffeomorphic registration using geodesic shooting and gauss-newton optimisation. Neuroimage **55**, 954–967 (2011)
11. Broit, C.: Optimal registration of deformed images. University of Pennsylvania (1981)
12. Rueckert, D., et al.: Diffeomorphic registration using B-splines. Med. Image Comput. Comput. Assist. Interv. **9**(Pt 2), 702–709 (2006)
13. Hello and Goodbye: Good Evening. Journal 67 (2019)
14. Thirion, J.-P.: Image matching as a diffusion process: an analogy with Maxwell's demons. Med. Image Anal. **2**(3), 243–260 (1998)
15. Beg, M.F., et al.: Computing large deformation metric mappings via geodesic flows of diffeomorphisms. Int. J. Comput. Vision **61**, 139–157 (2005)
16. Hernandez, M., et al.: Registration of anatomical images using geodesic paths of diffeomorphisms parameterized with stationary vector fields. In: IEEE 11th International Conference on Computer Vision, pp. 1–8 (2007)
17. Lorenzi, M., et al.: LCC-Demons: a robust and accurate symmetric diffeomorphic registration algorithm. Neuroimage **81**, 470–483 (2013)
18. Pham, N., et al.: Spectral graph wavelet based nonrigid image registration. IEEE Trans. Pattern Anal. Mach. Intell., 3348–3352 (2018)
19. Reaungamornrat, S., et al.: MIND Demons: symmetric diffeomorphic deformable registration of MR and CT for image-guided spine surgery. IEEE Trans. Med. Imaging **35**(11), 2413–2424 (2016)
20. Christensen, T., et al.: Shoving model for viscous flow. World Sci. **12**, 375 (1981)
21. Vercauteren, T., et al.: Symmetric log-domain diffeomorphic Registration: a demons-based approach. Med. Image Comput. Comput. Assist. Interv. **11**(Pt 1), 754–761 (2008)
22. Arsigny, V., et al.: A fast and log-euclidean polyaffine framework for locally linear registration. [Research Report]RR-5885, INRIA (2006)
23. Arsigny, V., Commowick, O., Pennec, X., Ayache, N.: A log-euclidean framework for statistics on diffeomorphisms. In: Larsen, R., Nielsen, M., Sporring, J. (eds.) MICCAI 2006. LNCS, vol. 4190, pp. 924–931. Springer, Heidelberg (2006). https://doi.org/10.1007/11866565_113

24. Pennec, X.: Bi-invariant means on Lie groups with Cartan-Schouten connections. Geom. Sci. Inf., 59–67 (2013)
25. Yang, X., Li, Y., Reutens, D., Jiang, T.: Diffeomorphic metric landmark mapping using stationary velocity field parameterization. Int. J. Comput. Vision **115**(2), 69–86 (2015). https://doi.org/10.1007/s11263-015-0802-4
26. Ou, Y., et al.: Comparative evaluation of registration algorithms in different brain databases with varying difficulty: results and insights. IEEE Trans. Med. Imag. **33**, 2039–2065 (2014)

Deformable Image Registration Uncertainty Quantification Using Deep Learning for Dose Accumulation in Adaptive Proton Therapy

A. Smolders[1,2(✉)], T. Lomax[1,2], D. C. Weber[1], and F. Albertini[1]

[1] Paul Scherrer Institute, Center for Proton Therapy, Villigen, Switzerland
`andreas.smolders@psi.ch`
[2] Department of Physics, ETH Zurich, Zürich, Switzerland

Abstract. Deformable image registration (DIR) is a key element in adaptive radiotherapy (AR) to include anatomical modifications in the adaptive planning. In AR, daily 3D images are acquired and DIR can be used for structure propagation and to deform the daily dose to a reference anatomy. Quantifying the uncertainty associated with DIR is essential. Here, a probabilistic unsupervised deep learning method is presented to predict the variance of a given deformable vector field (DVF). It is shown that the proposed method can predict the uncertainty associated with various conventional DIR algorithms for breathing deformation in the lung. In addition, we show that the uncertainty prediction is accurate also for DIR algorithms not used during the training. Finally, we demonstrate how the resulting DVFs can be used to estimate the dosimetric uncertainty arising from dose deformation.

Keywords: Deformable image registration · Proton therapy · Adaptive planning · Uncertainty · Deep learning

1 Introduction

Due to their peaked depth-dose profile, protons deposit a substantially lower dose to the normal tissue than photons for a given target dose [19]. However, the location of the dose peak is highly dependent on the tissue densities along the beam path, which are subject to anatomical changes throughout the treatment. Target margins are therefore applied, reducing the advantage of proton therapy (PT) [19]. The need to account for anatomical uncertainties can be alleviated using daily adaptive PT (DAPT), where treatment is reoptimized based on a daily patient image [1]. DAPT yields a series of dose maps, each specific to a daily anatomy. One important step of DAPT is to rely on the accurate accumulation of these doses for quality assurance (QA) of the delivered treatment and to trigger further adaptation [3,7,12,13]. To this end, the daily scans are registered to a reference and their corresponding doses are deformed before summation. In

© The Author(s), under exclusive license to Springer Nature Switzerland AG 2022
A. Hering et al. (Eds.): WBIR 2022, LNCS 13386, pp. 57–66, 2022.
https://doi.org/10.1007/978-3-031-11203-4_7

the presence of deforming anatomy, deformable image registration (DIR) is used [7,22,25]. However, DIR is ill-posed [4], which results in dosimetric uncertainty after deformation. Substantial work has been performed to quantify this, summarized in [7], but there remains a clear need for methods predicting uncertainty associated with DIR and its effect on dose deformation [4,18].

In this work, an unsupervised deep learning (DL) method is presented to predict the uncertainty associated with a DIR result. Section 2 describes our method. The results of hyperparameter tuning on the predicted registration uncertainty are presented in Sect. 3, followed by the effect on the dosimetric uncertainty arising from dose deformation. Section 4 provides a discussion and conclusions are stated in Sect. 5.

2   Methods

Our work aims to estimate the uncertainty of the solution of an existing DIR algorithm. It is based upon a probabilistic unsupervised deep neural network for DIR called VoxelMorph [8]. The main equations from [8] are first summarized, after which the changes are described.

2.1   Probabilistic VoxelMorph

With f and m respectively a fixed and a moving 3D volume, here CT images, a neural network learns z, the latent variable for a parameterized representation of a deformable vector field (DVF) Φ_z. The network aims to estimate the conditional probability $p(z|f, m)$, by assuming a prior probability $p(z) = \mathcal{N}(0, \Sigma_z)$, with $\Sigma_z^{-1} = \Lambda_z = \lambda(D - A)$, λ a hyperparameter, D the graph degree matrix and A the adjacency matrix. Further, f is assumed to be a noisy observation of the warped moving image with noise level σ_I^2, $p(f|m, z) = \mathcal{N}(m \circ \Phi_z, \sigma_I^2 I)$. With these assumptions, calculation of $p(z|f, m)$ is intractable. Instead, $p(z|f, m)$ is modelled as a multivariate Gaussian

$$q_\Psi(z|f, m) = \mathcal{N}(\mu_{z|f,m}, \Sigma_{z|f,m}) \tag{1}$$

with Ψ the parameters of the network which predicts $\mu_{z|f,m}$ and $\Sigma_{z|f,m}$ (Fig. 1). The parameters Ψ are optimized by minimizing the KL divergence between $p(z|f, m)$ and $q_\Psi(z|f, m)$, yielding, for K samples $z_k \sim q_\Psi(z|f, m)$, a loss function

$$\mathcal{L}(\Psi, f, m) = \frac{1}{2\sigma_I^2 K} \sum_k ||f - m \circ \Phi_z||^2 + \frac{\lambda}{4} \sum_{i=1}^m \sum_{j \in N(i)} (\mu_i - \mu_j)^2$$

$$+ tr(\frac{\lambda}{2}(D - A)\Sigma_{z|f,m}) - \frac{1}{2}log(|\Sigma_{z|f,m}|) + cte \tag{2}$$

with $N(i)$ the neighboring voxels of voxel i. When $\Sigma_{z|f,m}$ is diagonal, the last two terms of Eq. 2 reduce to $\frac{1}{2}tr(\lambda D \Sigma_{z|f,m} - log(\Sigma_{z|f,m}))$.

2.2   Combining Deep Learning with Existing DIR Software

Because the performance of DL based DIR is generally below conventional methods [9,10,24], our network aims to predict the uncertainty associated with a DVF generated by another algorithm without predicting the DVF itself. We therefore extend the VoxelMorph architecture to include the output DVF of an existing DIR algorithm (Fig. 1). First, an existing algorithm is ran on f and m, after which the resulting DVF is concatenated to f and m as network input. The network only predicts a diagonal matrix G, which is used to calculate $\Sigma_{z|f,m}$ (see Sect. 2.3), and the mean field $\mu_{z|f,m}$ is taken as the output of the DIR algorithm.

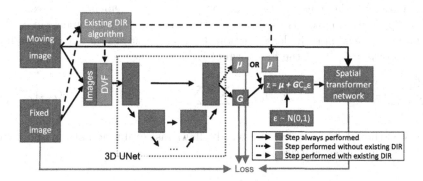

Fig. 1. Schematic network architecture. In case an existing DIR method is used, the resulting DVF of this algorithm is concatenated with the fixed and moving image, resulting in a $5 \times H \times W \times D$ tensor as network input. A 3D UNet predicts a diagonal matrix G, and taking the DVF of the existing DIR as mean field μ, DVF samples are generated with the reparametrization trick as $z = \mu + GC_{\sigma_c}\epsilon$ (see Sect. 2.3) [15]. Contrarily if no existing DIR is used, the network only receives the fixed and moving image as input and predicts a mean DVF besides G.

2.3   Non-diagonal Covariance Matrix

Dosimetric uncertainty will be estimated by sampling $q_\Psi(z|f,m)$, requiring spatially smooth samples. Nearby vectors can be correlated with a non-diagonal covariance matrix. However, a full covariance matrix cannot be stored in memory because it would require storing $(3 \times H \times W \times D)^2$ entries, which for a 32 bit image of $256 \times 265 \times 96$ requires 633 TB, compared to 25 MB for the diagonal elements. In [8] a non-diagonal $\Sigma_{z|f,m}$ is proposed by Gaussian smoothing of a diagonal matrix G, i.e. $\Sigma_{z|f,m} = C_{\sigma_c}GG^T C_{\sigma_c}^T$, but it is shown that this is unnecessary because the implemented diffeomorphic integration smooths the samples sufficiently. Because the existing DIR solutions are not necessarily diffeomorphic, we do not apply integration, which implies the need for a non-diagonal $\Sigma_{z|f,m}$.

Similar to [8], we apply Gaussian smoothing but invert the order $\Sigma_{z|f,m} = GC_{\sigma_c}C_{\sigma_c}^T G^T$ which yields a fixed correlation matrix $\rho = C_{\sigma_c}C_{\sigma_c}^T$. This has the advantage that the variance of the vector magnitude at voxel i is only dependent on the corresponding diagonal element $G_{i,i}$ and not on its neighbors. Furthermore, it allows to simplify the calculation of the loss terms in Eq. 2. Rewriting the last two terms of Eq. 2 with $\Sigma_{z|f,m} = GC_{\sigma_c}C_{\sigma_c}^T G^T$ results in

$$
\begin{aligned}
&tr(\frac{\lambda}{2}(D-A)\Sigma_{z|f,m}) - \frac{1}{2}log(|\Sigma_{z|f,m}|) \\
&= \sum_{i=1}^{m}\sum_{j=1}^{m}(\frac{\lambda}{2}(D-A)_{i,j}(\Sigma_{z|f,m})_{i,j}) - \frac{1}{2}log(|GG^T|) - \frac{1}{2}log(|C_{\sigma_c}C_{\sigma_c}^T|)
\end{aligned}
\tag{3}
$$

with i and j respectively the row and column indices, m the number voxels and $log(|C_{\sigma_c}C_{\sigma_c}^T|)$ a constant which can be excluded from the loss function. For each row (or voxel) i, the matrix $(D-A)$ has only 7 non-zero elements (the voxel itself and its 6 neighboring voxels), so that only the corresponding 7 elements in $\Sigma_{z|f,m}$ are needed to evaluate the loss function. By precomputing the 7 corresponding elements of $\rho = C_{\sigma_c}C_{\sigma_c}^T$, the first term of Eq. 3 becomes

$$
\sum_{i=1}^{m}\sum_{j=1}^{m}(\frac{\lambda}{2}(D-A)_{i,j}(\Sigma_{z|f,m})_{i,j}) = \sum_{i=1}^{m}\sum_{j\in N(i)}(\frac{\lambda}{2}(D-A)_{i,j}\rho_{i,j}G_{i,i}G_{j,j})
\tag{4}
$$

with $N(i)$ the neighbors of voxel i, which allows fast evaluation of $\mathcal{L}$ without the need of storing large matrices.

2.4   Training

52 CT scan pairs from 40 different patients with various indications treated at the Centre for Proton Therapy (CPT) in Switzerland are used for training. The pairs consist of one planning and one replanning or control CT from a proton treatment, and are therefore representative of both daily and progressive anatomical variations in DAPT. Scans are rigidly registered using the Elastix toolbox [16] and resampled to a fixed resolution $1.95 \times 1.95 \times 2.00$ mm, most frequently occurring in the dataset. The Hounsfield units are normalized with $\frac{HU+1000}{4000}$. Patches with a fixed size $256 \times 256 \times 96$ are randomly cropped from the full CTs during training and axis aligned flipping is applied as data augmentation.

The network is implemented in Pytorch [20] and training is ran on GPUs with 11 GB VRAM. A 3D UNet is used [8] with an initial convolution creating 16 feature maps, which are doubled in each of the 3 consecutive downsampling steps. The features are upsampled 3 times to their original resolution. The parameters are optimized with Adam [14] with initial learning rate $2 \cdot 10^{-4}$, which is halved 6 times during 500 epochs. Gaussian smoothing of the diagonal covariance matrix has a fixed kernel size of 61 voxels and blur $\sigma_c = 15$.

We train networks to predict the uncertainty associated with three existing DIR algorithms: a b-spline and a demon implementation in Plastimatch and a non-diffeomorphic VoxelMorph predicting both $\mu_{z|f,m}$ and $\Sigma_{z|f,m}$. The parameters for b-spline and demon are taken from [2,17]. Furthermore, we verify whether these networks can be used to predict the uncertainty of other DIR algorithms by evaluating them on the results of a commercial DIR in Velocity.

2.5   Validation

The hyperparameters λ and σ_I^2 are tuned for each method by quantitatively evaluating the predicted uncertainty on the publicly available 4DCT DIRLAB lung deformation dataset [5,6]. It contains 10 CT scan pairs with each 300 annotated landmarks (LM). These scans are split equally in a validation and test set. We maximize the probability of observing the moving landmarks $\boldsymbol{x}_m$ given the predicted probabilistic vector field, which, for a given set of CTs, is calculated as

$$p(LMs) = \prod_i^{CTs} \prod_j^{LM} p(\boldsymbol{x}_{m,i,j}|DVF_i), \tag{5}$$

assuming for simplicity that each landmark is independent of the others, which is reasonable if the landmarks are sufficiently far apart. Note that the probability of observing exactly $\boldsymbol{x}_m$ is infinitesimally small because the variables are continuous. We therefore maximize the probability that $\boldsymbol{x}_m$ is observed within a cube of 1 mm^3 around it with a homogeneous probability density, which is the same as maximizing the probability density at $\boldsymbol{x}_m$. We discard the 1% least probable points because $p(LMs)$ is heavily affected by the outliers due to the extremely low probability density at the tails of a normal distribution. Furthermore, we maximize the mean log $p(LMs)$ to avoid that the absolute value is dependent on the number of landmarks.

3   Results

3.1   Hyper Parameter Tuning

The optimal hyperparameters are $\lambda = 10$ and $\sigma_I^2 = 10^{-4}$ for both b-spline and demon (Fig. 2). Further, using both the networks trained on demon and b-spline, we find that the network trained with b-spline and $\lambda = 5$ and $\sigma_I^2 = 10^{-4}$ yields the highest average log $p(LMs)$ for Velocity (not shown).

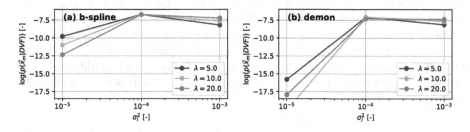

Fig. 2. Average log probability of observing moving landmarks $\boldsymbol{x}_m$ of the validation set for varying values of σ_I^2 and λ including an existing DIR output. Similar results were found for the test set (not shown).

For VoxelMorph, the hyperparameters influence both $\mu_{z|f,m}$ and $\Sigma_{z|f,m}$. Equation 2 shows that the trade off between similarity and smoothness is determined by the product $\lambda\sigma_I^2$. Therefore, we first minimize the target registration error (TRE) on the validation set by varying $\lambda\sigma_I^2$ (keeping $\lambda = 2$), which yields a minimum TRE around $\lambda\sigma_I^2 = 2 \cdot 10^{-3}$. Varying λ and σ_I^2 while keeping $\lambda\sigma_I^2 = 2 \cdot 10^{-3}$ results in a maximum $p(LMs)$ for $\sigma_I^2 = 5 \cdot 10^{-4}$. $p(LMs)$ is however lower than for the networks including the conventional (i.e. non deep learning) DIRs, indicating that these methods predict better probability distributions.

Figure 3 shows the uncertainty prediction for a lung CT in the DIRLAB dataset. As expected, the predicted uncertainty is low in regions with high contrast and high where contrast is low. Further, the Jacobian determinant is <0 for on average 0.01% of the voxels in sampled DVFs for the DIRLAB dataset, which, together with visual inspection, indicates that samples are sufficiently smooth.

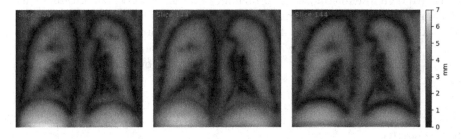

Fig. 3. Predicted uncertainty σ_p, i.e. the square root of the diagonal elements of $\Sigma_{z|f,m}$, in the sagittal (left), coronal (middle) and axial (right) direction for one example patient in the test set.

Comparing the target errors and their predictions for the tuned networks for all DIRLAB scans yields several conclusions (Fig. 4). First of all, our method is able to fairly accurately predict the uncertainty of multiple existing DIR algorithms. Secondly, the error prediction of Velocity shows that it is possible to predict the error from a DIR algorithm even if it was not used to train the network. Lastly, the average error is higher and the uncertainty prediction is worse for VoxelMorph than for the existing DIR algorithms, as expected from [9,10,24]. However, the performance can likely be improved by diffeomorphic integration, network adjustments or using more data, but this is not within the scope of the current study.

3.2 Dose Deformation

We create probabilistic dose maps by sampling the probabilistic DVF and warping the dose with the different samples. We focus here on the result of a single deformation to highlight the dosimetric uncertainty associated with warping. Even though the predicted DVFs have assumed to be Gaussian, the probabilistic dose maps are not. We therefore keep the individual samples and use a finite-sample distribution to approximate the probabilistic dose map.

The dose received by the tumor and organs at risk (OARs) is in PT frequently evaluated with dose volume histograms (DVHs). Probabilistic DVHs can be constructed from the probabilistic dose map. Here, the lower and upper bound of the DVH depict for each volume increment respectively the 5th and 95th percentile of all sampled doses (Fig. 5).

Verifying whether the dosimetric uncertainty is realistic is non-trivial. Previous work [2,17] quantified it by warping the dose with several DIR algorithms and calculating the dose differences between the results. Similarly, here we verify whether the warped dose with three conventional DIR algorithms falls in

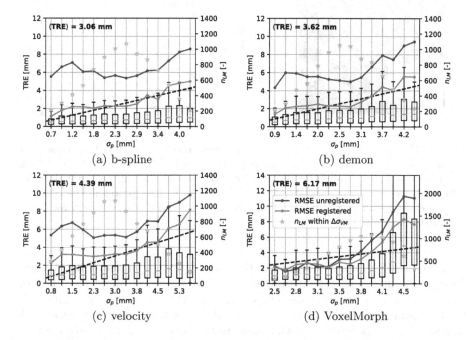

(a) b-spline

(b) demon

(c) velocity

(d) VoxelMorph

Fig. 4. TRE as a function of the predicted uncertainty σ_p for all DIRLAB scans. For each subplot, σ_p is divided into 15 equal intervals and the distribution of the TREs within each interval is plotted as a box, together with the unregistered and registered root mean squared error (RMSE). The number of landmarks n_{LM} within each interval is also shown (right axes). If the TREs were normally distributed and the networks had a perfect prediction, the registered RMSE would be exactly equal to the predicted σ_p (dashed line).

between our predicted lower and upper bound (Fig. 5). Using the same dataset of 7 lung cancer patients with each 9 repeated CTs as in [2,17], we find that the dose in on average 97% of the volume of the OARs (heart, esophagus and medulla) lies between the bounds predicted for b-spline. For the planning target volume (PTV) and gross tumor volume (GTV) it is on average 81%.

4   Discussion

Despite the promising preliminary results, more work is required before the method can be used in the clinic. Our approach should be verified on a dataset including typical deformations that occur during the course of six weeks of treatment, and not only during one breathing cycle. To that end, a dataset with typical anatomical deformations is currently being landmarked at the CPT.

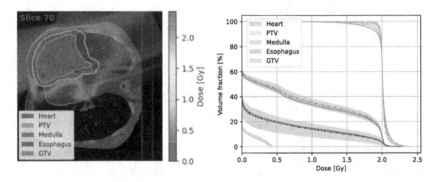

Fig. 5. Left: example of a deformed dose map with b-spline, overlayed with contours of the gross tumor volume (GTV), planning target volume (PTV) and three OARs. Right: corresponding probabilistic DVH as calculated with the optimal network for b-spline (shaded area). The dashed, dotted and dash-dotted lines represent the DVH for warped doses with three commercial DIR softwares, respectively Mirada, Raystation Anaconda and Velocity.

Even for the dataset under study, the error prediction is clearly not perfect. This can be due to several factors, among which imperfect annotation, lack of training data or inaccurate model assumptions. One important assumption is the Gaussian vector field. Although our results show that it is not unreasonable to assume that the errors are Gaussian, further research should look whether other probability distributions yield better results. Unfortunately, other analytical distributions are often mathematically more complex making exact treatment as in Eqs. 2 and 3 difficult. Learning a discretized posterior could resolve this [10,11,21,23].

The trained networks capture most of the dosimetric variations found in the OARs when running conventional DIRs. By contrast, for the GTV and PTV only 81% of the doses lie between the error bars, significantly below the expected

90% given the 5th and 95th percentile error bounds. However, we found that this value increases to 91% by simply adding a small margin to the error bounds (i.e. by increasing the upper and decreasing lower bound by only 0.1% of the dose). This indicates that the deviation from the error bounds is mostly very small.

5 Conclusion

In this work, a probabilistic unsupervised deep learning method for deformable image registration is presented to predict the uncertainty associated with DIR solutions. It is shown that the method can accurately predict the uncertainty of various conventional DIR algorithms and that the combination of deep learning with conventional DIR yields superior results than using deep learning alone.

Acknowledgments. This work has received funding from the European Union's Horizon 2020 Marie Skłodowska-Curie Actions under Grant Agreement No. 955956.

References

1. Albertini, F., Matter, M., Nenoff, L., Zhang, Y., Lomax, A.: Online daily adaptive proton therapy. Br. J. Radiol. **93**(1107), 20190594 (2020)
2. Amstutz, F., et al.: An approach for estimating dosimetric uncertainties in deformable dose accumulation in pencil beam scanning proton therapy for lung cancer. Phys. Med. Biol. **66**(10), 105007 (2021)
3. Brock, K.K., McShan, D.L., Ten Haken, R., Hollister, S., Dawson, L., Balter, J.: Inclusion of organ deformation in dose calculations. Med. Phys. **30**(3), 290–295 (2003)
4. Brock, K.K., Mutic, S., McNutt, T.R., Li, H., Kessler, M.L.: Use of image registration and fusion algorithms and techniques in radiotherapy: report of the AAPM radiation therapy committee task group no. 132. Med. Phys. **44**(7), e43–e76 (2017)
5. Castillo, E., Castillo, R., Martinez, J., Shenoy, M., Guerrero, T.: Four-dimensional deformable image registration using trajectory modeling. Phys. Med. Biol. **55**(1), 305 (2009)
6. Castillo, R., et al.: A framework for evaluation of deformable image registration spatial accuracy using large landmark point sets. Phys. Med. Biol. **54**(7), 1849 (2009)
7. Chetty, I.J., Rosu-Bubulac, M.: Deformable registration for dose accumulation. In: Seminars in Radiation Oncology, vol. 29, pp. 198–208. Elsevier (2019)
8. Dalca, A.V., Balakrishnan, G., Guttag, J., Sabuncu, M.R.: Unsupervised learning of probabilistic diffeomorphic registration for images and surfaces. Med. Image Anal. **57**, 226–236 (2019)
9. Hansen, L., Heinrich, M.P.: Tackling the problem of large deformations in deep learning based medical image registration using displacement embeddings. arXiv preprint arXiv:2005.13338 (2020)
10. Heinrich, M.P.: Closing the gap between deep and conventional image registration using probabilistic dense displacement networks. In: Shen, D., et al. (eds.) MICCAI 2019. LNCS, vol. 11769, pp. 50–58. Springer, Cham (2019). https://doi.org/10.1007/978-3-030-32226-7_6

11. Heinrich, M.P., Jenkinson, M., Papież, B.W., Brady, S.M., Schnabel, J.A.: Towards realtime multimodal fusion for image-guided interventions using self-similarities. In: Mori, K., Sakuma, I., Sato, Y., Barillot, C., Navab, N. (eds.) MICCAI 2013. LNCS, vol. 8149, pp. 187–194. Springer, Heidelberg (2013). https://doi.org/10. 1007/978-3-642-40811-3_24
12. Jaffray, D.A., Lindsay, P.E., Brock, K.K., Deasy, J.O., Tomé, W.A.: Accurate accumulation of dose for improved understanding of radiation effects in normal tissue. Int. J. Radiation Oncol.* Biol.* Phys. **76**(3), S135–S139 (2010)
13. Janssens, G., et al.: Evaluation of nonrigid registration models for interfraction dose accumulation in radiotherapy. Med. Phys. **36**(9Part1), 4268–4276 (2009)
14. Kingma, D.P., Ba, J.: Adam: a method for stochastic optimization. arXiv preprint arXiv:1412.6980 (2014)
15. Kingma, D.P., Salimans, T., Welling, M.: Variational dropout and the local reparameterization trick. Adv. Neural Inf. Process. Syst. **28** (2015)
16. Klein, S., Staring, M., Murphy, K., Viergever, M.A., Pluim, J.P.: Elastix: a toolbox for intensity-based medical image registration. IEEE Trans. Med. Imaging **29**(1), 196–205 (2009)
17. Nenoff, L., et al.: Deformable image registration uncertainty for inter-fractional dose accumulation of lung cancer proton therapy. Radiother. Oncol. **147**, 178–185 (2020)
18. Paganelli, C., Meschini, G., Molinelli, S., Riboldi, M., Baroni, G.: Patient-specific validation of deformable image registration in radiation therapy: overview and caveats. Med. Phys. **45**(10), e908–e922 (2018)
19. Paganetti, H.: Range uncertainties in proton therapy and the role of Monte Carlo simulations. Phys. Med. Biol. **57**(11), R99 (2012)
20. Paszke, A., et al.: PyTorch: an imperative style, high-performance deep learning library. Adv. Neural. Inf. Process. Syst. **32**, 8026–8037 (2019)
21. Rühaak, J.: Estimation of large motion in lung CT by integrating regularized keypoint correspondences into dense deformable registration. IEEE Trans. Med. Imaging **36**(8), 1746–1757 (2017)
22. Schultheiss, T.E., Tomé, W.A., Orton, C.G.: It is not appropriate to "deform" dose along with deformable image registration in adaptive radiotherapy. Med. Phys. **39**(11), 6531–6533 (2012)
23. Sedghi, A., Kapur, T., Luo, J., Mousavi, P., Wells, W.M.: Probabilistic image registration via deep multi-class classification: characterizing uncertainty. In: Greenspan, H., et al. (eds.) CLIP/UNSURE 2019. LNCS, vol. 11840, pp. 12–22. Springer, Cham (2019). https://doi.org/10.1007/978-3-030-32689-0_2
24. de Vos, B.D., Berendsen, F.F., Viergever, M.A., Sokooti, H., Staring, M., Išgum, I.: A deep learning framework for unsupervised affine and deformable image registration. Med. Image Anal. **52**, 128–143 (2019)
25. Zhong, H., Chetty, I.J.: Caution must be exercised when performing deformable dose accumulation for tumors undergoing mass changes during fractionated radiation therapy. Int. J. Radiat. Oncol. Biol. Phys. **97**(1), 182–183 (2016)

Distinct Structural Patterns
of the Human Brain: A Caveat
for Registration

Frithjof Kruggel[(✉)]

Department of Biomedical Engineering, University of California, Irvine, USA
fkruggel@uci.edu
http://sip.eng.uci.edu

Abstract. Current approaches for analyzing structural patterns of the human brain often implicitly assume that brains are variants of a single type, and use nonlinear registration to reduce the inter-individual variability. This assumption is challenged here. Regional anatomical and connection patterns cluster into statistically distinct types. An advanced analysis proposed here leads to a deeper understanding of the governing principles of cortical variability.

Keywords: Structural patterns · Connectivity · Human cortex

1 Introduction

Cortical structures of the human brain show a puzzling complexity and inter-individual variability. Numerous analytic approaches implicitly assume that structural properties of brains, represented in any high-dimensional space, form a single cluster and use nonlinear registration to reduce the inter-individual variability. We challenge this assumption. Depending on the features and similarity criteria involved in the registration process, the total variance is reduced by only 20–40%. Consider a simplifying analogy: Suppose we want to study structural properties of cars. We hardly doubt that a registration procedure can be designed that successfully matches gross car parts (e.g., the passenger and engine compartment, the trunk and wheels). However, when zooming into details, objects under study become distinct (e.g. a trunk of a truck vs. a sports car, a combustion engine vs. an electric motor). Here, we demonstrate here that structural variants of brain regions with distinctive properties exist in a population. Avoiding an arguable registration and embracing the actual variability leads to analytic procedures that actually *explain* sources of variability at a considerably larger proportion.

2 Methods

Data Source: We used anatomical and diffusion-weighted MRI data acquired in $nc = 1061$ subjects of the publicly available Human Connectome Project [2].

© The Author(s), under exclusive license to Springer Nature Switzerland AG 2022
A. Hering et al. (Eds.): WBIR 2022, LNCS 13386, pp. 67–71, 2022.
https://doi.org/10.1007/978-3-031-11203-4_8

Anatomical processing: We started out from triangulated meshes representing the white-gray matter interface of a hemisphere with a topological genus of zero. Using local curvature and geodesic depth, the surface was segmented into patches called *basins* that were centered around a locally deepest point, the *sulcal root*. A most isometric mapping was used to transfer and re-parameterize vertex-wise properties (e.g., basin label, depth, curvature) onto a common sphere with $nv = 163842$ vertices. Thus, we represented structural information as an image of $nc \times nv \times np$ properties. Refer to [4] for details.

Tractography: Diffusion-weighted data were corrected for subject motion and susceptibility distortions. Voxel-wise estimates of the orientation distribution function of water mobility were computed using the constrained spherical decon-volution method [3]. Probabilistic tracking [5] from basin-labeled surface seeds was performed to determine connectivity between basins. Results were kept in hemisphere-wise connectivity matrices C, where each element $C(i,j)$ corresponded to the probability of connecting basin i to j. Thus, C can be regarded as a discrete, empirical PDF of basin connectivity.

Distance Metrics: We computed a co-occurence matrix M of the basin labeling in hemispheres a, b and expressed the their structural distance by $D_M = 1 - \mathrm{NMI}(M_{a,b})$. For connectivity, we selected the Hellinger distance metric by experimentation:

$$\mathrm{D}_C(a,b) = \sqrt{2\left(1 - \sqrt{2\sum_i\sum_j\sqrt{C_a(i,j)\,C_b(i,j)}}\right)}. \tag{1}$$

Statistical Assessment: We computed the distance metrics for all hemisphere pairs a, b and compiled them in matrices for structure D_M and connectivity D_C of dimensions $nc \times nc$. Both matrices were mapped into a low-dimensional space using the ISOMAP algorithm [6], with a target dimension of $nd = 4$ estimated by the Grassberger-Procaccia method [1]. Thus, structural and connectivity of a hemisphere were represented by a point in an 8-dimensional space. We used a Gaussian mixture model to cluster into groups, where the number of classes was determined from the maximal Bayesian information criterion and silhouette coefficient. Note that this analysis can be restricted to any sub-region of the whole hemisphere.

3   Results

Due to space limitations, we provided results for the central sulcus (CS) only. For each dimension of the structural and connectivity matrices, we analyzed their dependence on several variables using linear regression (Tab. 1). Dimensions and their amount of represented variance were compiled in the second column. The first dimension represented more than 50% of the variance, and corresponded to the "regularity" of the sulcus structure. Regular sulci were straight, deep, and consisted of relatively few basins, in contrast to tortuous, shallow sulci with a larger number of basins (Fig. 1). Considering the number of basins as a proxy for structural regularity, we found that between 25% and 41% of the variance (R^2) were addressed to regularity. About 10% of the overall variance were explained by subject sex, handedness, and brain volume.

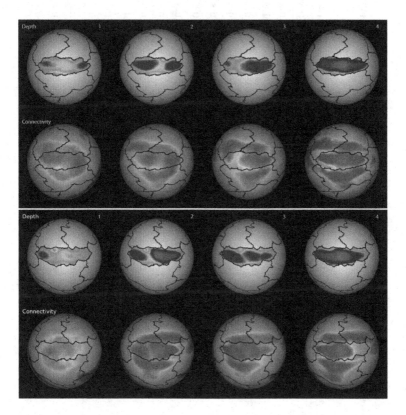

Fig. 1. Clustering of the central sulcus (CS) into four distinct, mirror-symmetric configurations on the left (top) and right (below) side. Rows 1, 3 show geodesic depth (increasing from red to magenta). Rows 2, 4 show the connection strength (increasing from magenta to red). (Color figure online)

Significant influences of subject sex, handedness, and brain volume were typically found for the second structural dimension and the third connectivity dimension. We assessed the absolute difference of scores within subject pairs grouped by genetic similarity. This heritability was typically reflected in the second dimension, representing between 2% and 6% of the total variance.

Clustering yielded four distinct structural and connectivity patterns (Fig. 1), with mirror-symmetric patterns on the left (top panel) and right side (below). Patterns were sorted by increasing regularity from left to right, as determined

Table 1. Analysis of dimensions obtained from domain decomposition of distance matrices for the central sulcus on the left and right side. The relevance of dimensions 1–4 was assessed in relation to the number of basins in this sulcus, demographic variables sex, handedness, and heritability.

Model	Dimension		# of Basins		R^2	Brain Volume		Sex		Handedness		Heritability	
		Exp. var.	p-value	Code		p-value	Code	p-value	Code	p-value	Code	p-value	Code
Structure left	1	0.559	< 2e-16	***	0.337	n.s	–	n.s	–	n.s	–	n.s	–
	2	0.123	8.13e-4	***	0.026	1.95e-5	***	n.s	–	n.s	–	1.05e-4	***
	3	0.084	4.29e-6	***	0.037	5.51e-3	**	1.54e-3	**	0.0149	*	n.s	–
	4	0.045	< 2e-16	***	0.127	n.s	–	n.s	–	n.s	–	n.s	–
Structure right	1	0.588	< 2e-16	***	0.255	n.s	–	n.s	–	0.0413	*	n.s	–
	2	0.098	0.0170	*	0.038	2.48e-8	***	1.13e-4	**	n.s	–	0.0349	*
	3	0.058	1.49e-13	***	0.054	0.016	*	n.s	–	n.s	–	n.s	–
	4	0.046	< 2e-16	***	0.194	n.s	–	n.s	–	n.s	–	n.s	–
Connectivity left	1	0.562	< 2e-16	***	0.406	n.s	–	7.40e-3	**	n.s	–	n.s	–
	2	0.230	< 2e-16	***	0.139	n.s	–	n.s	–	n.s	–	4.44e-3	**
	3	0.106	0.0378	*	0.020	2.69e-3	**	8.35e-3	**	6.29e-3	**	n.s	–
	4	0.030	n.s	–	0.024	5.45e-5	***	n.s	–	n.s	–	0.0451	*
Connectivity right	1	0.528	< 2e-16	***	0.380	n.s	–	n.s	–	n.s	–	n.s	–
	2	0.253	< 2e-16	***	0.111	n.s	–	n.s	–	n.s	–	n.s	–
	3	0.119	1.52e-5	***	0.025	n.s	–	0.0122	*	n.s	–	n.s	–

from scores of the first dimension above. The first pattern (column 1) showed a low regularity, consisting of two shallow centers of low variability. Patterns 2 and 3 revealed two stronger centers, in pattern 2 more prominent in the upper CS, in pattern 3 more prominent in the lower CS. Finally, pattern 4 showed a straight and deep sulcus with high regularity. Interestingly, more regular sulcal patterns were related to a stronger, more distinctive connectivity (rows 2 and 4). Note that connection strength closely followed a lower basin variability not only in the central sulcus, but also adjacent regions in the pre- and post-central sulcus, and the mid-posterior insula on both sides.

4   Conclusion

By this short demonstration, we wanted to illustrate two points: (1) Structural and connectivity patterns of the human brain do not originate from a continuum, but show distinct properties, at least at the regional level. This finding renders the application of registration processes as arguable, at least at the hemispheric level. (2) Instead of attempting to reduce the inter-individual variability by registration, we suggest to embrace this variability and to analyze and identify their sources. As demonstrated here, up to 80% of the total variance can be explained by identifiable factors.

References

1. Grassberger, P., Procaccia, I.: Characterization of strange attractors. Physica D: Nonlinear Phenomena **9**, 189–208 (1983)
2. Human Connectome Project: 1200 Subjects Data Release Reference Manual. https://www.humanconnectome.org/study/hcp-young-adult/document/1200-subjects-data-release, Accessed 17 Apr 2022
3. Jeurissen, B., Tournier, J.D., Dhollander, T., Connelly, A., Sijbers, J.: Multi-tissue constrained spherical deconvolution for improved analysis of multi-shell diffusion MRI data. NeuroImage **103**, 411–426 (2014)
4. Kruggel, F.: The macro-structural variability of the human neocortex. NeuroImage **172**, 620–630 (2018)
5. Smith, R.E., Tournier, J.D., Calamante, F., Connelly, A.: Anatomically-constrained tractography: improved diffusion MRI streamlines tractography through effective use of anatomical information. NeuroImage **62**, 1924–1938 (2016)
6. Tenenbaum, J.B., de Silva, V., Langford, J.C.: A global geometric framework for nonlinear dimensionality reduction. Science **290**, 2319–2323 (2000)

Architectures

A Multi-organ Point Cloud Registration Algorithm for Abdominal CT Registration

Samuel Joutard[1,3(✉)], Thomas Pheiffer[3], Chloe Audigier[2], Patrick Wohlfahrt[2], Reuben Dorent[1], Sebastien Piat[3], Tom Vercauteren[1], Marc Modat[1], and Tommaso Mansi[3]

[1] King's College London, London, UK
`samuel.joutard@kcl.ac.uk`
[2] Siemens Healthineers, Erlangen, Germany
[3] Siemens Healthineers, Princeton, USA

Abstract. Registering CT images of the chest is a crucial step for several tasks such as disease progression tracking or surgical planning. It is also a challenging step because of the heterogeneous content of the human abdomen which implies complex deformations. In this work, we focus on accurately registering a subset of organs of interest. We register organ surface point clouds, as may typically be extracted from an automatic segmentation pipeline, by expanding the Bayesian Coherent Point Drift algorithm (BCPD). We introduce MO-BCPD, a multi-organ version of the BCPD algorithm which explicitly models three important aspects of this task: organ individual elastic properties, inter-organ motion coherence and segmentation inaccuracy. This model also provides an interpolation framework to estimate the deformation of the entire volume. We demonstrate the efficiency of our method by registering different patients from the LITS challenge dataset. The target registration error on anatomical landmarks is almost twice as small for MO-BCPD compared to standard BCPD while imposing the same constraints on individual organs deformation.

1 Introduction

Registering CT images of the chest is an important step for several pipelines such as surgical planning for liver cancer resection or disease progression tracking [1,2,10,15]. This step is both crucial and challenging as the deformations involved are large and may contain complex patterns such as sliding motion between organs. While traditional registration methods tend to fail on this task, learning approaches such as [5,6,12] obtained promising results at the Learn2Reg 2020 challenge, task 3 [7]. Yet, traditional and learning approaches both aims at

---

Supplementary Information The online version contains supplementary material available at https://doi.org/10.1007/978-3-031-11203-4_9.

A. Hering et al. (Eds.): WBIR 2022, LNCS 13386, pp. 75–84, 2022.
https://doi.org/10.1007/978-3-031-11203-4_9

registering the whole image content instead of focusing on the relevant structures of interests. This introduces undesired noise and complexity to the registration process. To tackle this issue, we propose to exploit the recent availability of high quality automatic segmentation pipelines such as [3,17] and register the segmented structures. Specifically, structures are registered using their surface point cloud representation, allowing for exploiting meaningful geometric information of the different organs and finely modeling their dynamic properties. We also stress that surface point clouds are easy to derive from segmentation masks and are a lightweight representation of the structures of interest.

The Coherent Point Drift [13] (CPD) algorithm is one of the most popular method for deformable point cloud registration considered as state of the art [11]. A recent work [9] extended this framework using a Bayesian formulation and obtained more robust performances. CDP and BCPD both assume that points move coherently as a group to preserve the structure coherence. This is mainly because these frameworks are designed to register point clouds representing a single object. Consequently, [9,13] are not adapted for registering multi-organ points clouds. In particular, the coherency assumption doesn't stand for organs registration as each organ-specific point cloud may move independently to its neighbour, especially if we aim at registering inter-patient images.

In this work, we introduce a Multi-Organ Bayesian Coherent Point Drift algorithm (MO-BCPD) that models independent coherent structures. The contribution of this work is four-fold. Firstly, we extend the Bayesian formulation of CPD to model more complex structures interactions such as organ motion independence. Secondly, given that points clouds are obtained using automated segmentations, the proposed framework models partial segmentation errors allowing MO-BCPD to recover them. Thirdly, we model individual organ elasticity as part of the formulation. Fourthly, extensive experiments on 104 patients (10,712 pairs of patients) from the LiTS public dataset [4] demonstrate the effectiveness of our approach compared to BCPD. In particular, our method achieves an average target registration error on anatomical landmarks of 13 mm compared to 22 mm for the standard BCPD.

2   Method

In this section, we present our Multi-Organ Bayesian Coherent Point Drift algorithm. Let $\mathbf{y} = [\mathbf{y}_m]_{m \in \{1...M\}} \in \mathbb{R}^{M,3}$ be the source point cloud and $\mathbf{x} = [\mathbf{x}_n]_{n \in \{1...N\}} \in \mathbb{R}^{N,3}$ be the target point cloud where N and M are respectively the number of source and target points. We aim at finding the transformation $\mathcal{T}$ that realistically aligns these point clouds. In particular here, unlike in [9,13], the considered point clouds both represent a set of organ surfaces. Hence, each point is associated with an organ. Let $\mathbf{l}^y = [l_m^y]_{m \in \{1...M\}} \in \{1...L\}^M$ be the organ labels of the source point cloud and $\mathbf{l}^x = [l_n^x]_{n \in \{1...N\}} \in \{1...L\}^N$ be the organ labels of the target point cloud. L is the number of organs.

Transformation Model. Similarly to the BCPD, the Multi-Organ Bayesian Coherent Point Drift (MO-BCPD), decomposes the motion in two components: a sim-

Algorithm 1: Multi-Organ BCPD $(\mathbf{y}, \mathbf{x}, \omega, \Lambda, B, S, U, \kappa, \gamma, \epsilon)$

$\mathbf{v} \leftarrow 0_{M,3},\ \Sigma \leftarrow Id_M,\ s \leftarrow 1,\ R \leftarrow Id_3,\ t \leftarrow 0_3,\ <\alpha_m> \leftarrow \frac{1}{M},$

$\sigma^2 \leftarrow \frac{\gamma}{D \sum_{m,n} u_{l_m^y,l_n^x}} \sum_{m,n} u_{l_m^y,l_n^x} \|x_n - y_m\|^2,\ \theta \leftarrow (\mathbf{v}, \alpha, \mathbf{c}, \mathbf{e}, \rho, \sigma^2),\ P \leftarrow \frac{1}{M}\mathbf{1}_{M,N}$

$\nu' \leftarrow \mathbf{1}_N,\ q_1(.,.) \leftarrow D^{\kappa \mathbf{1}_M} \phi^{0,\Sigma},$

$q_2(c,e) \leftarrow \prod\limits_{n=1}^{N} (1 - \nu'_n)^{1-c_n} \left(\nu'_n \prod\limits_{m=1}^{M} \left(\frac{p_{mn}}{\nu'_n}\right)^{\delta_n(e_m)} \right)^{c_n},\ q_3(.,.) \leftarrow \delta_\rho \delta_{\sigma^2}$

while $L(q_1 q_2 q_3)$ *increases more than* ϵ **do**

    Update P and related terms:

    $\forall m,n\ \phi_{m,n} \leftarrow u_{l_m^y,l_n^x} \phi^{y'_m, \sigma^2 Id_3}(x_n) \exp -\frac{3s^2 \Sigma_{m,m}}{2\sigma^2},$

    $\forall m,n\ p_{m,n} \leftarrow \dfrac{(1-\omega)<\alpha_m>\phi_{m,n}}{\omega p_{out}(x_n)+(1-\omega)\sum\limits_{m'}<\alpha_{m'}>\phi_{m',n}},\ \nu \leftarrow P.\mathbf{1}_N,\ \nu' \leftarrow P^T.\mathbf{1}_M,$

    $\hat{N} \leftarrow \nu^T.\mathbf{1}_M,\ \hat{\mathbf{x}} \leftarrow \Delta(\nu)^{-1}.P.\mathbf{x},$

    Update displacement field and related terms:

    $\Sigma \leftarrow \left(G^{-1} + \frac{s^2}{\sigma^2}\Delta(\nu)\right),\ \forall d \in \{1,2,3\}\ \mathbf{v}^d \leftarrow \frac{s^2}{\sigma^2}\Sigma\Delta(\nu)(\rho^{-1}(\hat{\mathbf{x}}^d) - \mathbf{y}^d),$

    $\mathbf{u} \leftarrow \mathbf{y} + \mathbf{v},\ <\alpha_m> \leftarrow exp\{\psi(\kappa + \nu_m) - \psi(\kappa M + \hat{N})\}$

    Update ρ and related terms: $\bar{x} \leftarrow \frac{1}{\hat{N}} \sum\limits_{m=1}^{M} \nu_m \hat{x}_m,\ \bar{\sigma}^2 \leftarrow \frac{1}{\hat{N}} \sum\limits_{m=1}^{M} \nu_m \sigma_m^2,$

    $\bar{u} \leftarrow \frac{1}{\hat{N}} \sum\limits_{m=1}^{M} \nu_m u_m,\ S_{xu} \leftarrow \frac{1}{\hat{N}} \sum\limits_{m=1}^{M} (\hat{x}_m - \bar{x})(u_m - \bar{u})^T,$

    $S_{uu} \leftarrow \frac{1}{\hat{N}} \sum\limits_{m=1}^{M} (u_m - \bar{u})(u_m - \bar{u})^T + \bar{\sigma}^2 Id_3,\ \Phi S'_{xu}\Psi^T \leftarrow svd(S_{xu}),$

    $R \leftarrow \Phi d(1,\ldots,1,|\Phi\Psi|)\Psi^T,\ s \leftarrow \frac{Tr(RS_{xu})}{Tr(S_{uu})},\ t \leftarrow \bar{x} - sR\bar{u},\ \mathbf{y}' \leftarrow \rho(\mathbf{y} + \mathbf{v})$

    $\sigma^2 \leftarrow \frac{1}{3\hat{N}} \sum\limits_{d=1}^{3} \left((\mathbf{x}^d)^T \Delta(\nu')\mathbf{x}^d - 2\mathbf{x}^d P^T \mathbf{y}'^d + (\mathbf{y}'^d)^T \Delta(\nu)\mathbf{y}'^d\right) + s^2\bar{\sigma}^2$

    Update q: $q_1(.,.) \leftarrow D^{\kappa \mathbf{1}_M} \phi^{\mathbf{v},\Sigma},$

    $q_2(c,e) \leftarrow \prod\limits_{j=1}^{N} (1 - \nu'_j)^{1-c_j} \left(\nu'_j \prod\limits_{i=1}^{M} \left(\frac{p_{ij}}{\nu'_j}\right)^{\delta_i(e_j)} \right)^{c_j},\ q_3(.,.) \leftarrow \delta_\rho \delta_{\sigma^2}$

end

---

ilarity transform $\rho : \mathbf{p} \longrightarrow s\mathbf{R}\mathbf{p} + \mathbf{t}$ and a dense displacement field $\mathbf{v}$. Hence the deformed source point could is $[\mathcal{T}(\mathbf{y}_m)]_{m\in\{1...M\}} = [\rho(\mathbf{y}_m + \mathbf{v}_m)]_{m\in\{1...M\}}$. While this parametrization is redundant, [9] has shown that this makes the algorithm more robust to target rotations. Moreover, it is equivalent to performing a rigid alignment followed by a non-rigid refinement which corresponds to the common practice in medical image registration.

Generative Model. As in [9], MO-BCPD assumes that all points from the target point cloud $[\mathbf{x}_n]_{n\in\{1...N\}}$ are sampled independently from a generative model. A point x_n from the target point cloud is either an outlier or an inlier which is indicated by a hidden binary variable c_n. We note the probability for a point to be an outlier ω (i.e. $\mathcal{P}(c_n = 0) = \omega$). If x_n is an outlier, it is sampled from an outlier distribution of density p_{out} (typically, a uniform distribution over a volume containing the target point cloud). If x_n is an inlier ($c_n = 1$), x_n is associated with a point $\mathcal{T}(y_m)$ in the deformed source point cloud. Let e_n be a multinomial variable indicating the index of the point of the deformed source point cloud with which x_n is associated (i.e. $e_n = m$ in our example). Let α_m be the probability of selecting the point $T(y_m)$ to generate a point of the target point cloud (i.e. $\forall n \; \mathcal{P}(e_n = m|c_n = 1) = \alpha_m$). x_n is then sampled from a Gaussian distribution with covariance-matrix $\sigma^2 Id_3$ (Id_3 is the identity matrix of $\mathbb{R}^3$) centered on $T(\mathbf{y}_m)$. Finally, the organ label l_n^x is sampled according to the label transition distribution $\mathcal{P}(l_n^x|l_m^y) = u_{l_n^x, l_m^y}$. The addition of the label transition term is our contribution to the original generative model [9]. This term encourages to map corresponding organs between the different anatomies while allowing to recover from partial segmentation errors from the automatic segmentation tool.

We can now write the following conditional probability density:

$$p^e(x_n, l_n^x, c_n, e_n|\mathbf{y}, \mathbf{l}^y, \mathbf{v}, \alpha, \rho, \sigma^2)$$

$$= (\omega p_{out}(x_n))^{1-c_n} \left((1 - \omega) \prod_{m=1}^{M} \left(\alpha_m u_{l_m^y, l_n^x} \phi^{\mathbf{y}'_m, \sigma^2 Id_3}(\mathbf{x}_n) \right)^{\delta_{e_n = m}} \right)^{c_n} \quad (1)$$

where $\phi^{\mu, \Sigma}$ is the density of a multivariate Gaussian distribution $\mathcal{N}(\mu, \Sigma)$ and δ is the Kronecker symbol.

Prior Distributions. MO-BCPD also relies on prior distributions in order to regularize the registration process and obtain realistic solutions. As in [9], MO-BCPD defines two prior distributions: $p^v(\mathbf{v}|\mathbf{y}, \mathbf{l}_y)$ that regularizes the dense displacement field and $p^\alpha(\alpha)$ that regularizes the parameters α of the source point cloud selection multinomial distribution mentioned in the generative model. The prior on α follows a Dirichlet distribution of parameter $\kappa \mathbf{1}_M$. In practice, κ is set to a very high value which forces $\alpha_m \approx 1/M$ for all m. To decouple motion characteristics within and between organs, we propose a novel formulation of the displacement field prior p^v. Specifically, we introduce 3 parameters: a symmetric matrix $S = [s_{l,l'}]_{l,l' \in \{0...L\}}$ and two vectors $\Lambda = [\Lambda_l]_{l \in \{0...L\}}$ and $B = [B_l]_{l \in \{0...L\}}$. The matrix S parametrizes the motion coherence inter-organs. The vectors Λ and B respectively characterizes the variance of the deformation magnitude and motion coherence bandwidth within each organ. We define the displacement field prior for the MO-BCPD as:

$$p^v(\mathbf{v}|\mathbf{y}, \mathbf{l}^y) = \phi^{0,G}(\mathbf{v}^1)\phi^{0,G}(\mathbf{v}^2)\phi^{0,G}(\mathbf{v}^3) \quad (2)$$

$$G = \left[\Lambda_{l_i^y} \Lambda_{l_j^y} S_{l_i^y, l_j^y} \exp -\frac{\|y_i - y_j\|^2}{2 B_{l_i^y} B_{l_j^y}} \right]_{i,j \leq M} \quad (3)$$

Note that G must be definite-positive, leading to strictly positive values for variance of the displacement magnitude Λ_l and mild constraints on S.

Learning. Combining equations (1) and (2), the joint probability distribution of the variables $\mathbf{y}, \mathbf{l}^y, \mathbf{x}, \mathbf{l}^x, \theta$, where $\theta = (\mathbf{v}, \alpha, \mathbf{c}, \mathbf{e}, \rho, \sigma^2)$ is defined as:

$$p(\mathbf{x}, \mathbf{l}^x, \mathbf{y}, \mathbf{l}^y, \theta) \propto p^v(\mathbf{v}|\mathbf{y}, \mathbf{l}_y) p^\alpha(\alpha) \prod_{n=1}^{N} p^e(x_n, l_n^x, c_n, e_n|\mathbf{y}, \mathbf{l}_y, \mathbf{v}, \alpha, \rho, \sigma^2) \quad (4)$$

As in [9], we use variational inference to approximate the posterior distribution $p(\theta|\mathbf{x}, \mathbf{y})$ with a factorized distribution $q(\theta) = q_1(\mathbf{v}, \alpha)q_2(\mathbf{c}, \mathbf{e})q_3(\rho, \sigma^2)$ so that $q = \underset{q_1, q_2, q_3}{\arg\min} KL(q|p(.|\mathbf{x}, \mathbf{y}))$ where KL is the Kullback-Leibler divergence. Similarly to [9], we derive the MO-BCPD algorithm presented in algorithm 1. The steps detailed in Algorithm 1 perform coordinate ascent on the evidence lower bound $L(\theta) = \int_\theta q(\theta) \ln \frac{p(\mathbf{x}, \mathbf{y}, \theta)}{q(\theta)} d\theta$. In algorithm 1, γ is a hyper-parameter used to scale the initial estimation of σ^2 and ϵ is used for stopping criteria. We note $\Delta(\nu)$ the diagonal matrix with diagonal entries equal to ν.

Hyper-parameter Setting. The model has a large number of hyper-parameters which can impact the performance of the algorithm. Regarding κ, the parameter of the prior distribution p^α, and γ, the scaling applied to the initial estimation of σ^2, we followed the guidelines in [9]. ω is set based on an estimate of the proportion of outliers on a representative testing set. Regarding B and Λ, respectively the vector of organ-specific motion coherence bandwidth and expected deformation magnitude, they characterise organs elastic properties. Concretely, a larger motion coherence bandwidth B_l increases the range of displacement correlation for organ l (points that are further away are encouraged to move in the same direction). A larger expected deformation magnitude Λ_l increases the probability of larger displacements for organ l. These are physical quantities expressed in mm that could be set based on organs physical properties. The inter-organ motion coherence matrix S should be a symmetric matrix containing values between 0 and 1. $S_{l,l} = 1$ for all organs $l \in \{1 \ldots L\}$ and $S_{l,l'}$ is closer to 0 if organs l and l' can move independently.

$u_{l,l'}$ is the probability that a point with label l' generates a point with label l. As points labels are in practice obtained from an automatic segmentation tool, we note $[g^y_m]_{m \in \{1 \ldots M\}}$ and $[g^x_n]_{n \in \{1 \ldots n\}}$ respectively the unknown true organ labels of the source and target point clouds (as opposed to the estimated ones $[l^y_m]_{m \in \{1 \ldots M\}}$ and $[l^y_m]_{m \in \{1 \ldots M\}}$). We assume that points from the deformed source point cloud generate points with the same true labels (i.e. $\mathcal{P}(g^x_n = g^y_m|e_n = m) = 1$). Hence, the probability for y_m, with estimated organ label l^y_m to generate a point with label l^x_n is given by: $u_{l^x_n, l^y_m} = \sum_k p(g^y_m = k|l^y_m)p(l^x_n|g^x_n = k)$ where $p(g^y_m = k|l^y_m)$ is the probability that a point labelled l^y_m by the automatic segmentation tool has true label k and $p(l^x_n|g^x_n = k)$ is the probability that the automatic segmentation tool predicts the organ label l^x_n for a point with true label k. These probabilities need to be estimated on a representative testing set. We note that if the segmentation is error-free, the formula above gives $U = Id_L$. Indeed, in that case the points organ labels correspond exactly to the organ true labels so a point belonging to a certain organ can only generate a point from the same organ in the target anatomy. Properly setting the organ label transition probability matrix U is crucial to recover from potential partial segmentation errors. Figure 1 illustrates with a toy example a situation where the algorithm converges to an undesired state if the segmentation error is not modeled properly.

Interpolation. Once the deformation on the organ point clouds is known, one might want to interpolate the deformation back to image space in order to

Fig. 1. Toy example registering a pair of organs (a blue and an orange organ) with ~10% segmentation error (corrupted input labels). Both organs (orange and blue) of the target point cloud are shown in (a) in transparent while the source point cloud is shown in opaque. The blue (orange) dots on the left (right) of the figure corresponds to simulated segmentation errors. (b) shows the registered point cloud without modeling the inter-organ segmentation error, (c) shows the registered point cloud with segmentation error modelization

resample the whole volume. As in [8] we propose to use Gaussian process regression to interpolate the deformation obtained by the MO-BCPD algorithm. This interpolation process can also be used to register sub-sampled point clouds to decrease computation time as in [8].

Given a set of points $\tilde{\mathbf{y}} = [\tilde{y}_i]_{i \in \{1...\tilde{M}\}}$ with labels $\mathbf{l}^{\tilde{y}} = [l_i^{\tilde{y}}]_{i \in \{1...\tilde{M}\}}$. We compute the displacement for the set of points $\tilde{y}$ as:

$$\mathbf{v}^{\tilde{y}} = G^{int}(\tilde{\mathbf{y}}, \mathbf{l}^{\tilde{y}}, \mathbf{y}, \mathbf{l}^y, B, \Lambda, S).G^{-1}.v \tag{5}$$

$$G^{int}(\tilde{\mathbf{y}}, \mathbf{l}^{\tilde{y}}, \mathbf{y}, \mathbf{l}^y, B, \Lambda, S)_{i,j} = \Lambda_{l_i^{\tilde{y}}} \Lambda_{l_i^y} S_{l_i^{\tilde{y}}, l_i^y} \exp - \frac{\|\tilde{y}_i - y_j\|^2}{2 \beta_{l_i^{\tilde{y}}} \beta_{l_j^y}} \tag{6}$$

Acceleration. The speed ups strategies mentioned in [9] are fully transferable to the MO-BCPD pipeline. In our experiments though, the main improvement, by far, came from performing a low rank decomposition of G at the initialization of the algorithm. Indeed, this yielded consistent reliable ×10 speed-ups with negligible error when using ≥ 20 eigen values. The Nystrom methods to approximate P sometimes implied large error due to the stochasticity of the method while yielding up to ×2 speed-ups which is why we did not use it. This allows MO-BCPD to be run in a few seconds with $M, N \approx 5000$.

3   Experiments

We evaluate the MO-BCPD algorithm by performing inter-patient registration from the LITS challenge training dataset [4] which contains 131 chest-CT patient images. 27 patients were removed due to different field of view, issues with the segmentation or landmark detection. In total, 10,712 registrations were performed on all the pairs of remaining patients. The segmentation was automatically performed using an in-house tool derived from [17] which also provides a

Table 1. Target registration error on landmarks. Results in mm (std).

	Sim	BCPD	GMC-MO-BCPD	OMC-MO-BCPD
Bladder	257 (26)	29 (15)	30 (15)	**26** (15)
Left kidney bottom	128 (18)	23 (11)	22 (10)	**8** (4)
Left kidney center	107 (13)	18 (10)	15 (8)	**6** (3)
Left kidney top	101 (15)	23 (12)	21 (10)	**9** (4)
Liver bottom	114 (15)	29 (14)	28 (14)	**24** (13)
Liver center	65 (10)	12 (7)	12 (7)	**11** (7)
Liver top	123 (15)	**24** (12)	25 (12)	26 (14)
Right kidney bottom	98 (15)	26 (13)	24 (11)	**10** (6)
Right kidney center	65 (11)	21 (12)	17 (9)	**5** (3)
Right kidney top	64 (15)	25 (13)	21 (11)	**9** (4)
Round ligament of liver	95 (20)	27 (14)	27 (13)	**25** (13)

set of anatomical landmarks for each image which were used for evaluation. We considered five organs of interest: the liver, the spleen, the left and right kidneys and the bladder. We compared 4 different algorithms: registration of the point clouds with a similarity transform (Sim), BCPD, GMC-MO-BCPD which is MO-BCPD with global motion coherence ($S = 1_{L,L}$) and OMC-MO-BCPD which is MO-BCPD with intra-organ motion coherence only ($S = Id_L$). As the segmentation tool performed very well on the considered organs, we set $U = Id_L$ and $\omega = 0$ for both MO-BCPD versions ($\omega = 0$ for BCPD as well). We used for all organs the same values for Λ_l and B_l respectively 10 mm and 30 mm as a trade off between shape matching and preservation of individual organs appearance ($\beta = 30$ and $\lambda = 0.1$ for BCPD which is the equivalent configuration). We also set, $\gamma = 1$ and $\epsilon = 0.1$. We compared those algorithms by computing the registration error on the anatomical landmarks belonging to those organs. We chose this generic, relatively simple setting (same rigidity values for all organs, no outlier modeling, only two extreme configurations for S) in order to perform large scale inter-patient registration experiments but we would like to stress that further fine tuning of these parameters for a specific application or even for a specific patient would further improve the modeling and hence the registration outcome. Results are presented in Table 1. We observe that while GMC-MO-BCPD induces some marginal improvements with respect to BCPD, OMC-MO-BCPD allows a much more precise registration. As illustrated in Fig. 2, the main improvement from BCPD to GMC-MO-BCPD is that different organs no longer overlap. Indeed, as highlighted by the green and blue ellipses, the spleen and the right kidney from the deformed source patient overlap the liver of the target patient when using BCPD. OMC-MO-BCPD properly aligns the different organs while preserving their shape (see orange and purple ellipses for instance).

Fig. 2. Qualitative comparison of registration output for the liver, left/right kidneys and spleen. From left to right, BCPD, GMC-MO-BCPD, OMC-MO-BCPD. The target organs are shown in transparency while the deformed point cloud are shown in opaque.

4   Conclusion

We introduced MO-BCPD, an extension of the BCPD algorithm specifically adapted to abdominal organ registration. We identified three limitations of the original work [9] on this task and proposed solutions to model: the segmentation error between neighboring organs of interest, the heterogeneous elastic properties of the abdominal organs and the complex interaction between various organs in terms of motion coherence. We demonstrated significant improvements over BCPD on a large validation set (N=10,712).

Moreover, we would like to highlight that segmentation error could also be taken into account by tuning the outlier probability distribution p_{out} and the probability of being an outlier ω. When the point is estimated as a potential outlier by the algorithm, its contribution to the estimation of the transformation $\mathcal{T}$ is lowered. Hence, the segmentation error modelled by p_{out} and ω corresponds to over/under segmentation, i.e. when there is a confusion between an organ and another class we don't make use of in MO-BCPD (e.g. background). Hence, MO-BCPD introduces a finer way of handling segmentation error by distinguishing two types of errors: mis-labeling between classes of interest which is modelled by U and over/under-segmentation of classes of interest modeled by ω and p_{out}.

In this manuscript, we focused on highlighting the improvements yielded by the MO-BCPD formulation specifically designed for multi-organ point cloud registration. That being said, some clinical applications would require the deformation on the whole original image volume. Hence, we present in supplementary material preliminary results on a realistic clinical use case. In Figs. 3 and 4 we see that MO-BCPD coupled with the proposed interpolation framework obtain better alignment on the structures of interest than traditional intensity-based baselines. It is also interesting to note that MO-BCPD also better aligns structures that are particularly challenging to align such as the hepatic vein while not using these structures in the MO-BCPD.

Future work will investigate how MO-BCPD could be used as a fast, accurate initialization for image-based registration algorithms. From a modeling standpoint, we would also like to further work on segmentation error modeling (in particular over/under segmentation) with a more complex organ specific outlier distribution.

References

1. Carrillo, A., Duerk, J., Lewin, J., Wilson, D.: Semiautomatic 3-D image registration as applied to interventional MRI liver cancer treatment. IEEE Trans. Med. Imaging **19**(3), 175–185 (2000). https://doi.org/10.1109/42.845176
2. Cash, D.M., et al.: Concepts and preliminary data toward the realization of image-guided liver surgery. J. Gastrointest. Surg. **11**(7), 844–59 (2007). https://www.proquest.com/scholarly-journals/concepts-preliminary-data-toward-realization/docview/1112236808/se-2?accountid=11862. copyright - The Society for Surgery of the Alimentary Tract 2007; Dernière mise á jour - 2014-03-30
3. Chen, X., et al.: A deep learning-based auto-segmentation system for organs-at-risk on whole-body computed tomography images for radiation therapy. Radiother. Oncol. **160**, 175–184 (2021). https://doi.org/10.1016/j.radonc.2021.04.019
4. Christ, P.F., et al.: The liver tumor segmentation benchmark (LiTS). CoRR abs/1901.04056 (2019). http://arxiv.org/abs/1901.04056
5. Estienne, T., et al.: Deep learning based registration using spatial gradients and noisy segmentation labels. CoRR abs/2010.10897 (2020). https://arxiv.org/abs/2010.10897
6. Heinrich, M.P.: Closing the gap between deep and conventional image registration using probabilistic dense displacement networks. In: Shen, D., et al. (eds.) MICCAI 2019. LNCS, vol. 11769, pp. 50–58. Springer, Cham (2019). https://doi.org/10.1007/978-3-030-32226-7_6
7. Hering, A., et al.: Learn2Reg: comprehensive multi-task medical image registration challenge, dataset and evaluation in the era of deep learning. arXiv preprint arXiv:2112.04489 (December 2021)
8. Hirose, O.: Acceleration of non-rigid point set registration with downsampling and Gaussian process regression. IEEE Trans. Pattern Anal. Mach. Intell. **43**(8), 2858–2865 (2021). https://doi.org/10.1109/TPAMI.2020.3043769
9. Hirose, O.: A Bayesian formulation of coherent point drift. IEEE Trans. Pattern Anal. Mach. Intell. **43**(7), 2269–2286 (2021). https://doi.org/10.1109/TPAMI.2020.2971687
10. Lange, T., et al.: Registration of portal and hepatic venous phase of MR/CT data for computer-assisted liver surgery planning. In: International Congress Series, vol. 1281, pp. 768–772 (2005). https://doi.org/10.1016/j.ics.2005.03.332
11. Maiseli, B., Gu, Y., Gao, H.: Recent developments and trends in point set registration methods. J. Vis. Commun. Image Represent. **46**, 95–106 (2017). https://doi.org/10.1016/j.jvcir.2017.03.012, https://www.sciencedirect.com/science/article/pii/S1047320317300743
12. Mok, T.C.W., Chung, A.C.S.: Large deformation diffeomorphic image registration with Laplacian pyramid networks. ArXiv abs/2006.16148 (2020)
13. Myronenko, A., Song, X.: Point set registration: coherent point drift. IEEE Trans. Pattern Anal. Mach. Intell. **32**(12), 2262–2275 (2010). https://doi.org/10.1109/TPAMI.2010.46
14. Papież, B.W., Franklin, J.M., Heinrich, M.P., Gleeson, F.V., Brady, M., Schnabel, J.A.: Gifted demons: deformable image registration with local structure-preserving regularization using supervoxels for liver applications. J. Med. Imaging **5**, 024001 (2018). https://doi.org/10.1117/1.JMI.5.2.024001
15. Robu, M.R., et al.: Global rigid registration of CT to video in laparoscopic liver surgery. Int. J. Comput. Assist. Radiol. Surg. **13**, 947–956 (2018)

16. Thirion, J.P.: Image matching as a diffusion process: an analogy with Maxwell's demons. Med. Image Anal. **2**(3), 243–260 (1998). https://doi.org/10.1016/S1361-8415(98)80022-4, https://www.sciencedirect.com/science/article/pii/S1361841598800224

17. Yang, D., et al.: Automatic liver segmentation using an adversarial image-to-image network. In: Descoteaux, M., Maier-Hein, L., Franz, A., Jannin, P., Collins, D.L., Duchesne, S. (eds.) MICCAI 2017. LNCS, vol. 10435, pp. 507–515. Springer, Cham (2017). https://doi.org/10.1007/978-3-319-66179-7_58

Voxelmorph++

Going Beyond the Cranial Vault with Keypoint Supervision and Multi-channel Instance Optimisation

Mattias P. Heinrich[(✉)] and Lasse Hansen

Institute of Medical Informatics, University of Lübeck, Lübeck, Germany
{heinrich,hansen}@imi.uni-luebeck.de

Abstract. The majority of current research in deep learning based image registration addresses inter-patient brain registration with moderate deformation magnitudes. The recent Learn2Reg medical registration benchmark has demonstrated that single-scale U-Net architectures, such as VoxelMorph that directly employ a spatial transformer loss, often do not generalise well beyond the cranial vault and fall short of state-of-the-art performance for abdominal or intra-patient lung registration. Here, we propose two straightforward steps that greatly reduce this gap in accuracy. First, we employ keypoint self-supervision with a novel network head that predicts a discretised heatmap and robustly reduces large deformations for better robustness. Second, we replace multiple learned fine-tuning steps by a single instance optimisation with hand-crafted features and the Adam optimiser. Different to other related work, including FlowNet or PDD-Net, our approach does not require a fully discretised architecture with correlation layer. Our ablation study demonstrates the importance of keypoints in both self-supervised and unsupervised (using only a MIND metric) settings. On a multi-centric inspiration-exhale lung CT dataset, including very challenging COPD scans, our method outperforms VoxelMorph by improving nonlinear alignment by 77% compared to 19% - reaching target registration errors of 2 mm that outperform all but one learning methods published to date. Extending the method to semantic features sets new stat-of-the-art performance on inter-subject abdominal CT registration.

Keywords: Registration · Heatmaps · Deep learning

1 Introduction

Medical image registration aims at finding anatomical and semantic correspondences between multiple scans of the same patient (intra-subject) or across a population (inter-subject). The difficulty of this task with plentiful clinical applications lies in discriminating between changes in intensities due to image appearance changes (acquisition protocol, density difference, contrast, etc.) and nonlinear deformations. Advanced similarity metrics may help in finding a good

© The Author(s), under exclusive license to Springer Nature Switzerland AG 2022
A. Hering et al. (Eds.): WBIR 2022, LNCS 13386, pp. 85–95, 2022.
https://doi.org/10.1007/978-3-031-11203-4_10

contrast-invariant description of local neighbourhoods, e.g. normalised gradient fields [7] or MIND [14]. Due to the ill-posedness of the problem some form of regularisation is often employed to resolve the disambiguity between several potential local minima in the cost function. Powerful optimisation frameworks that may comprise iterative gradient descent, discrete graphical models or both (see [28] for an overview) aim at solving for a global optimum that best aligns the overall scans (within the respective regions of interest). Many deep learning (DL) registration frameworks (e.g. DLIR [30] and VoxelMorph [1]) rely on a spatial transformer loss that may be susceptible to ambiguous optimisation landscapes - hence multiple resolution or scales levels need to be considered. The focus of this work is to reflect such local minima more robustly in the loss function of DL-registration. We propose to predict probabilistic displacement likelihoods as heatmaps, which can better capture multiple scales of deformation within a single feed-forward network.

Related Work: Addressing large deformations with learning based registration is generally approached by either multi-scale, label-supervised networks [15,21,22] or by employing explicitly discretised displacements [9,11]. Many variants of U-Net like architectures have been proposed that include among others, dual-stream [18], cascades [32] and embeddings [5]. Different to those works, we do neither explicitly model a discretised displacement space, multiple scales or warps, nor modify the straightforward feed-forward U-Net of DLIR or VoxelMorph. Other works aimed at learning lung deformations through simulated transformations [4,26].

Learning based point-cloud registration is another research field of interest (FlowNet3d [20]) that has however so far been restricted to lung registration in the medical domain [8]. For a comparison of classical approaches using thin-plate splines or expectation-maximisation e.g. coherent point drift to learning-based ones the reader is referred to comparison experiments in [8]. Stacked hourglass networks that predict discretised heatmaps of well-defined anatomical landmarks are commonly used in human pose estimation [24]. They are, however, restricted to datasets and registration applications where not only pairwise one-to-one correspondences can be obtained as training objective but generic landmarks have to be found across all potential subjects. Due to anatomical variations this restriction often prevents their use in medical registration. DRAMMS [25] also aims at matching keypoints across scans and learns discriminative features based on mutual saliency maps but does not offer the benefits of fast feed-forward prediction of displacements.

Combining an initial robust deformation prediction with instance optimisation or further learning fine-tuning steps (Learn-to-optimise, cf. [29]) has become a new trend in learning-based registration, e.g. [16,27]. This paradigm, which expects coarse but large displacements from a feed-forward prediction, can shift the focus away from sub-pixel accuracy and towards avoiding failure cases in the global transformation due to local minima. It is based on the observation that local iterative optimisers are very precise when correctly initialised and have become extremely fast due to GPU acceleration.

Contributions: We demonstrate that a single-scale U-Net without any bells and whistles in conjunction with a fast MIND-based instance optimisation can achieve or outperform state-of-the-art in large-deformation registration. This is achieved by focussing on coarse scale global transformation by introducing a novel heatmap prediction network head. In our first scenario we employ weak self-supervision, through automatic keypoint correspondences [12]. Here, the heatmap enables a discretised integral regression of displacements to directly and explicitly match the keypoint supervision. Second, we incorporate the heatmap prediction into a non-local unsupervised metric loss. This enables a direct comparison within the same network architecture to the commonly used spatial transformer (warping) loss in unsupervised DL registration and highlights the importance of providing better guidance to avoid local minima. Our extensive ablation experiments with and without instance optimisation on a moderately large and challenging inspiration-exhale lung dataset demonstrate state-of-the-art performance.

Our code is publicly available at:
https://www.github.com/mattiaspaul/VoxelMorphPlusPlus

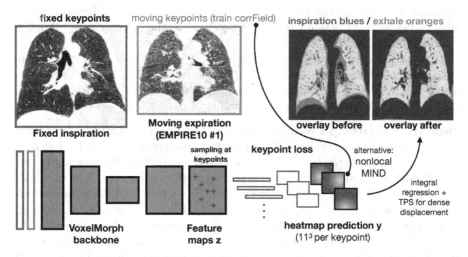

Fig. 1. Overview of method and qualitative result for held-out case #1 from [23]. The key new element of our approach is the heatmap prediction head that is appended to a standard VoxelMorph. It helps overcome local minima through a probabilistic loss with either automatic keypoint correspondences or non-locally weighted MIND features.

2 Method

Keypoint correspondences are an excellent starting point to explore the benefits of incorporating a heatmap prediction within DL-registration. Our method can be either trained when those automatically computed displacements at around $|K| \approx 2000$ locations per scan are available for training data, or we can use a non-local unsupervised metric loss (see details below). In both scenarios, we use Förstner keypoints in the fixed scan to focus on distinct locations within a region of interest. We will first describe the baseline single-scale U-Net backbone, followed by our novel heatmap prediction head and the non-local MIND loss (which is an extension of [13] to 3D).

Baseline Backbone: Given two input CT scans, fixed and moving image $F, M :$ $\mathbb{R}^3 \to \mathbb{R}$ and a region of interest $\Omega \in \mathbb{R}^3$, we firstly define a feed-forward U-Net [6] $\Theta(F, M, \Omega, \theta) \to \mathbb{R}^C$ with trainable parameters θ that maps the concatenated input towards a shared C-dimensional feature representation $\mathbf{z}$ (we found $C \approx 64$ is expressive enough to represent displacements). This representation may have a lower spatial resolution than F or M and is the basis for predicting a (sparse) displacement field φ that spatially aligns F and M within Ω. Θ comprises in our implementation a total of eleven 3D convolution blocks, each consisting of a $3 \times 3 \times 3$ convolution, instance normalisation (IN), ReLU, a $1 \times 1 \times 1$ convolution, and another IN+ReLU. Akin to VoxelMorph, we use $2 \times 2 \times 2$ max-pooling after each of the four blocks in the encoder and nearest neighbour upsampling to restore the resolution in the decoder, but use a half-resolution output. The network has up to $C = 64$ hidden feature channels and 901'888 trainable parameters.

Due to the fact that this backbone already contains several convolution blocks on the final resolution at the end of the decoder, it is directly capable of predicting a continuous displacement field φ by simply appending three more $1 \times 1 \times 1$ convolutions (and IN + ReLU) with a number of output channels equal to 3.

Discretised Heatmap Prediction Head: The aim of the heatmap prediction head is to map a C-dimensional feature vector (interpreted as a $1 \times 1 \times 1$ spatial tensor with $|K|$ being the batch dimension) into a discretised displacement tensor $y \in \mathcal{Q}$ with predefined size and spatial range R (see Fig. 1). Here we chose $R = 0.3$, in a coordinate system that ranges from -1 to $+1$, which captures even large lung motion between respiratory states. We define $\mathcal{Q}$ to be a discretised map of size $11 \times 11 \times 11$ to balance computational complexity and limit quantisation effects. This means we need to design another *nested* decoder that increases the spatial resolution from 1 to 11. Our heatmap network comprises a transpose 3D convolution with kernel size $7 \times 7 \times 7$, six further 3D convolution blocks (kernel size $7 \times 7 \times 7$ and IN+ReLU) once interleaved with a single trilinear upsampling to $11 \times 11 \times 11$. It has 462'417 trainable parameters and its number of output channels is equal to 1.

Next, we can define a probabilistic displacements tensor $\mathcal{P}$ with a dimensionality of 6 (3 spatial and 3 displacement dimensions) using a softmax operation along the combined displacement dimensions as:

$$\mathcal{P}(\mathbf{x}, \Delta\mathbf{x}) = \frac{\exp(y(\mathbf{x}, \Delta\mathbf{x}))}{\sum_{\Delta\mathbf{x}} \exp(y(\mathbf{x}, \Delta\mathbf{x}))}, \tag{1}$$

where $\mathbf{x}$ are global spatial 3D coordinates and $\Delta\mathbf{x}$ local 3D displacements. In order to define a continuous valued displacement field, we apply a weighted sum:

$$\varphi(\mathbf{x}) = \sum_{\Delta\mathbf{x}} \mathcal{P}(\mathbf{x}, \Delta\mathbf{x}) \cdot \mathcal{Q}(\Delta\mathbf{x}) \tag{2}$$

This output is used during training to compute a mean-squared error between predicted and pre-computed keypoint displacements. Since, the training correspondences are regularised using a graphical model, we require no further penalty.

Non-local MIND Loss: To avoid the previously described pitfalls of directly employing a spatial transformer (warping) loss, we can better employ the probabilistic heatmap prediction and compute the discretely warped MIND vectors of the moving scan implicitly by a weighted average of the underlying features within pre-defined capture region (where c describes one of the 12 MIND channels) as:

$$MIND_{warped}(c, \mathbf{x}) = \sum_{\Delta\mathbf{x}} \mathcal{P}(\mathbf{x}, \Delta\mathbf{x}) \cdot MIND(c, \mathbf{x} + \Delta\mathbf{x}) \tag{3}$$

Implementation Details: Note that the input to both the small regression network (baseline) and our proposed are feature vectors sampled at the keypoint locations, which already improves the baseline architecture slightly. We use trilinear interpolation in all cases where the input and output grids differ in size to obtain off-grid values. All predicted sparse displacements φ are extrapolated to a dense field using thin-plate-splines with $\lambda = 0.1$ that yields φ^*.

For the baseline regression setup (VoxelMorph) we employ a common MIND warping loss and a diffusion regularisation penalty that is computed based on the Laplacian of a kNN-graph ($k = 7$) between the fixed keypoints. The weighting of the regularisation was empirically set to $\alpha = 0.25$. We found that using spatially aggregated CT and MIND tensors the former using average pooling with kernel size 2, the latter two of those pooling steps, leads to stabler training in particular for the regression baseline.

Multi-channel Instance Optimisation: We directly follow the implementation described in [27][1]. It is initialised with φ^*, runs for 50 iterations, employs a combined B-spline and diffusion regularisation coupled with a MIND metric loss and a grid spacing of 2. This step is extremely fast, but relies on a robust initialisation as we will demonstrate in our experiments. The method can also be employed when semantic features, e.g. segmentation predictions from an nnUNet [19], are available in the form of one-hot tensors.

3   Experiments and Results

We perform extensive experiments on inspiration-exhale registration of lung CT scans - arguably one of the most challenging tasks in particular for learning-based

---

[1] https://github.com/multimodallearning/convexAdam.

registration [16]. A dataset of 30 scan pairs with large respiratory differences is collected from EMPIRE10 (8 scan pairs #1, #7, #8, #14, #18, #20, #21 and #28) [23], Learn2Reg Task 2 [17] and DIR-Lab COPD [2] (10 pairs). The exhale and inspiration scans are resampled to $1.75 \times 1.25 \times 1.75$ mm and $1.75 \times 1.00 \times 1.25$ mm respectively to account for average overall volume scaling and a fixed region with dimensions $192 \times 192 \times 208$ voxels was cropped that centres the mass of automatic lung masks. Note that this pre-processing approximately halves the initial target registration error (TRE) of the COPD dataset. Lung masks are also used to define a region-of-interest for the loss evaluation and to mask input features for the instance optimisation. We split the data into five folds for cross-validation that reflect the multi-centric data origin (i.e. approx. two scans per centre are held out for validation each).

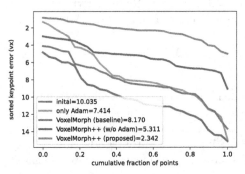

Table 1. Results of ablation study on lung CT: VOXELMORPH++ improves error reduction of nonlinear alignment from 18% to 77%.

	UNet	Heatmap	Keypoints	Error w/o Adam	Error w/Adam
Initial/Adam				10.04 vx	7.41 vx
VoxelMorph	✓			8.17 vx	4.80 vx
VM + Heatmap	✓	✓		6.49 vx	3.18 vx
VM + Keypoints	✓		✓	6.30 vx	2.79 vx
VoxelMorph++	✓	✓	✓	5.31 vx	**2.34 vx**

Fig. 2. Cumulative keypoint error of proposed model compared to a VoxelMorph baseline and using only Adam instance optimisation with MIND.

Keypoint Self-supervision: To create correspondence as self-supervision for our proposed VoxelMorph++ method, we employ the **corrField** [12][2], which is designed for lung registration and based on a discretised displacement search and a Markov random field optimisation with multiple task specific improvements. It runs within a minute per scan pair and creates $\approx 2000 = |K|$ highly accurate (≈ 1.68 mm) correspondences at Förstner keypoints.

As additional experiments we also apply the same technique to the popular (but less demanding) DIR-Lab 4DCT lung CT benchmark and we extend our method to the inter-subject alignment of 30 abdominal CTs [31] that was also part of the Learn2Reg 2020 challenge (Task 3) and provides 13 difficult anatomical organ labels for training and validation.

Ablation Study: We consider a five-fold cross-validation for all ablation experiments with an initial error of 10.04 vx (after translation and scaling) across 30 scan pairs computed based on keypoint correspondences. Employing only the

---

[2] http://www.mpheinrich.de/code/corrFieldWeb.zip.

Adam instance optimisation with MIND features results in an error of 8.17 vx, with default settings of grid spacing = 2 voxels, 50 iterations and $\lambda_{Adam} = 0.65$. Note that a dense displacement is estimated with a parametric B-spline model. We start from the slightly improved VoxelMorph **baseline** with MIND loss, diffusion regularisation and increased number of convolution operations described above. This yields a keypoint error of **8.17 vx** that represents an error reduction of 19% and can be further improved to 4.80 vx when adding instance optimisation. A weighting parameter $\lambda = 0.75$ for diffusion regularisation was empirically found with $k = 7$ for the sparse neighbourhood graph of keypoints. Replacing the traditional spatial transformer loss with our proposed heatmap prediction head that uses the nonlocal MIND loss much improves the performance to 6.49 vx and 3.18 vx (with and without Adam respectively). But the key improvement can be gained when including the self-supervised keypoint loss. Using our baseline VoxelMorph architecture that regresses continuous 3D vectors, we reach 6.30 and **2.79 vx**. See Table 1 and Fig. 2 for numerical and cumulative errors. Our heatmap-based network and the instance optimisation require around 0.43 and 0.41 s inference time, respectively. The complexity of transformations measured as standard deviation of log-Jacobian determinants is on average 0.0554. The number of negative values is zero (no folding) in 8 out of 10 COPD cases and negligible ($<10^{-4}$) in the others.

Comparison to State-of-the-Art on DIR-Lab: When evaluation the target registration error (TRE) in mm for the 10 pairs of DIR-Lab COPD [2] one of the most challenging benchmarks in medical registration with an initial misalignment of 23.36 mm (and 12.0 mm after pre-alignment), we reach 2.16 mm. This compares very favourable to VoxelMorph+ with 7.98 mm and LapIRN with 3.68 mm (see Table 2). Of all published DL-methods only GraphRegNet [9] is superior with 1.34 mm. The high visual quality of our registration is shown in Fig. 3.

Table 2. Target registration error in mm for 10 pairs each of DIR-Lab 4D lung-CT [3] and COPD [2] datasets in comparison to a selection of other published methods.

Method (citation)	Before	[26]	[15]	[4]	[21][a]	[1]	[9]	Ours
TRE (4DCT)	8.46 ± 6.6	2.52 ± 3.0	1.14 ± 0.8	3.68 ± 3.3	1.60	1.71 ± 2.9	1.39 ± 1.3	**1.33**
TRE (COPD)	23.36 ± 11.9	–	–	–	3.83	7.98 ± 3.8	1.34 ± 1.3	**2.16**

[a]own experiments including instance optimisation

Limitations and Further Potential: We have not yet considered more advanced network architectures as backbone, e.g. two-stream or multi-level, which are likely to yield further improvements. However, based on our experiments we expect that it could merely reduce the reliance on instance optimisation.

Inter-subject Abdominal CT Registration: We apply our proposed VoxelMorph++ model with nonlocal loss and no architectural modifications to

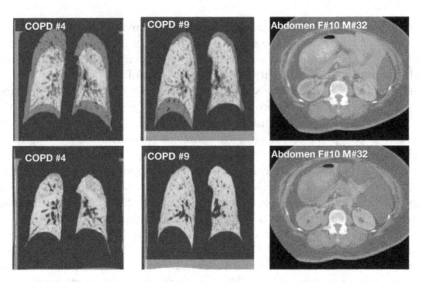

Fig. 3. Exemplary results of our proposed method before (top row) and after (bottom row) registration. Fixed exhale scans are shown in blue and inspiration in orange shades respectively (adding up to grayscale when aligned). For abdominal alignment transformed segmentation labels are shown, here: right kidney ▨, left kidney ■, gallbladder ▨, liver ■, stomach ■, aorta ■ and pancreas ▨ are visible. (Color figure online)

another challenging task of inter-subject abdominal CT registration with initial Dice overlap for 13 organs of only 25.9% and weakly supervised learning (45 registration pairs). Following [10], we decouple the semantic feature extraction and directly train an nnUNet model [19]. The best published VoxelMorph model that was trained with label-supervision and extended to a two-stream architecture reached 43.9% [27], the two top-ranked methods of the Learn2Reg challenge yield 65.7% (ConvexAdam [27]) and 67% (LapIRN [22]) respectively. Directly employing instance optimisation with 25 iterations on the nnUNet features achieves 62.9%. We use 2048 keypoints that are sampled inversely proportional to the predicted label maps and employ two warps (and inverse consistency for the first of them). Our model substantially outperforms VoxelMorph with 52.3% and sets a new state-of-the-art performance after instance optimisation reaching 69.6% with a total run time of less than a second.

4   Discussion and Conclusions

Our results demonstrate that contrary to previous belief, a simple single-scale U-Net architecture can provide large deformation estimation that is robust enough to reach high accuracy with a subsequent instance optimisation. The key insight of our work is the importance to predict a discretised heatmap to alleviate the problematic direct regression and use strong self-supervision either using automatic keypoint correspondences or a nonlocal multichannel loss together with a straightforward instance optimisation. Our work is related to mlVIRN [15], which also incorporates a keypoint loss for lung registration in addition to lobe segmentations, but has to be trained with several hundreds of paired CTs and did not report TRE values for DIRlab-COPD. Our network can be trained within 17 min on a single RTX A4000 requiring less than 2 GB of VRAM, indicating the improved training efficiency with fewer scans when using heatmaps. GraphReg-Net [9] is similar in that it also employs heatmaps (integral regression) but more explicitly by defining the exact same discretised displacement grid beforehand and computing an SSD cost tensor based on hand-crafted features as input. While it outperforms our method with a TRE of 1.34 mm it appears to be more tailored towards the specific task and might not be easily extendable to end-to-end feature learning or abdominal registration.

References

1. Balakrishnan, G., Zhao, A., Sabuncu, M.R., Guttag, J., Dalca, A.V.: VoxelMorph: a learning framework for deformable medical image registration. IEEE Trans. Med. Imaging **38**(8), 1788–1800 (2019)
2. Castillo, R., et al.: A reference dataset for deformable image registration spatial accuracy evaluation using the copdgene study archive. Phys. Med. Biol. **58**(9), 2861 (2013)
3. Castillo, R., et al.: A framework for evaluation of deformable image registration spatial accuracy using large landmark point sets. Phys. Med. Biol. **54**(7), 1849 (2009)
4. Eppenhof, K.A., Lafarge, M.W., Veta, M., Pluim, J.P.: Progressively trained convolutional neural networks for deformable image registration. IEEE Trans. Med. Imaging **39**(5), 1594–1604 (2019)
5. Estienne, T., et al.: MICS: multi-steps, inverse consistency and symmetric deep learning registration network (2021)
6. Falk, T., et al.: U-Net: deep learning for cell counting, detection, and morphometry. Nat. Methods **16**(1), 67–70 (2019)
7. Haber, E., Modersitzki, J.: Intensity gradient based registration and fusion of multi-modal images. Methods Inf. Med. **46**(03), 292–299 (2007)
8. Hansen, L., Dittmer, D., Heinrich, M.P.: Learning deformable point set registration with regularized dynamic graph CNNs for large lung motion in COPD patients. In: Zhang, D., Zhou, L., Jie, B., Liu, M. (eds.) GLMI 2019. LNCS, vol. 11849, pp. 53–61. Springer, Cham (2019). https://doi.org/10.1007/978-3-030-35817-4_7
9. Hansen, L., Heinrich, M.P.: GraphregNet: deep graph regularisation networks on sparse keypoints for dense registration of 3d lung CTS. IEEE Trans. Med. Imaging **40**(9), 2246–2257 (2021)
10. Hansen, L., Heinrich, M.P.: Revisiting iterative highly efficient optimisation schemes in medical image registration. In: de Bruijne, M., et al. (eds.) MICCAI 2021. LNCS, vol. 12904, pp. 203–212. Springer, Cham (2021). https://doi.org/10.1007/978-3-030-87202-1_20
11. Heinrich, M.P.: Closing the gap between deep and conventional image registration using probabilistic dense displacement networks. In: Shen, D., et al. (eds.) MICCAI 2019. LNCS, vol. 11769, pp. 50–58. Springer, Cham (2019). https://doi.org/10.1007/978-3-030-32226-7_6

12. Heinrich, M.P., Handels, H., Simpson, I.J.A.: Estimating large lung motion in COPD patients by symmetric regularised correspondence fields. In: Navab, N., Hornegger, J., Wells, W.M., Frangi, A.F. (eds.) MICCAI 2015. LNCS, vol. 9350, pp. 338–345. Springer, Cham (2015). https://doi.org/10.1007/978-3-319-24571-3_41

13. Heinrich, M.P., Hansen, L.: Highly accurate and memory efficient unsupervised learning-based discrete CT registration using 2.5D displacement search. In: Martel, A.L., et al. (eds.) MICCAI 2020. LNCS, vol. 12263, pp. 190–200. Springer, Cham (2020). https://doi.org/10.1007/978-3-030-59716-0_19

14. Heinrich, M.P., et al.: Mind: modality independent neighbourhood descriptor for multi-modal deformable registration. Med. Image Anal. **16**(7), 1423–1435 (2012)

15. Hering, A., Häger, S., Moltz, J., Lessmann, N., Heldmann, S., van Ginneken, B.: CNN-based lung CT registration with multiple anatomical constraints. Med. Image Anal., 102139 (2021)

16. Hering, A., et al.: Learn2Reg: comprehensive multi-task medical image registration challenge, dataset and evaluation in the era of deep learning (2021)

17. Hering, A., Murphy, K., van Ginneken, B.: Learn2Reg challenge: CT lung registration - training data, May 2020. https://doi.org/10.5281/zenodo.3835682

18. Hu, X., Kang, M., Huang, W., Scott, M.R., Wiest, R., Reyes, M.: Dual-stream pyramid registration network. In: Shen, D., et al. (eds.) MICCAI 2019. LNCS, vol. 11765, pp. 382–390. Springer, Cham (2019). https://doi.org/10.1007/978-3-030-32245-8_43

19. Isensee, F., Jaeger, P.F., Kohl, S.A., Petersen, J., Maier-Hein, K.H.: NNU-Net: a self-configuring method for deep learning-based biomedical image segmentation. Nat. Methods **18**(2), 203–211 (2021)

20. Liu, X., Qi, C.R., Guibas, L.J.: FlowNet3d: learning scene flow in 3d point clouds. In: Proceedings of the IEEE/CVF Conference on Computer Vision and Pattern Recognition, pp. 529–537 (2019)

21. Mok, T.C.W., Chung, A.C.S.: Conditional deformable image registration with convolutional neural network. In: de Bruijne, M., et al. (eds.) MICCAI 2021. LNCS, vol. 12904, pp. 35–45. Springer, Cham (2021). https://doi.org/10.1007/978-3-030-87202-1_4

22. Mok, T.C.W., Chung, A.C.S.: Large deformation diffeomorphic image registration with Laplacian pyramid networks. In: Martel, A.L., et al. (eds.) MICCAI 2020. LNCS, vol. 12263, pp. 211–221. Springer, Cham (2020). https://doi.org/10.1007/978-3-030-59716-0_21

23. Murphy, K., et al.: Evaluation of registration methods on thoracic CT: the empire10 challenge. IEEE Trans. Med. Imaging **30**(11), 1901–1920 (2011)

24. Newell, A., Yang, K., Deng, J.: Stacked hourglass networks for human pose estimation. In: Leibe, B., Matas, J., Sebe, N., Welling, M. (eds.) ECCV 2016. LNCS, vol. 9912, pp. 483–499. Springer, Cham (2016). https://doi.org/10.1007/978-3-319-46484-8_29

25. Ou, Y., Sotiras, A., Paragios, N., Davatzikos, C.: DRAMMS: deformable registration via attribute matching and mutual-saliency weighting. Med. Image Anal. **15**(4), 622–639 (2011)

26. Sang, Y., Ruan, D.: Scale-adaptive deep network for deformable image registration. Med. Phys. **48**(7), 3815–3826 (2021)

27. Siebert, H., Hansen, L., Heinrich, M.P.: Fast 3d registration with accurate optimisation and little learning for learn2reg 2021 (2021)

28. Sotiras, A., Davatzikos, C., Paragios, N.: Deformable medical image registration: a survey. IEEE Trans. Med. Imaging **32**(7), 1153–1190 (2013)

29. Teed, Z., Deng, J.: RAFT: recurrent all-pairs field transforms for optical flow. In: Vedaldi, A., Bischof, H., Brox, T., Frahm, J.-M. (eds.) ECCV 2020. LNCS, vol. 12347, pp. 402–419. Springer, Cham (2020). https://doi.org/10.1007/978-3-030-58536-5_24

30. de Vos, B.D., et al.: A deep learning framework for unsupervised affine and deformable image registration. Med. Image Anal. **52**, 128–143 (2019)

31. Xu, Z., et al.: Evaluation of six registration methods for the human abdomen on clinically acquired CT. IEEE Trans. Biomed. Eng. **63**(8), 1563–1572 (2016)

32. Zhao, S., Lau, T., Luo, J., Eric, I., Chang, C., Xu, Y.: Unsupervised 3d end-to-end medical image registration with volume tweening network. IEEE J. Biomed. Health Inform. **24**(5), 1394–1404 (2019)

Unsupervised Learning of Diffeomorphic Image Registration via TransMorph

Junyu Chen$^{(\boxtimes)}$, Eric C. Frey, and Yong Du

Russell H. Morgan Department of Radiology and Radiological Science, Johns Hopkins Medical Institutes, Baltimore, MD, USA
jchen245@jhmi.edu

Abstract. In this work, we propose a learning-based framework for unsupervised and end-to-end learning of diffeomorphic image registration. Specifically, the proposed network learns to produce and integrate time-dependent velocity fields in an LDDMM setting. The proposed method guarantees a diffeomorphic transformation and allows the transformation to be easily and accurately inverted. We also showed that, without explicitly imposing a diffeomorphism, the proposed network can provide a significant performance gain while preserving the spatial smoothness in the deformation. The proposed method outperforms the state-of-the-art registration methods on two widely used publicly available datasets, indicating its effectiveness for image registration. The source code of this work is available at: https://bit.ly/3EtYUFN.

Keywords: Image registration · Transformer · Deep neural networks

1 Introduction

Deformable image registration functions by establishing the spatial correspondence between the moving and the fixed images. Traditionally, image registration has been accomplished by optimizing a pair-wise objective function iteratively [3,5,9,18]. Over the last decade, deep learning has emerged as a major area of research in the field of medical image analysis, including registration [4,7,8,12,15,16,20]. Learning-based registration models optimize a global functional for a dataset during training, thereby obviating the time-consuming and computationally expensive per-image optimization during inference.

Diffeomorphic image registration is appealing in many medical imaging applications, owing to its properties like topology preservation and transformation invertibility. A diffeomorphic transformation can be achieved via the time integration of sufficiently smooth time-stationary [1,2,11] or time-dependent velocity fields [3,5]. Almost all existing *end-to-end* learning-based registration models adopt stationary velocity fields because of their ease of implementation and relatively low computational cost [8,15,16]. In this work, however, we demonstrate how time-dependent velocity fields can be efficiently incorporated into an *end-to-end* deep neural network framework, which results in diffeomorphisms (an illustrative example is shown in Fig. 1) and improved registration performance.

A. Hering et al. (Eds.): WBIR 2022, LNCS 13386, pp. 96–102, 2022.
https://doi.org/10.1007/978-3-031-11203-4_11

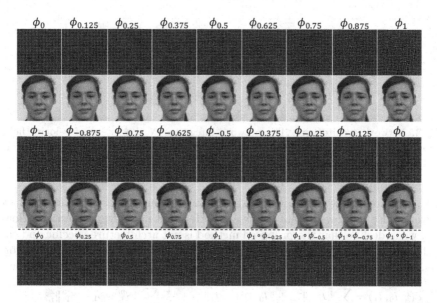

Fig. 1. Inversion and composition of the deformation fields using the proposed method. A neural network learns to generate time-dependent velocity fields for 8 time-steps.

2 Background on LDDMM

In the LDDMM setting [5], the transformation ϕ_t is computed as the flow of a time-dependent velocity field v_t, specified by the ODE: $\frac{d\phi}{dt} = v_t(\phi_t)$ with $t \in [0, 1]$. The final transformation at $t = 1$ is gained by integrating the velocity fields in time: $\phi_1 = \phi_0 + \int_0^1 v_t(\phi_t)dt$ with $\phi_0 = Id$. Then, the optimal transformation is formulated as a variational problem of the form:

$$v^* = \arg\min_v \left(\lambda \int_0^1 \|v_t\|_V^2 dt + \|I_0 \circ \phi_1 - I_1\|_{L^2}^2 \right), \tag{1}$$

where $\| \cdot \|_{L^2}$ denotes the standard L_2-norm, $\|f\|_V = \|Lf\|_{L^2}$ and L is a differential operator of the type $(-\alpha\Delta + \gamma)^\beta Id$ with $\beta > 1.5$, and I_0 and I_1 are the moving and fixed images, respectively. With sufficiently smooth v, a dffieomorphism is guaranteed in this setting.

3 Methods

In this work, a neural network was used to generate velocity fields with a predetermined discretized number of time-steps, specified by N (as shown in Fig. 2). Then, the field integration layer integrates the generated velocity fields to form the transformation at the end-point, i.e., $\phi_1 \approx Id + \sum_{t=1}^{N} v_t \circ \phi_t$, and the inverse transformation ϕ_{-1} is computed as $Id - \sum_{t=1}^{N} v_t \circ \phi_t$. The proposed network may be trained self-supervisedly, end-to-end, using moving and fixed image pairs. We

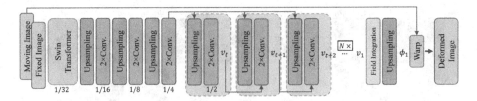

Fig. 2. Network architecture. The network integrates N time-steps of velocity fields to form a final deformation field. Note that skip connections and activation functions were omitted for visualization.

chose our previously developed `TransMorph` [6] (denoted as TM) as the base network since it showed state-of-the-art performance on several datasets. However, we underline that the proposed method is not architecture-specific and can readily be integrated into any architecture. The loss function was derived from Eq. 1 with an additional term to account the available label map information:

$$\mathcal{L}(v, I_0, I_1) = \sum_t \|Lv_t\|_{L^2}^2 + \|I_0 \circ \phi_1 - I_1\|_{L^2}^2 + \frac{1}{M} \sum_m \|S_0^m \circ \phi_1 - S_1^m\|_{L^2}^2, \qquad (2)$$

where S_0 and S_1 denote the M-channel label maps of the moving and fixed images, respectively, where each channel corresponds to the label map of an anatomical structure. We denote the model trained using this loss function as TM-TVF$_{LDDMM}$.

As a consequence of imposing a diffeomorphic transformation, excessive regularization may lead to a suboptimal registration accuracy measured by image similarity or segmentation overlap. Here, we demonstrate that by integrating time-dependent velocity fields, we could implicitly enforce transformation smoothness and improve performance without explicitly imposing a diffeomorphism. In this setting, we used a diffusion regularizer to regularize *only* the velocity field at the end-point:

$$\mathcal{L}(v, I_0, I_1) = \|\nabla v_1\|_{L^2}^2 + NCC(I_0 \circ \phi_1, I_1) + Dice(S_0 \circ \phi_1, S_1), \qquad (3)$$

where ∇v is the spatial gradient operator applied to v, $NCC(\cdot)$ denotes normalized cross-correlation, and $Dice(\cdot)$ denotes Dice loss. We denote the model trained using this loss function as TM-TVF.

4   Experiments and Results

We validated the proposed method using two publicly available datasets, one in 2D and one in 3D. The 2D dataset is the Radboud Faces Database (RaFD) [13], and it comprises eight distinct facial expression images for each of 67 subjects. We randomly divided the subjects into 53, 7, and 7 subjects, and used face images of subjects glancing in the direction of the camera. A total of 2968, 392, and 392 image pairs were used for training, validation, and testing. The images were cropped then resized into 256×256. The 3D dataset is the OASIS dataset

Table 1. SSIM [19] and FSIM [21] comparisons between the proposed method and the others on the RaFD dataset.

	VM-2 [4]	VM-diff [8]	CycleMorph [12]	TM [6]	TM-TVF$_{LDDMM}$	TM-TVF
SSIM↑	0.858 ± 0.038	0.805 ± 0.044	0.875 ± 0.038	0.899 ± 0.035	0.829 ± 0.049	**0.910 ± 0.028**
FSIM↑	0.669 ± 0.039	0.613 ± 0.041	0.687 ± 0.042	0.716 ± 0.043	0.620 ± 0.053	**0.734 ± 0.033**
% of $|J_\phi| \le 0$ ↓	0.798 ± 0.812	<0.001	0.092 ± 0.163	0.190 ± 0.194	<0.001	0.062 ± 0.107
SDlogJ↓	0.086 ± 0.022	0.051 ± 0.011	0.059 ± 0.014	0.065 ± 0.016	**0.046 ± 0.010**	0.057 ± 0.013

Table 2. Validation and test results for the OASIS dataset from the 2021 Learn2Reg challenge [10]. The validation results came from the challenge's leaderboard, whereas the test results came directly from the challenge's organizers.

	Validation			Test		
	Dice↑	SDlogJ↓	HdDist95↓	Dice↑	SDlogJ↓	HdDist95↓
ConvexAdam [17]	0.846 ± 0.016	**0.067 ± 0.005**	1.500 ± 0.304	0.81	**0.07**	1.63
LapIRN [16]	0.861 ± 0.015	0.072 ± 0.007	1.514 ± 0.337	0.82	**0.07**	1.67
TM [6]	0.862 ± 0.014	0.128 ± 0.021	1.431 ± 0.282	0.820	0.124	1.656
TM-TVF$_{LDDMM}$	0.833 ± 0.016	0.090 ± 0.005	1.630 ± 0.353	–	–	–
TM-TVF	**0.869 ± 0.014**	0.094 ± 0.018	**1.396 ± 0.297**	**0.824**	0.090	**1.633**

[14] obtained from the 2021 Learn2Reg challenge [10]. This dataset comprises a total of 451 brain T2 MRI images, with 394, 19, and 38 images being used for training, validation, and testing, respectively. We trained the proposed method for 500 epochs using a learning rate of $1e^{-4}$. The number of time-steps, N, was empirically set to 8. We set $\alpha = 0.01$, $\gamma = 0.01$, and $\beta = 2$ for RaFD dataset, and $\alpha = 0.01$, $\gamma = 0.001$, and $\beta = 2$ for OASIS dataset. Note that due to the absence of segmentation in the RaFD dataset, the segmentation losses in Eqs. 1 and 2 were omitted.

Table 1 and 2 show quantitative results of the proposed models on the RaFD and OASIS datasets. On both datasets, the proposed TM-TVF yielded the highest performance against all other methods, including the first-ranking method (LapIRN [16]) from the Learn2Reg challenge. Specifically, TM-TVF outperformed its base network TM in image similarity and segmentation overlap on the two datasets, with p values < 0.0001 from paired t-tests. Although, a diffeomorphism was not explicitly guaranteed in TM-TVF, it still produced much smoother transformations than TM and VM measured by SDlogJ and the percentage of non-positive Jacobian determinant. On the other hand, although TM-TVF$_{LDDMM}$ guarantees a diffeomorphic transformation (as shown in Fig. 1, 3, and 4), it results in relatively poor registration performance, which is most likely owing to the excessive regularization imposed on the transformation.

5   Conclusion

In conclusion, we have proposed a learning-based framework for learning to generate time-dependent velocity fields in the LDDMM setting. The quantitative results show that the framework outperformed state-of-the-art registration models, indicating the effectiveness of the proposed method. Moreover, the proposed method is not architecture-specific and may be easily incorporated to improve registration performance in any network architecture.

Acknowledgment. This work was supported by a grant from the National Cancer Institute, U01-CA140204.

Appendix A. Additional Qualitative Results

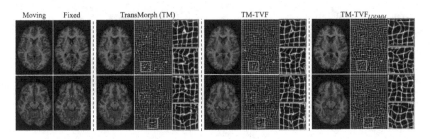

Fig. 3. Qualitative comparisons of the deformation field smoothness. TM yielded a deformation field with noticeable folded voxels, but TM-TVF generated a smoother field with state-of-the-art registration accuracy (as seen in Tables 1 and 2). TM-TVF$_{LDDMM}$ generated a highly regularized deformation field with nearly no visible folded voxels.

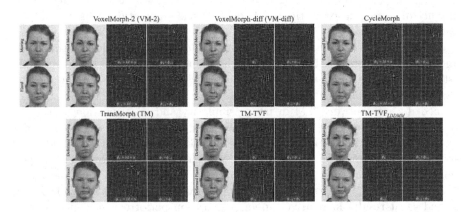

Fig. 4. Qualitative comparisons of facial expression registration. TM-TVF$_{LDDMM}$ produced a smooth and invertible transformation, but all other transformations were not. Additionally, TM-TVF yielded the best qualitative results for both forward and backward registration. Note that the transformation inversions for VM-2, CycleMorph, and TM were approximated using $Id - u$, where u denotes the displacement field.

References

1. Arsigny, V., Commowick, O., Pennec, X., Ayache, N.: A log-euclidean framework for statistics on diffeomorphisms. In: Larsen, R., Nielsen, M., Sporring, J. (eds.) MICCAI 2006. LNCS, vol. 4190, pp. 924–931. Springer, Heidelberg (2006). https://doi.org/10.1007/11866565_113

2. Ashburner, J.: A fast diffeomorphic image registration algorithm. Neuroimage **38**(1), 95–113 (2007)

3. Avants, B.B., Epstein, C.L., Grossman, M., Gee, J.C.: Symmetric diffeomorphic image registration with cross-correlation: evaluating automated labeling of elderly and neurodegenerative brain. Med. Image Anal. **12**(1), 26–41 (2008)

4. Balakrishnan, G., Zhao, A., Sabuncu, M.R., Guttag, J., Dalca, A.V.: Voxelmorph: a learning framework for deformable medical image registration. IEEE Trans. Med. Imaging **38**(8), 1788–1800 (2019)

5. Beg, M.F., Miller, M.I., Trouvé, A., Younes, L.: Computing large deformation metric mappings via geodesic flows of diffeomorphisms. Int. J. Comput. Vision **61**(2), 139–157 (2005)

6. Chen, J., Frey, E.C., He, Y., Segars, W.P., Li, Y., Du, Y.: Transmorph: transformer for unsupervised medical image registration (2021). https://arxiv.org/abs/2111.10480

7. Chen, J., He, Y., Frey, E.C., Li, Y., Du, Y.: ViT-V-Net: vision transformer for unsupervised volumetric medical image registration. In: Medical Imaging with Deep Learning (2021)

8. Dalca, A.V., Balakrishnan, G., Guttag, J., Sabuncu, M.R.: Unsupervised learning of probabilistic diffeomorphic registration for images and surfaces. Med. Image Anal. **57**, 226–236 (2019)

9. Heinrich, M.P., Jenkinson, M., Brady, M., Schnabel, J.A.: MRF-based deformable registration and ventilation estimation of lung CT. IEEE Trans. Med. Imaging **32**(7), 1239–1248 (2013)

10. Hering, A., et al.: Learn2reg: comprehensive multi-task medical image registration challenge, dataset and evaluation in the era of deep learning. arXiv preprint arXiv:2112.04489 (2021)

11. Hernandez, M., Bossa, M.N., Olmos, S.: Registration of anatomical images using paths of diffeomorphisms parameterized with stationary vector field flows. Int. J. Comput. Vision **85**(3), 291–306 (2009)

12. Kim, B., et al.: CycleMorph: cycle consistent unsupervised deformable image registration. Med. Image Anal. **71**, 102036 (2021)

13. Langner, O., et al.: Presentation and validation of the radboud faces database. Cogn. Emot. **24**(8), 1377–1388 (2010)

14. Marcus, D.S., Wang, T.H., Parker, J., Csernansky, J.G., Morris, J.C., Buckner, R.L.: Open access series of imaging studies (OASIS): cross-sectional MRI data in young, middle aged, nondemented, and demented older adults. J. Cogn. Neurosci. **19**(9), 1498–1507 (2007)

15. Mok, T.C., Chung, A.: Fast symmetric diffeomorphic image registration with convolutional neural networks. In: Proceedings of the IEEE/CVF Conference on Computer Vision and Pattern Recognition, pp. 4644–4653 (2020)

16. Mok, T.C.W., Chung, A.C.S.: Conditional deformable image registration with convolutional neural network. In: de Bruijne, M., et al. (eds.) MICCAI 2021. LNCS, vol. 12904, pp. 35–45. Springer, Cham (2021). https://doi.org/10.1007/978-3-030-87202-1_4

17. Siebert, H., Hansen, L., Heinrich, M.P.: Fast 3d registration with accurate optimisation and little learning for learn2reg 2021. arXiv preprint arXiv:2112.03053 (2021)
18. Vercauteren, T., Pennec, X., Perchant, A., Ayache, N.: Diffeomorphic demons: efficient non-parametric image registration. Neuroimage 45(1), S61–S72 (2009)
19. Wang, Z., Bovik, A.C., Sheikh, H.R., Simoncelli, E.P.: Image quality assessment: from error visibility to structural similarity. IEEE Trans. Image Process. 13(4), 600–612 (2004)
20. Yang, X., Kwitt, R., Styner, M., Niethammer, M.: Quicksilver: fast predictive image registration-a deep learning approach. Neuroimage 158, 378–396 (2017)
21. Zhang, L., Zhang, L., Mou, X., Zhang, D.: FSIM: a feature similarity index for image quality assessment. IEEE Trans. Image Process. 20(8), 2378–2386 (2011)

SuperWarp: Supervised Learning and Warping on U-Net for Invariant Subvoxel-Precise Registration

Sean I. Young[1,2(✉)], Yaël Balbastre[1,2], Adrian V. Dalca[1,2], William M. Wells[1,2], Juan Eugenio Iglesias[1,2], and Bruce Fischl[1,2]

[1] MGH/HST Martinos Center for Biomedical Imaging, Boston, USA
{siyoung,adalca}@mit.edu, ybalbastre@mgh.harvard.edu
[2] Massachusetts Institute of Technology, Cambridge, USA

Abstract. In recent years, learning-based image registration methods have gradually moved away from direct supervision with target warps to self-supervision using segmentations, producing promising results across several benchmarks. In this paper, we argue that the relative failure of supervised registration approaches can in part be blamed on the use of regular U-Nets, which are jointly tasked with feature extraction, feature matching, and estimation of deformation. We introduce one simple but crucial modification to the U-Net that disentangles feature extraction and matching from deformation prediction, allowing the U-Net to warp the features, across levels, as the deformation field is evolved. With this modification, direct supervision using target warps begins to outperform self-supervision approaches that require segmentations, presenting new directions for registration when images do not have segmentations. We hope that our findings in this preliminary workshop paper will re-ignite research interest in supervised image registration techniques. Our code is publicly available from https://github.com/bal basty/superwarp.

Keywords: Image registration · Optical flow · Supervised learning

1 Introduction

In recent years, fully convolutional networks (FCNs) have become a universal framework for tackling an array of problems in medical imaging, ranging from image denoising and super-resolution [1, 2] to semantic segmentation [3–5], and from style transfer [6, 7] to image registration. Among these, image registration methods have benefitted immensely from FCNs, allowing methods to transition from an optimization-based paradigm to a learning-based one and to accelerate the alignment of images with different contrasts or modalities, for example.

An overwhelming majority of recent image registration networks [8–11] are trained unsupervised, in the sense that ground-truth deformation fields are not required in the supervision of these networks. Instead, a surrogate photometric loss is used to maximize the similarity between the fixed image and the moving one—warped by the predicted

© The Author(s), under exclusive license to Springer Nature Switzerland AG 2022
A. Hering et al. (Eds.): WBIR 2022, LNCS 13386, pp. 103–115, 2022.
https://doi.org/10.1007/978-3-031-11203-4_12

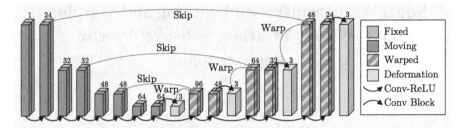

Fig. 1. The SuperWarp U-Net for image registration (first four levels shown). The fixed and moving images are concatenated along the batch axis and processed through the network. The two image features are reconcatenated along the channel axis at each level of the U-Net's upward path to be processed into a residual deformation, used to warp the moving features, and scaled and summed to produce the final deformation.

deformation field—in lieu of a loss that penalizes the differences between the predicted and ground-truth deformation fields. Since images typically contain large untextured regions as well as different contrasts and voxel intensities, merely minimizing differences in the fixed image and the moving one is insufficient to recover the ground truth deformation, even when a smoothness prior (or regularization) is imposed on the predicted deformation field. While supervised [12–14] and self-supervised approaches [9, 10]—based on segmentations, for example—produce excellent results, direct supervision using target warps is still desirable in many cases especially if the images do not have segmentations. However, supervised registration has not been as successful for many applications due to severe optimization difficulties faced—the network is jointly tasked with feature extraction and matching in addition to deformation estimation, which is not handled well by a fully convolutional network.

In this work, we will propose SuperWarp, a supervised learning approach to medical image registration. We first re-visit the classic optical flow equation of Horn and Schunck [15] to analyze its implications for supervised registration—the duality of intensity-invariant feature extraction and deformation estimation and the need for multi-scale warping. With such implications in mind, we make one simple but critical modification to the segmentation U-Net that repurposes it for subvoxel- (or subpixel-) accurate supervised image registration. With this modification, direct supervision using target warps outperforms self-supervised registration requiring segmentations. The network, shown in Fig. 1, is strikingly similar to a segmentation U-Net except for warping and deformation extraction layers, allowing U-Net to warp the features as the deformation field is evolved.

2   Related Work

SuperWarp is heavily inspired by previous work on optical flow estimation, the aim of which is to recover apparent motion from an image pair [15]. Expressed mathematically, however, optical flow estimation and image registration are in fact identical problems possibly except for the notion of regularity in each—an optical flow field for the former is typically assumed differentiable a.e. whereas a deformation field for the latter infinitely differentiable or diffeomorphic. This subtle distinction between the two problems does however disappear under the supervised learning paradigm since the type of regularity desired is reflected in the ground-truth optical flow (or deformation) fields of the training data.

2.1 Optical Flow Estimation

Here, we briefly recap development in classical and learning-based optical flow estimation methods—see e.g. [16] for a review. In their seminal work, Horn and Schunck [15] formulated optical flow estimation via a regularized optimization problem, noting that the problem is generally ill-posed in the absence of local smoothness priors. Several works extend the original Horn–Schunck model [15] using sub-quadratic regularization and data fidelity terms [17–21] that mitigate the deleterious effects of occlusions on flow estimation. Oriented regularization terms [21–25] regularize the flow only along the direction tangent to the image gradient while non-local terms [26–29] regularize flow even across disconnected pixels subject to similar motion. Median filtering of intermediate flows [23, 26] achieves similar effects to non-local regularity terms. Higher-order regularizers [28, 30] assign zero penalty to affine trends in the flow to encourage piecewise-linear flow predictions. Despite the advances, designing a regularizer is highly domain-specific, suggesting that it can be alleviated via supervised learning.

Orthogonally to the choice of regularizers, multi-scale schemes [31–34] have been used to estimate larger flows. Descriptor matching [31, 32] introduces an extra data fidelity term that penalizes misalignment of scale-invariant features (e.g. SIFT), overcoming the deterioration of the conventional data fidelity term at large scales due to the loss of small image structures. Since the optical flow equation no longer holds in the presence of a global brightness change, several authors propose to attenuate the brightness component of the images as a first step using high-pass filters [18, 24, 35], structure-texture decomposition [27, 36] or color space transforms [24]. Thus, in traditional approaches, both multi-scale processing and brightness-invariant transforms require us to handcraft suitable pre-processing filters, which can be highly time-consuming owing to the image-dependent nature of such filters. As we will see, the U-Net architecture used in the SuperWarp obviates the need to handcraft such filters, allowing the U-Net to learn them directly from the training data, end-to-end, to enable brightness-invariant image registration with exceptional generalization ability.

Fischer et al. [37] formulate optical flow estimation as a supervised learning problem. They train a U-Net model to output the optical flow field directly for a pair of input images, supervising the training using the ground-truth optical flow field as the target. Later works extend [37], cascading multiple instances of the network with warping [38], introducing a warping layer [39] or using a fixed image pyramid [40] to improve the accuracy of flow prediction [38, 39] as well as reduce the model size. Some authors propose to tackle optical flow estimation as an unsupervised learning task [41, 42], using a photometric loss to penalize the intensity differences across the fixed and moved images. Recent extensions in this unsupervised direction include occlusion-robust losses [43, 44] based on forward-backward consistency, and self-supervision losses [42, 45]. These are also the building blocks of unsupervised image registration methods [9–11].

3   Mathematical Framework

3.1   Optical Flow Estimation and Duality Principle

Under a sufficiently high temporal sampling rate, we can relate the intensities of a successive pair of three-dimensional images (f_0, f_1) to components (u, v, w) of the displacement between the two images using the optical flow equation

$$(\partial f_1 / \partial x) \cdot u + (\partial f_1 / \partial y) \cdot v + (\partial f_1 / \partial z) \cdot w = f_0 - f_1 \qquad (1)$$

[15], where $(\partial/\partial x, \partial/\partial y, \partial/\partial z)$ denotes the 3D spatial gradient operator. PDE (1) can also be seen as a linearization of the small deformation model in image registration [46]. Since (1) involves three unknowns for every equation, finding (u, v, w) given (f_0, f_1) is an ill-posed inverse problem. Smoothness assumptions are therefore made in optimization-based flow estimation [17–21] to render the inverse problem well-posed again similar to image registration [9–11].

A global change in the brightness or contrast across the image pair (f_0, f_1) introduces an additive bias in the right-hand side of (1) such that the equation no longer holds. Compensating for this change in pre-processing would require knowledge of the displacement field (u, v, w) that we seek in the first place. A similar issue is often met in medical image registration, with different imaging modalities across f_0 and f_1 injecting additive and multiplicative biases in (1). If however we knew the ground-truth displacement (u, v, w), harmonizing (f_0, f_1) in a normalized intensity space is readily achieved via (1). Conversely, given a harmonized image pair, the displacement field can be recovered using (1).

Image segmentation [47] is the ultimate form of image harmonization, since it removes brightness and contrast from images altogether and turns them into piecewise smooth (constant) signals by construction. This suggests that the use segmentation maps to supervise registration [9, 10] can be beneficial. However, many types of images do not have segmentations available or lack the notion of segmentation altogether, e.g. fMRI activations, so supervision using the ground-truth warps instead can be an expedient way of learning to register.

In practice, images (f_0, f_1) are acquired at a low temporal sampling rate so (1) holds only over regions where both image intensities are linear functions of their spatial coordinates [15]. Equivalently, (1) holds in the general case only if the magnitudes of the components (u, v, w) are less than one voxel. Since this can pose a major limitation for practical applications, multi-scale processing is used to linearize the images at gradually smaller scales, with the displacement field estimated at the larger scale used to initialize the residual flow estimation at the smaller scale. Linearizing images at larger scales, however, results in the loss of small structures due to the smoothing filters. Handcrafting filters that have an optimum tradeoff between linearization and preservation of image features at every scale is image-dependent and can be time-consuming, implying that learning such filters end-to-end can be beneficial for generalization ability.

3.2   Supervised Learning and Multi-scale Warping

SuperWarp exploits the duality principle (1) to supervise an image registration network equipped with multi-scale warping to estimate large deformations. We train a U-Net model on pairs of images with different intensities related via our smoothly synthesized ground-truth deformation fields. The downward path of the U-Net model first extracts intensity-invariant features from the two images separately. The upward path then extracts from the feature pair a deformation that minimizes the differences with respect to the ground-truth target.

SuperWarp makes one important modification to the registration U-Net for large displacement estimation. At each level of the network's upward path, the features of the moving image are first warped using the deformation field from the previous level, such that only the residual deformation, less than a voxel in magnitude, need be extracted at the current level. Processing the two images jointly as a single multi-channel image through U-Net, as done in [9, 10], would entangle the features of the fixed and the moving images, so that warping only the features of the moving one post hoc is not feasible. Instead, we process the two images as a batch with the image pair interacting only during deformation extraction, where the two image features are reconcatenated along the channels axis and processed into deformation field using a convolution block.

Note from the left and the right-hand sides of (1) that it is $(\mathbf{f}_1, \mathbf{f}_0 - \mathbf{f}_1)$, not $(\mathbf{f}_1, \mathbf{f}_0)$, which needs to be processed for displacement estimation. This suggests that feeding the features of $\mathbf{f}_1$ and pre-computed feature differences between $\mathbf{f}_0$ and (warped) $\mathbf{f}_1$ into deformation blocks can yield a saving of one convolution layer per block, which is substantial given that these blocks typically have no more than three convolution layers in total. In practice, we reparameterize the input further to the features of $(\mathbf{f}_0 + \mathbf{f}_1, \mathbf{f}_0 - \mathbf{f}_1)$ to help the extraction blocks average the spatial derivatives of the features across the two images, similarly to the practice in optimization-based approaches [16]. This reparameterization can be seen as a Hadamard transform [48] across the two image feature sets.

Table 1. Parameter ranges and probabilities used for random spatial transformation and intensity augmentation of the image pair. Applied separately to each image in the pair.

	Spatial Transformation					Intensity Augmentation			
	Translate	Scale	Rotate	Shear	Elastic	Noise Std	Multiply	Contrast	Gamma
Range	±12	[0.75,1.25]	±30°	±0.012	±4 (256²)	[0,0.05]	[0.75,1.25]	[0.75,1.25]	[0.70,1.50]
Prob.	1.0	1.0	1.0	1.0	1.0	0.5	0.5	0.5	0.5

3.3 Deep Supervision, Data Augmentation and Training

Following the approach of deep supervision for semantic segmentation [49], we supervise the deformation block at each level of the U-Net's upward path with a displacement target. We use the MSE loss between the predicted $(\mathbf{u}, \mathbf{v}, \mathbf{w})$ and the target $(\mathbf{p}, \mathbf{q}, \mathbf{r})$ to minimize $E(\mathbf{u}, \mathbf{v}, \mathbf{w}) = \|(\mathbf{u}, \mathbf{v}, \mathbf{w}) - (\mathbf{p}, \mathbf{q}, \mathbf{r})\|_2^2$. The loss is summed across levels without weighting to produce the final training loss. The deformation block at each level is supervised using the target ground-truth field down-sampled to the spatial dimensions of its predictions. For evaluation, more forgiving EPE loss $E_{\text{EPE}}(\mathbf{u}, \mathbf{v}, \mathbf{w}) = \|(\mathbf{u}, \mathbf{v}, \mathbf{w}) - (\mathbf{p}, \mathbf{q}, \mathbf{r})\|_{2,1}$ is used instead.

To generate training pairs of images with their corresponding ground-truth deformation targets, we sample an image $\mathbf{f}$ from the training set and synthesize two different smooth displacements $(\mathbf{p}_0, \mathbf{q}_0, \mathbf{r}_0)$ and $(\mathbf{p}_1, \mathbf{q}_1, \mathbf{r}_1)$ that warp $\mathbf{f}$ and produce $\mathbf{f}_0$ and $\mathbf{f}_1$, respectively. The ground-truth displacement is given by

$$(\mathbf{p}, \mathbf{q}, \mathbf{r}) = (\mathbf{Id} + (\mathbf{p}_1, \mathbf{q}_1, \mathbf{r}_1))^{-1}(\mathbf{Id} + (\mathbf{p}_0, \mathbf{q}_0, \mathbf{r}_0)) - \mathbf{Id}, \qquad (2)$$

in which the identity $\mathbf{Id}$ denotes the (vectorization) of the grid coordinates. To facilitate computation, we restrict $(\mathbf{p}_1, \mathbf{q}_1, \mathbf{r}_1)$ to affine fields so that the inverse coordinate mapping $(\cdot)^{-1}$ (2) can be computed by inverting a 4×4 matrix. We apply a small elastic deformation on $(\mathbf{p}_0, \mathbf{q}_0, \mathbf{r}_0)$ to approximate a higher-order (non-affine) component of the spatial distortion typically seen in MR scans. We then transform the voxel intensities of $\mathbf{f}_0$ and $\mathbf{f}_1$ using a standard augmentation pipeline (Gaussian noise, brightness multiplication, contrast augmentation, and gamma transform); see Table 1 for the hyperparameters of these transforms.

For training, we use a batch size of 1, which actually becomes 2 because the moving and fixed images are concatenated along the batch axis. The Adam [50] optimizer is used with an initial learning rate of 10^{-4}, linearly reduced to 10^{-6} across 200,000 iterations. We find it beneficial to initially train the network for 20,000 iterations on training examples with zero displacement and deformation but still with intensity augmentations to enable the network to learn to extract contrast-invariant features, then introducing deformations to train the network to predict deformations with brightness change across the image pair.

In Fig. 2, we plot validation Dice and end-point error curves of SuperWarp U-Net (ours) and a VoxelMorph-like U-Net baseline for the registration of MR brain scans. In the Dice-supervised case, we train the networks to minimize the regularized Dice loss between the segmentations of the fixed and moving images

$$E_{\text{Dice}}(\mathbf{u}, \mathbf{v}, \mathbf{w}) = D_{\text{Dice}}(\mathbf{f}_1 \circ (\mathbf{Id} + (\mathbf{u}, \mathbf{v}, \mathbf{w})), \mathbf{f}_0) + R(\mathbf{u}, \mathbf{v}, \mathbf{w}), \qquad (3)$$

in which R penalizes the (squared) Laplacian of the components $\mathbf{u}, \mathbf{v}, \mathbf{w}$. In the MSE-supervised case, the same networks are trained to minimize the MSE in the predicted and target deformations. Regardless of the training objective, our SuperWarp U-Net outperforms the baseline U-Net and also trains significantly faster, requiring only 20 iterations to reach maximum accuracy in the case where the Dice loss is used. Moreover, SuperWarp U-Net trained using the MSE loss (no segmentations) outperforms the Dice baseline requiring segmentations.

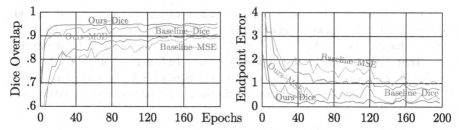

Fig. 2. Validation registration accuracy. In both self-supervised (MSE) and supervised (Dice) cases, the SuperWarp U-Net leads to better mean Dice and endpoint error than the baseline (similar to VoxelMorph) and trains faster, requiring only 40 epochs to reach the final accuracy, which are 0.954, 0.152 (Ours–Dice), 0.906, 0.711 (Baseline–Dice).

4 Experimental Evaluation

We validate our proposed SuperWarp approaches on two datasets—a set of 2D brain magnetic resonance (MR) scans, as well as the Flying Chairs [37] optical flow dataset widely used in computer vision. The brain image registration task allows us to benchmark the performance of SuperWarp against related work in medical image registration [9, 10] while Flying Chairs allows us to compare the SuperWarp U-Net with the state-of-the-art optical flow estimation networks. In addition to Dice scores between fixed and moved images, we also use the mean EPE to evaluate the accuracy of the displacements. All U-Nets have 7 levels of [24, 32, 48, 64, 96, 128, 192] features and two convolution layers at each level.

4.1 Invariant Registration of Brain MR Images

Here, we apply SuperWarp to deformable registration of 2D brain scans within a subject. Obviously, SuperWarp could be applied to the cross-subject setting too but the accuracy of predicted deformations is easier to assess in the within-subject case and facilitates comparisons with other methods. We use the whole brain dataset of [51] containing 40 T1-weighted brain MR scans, along with the corresponding segmentations produced using FreeSurfer [51]. For test, we use a collection of 500 T1-weighted brain MR scans curated from: OASIS, ABIDE-I and -II, ADHD, COBRE, GSP, MCIC, PPMI, and UK Bio. The scan pairs are generated as described in Sect. 3.3. We do not perform linear registration of the images as a preprocessing step in any of the methods since the displacements are rather small (Table 1) and this provides better insights into their behavior.

To show the improvement in the accuracy of the deformation field recovered using our methods, we plot statistics of the validation end-point error and Dice scores produced by all methods including the baseline—similar to VoxelMorph [9]—in Fig. 3. While our Dice scores are higher than those of the baseline only by 0.04, our end-point errors are more significantly reduced from the baselines (by 80%, on average across, foreground pixels). Figure 4 shows the displacements predicted by our method, comparing them with those from the baseline.

To better understand the sources of improvement between the baseline and our approach, we conduct an extensive set of ablation studies on SuperWarp as listed in Table
2. We see that the multi-scale loss used in [49] can actually hurt accuracy for this experiment. Training with the EPE loss produces a worse EPE than training with the MSE
loss likely due to numerical instability at zero. The number of U-Net levels should also
be high enough (seven) to cover the largest displacements (about ± 64) at the coarsest
level of the U-Net.

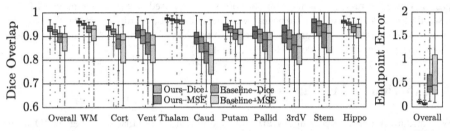

Fig. 3. Test Dice (left) and endpoint error (right) statistics on 10 structures across 500 T1w brain
images. Regardless of the choice of the training loss function, the SuperWarp produces better Dice
and endpoint error than the baseline (similar to VoxelMorph). Note that Ours–MSE does not need
or use segmentation information.

4.2   Optical Flow Estimation

To further benchmark the network architecture used by the SuperWarp, we run additional experiments on the Flying Chairs optical flow dataset [37], popularly used by
the computer vision community. To facilitate a fair comparison, we set our network
and training hyperparameters very similarly to [39]: 7 U-Net levels for a total of 6.9M
learnable parameters, 1M steps, EPE loss for training, multi-scale loss (but weight all
scales equally) with the Adam optimizer. Table 3 lists the validation EPE of flow fields
predicted using the SuperWarp and other well-performing models.

Both PWC-Net [39] and FlowNet-C [37] attribute their good performance to the
use of the cost-volume layer but we find cost volumes to be unnecessary to achieve
a good accuracy at least on this dataset. While SPY-Net [40] also uses a multi-scale
warping strategy, it is based on a fixed image pyramid. This helps to bring down the
number of trainable parameters but can also lead to a loss of image structure at coarser
levels. FlowNet2 cascades multiple FlowNet models and warps in between, while the
SuperWarp U-Net incorporates warps directly in the model, significantly reducing the
model size with a comparable accuracy.

5   Discussion

In this paper we have shown that supervising an image registration network with a target warp can achieve state-of-the-art accuracy. Our approach outperforms previous supervised ones due to the multi-scale nature of our prediction, where deformations are composed across the upward path of the U-Net and applied to the features of the moving image. This way, each spatial scale receives a moving image as input that has

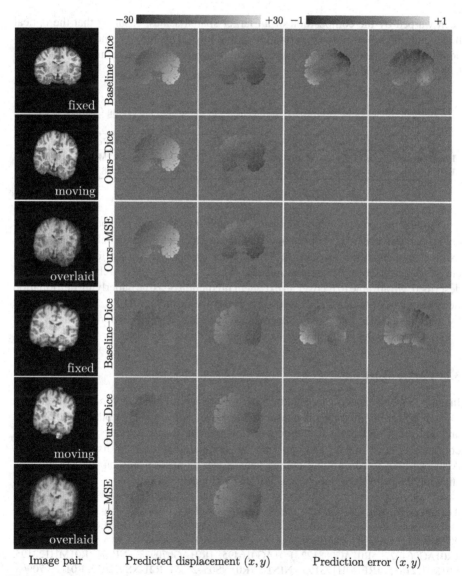

Fig. 4. Visualization of predicted displacement fields (test). Both Dice and MSE variants of the SuperWarp can produce highly accurate displacements (in the first example, 0.08 and 0.06 mm, respectively) whereas the baseline (similar to VoxelMorph) prediction has larger errors (0.41 mm). Images are 2D, 256 × 256, 1 mm isotropic. Cf. Fig. 3 (right).

Table 2. Ablation of network and training hyperparameters used and their influence on the best epoch validation accuracy. Default hyperparameter: (7, MSE, False, True).

	Number of levels		Training loss function			Multi-scale loss		Multi-scale warp	
	6	7	Dice	EPE	MSE	True	False	True	False
Dice	0.927	**0.947**	**0.954**	0.942	0.947	0.939	**0.947**	**0.947**	0.903
EPE	0.450	**0.122**	**0.103**	0.195	0.122	0.270	**0.122**	**0.122**	0.738

been warped by the composition of all larger spatial scales, ensuring that the optical flow condition holds for the deformation at that level. This recovers the accuracy of the deformation estimation that was likely lost in previous supervised techniques due to the lack of multi-scale warping.

Table 3. Mean EPE achieved by various network models on the Flying Chairs test set.

	PWC-Net	SPY-Net	FlowNetS	FlowNetC	FlowNet2	Ours–EPE
Parameters	8.75M	**1.20M**	32.1M	32.6M	64.2M	6.9M
EPE	2.00	2.63	2.71	2.19	**1.78**	1.82

While segmentation accuracy is itself of course important, we also point out that there are instances in which it is important to recover an exact deformation field. In these cases, using a segmentation loss leads to inaccuracies when there are too few segmentation classes to guide thedeformation estimation. We show that using the architecture we have described, we are able to recover an excellent prediction of a true underlying deformation field. Uses cases include distortion estimation and removal in MRI, such as those caused by inhomogeneities in the main magnetic field (B0) and image distortions induced by nonlinearities in the gradient coils used to encode spatial location.

5.1 Future Work

In this workshop paper, we have addressed only one type of invariance, namely invariance to intensity (or illumination) change across images. In the sequel, we plan to add contrast and distortion invariance to the network by training it on synthetic scans of various contrasts as done in [52] and applying the synthetic approach to distortions as well. Also, we plan to run a more comprehensive set of experiments on 3D MR images, showing the benefits of our approach in many clinical applications.

Acknowledgments. Support for this research provided in part by the BRAIN Initiative Cell Census Network grant U01MH117023, NIBIB (P41EB015896, 1R01EB023281, R01EB-006758, R21EB018907, R01EB019956, P41EB030006, P41EB028741), NIA (1R-56AG064027, 1R01AG064027, 5R01AG008122, R01AG016495, 1R01AG070988), NIMH (R01MH123195, R01MH121885, 1RF1MH123195), NINDS (R01NS05–25851, R21-NS072652, R01NS070963, R01NS083534, 5U01NS086625, 5U24NS-10059103, R01NS105820), ARUK (IRG2019A-003), and was made possible by resources from Shared Instrumentation Grants 1S10RR023401,

1S10RR019307, and 1S10-RR023043. Additional support was provided by the NIH Blueprint for Neuroscience Research (5U01-MH093765), part of the multi-institutional Human Connectome Project.

References

1. Zhang, K., Zuo, W., Chen, Y., Meng, D., Zhang, L.: Beyond a Gaussian denoiser: residual learning of deep CNN for image denoising. IEEE Trans. Image Process. **26**, 3142–3155 (2017)
2. Dong, C., Loy, C.C., He, K., Tang, X.: Image super-resolution using deep convolutional networks. IEEE Trans. Pattern Anal. Mach. Intell. **38**, 295–307 (2016)
3. Ronneberger, O., Fischer, P., Brox, T.: U-Net: convolutional networks for biomedical image segmentation. In: Navab, N., Hornegger, J., Wells, W.M., Frangi, A.F. (eds.) MICCAI 2015. LNCS, vol. 9351, pp. 234–241. Springer, Cham (2015). https://doi.org/10.1007/978-3-319-24574-4_28
4. Long, J., Shelhamer, E., Darrell, T.: Fully convolutional networks for semantic segmentation. In: Proceedings of CVPR, pp. 3431–3440 (2015)
5. Noh, H., Hong, S., Han, B.: Learning deconvolution network for semantic segmentation. In: Proceedings of CVPR (2015)
6. Gatys, L.A., Ecker, A.S., Bethge, M.: Image style transfer using convolutional neural networks. In: Proceedings of CVPR, pp. 2414–2423 (2016)
7. Johnson, J., Alahi, A., Fei-Fei, L.: Perceptual losses for real-time style transfer and super-resolution. In: Leibe, B., Matas, J., Sebe, N., Welling, M. (eds.) ECCV 2016. LNCS, vol. 9906, pp. 694–711. Springer, Cham (2016). https://doi.org/10.1007/978-3-319-46475-6_43
8. de Vos, B.D., Berendsen, F.F., Viergever, M.A., Staring, M., Išgum, I.: End-to-end unsupervised deformable image registration with a convolutional neural network. In: Cardoso, M.J., et al. (eds.) DLMIA/ML-CDS -2017. LNCS, vol. 10553, pp. 204–212. Springer, Cham (2017). https://doi.org/10.1007/978-3-319-67558-9_24
9. Balakrishnan, G., Zhao, A., Sabuncu, M.R., Guttag, J., Dalca, A.V.: An unsupervised learning model for deformable medical image registration. In: Proceedings of CVPR (2018)
10. Dalca, A.V., Balakrishnan, G., Guttag, J., Sabuncu, M.R.: Unsupervised learning for fast probabilistic diffeomorphic registration. In: Frangi, A.F., Schnabel, J.A., Davatzikos, C., Alberola-López, C., Fichtinger, G. (eds.) MICCAI 2018. LNCS, vol. 11070, pp. 729–738. Springer, Cham (2018). https://doi.org/10.1007/978-3-030-00928-1_82
11. Mok, T.C.W., Chung, A.C.S.: Fast symmetric diffeomorphic image registration with convolutional neural networks. In: Proceedings of CVPR (2020)
12. Cao, X., et al.: Deformable image registration based on similarity-steered CNN regression. In: Descoteaux, M., Maier-Hein, L., Franz, A., Jannin, P., Collins, D.L., Duchesne, S. (eds.) MICCAI 2017. LNCS, vol. 10433, pp. 300–308. Springer, Cham (2017). https://doi.org/10.1007/978-3-319-66182-7_35
13. Rohé, M.-M., Datar, M., Heimann, T., Sermesant, M., Pennec, X.: SVF-Net: learning deformable image registration using shape matching. In: Descoteaux, M., Maier-Hein, L., Franz, A., Jannin, P., Collins, D.L., Duchesne, S. (eds.) MICCAI 2017. LNCS, vol. 10433, pp. 266–274. Springer, Cham (2017). https://doi.org/10.1007/978-3-319-66182-7_31
14. Yang, X., Kwitt, R., Styner, M., Niethammer, M.: Quicksilver: fast predictive image registration – a deep learning approach. Neuroimage **158**, 378–396 (2017)
15. Horn, B.K.P., Schunck, B.G.: Determining optical flow. Artif. Intell. **17**, 185–203 (1981)
16. Sun, D., Roth, S., Black, M.J.: A quantitative analysis of current practices in optical flow estimation and the principles behind them. Int. J. Comput. Vis. **106**, 115–137 (2014)

17. Black, M.J., Anandan, P.: The robust estimation of multiple motions: parametric and piecewise-smooth flow fields. Comput. Vis. Image Underst. **63**, 75–104 (1996)
18. Papenberg, N., Bruhn, A., Brox, T., Didas, S., Weickert, J.: Highly accurate optic flow computation with theoretically justified warping. Int. J. Comput. Vis. **67**, 141–158 (2006)
19. Roth, S., Lempitsky, V., Rother, C.: Discrete-continuous optimization for optical flow estimation. In: Cremers, D., Rosenhahn, B., Yuille, A.L., Schmidt, F.R. (eds.) Statistical and Geometrical Approaches to Visual Motion Analysis. LNCS, vol. 5604, pp. 1–22. Springer, Heidelberg (2009). https://doi.org/10.1007/978-3-642-03061-1_1
20. Zach, C., Pock, T., Bischof, H.: A duality based approach for realtime TV-L 1 optical flow. In: Hamprecht, F.A., Schnörr, C., Jähne, B. (eds.) Pattern Recognition, pp. 214–223. Springer, Heidelberg (2007). https://doi.org/10.1007/978-3-540-74936-3_22
21. Sun, D., Roth, S., Lewis, J.P., Black, M.J.: Learning Optical Flow. In: Forsyth, D., Torr, P., Zisserman, A. (eds.) ECCV 2008. LNCS, vol. 5304, pp. 83–97. Springer, Heidelberg (2008). https://doi.org/10.1007/978-3-540-88690-7_7
22. Nagel, H., Enkelmann, W.: An investigation of smoothness constraints for the estimation of displacement vector fields from image sequences. IEEE Trans. Pattern Anal. Mach. Intell. **8**, 565–593 (1986)
23. Wedel, A., Cremers, D., Pock, T., Bischof, H.: Structure- and motion-adaptive regularization for high accuracy optic flow. In: Proceedings of ICCV, pp. 1663–1668 (2009)
24. Zimmer, H., Bruhn, A., Weickert, J.: Optic flow in harmony. Int. J. Comput. Vis. **93**, 368–388 (2011)
25. Zimmer, H., et al.: Complementary optic flow. In: Cremers, D., Boykov, Y., Blake, A., Schmidt, F.R. (eds.) EMMCVPR 2009. LNCS, vol. 5681, pp. 207–220. Springer, Heidelberg (2009). https://doi.org/10.1007/978-3-642-03641-5_16
26. Sun, D., Roth, S., Black, M.J.: Secrets of optical flow estimation and their principles. In: Proceedings of CVPR, pp. 2432–2439 (2010)
27. Werlberger, M., Pock, T., Bischof, H.: Motion estimation with non-local total variation regularization. In: Proceedings of CVPR, pp. 2464–2471 (2010)
28. Ranftl, R., Bredies, K., Pock, T.: Non-local total generalized variation for optical flow estimation. In: Fleet, D., Pajdla, T., Schiele, B., Tuytelaars, T. (eds.) ECCV 2014. LNCS, vol. 8689, pp. 439–454. Springer, Cham (2014). https://doi.org/10.1007/978-3-319-10590-1_29
29. Krähenbühl, P., Koltun, V.: Efficient nonlocal regularization for optical flow. In: Fitzgibbon, A., Lazebnik, S., Perona, P., Sato, Y., Schmid, C. (eds.) ECCV 2012. LNCS, vol. 7572, pp. 356–369. Springer, Heidelberg (2012). https://doi.org/10.1007/978-3-642-33718-5_26
30. Bredies, K., Kunisch, K., Pock, T.: Total generalized variation. SIAM J. Imaging Sci. **3**, 492–526 (2010)
31. Liu, C., Yuen, J., Torralba, A.: Sift flow: dense correspondence across scenes and its applications. IEEE Trans. Pattern Anal. Mach. Intell. **33**, 978–994 (2011)
32. Brox, T., Malik, J.: Large displacement optical flow: descriptor matching in variational motion estimation. IEEE Trans. Pattern Anal. Mach. Intell. **33**, 500–513 (2011)
33. Weinzaepfel, P., Revaud, J., Harchaoui, Z., Schmid, C.: DeepFlow: large displacement optical flow with deep matching. In: ICCV, pp. 1385–1392 (2013)
34. Hu, Y., Song, R., Li, Y.: Efficient coarse-to-fine patchmatch for large displacement optical flow. In: Proceedings of CVPR, pp. 5704–5712 (2016)
35. Lempitsky, V., Rother, C., Roth, S., Blake, A.: Fusion moves for markov random field optimization. IEEE Trans. Pattern Anal. Mach. Intell. **32**, 1392–1405 (2010)
36. Wedel, A., Pock, T., Zach, C., Bischof, H., Cremers, D.: An improved algorithm for TV-L1 optical flow. In: Proceedings of Statistical and Geometrical Approaches to Visual Motion Analysis, pp. 23–45 (2009)
37. Dosovitskiy, A., et al.: FlowNet: learning optical flow with convolutional networks. In: Proceedings of CVPR (2015)

38. Ilg, E., Mayer, N., Saikia, T., Keuper, M., Dosovitskiy, A., Brox, T.: FlowNet 2.0: evolution of optical flow estimation with deep networks. In: Proceedings of CVPR (2017)
39. Sun, D., Yang, X., Liu, M.-Y., Kautz, J.: PWC-Net: CNNs for optical flow using pyramid, warping, and cost volume. In: Proceedings of CVPR (2018)
40. Ranjan, A., Black, M.J.: Optical flow estimation using a spatial pyramid network. In: Proceedings of CVPR (2017)
41. Yu, J.J., Harley, A.W., Derpanis, K.G.: Back to basics: unsupervised learning of optical flow via brightness constancy and motion smoothness. In: Hua, G., Jégou, H. (eds.) ECCV 2016. LNCS, vol. 9915, pp. 3–10. Springer, Cham (2016). https://doi.org/10.1007/978-3-319-494 09-8_1
42. Liu, P., Lyu, M., King, I., Xu, J.: SelFlow: self-supervised learning of optical flow. In: Proceedings of CVPR (2019)
43. Wang, Y., Yang, Y., Yang, Z., Zhao, L., Wang, P., Xu, W.: Occlusion aware unsupervised learning of optical flow. In: Proceedings of CVPR (2018)
44. Hur, J., Roth, S.: Iterative residual refinement for joint optical flow and occlusion estimation. In: Proceedings of CVPR (2019)
45. Liu, P., King, I., Lyu, M.R., Xu, J.: DDFlow: learning optical flow with unlabeled data distillation. In: Proceedings of AAAI Conference on Artificial Intelligence, vol. 33, pp. 8770–8777 (2019)
46. Ashburner, J.: A fast diffeomorphic image registration algorithm. Neuroimage **38**, 95–113 (2007)
47. Blake, A., Zisserman, A.: Visual Reconstruction. MIT Press, Cambridge (1987)
48. Pratt, W.K., Kane, J., Andrews, H.C.: Hadamard transform image coding. Proc. IEEE. **57**, 58–68 (1969)
49. Lee, C.-Y., Xie, S., Gallagher, P., Zhang, Z., Tu, Z.: Deeply-supervised nets. In: Proceedings of MLR, pp. 562–570 (2015)
50. Kingma, D.P., Ba, J.: Adam: a method for stochastic optimization. In: Proceedings of ICLR (2017)
51. Fischl, B.: FreeSurfer. NeuroImage. **62**, 774–781 (2012)
52. Hoffmann, M., Billot, B., Greve, D.N., Iglesias, J.E., Fischl, B., Dalca, A.V.: SynthMorph: learning contrast-invariant registration without acquired images. IEEE Trans. Med. Imaging **41**, 543–558 (2021)

Optimisation

Learn to Fuse Input Features for Large-Deformation Registration with Differentiable Convex-Discrete Optimisation

Hanna Siebert[(✉)][iD] and Mattias P. Heinrich[iD]

Institute of Medical Informatics, Universität zu Lübeck, Lübeck, Germany
{siebert,heinrich}@imi.uni-luebeck.de

Abstract. Hybrid methods that combine learning-based features with conventional optimisation have become popular for medical image registration. The ConvexAdam algorithm that ranked first in the comprehensive Learn2Reg registration challenges completely decouples semantic and/or hand-crafted feature extraction from the estimation of the transformation due to the difficulty of differentiating the discrete optimisation step. In this work, we propose a simple extension that enables backpropagation through discrete optimisation and learns to fuse the semantic and hand-crafted features in a supervised setting. We demonstrate state-of-the-art performance on abdominal CT registration.

Keywords: Large deformation registration · Convex optimisation · End-to-end learning

1 Introduction and Related Work

While end-to-end learning of fully-convolutional networks is the method of choice for semantic segmentation, image registration continues to benefit from integrating conventional optimisation steps, e.g. pairwise instance optimisation [7], a discretised search of displacements [1] or iterative recurrent updates [9]. Discrete optimisation has been shown to yield excellent registration quality for numerous tasks [2,5,7] but does rely on non-differentiable steps which would prevent its use in end-to-end learning. We aim for a method that offers the possibility to use discrete optimisation in an end-to-end learning setting. Therefore, we introduce a differentiable convex discrete optimisation approach that is able to align images with large deformations. This differentiable optimisation is used to learn the fusion of semantic and hand-crafted image features.

2 Method

Figure 1 gives an overview of our method: First, hand-crafted and semantic features are extracted from the input images, concatenated and passed to a small

A. Hering et al. (Eds.): WBIR 2022, LNCS 13386, pp. 119–123, 2022.
https://doi.org/10.1007/978-3-031-11203-4_13

network comprising layers for feature fusion. The fixed and moving features output from this network are then used for our differentiable discretised convex optimisation method to align images with large deformations.

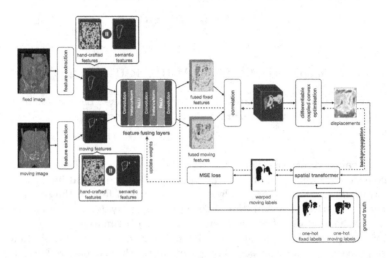

Fig. 1. Overview of our method: Hand-crafted and semantic features are concatenated and fused with feature fusing network layers. The fused features are used for our differentiable optimisation method to compute displacements. For backpropagation, warped moving and fixed labels are passed to a MSE loss function to update the feature fusing network's weights whereas the feature extraction part of the framework remains frozen.

2.1   Differentiable Convex-Discrete Optimisation

For pairwise deformable image registration, a deformation field $\mathbf{u}$ is sought that minimises the cost function $E(I_F, I_M, \mathbf{u})$ to align a fixed image I_F and a moving image I_M. In [3], a non-differentiable convex-discrete method has been proposed to find a deformation field $\mathbf{u}$ by solving a combined cost function

$$E(\mathbf{v}, \mathbf{u}) = DSV(\mathbf{v}) + \frac{1}{2\theta}(\mathbf{v} - \mathbf{u})^2 + \alpha|\nabla \mathbf{u}|^2 \qquad (1)$$

that ensures similarity and smoothness optimisation. In this function, $\mathbf{v}$ is an auxiliary second deformation field used to compute the displacement space volume DSV. The regularisation parameter α controls the smoothness of the deformation field and the parameter θ models the coupling between similarity and regularisation penalty and is decreased during iterative solving of the equation. The optimal selection of $\mathbf{v}$ with respect to the similarity term can be performed globally optimal using local cost aggregation [3].

In this work, we introduce a differentiable discretised convex optimisation by replacing argmin operators with their corresponding softmin counterparts and make suitable adjustments to hyper-parameters that reduce memory require-ments for end-to-end learning. Coupled-convex discrete optimisation [3] approx-imates more complex MRF-solutions by the following steps:

(0) initialisation of the current displacement field to zeros
(1) computation of a correlation volume based on sum of squared differences of feature tensors (the volume comprises 6 dimensions, 3 spatial dimension and 3 displacement dimensions)
(2a) a regularising coupling term that adds 3D parabolas in displacement dimen-sions that are rooted at the current displacement solution
(2b) the argmin operator (across all possible displacements) that defines a new regularised displacement field
(2c) a spatial smoothing step (e.g. a box-filter)

The correlation volume (step (1)) directly depends on the feature maps obtained from fixed and moving scans. By defining a large enough capture range and correspondingly a discrete mesh grid of relative displacements the method can robustly find a near global optimum without multiple warping steps or cas-caded architectures. Steps (2a)–(2c) are iteratively repeated with a continuously increasing weight for the coupling term that helps to ensure convergence of the optimisation. Step (2b), which takes the argmin is not differentiable and will be replaced with a softmin operator along the displacement dimension followed by a point-wise multiplication with the relative displacements of the predefined discrete mesh grid and subsequent reduction.

2.2 Learning of Input Feature Fusion

Previous work [3,7] has shown that hand-crafted MIND features [4] or automatic nnU-Net segmentations [6] can be used as input for a coupled convex optimisa-tion method for image registration. In this work, we combine hand-crafted and semantic features by fusing them with help of trainable feature fusing network layers comprising two $1 \times 1 \times 1$-convolutions followed by instance normalisations and ReLU activations. The first convolution increases the number of feature channels to 32 and a third $1 \times 1 \times 1$-convolution reduces the number of feature channels to 15. The resulting feature maps are then used to solve the differ-entiable convex-discrete optimisation problem described in Sect. 2.1 in order to compute the displacement fields that are then used to warp the moving label maps. One-hot representations of warped and fixed label maps weighted inversely proportional to the square root of the class frequency are passed to a MSE loss function that is used to train the feature fusing network's parameters whereas the feature extraction part of the framework stays frozen.

3 Experiments and Results

For our experiments we use the Learn2Reg-2020 challenge's (task 3) dataset con-taining 30 abdominal inter-patient CT scans with 13 manually labeled abdominal

Table 1. Left: Quantitative results: Accuracy is measured by the Dice similarity of segmentations and the 95% Hausdorff distance for segmentations. Plausibility of the deformations is measured by the standard deviation of the logarithmic Jacobian determinant. Right: example visualisation of fixed image and warped moving labels.

	Dice [%]	HD [mm]	SDlogJ
initial	25.14	40.21	–
MIND features	37.79	37.22	0.050
nnU-Net features	50.56	24.71	0.021
concatenated features	49.71	28.33	0.050
fused features	56.37	24.13	0.049

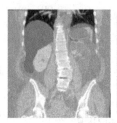

- spleen
- right kidney
- left kidney
- gallbladder
- esophagus
- liver
- stomach
- aorta
- inferior vena cava
- portal vein
- pancreas
- right adrenal gland
- left adrenal gland

organs and a resolution of $192 \times 160 \times 256$ [5,10]. The scans have been linearly pre-registered and split into 20 training cases and 10 test cases. For evaluation we consider all possible pairwise combinations of the test cases. From the image data, we extract MIND features (leading to 12 feature channels) and compute one-hot encoded label features by applying a nnU-Net trained on the 20 training cases (leading to 14 feature channels). We downsample the features to a resolution of $48 \times 40 \times 64$, concatenate them and pass the 26-channel input to our feature fusing network. The network's 15-channel output is then used for displacement computation with the differentiable convex optimisation method. Therefore, we use a displacement range that covers ~ 32 mm within the scanned abdominal region and scale the softmin operation's output (step (2b)) by half of the downsampled feature dimensions. The feature fusing network is trained for 50 epochs using Adam and a learning rate of 0.005.

For evaluation, we upsample the obtained displacement fields to the original image resolution. We compare our fused features with the direct use of MIND features, nnU-Net label features, and concatenation of MIND and nnU-Net features. The results given in Table 1 show that the fusion of MIND and nnU-Net features clearly outperforms the other investigated feature variants with an average Dice score of 56.37% compared to 50.56% when using only nnU-Net features. As using nnU-Net features yields to a deformation field that is optimised to warp the foreground structures, the SDlogJ value is lower than when MIND features are involved. We evaluated the potential problem of label bias with an experiment on additional structures (lumbar and thoracic vertebrae[1] [8]) unseen for the nnU-Net segmentation training and our fusion learning. While using only MIND features yields the highest accuracy we see great potential for the proposed feature fusion that only reduced the Dice score of the spine by 5% while the nnU-Net-based registration results in a drop of 42%. Hence the influence of label bias is substantially reduced.

---

[1] https://github.com/MIRACLE-Center/CTSpine1K.

4   Discussion and Conclusion

This work introduced a differentiable version of coupled convex discrete optimisation for image registration with large deformation. It has opened up the possibility of end-to-end feature learning and has well-performed for our feature fusing network. We show that the fusion of semantic label features and hand-crafted features based on image self-similarities leads to an improved registration performance compared to either using only semantic or only hand-crafted features or the simple concatenation of both.

References

1. Dosovitskiy, A., et al.: FlowNet: learning optical flow with convolutional networks. In: Proceedings of the IEEE International Conference on Computer Vision, pp. 2758–2766 (2015)
2. Heinrich, M.P., Jenkinson, M., Brady, M., Schnabel, J.A.: MRF-based deformable registration and ventilation estimation of lung CT. IEEE Trans. Med. Imag. **32**(7), 1239–1248 (2013)
3. Heinrich, M.P., Papież, B.W., Schnabel, J.A., Handels, H.: Non-parametric discrete registration with convex optimisation. In: Ourselin, S., Modat, M. (eds.) WBIR 2014. LNCS, vol. 8545, pp. 51–61. Springer, Cham (2014). https://doi.org/10.1007/978-3-319-08554-8_6
4. Heinrich, M.P., Jenkinson, M., Papież, B.W., Brady, S.M., Schnabel, J.A.: Towards realtime multimodal fusion for image-guided interventions using self-similarities. In: Mori, K., Sakuma, I., Sato, Y., Barillot, C., Navab, N. (eds.) MICCAI 2013. LNCS, vol. 8149, pp. 187–194. Springer, Heidelberg (2013). https://doi.org/10.1007/978-3-642-40811-3_24
5. Hering, A., et al.: Learn2reg: comprehensive multi-task medical image registration challenge, dataset and evaluation in the era of deep learning. arXiv preprint arXiv:2112.04489 (2021)
6. Isensee, F., Jaeger, P.F., Kohl, S.A., Petersen, J., Maier-Hein, K.H.: nnU-Net: a self-configuring method for deep learning-based biomedical image segmentation. Nat. Methods **18**(2), 203–211 (2021)
7. Siebert, H., Hansen, L., Heinrich, M.P.: Fast 3D registration with accurate optimisation and little learning for Learn2Reg 2021. In: Aubreville, M., Zimmerer, D., Heinrich, M. (eds.) Biomedical Image Registration, Domain Generalisation and Out-of-Distribution Analysis. MICCAI 2021. LNCS, vol. 13166, pp. 174–178. Springer, Cham (2021). https://doi.org/10.1007/978-3-030-97281-3_25
8. Smith, K., et al.: Data from CT_colonography. Cancer Imag. Arch. (2015). https://doi.org/10.7937/K9/TCIA.2015.NWTESAY1
9. Teed, Z., Deng, J.: RAFT: recurrent all-pairs field transforms for optical flow. In: Vedaldi, A., Bischof, H., Brox, T., Frahm, J.-M. (eds.) ECCV 2020. LNCS, vol. 12347, pp. 402–419. Springer, Cham (2020). https://doi.org/10.1007/978-3-030-58536-5_24
10. Xu, Z., et al.: Evaluation of six registration methods for the human abdomen on clinically acquired CT. IEEE Trans. Biomed. Eng. **63**(8), 1563–1572 (2016)

Multi-magnification Networks for Deformable Image Registration on Histopathology Images

Oezdemir Cetin[1], Yiran Shu[1], Nadine Flinner[2], Paul Ziegler[2], Peter Wild[2], and Heinz Koeppl[1(✉)]

[1] Department of Electrical Engineering and Information Technology, Technische Universität Darmstadt, Darmstadt, Germany
heinz.koeppl@tu-darmstadt.de

[2] Senckenberg Institute of Pathology, University Hospital Frankfurt, Frankfurt, Germany

Abstract. We present an end-to-end unsupervised deformable registration approach for high-resolution histopathology images with different stains. Our method comprises two sequential registration networks, where the local affine network can handle small deformations, and the non-rigid network is able to align texture details further. Both networks adopt the multi-magnification structure to improve registration accuracy. We train the proposed networks separately and evaluate them on the dataset provided by the University Hospital Frankfurt, which contains 41 multi-stained histopathology whole-slide images. By comparing with methods using the single-magnification structure, we confirm that the proposed multi-view architecture can significantly improve the performance of the local affine registration algorithm. Moreover, the proposed method achieves high registration accuracy of contents at the cell level and is potentially applicable to other medical image alignment tasks.

Keywords: Histopathological image · Affine transformation · Non-rigid registration · Unsupervised learning · Multi-magnification network

1 Introduction

Histopathological whole slide images, i.e., digital tissue slides produced by scanning conventional glass slides under high-resolution microscopy, are vital for modern histopathology analysis [15]. Standard whole slide images employ the pyramid structure to support different resolutions, making it easy for pathologists to observe by zooming. Each layer of the pyramid corresponds to a resolution level, with the bottom being the highest resolution information. In general, histopathologists utilise various staining techniques based on chemical features of the tissue, e.g. Hematoxylin-Eosin (H&E), periodic-acid Schiff (PAS) or elastic-van Gieson (EvG). In addition, antibody-mediated visualization of specific proteins, termed immunohistochemistry, is widely used in modern histopathology.

© The Author(s), under exclusive license to Springer Nature Switzerland AG 2022
A. Hering et al. (Eds.): WBIR 2022, LNCS 13386, pp. 124–133, 2022.
https://doi.org/10.1007/978-3-031-11203-4_14

As tissue specimens are prepared by approx. 3 μm-thin cuts each specimen represents an unique sample and slides obtained from directly adjacent tissue differ slightly in their morphology. Even when the same tissue slide is used for multiple staining, e.g. by bleaching and re-staining, shifts and/or deformations inevitably occur. These digital multi-stained histopathology images that are not aligned accurately pose obstacles to the diagnosis or further processing, thus need to be registered first.

Image registration is the process of matching two images geometrically so that corresponding coordinate points in both images correspond to the same physical region of the scene being imaged [21]. Biomedical image registration constitutes one of the key research areas for medical analysis that has been extensively studied. Traditional registration methods search for spatial transformation that brings the defined similarity metric to be optimum by an iterative optimization algorithm [1]. Nevertheless, the superiority of accuracy and robustness of classical approaches come at the cost of time, which becomes the main bottleneck in archiving desirable performance for practical applications. With the revival of deep learning, attempts have been made to develop learning-based approaches to implement faster registration, which can be grouped into three main categories [5,6]: (i) deep iterative registration, which follows the framework of traditional methods but instead adopts similarity metrics learned by deep neural networks [16,17], (ii) supervised transformation prediction, utilizing the known ground truth transformations to define the cost function [9,18], (iii) Unsupervised transformation prediction, where a spatial transformation network is applied to calculate the error of the given metric(s) with an appropriate regularization term [2,19]. The first class of methods inherits the time-consuming drawback of conventional approaches due to the iterative process, whereas the supervised training requires a large amount of data with annotations. In contrast, unsupervised transformation approaches produce the supervisory signals required for training directly by data and can achieve real-time registration during prediction. Therefore, we focused on the unsupervised methods in this work.

An obstacle to applying learning-based methodologies to histopathology images concerns their ultra-high resolution. Some studies have resampled images down to an acceptable memory limit before deformation estimation [3,15]. However, such detailed information as the cell morphological structure is almost impossible to observe on low-resolution images, becoming a key hamper in improving alignment accuracy. An alternative solution is to perform registration on smaller patches [8,12]. The shortcoming of this approach is the irreversible loss of neighboring information when splitting the images, resulting in the narrow field-of-view. In this work, we propose two deep multi-magnification network architectures for patch-based affine and non-rigid registration. The proposed local affine algorithm can effectively deal with imperceptible collective shifts of cell nuclei in the low-resolution pattern, and non-rigid registration is able to align further the cell components that are slightly altered in the morphological structure. We train the presented networks unsupervised and yield higher registration accuracy than the methods using only ordinary single-magnification networks. The result reaches precise alignment at the cellular level under the maximum resolution of histopathology WSIs, which significantly contributes to the manual/automatic pathological diagnosis on the differently stained tissue sections.

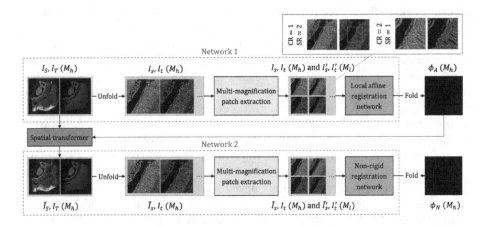

Fig. 1. Overview of the proposed algorithm: Both networks take as input concatenated patches I_s, I_t for M_h (high) magnification, and concatenated patches $I_s{}'$, $I_t{}'$ for M_l (low) magnification. An example in the upper right corner illustrates the construction process of a patch set, where the cropping rate (CR) and sampling factor (SR) used to build patches for each magnification level are given. The red boxes denote the corresponding regions at different magnifications. (Color figure online)

2 Methods

Let I_S, I_T: $\Omega \rightarrow \mathbb{R}$ represent the whole slide source and target images, defined in the spatial domain $\Omega \subset \mathbb{R}^d$, where d denotes ($d = 2$ in this study) spatial dimensionality of the given data. Similarly, I_s, I_t: $\omega \rightarrow \mathbb{R}$ with $\omega \subset \Omega$ represent the patch-wise source and target images, extracted from I_S and I_T. Assuming that the image pairs to be registered are pre-aligned well, we aim to find two deformation fields ϕ_A, ϕ_N: $\Omega \rightarrow \Omega$ to deform the source image such that:

$$I_S(\phi_N \circ \phi_A(x)) \approx I_T, \forall x \in \Omega. \tag{1}$$

Here "$\circ$" represents the composition of deformations and $I(\phi)$ indicates I deformed by ϕ. The deformations ϕ_A, ϕ_N are defined as a patch-wise affine deformation and a pixel-wise non-rigid deformation, respectively. They are obtained by aggregating the local deformations ϕ_a^p, ϕ_n^p : $\omega \rightarrow \omega$ of image patches (I_s, I_t) extracted from (I_S, I_T), where p indicates the index of the patch on the whole slide image. Two convolutional neural networks f_a and f_n are used to realize the affine registration $\phi_a = f_a(I_s, I_t)$ and non-rigid registration $\phi_n = f_n(I_s(\phi_a), I_t)$, respectively.

An affine registration network is leveraged to learn the affine transformation $\phi_a := Tx$, where $T \in \mathbb{R}^{d \times m}$ with $m = d + 1$. Next, the affinely registered images are fed into the non-rigid registration network to learn the displacement field $u(x)$ with $\phi_n := x + u(x)$, which represents the displacements for $\forall x \in \omega$ in the vertical and horizontal directions.

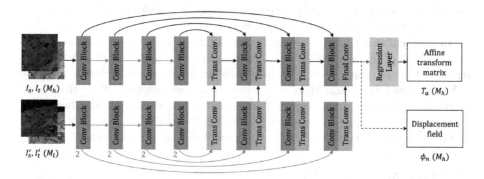

Fig. 2. Architecture of the local affine and non-rigid registration networks: *Conv Block* includes two sets, each consisting of a 3×3 convolution layer with group normalization (GN), activated by PReLU. *Trans Block* comprises a 2×2 transposed convolution layer with a stride of 2 followed by GN and PReLU activation. The green and red arrows indicate maximum pooling and average pooling, respectively. The center cropping operations are denoted by brown arrows with the cropping rates written in brown. Other blocks are described in the text. (Color figure online)

The input of both networks is a set of image patches with different magnifications, providing multiple field-of-views to the networks. Figure 1 offers an overview of the proposed registration algorithm for the case of two magnification levels. The strategy adopted for extracting multi-magnification patches in this work is described as follows: In a multi-magnification set, all other patches are obtained by center-cropping the base image with different cropping rates. Then, the patches are downsampled with the corresponding sampling factors to uniform the patch size. The downsampled base image is the one with the lowest magnification level in the set. Registration networks take the patch set as input and predict the local affine transform matrix/displacement field corresponding to the patch with the highest magnification level, as details described in the next section. According to Eq. 1, the final deformation for the given images I_S and I_T is obtained by composing the folded ϕ_A and ϕ_N.

2.1 Network Architectures

The proposed networks are inspired from [7], which contains multiple magnification layers that obtain more information from different field-of-views. Considering that the architectures of both networks are quite similar, they are shown in one figure for brevity, as visualized in Fig. 2. The concatenation of the high-magnification patches I_s and I_t is fed into the target magnification layer based on the U-Net [13], to extract the higher magnification feature maps. During reconstruction, these feature maps are concatenated with the corresponding lower magnification feature maps extracted from the lower-magnification patches I_s' and I_t' in another magnification layer. To limit the usage of feature maps from cropped boundary areas in a wide field-of-view, the lower magnification feature maps are center-cropped with a given cropping rate followed by up-sampling utilizing transpose convolution to match the size.

In the local affine network, $FinalBlock$ has the same structure as $ConvBlock$ but a stride of 2, followed by an adaptive average pooling layer. The reconstructed feature maps are transformed into six numeric parameters through a fully-connected layer and then rearranged into the resulting affine transform matrix T in the regression layer. Whereas, in the non-rigid network, the reconstructed feature maps are compressed utilizing $Final\ Block$, a stack of a 3×3 and a 1×1 convolution layer, into two-channel displacement field $u(x)$.

2.2 Loss Function

Assume that $\phi : \omega \rightarrow \omega$ is the local deformation field estimated by networks with image patches I_s and I_t as input, the loss function can be described as

$$\mathcal{L}\left(I_s, I_t, \phi\right) = \mathcal{L}_S\left(I_s(\phi), I_t\right) + \lambda\mathcal{L}_R\left(\phi\right), \tag{2}$$

where the first term $\mathcal{L}_S$ measures the similarity between the warped source and the target patches, and $\mathcal{L}_R$ is a regularization term considered only in the non-rigid network. Parameter λ controls the trade-off between these two terms as a hyperparameter in the training process.

We choose the normalized cross-correlation (NCC) [10] as the similarity metric $\mathcal{L}_S$. Let I_1, I_2 be two images then this similarity can be computed as

$$NCC(I_1, I_2) = \frac{1}{N-1} \sum_{x \in \omega} \frac{(I_1(x) - \bar{I}_1)(I_2(x) - \bar{I}_2)}{\sigma_{I_1}\sigma_{I_2}}, \tag{3}$$

where N indicates the number of non-zero pixels, $\bar{I}$ and σ_I represent the mean and standard deviation of the intensities in image I, respectively. The negative normalized cross-correlation (NCC) is used in training to minimize the loss function, while a higher NCC value corresponds to a higher similarity between images.

Under the intuition that a desirable deformation field should not vary too much between nearby points, the curvature regularization [4] is used to constrain the geometric smoothness of the displacement field ϕ predicted by the non-rigid network, i.e.,

$$\mathcal{L}_R(\phi) = \sum_{x \in \omega} \| \nabla\phi(x) \|^2. \tag{4}$$

3   Experiments

The University Hospital Frankfurt (UKF) provided the images used in this study to evaluate the proposed algorithm, with clinical data removed and completely anonymized. The UKF dataset comprises two parts: The first part offers 36 histopathological WSIs, where every two images are from the same tissue section, respectively stained with H&E and IHC-CD8. The second part consists of 5 WSIs obtained from two staining experiments in which multiple staining was performed on the tissue slides from one tissue in different orders. All WSIs are provided as .mrxs files with a unified specification. Each of them contains images at nine resolutions with a downsampling factor of 2, where the full resolution exceeds $180k \times 90k$ pixels in size. We generated 18 and 5 image pairs respectively from two parts of the UKF dataset for training and evaluation. The experiment details are presented next.

3.1   Experimental Settings

Data Preprocessing. We removed large background areas in the raw data by a boundary detection algorithm and then converted them into single-channel grayscale images. The rigid alignment method derived from [20] was adopted to handle the large misalignment of the image pairs.

Technical Details. The proposed algorithm was implemented by modifying and extending the DeepHistReg framework [20]. Unsupervised methods were trained on the resolution-level 4 images whose size varies from $3k$ to $7k$ pixels in one dimension. The images are split into overlapping patches, followed by extracting 224×224 patches of different magnification levels as the input to the networks. We trained both presented networks with a batch size of 4 using Nvidia Tesla P100 (PCIe). The Adam optimizer with an initial learning rate of $1e-3$ and a decay rate of 0.95 was adopted to update the network parameters. The constraint coefficient λ for the non-rigid network training was chosen to be 60.

Baseline Methods. We built two single-magnification networks for local affine and non-rigid registration as the baseline models for comparison. The architecture of both networks inherited the target magnification layer of the corresponding multi-amplification network with some adaptations. The training settings were the same as the proposed methods.

3.2   Evaluation Metrics

We quantified the registration accuracy by several similarity metrics since no ground truth such as landmarks or segmentation maps are provided for the UKF dataset. Except for the metric NCC used as the objective function during network training, the quality of the deformation fields was also evaluated by the

Mean-Squared-Error (MSE) [11] and the normalized Mutual-Information (NMI) [14], which are respectively defined as

$$MSE(I_1, I_2) = \frac{1}{N} \sum_{x \in \omega} [I_1(x) - I_2(x)]^2, \tag{5}$$

$$NMI(I_1, I_2) = \frac{2 \cdot H(I_1, I_2)}{H(I_1) + H(I_2)}, \tag{6}$$

where H indicates Shannon's entropy and $H(I_1, I_2)$ represents the dependence of variables (images) I_1 and I_2.

Table 1. Comparison among methods with single/multi-magnification registration networks, containing the average inference time and performance quantified by the similarity metrics NCC, MSE, and NMI (arrows indicate the trend of the increased similarity): The methods are named according to the adopted network architectures, where S/M stands for networks with the single/multi-magnification structure, and A/N denotes the local affine transformation and non-rigid deformation. For example, MASN refers to combining a multi-magnification local affine network and a single-magnification non-rigid network. Besides, an iterative approach is applied based on the presented method, with the number of iterations denoted in parentheses.

Metric	Initial	SASN	SAMN	MASN	MAMN	MASN(3)
NCC ↑	0.6828	0.7123	0.7060	**0.7461**	0.7443	**0.7728**
MSE ↓	0.0403	0.0376	0.0382	**0.0336**	0.0338	**0.0305**
NMI ↑	0.1670	0.1781	0.1756	0.1952	**0.1954**	**0.2038**
Time (sec)	–	25.48	28.96	28.95	32.31	45.29

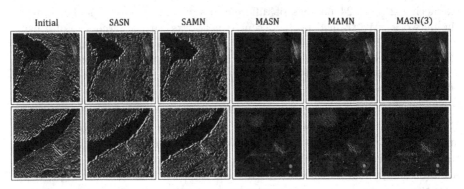

Fig. 3. Local subtractions of a high-resolution image pair registered by different methods: The non-overlapping regions appear as fluorescent green due to the nature of stains. For visibility, the contrast/brightness of images has been increased by 50%. (Color figure online)

4   Results

Table 1 summarizes the overall performance of our proposed algorithm in comparison to approaches containing one or more baseline models. All of them take images pre-aligned by rigid alignment as input.

As shown in Table 1, the proposed multi-magnification structures outperformed the ordinary single-magnification architecture for the local affine algorithm with remarkable benefits, whereas yielding almost no improvement in the performance of the non-rigid network. The increase in runtime due to the multi-magnification structure is not significant compared to the base runtime (SASN). According to the proposed algorithm, the difference in time will decrease exponentially for smaller image pairs. By iterating the prediction on the previous result by the same network, we obtained registration results with significantly higher accuracy.

We upsampled the predicted deformation fields for generating the registered images at a higher resolution. By performing local subtraction between the deformed source and target images, we evaluated the registration performance of different methods at the cellular level, as illustrated in Fig. 3. It can be observed that the local affine network improved by the multi-magnification structure is crucial for the enhancement of the overall performance. The cell nuclei can overlap completely in the best cases.

5   Discussion and Conclusion

In modern histopathology multiple staining techniques are used to detect specific structures within biological tissues. Each technique highlights different characteristics of the tissue and proper analysis needs to address the spatial distribution of these characteristics. In this context, we developed two novel deep networks with the multi-magnification structure for patch-based image registration, which can learn peripheral information outside the patches as auxiliary information to improve network performance. The presented method is of great importance for biomedical image registration since studies for them can often be performed only on smaller patches due to the large image size. Moreover, the network architectures can be easily expanded with more magnification levels. Nevertheless, this expansion makes little sense since too many field-of-views may instead negatively affect the network performance, especially for cases with no apparent global misalignment.

Our experiments compared the impact of single- and multi-magnification networks on the overall alignment performance by different network combinations. The results revealed that the multi-magnification structure could significantly improve the performance of the patch-based affine registration network. However, it yielded little success on the local non-rigid network. This might mainly attribute to the transformation nature of these two registration methods. The lack of neighboring information can aggravate the estimation error of deformation for the whole image patch region by the local affine approach, while this error occurs only within the edge region of the image patches in the non-rigid method due to the dense prediction. Therefore, the enhancement of the non-rigid method by the multi-magnification structure was much less evident than that of the local affine approach. Besides, we introduced an iterative approach on the method with the best performance, which further improved the registration accuracy, with an acceptable growth of inference time. The proposed method has the potential to be applicable for other medical image registration tasks.

Acknowledgement. H.K. acknowledges support from the European Research Council (ERC) with the consolidator grant CONSYN (nr. 773196). O.C. is supported by the Alexander von Humboldt Foundation Philipp Schwartz Initiative.

References

1. Costin, H.N., Rotariu, C.: Registration of multimodal medical images. Comput. Sci. J. Moldova **51**(3), 231–254 (2009)
2. Dalca, A.V., Balakrishnan, G., Guttag, J., Sabuncu, M.R.: Unsupervised learning for fast probabilistic diffeomorphic registration. In: Frangi, A.F., Schnabel, J.A., Davatzikos, C., Alberola-López, C., Fichtinger, G. (eds.) MICCAI 2018. LNCS, vol. 11070, pp. 729–738. Springer, Cham (2018). https://doi.org/10.1007/978-3-030-00928-1_82
3. Feuerstein, M., Heibel, H., Gardiazabal, J., Navab, N., Groher, M.: Reconstruction of 3-D histology images by simultaneous deformable registration. In: Fichtinger, G., Martel, A., Peters, T. (eds.) MICCAI 2011. LNCS, vol. 6892, pp. 582–589. Springer, Heidelberg (2011). https://doi.org/10.1007/978-3-642-23629-7_71
4. Fischer, B., Modersitzki, J.: A unified approach to fast image registration and a new curvature based registration technique. Linear Algebra Appl. **380**, 107–124 (2004)
5. Fu, Y., Lei, Y., Wang, T., Curran, W.J., Liu, T., Yang, X.: Deep learning in medical image registration: a review. Phys. Med. Biol. **65**(20), 20TR01 (2020)
6. Haskins, G., Kruger, U., Yan, P.: Deep learning in medical image registration: a survey. Mach. Vis. Appl., 1–18 (2020). https://doi.org/10.1007/s00138-020-01060-x
7. Ho, D.J., et al.: Deep multi-magnification networks for multi-class breast cancer image segmentation. Comput. Med. Imaging Graph. **88**, 101866 (2021)
8. Lotz, J., et al.: Patch-based nonlinear image registration for gigapixel whole slide images. IEEE Trans. Biomed. Eng. **63**(9), 1812–1819 (2015)
9. Lv, J., Yang, M., Zhang, J., Wang, X.: Respiratory motion correction for free-breathing 3d abdominal MRI using CNN-based image registration: a feasibility study. Br. J. Radiol. **91**(xxxx), 20170788 (2018)
10. Modersitzki, J.: FAIR: flexible algorithms for image registration. SIAM (2009)
11. Pishro-Nik, H.: Introduction to Probability, Statistics, and Random Processes. Kappa Research, Athens (2016)
12. Pitiot, A., Bardinet, E., Thompson, P.M., Malandain, G.: Piecewise affine registration of biological images for volume reconstruction. Med. Image Anal. **10**(3), 465–483 (2006)
13. Ronneberger, O., Fischer, P., Brox, T.: U-Net: convolutional networks for biomedical image segmentation. In: Navab, N., Hornegger, J., Wells, W.M., Frangi, A.F. (eds.) MICCAI 2015. LNCS, vol. 9351, pp. 234–241. Springer, Cham (2015). https://doi.org/10.1007/978-3-319-24574-4_28

14. Schütze, H., Manning, C.D., Raghavan, P.: Introduction to Information Retrieval, vol. 39. Cambridge University Press, Cambridge (2008)
15. Schwier, M., Böhler, T., Hahn, H.K., Dahmen, U., Dirsch, O.: Registration of histological whole slide images guided by vessel structures. J. Pathol. Inform. 4(Suppl) (2013)
16. Simonovsky, M., Gutiérrez-Becker, B., Mateus, D., Navab, N., Komodakis, N.: A deep metric for multimodal registration. In: Ourselin, S., Joskowicz, L., Sabuncu, M.R., Unal, G., Wells, W. (eds.) MICCAI 2016. LNCS, vol. 9902, pp. 10–18. Springer, Cham (2016). https://doi.org/10.1007/978-3-319-46726-9_2
17. So, R.W., Chung, A.C.: A novel learning-based dissimilarity metric for rigid and non-rigid medical image registration by using Bhattacharyya distances. Pattern Recogn. **62**, 161–174 (2017)
18. Uzunova, H., Wilms, M., Handels, H., Ehrhardt, J.: Training CNNs for image registration from few samples with model-based data augmentation. In: Descoteaux, M., Maier-Hein, L., Franz, A., Jannin, P., Collins, D.L., Duchesne, S. (eds.) MICCAI 2017. LNCS, vol. 10433, pp. 223–231. Springer, Cham (2017). https://doi.org/10.1007/978-3-319-66182-7_26
19. Wodzinski, M., Müller, H.: Unsupervised learning-based nonrigid registration of high resolution histology images. In: Liu, M., Yan, P., Lian, C., Cao, X. (eds.) MLMI 2020. LNCS, vol. 12436, pp. 484–493. Springer, Cham (2020). https://doi.org/10.1007/978-3-030-59861-7_49
20. Wodzinski, M., Müller, H.: DeephistReg: unsupervised deep learning registration framework for differently stained histology samples. Comput. Methods Programs Biomed. **198**, 105799 (2021)
21. Zitova, B., Flusser, J.: Image registration methods: a survey. Image Vis. Comput. **21**(11), 977–1000 (2003)

Realtime Optical Flow Estimation on Vein and Artery Ultrasound Sequences Based on Knowledge-Distillation

Till Nicke[1,4]([⊠]), Laura Graf[2], Mikko Lauri[1], Sven Mischkewitz[3], Simone Frintrop[1], and Mattias P. Heinrich[2]

[1] Department of Informatics, University of Hamburg, Hamburg, Germany
{till.nicke,mikko.lauri,simone.frintrop}@uni-hamburg.de
[2] Institute of Medical Informatics, University of Lübeck, Lübeck, Germany
{graf,heinrich}@imi.uni-luebeck.de
[3] ThinkSono GmbH, Potsdam, Germany
sven@thinksono.com
[4] Fraunhofer Institute for Image Computing MEVIS, Lübeck, Germany

Abstract. In this paper, we propose an approach for realtime optical flow estimation in ultrasound sequences of vein and arteries based on knowledge distillation. Knowledge distillation is a technique to train a faster, smaller model by learning from cues of larger models. Mobile devices with limited resources could be key in providing effective point-of-care healthcare and motivate the search of more lightweight solutions in the deep learning based image analysis. For ultrasound video analysis, motion correspondences of image contents (anatomies) have to be computed for temporal context and for real time application, fast solutions are required. We use a PWC-Net's [1] optical flow estimation output to create soft targets to train a PDD-Net [2] as lightweight optical flow estimator. We analyse the students' performance on the challenging task of fast segmentation propagation of vein and arteries in ultrasound images. Experiments show that even though we did not fine-tune the teachers on this task, a model trained with soft targets outperformed a model trained directly with labels and without a teacher.

Keywords: Knowledge distillation · Realtime video inference · Ultrasound images

1 Introduction

The analysis of objects in a sequence of images is a task that plenty of research has been done for, recently mostly in the deep learning field [3]. To achieve a coherent and accurate result over the different time points, it is important that

© The Author(s), under exclusive license to Springer Nature Switzerland AG 2022
A. Hering et al. (Eds.): WBIR 2022, LNCS 13386, pp. 134–143, 2022.
https://doi.org/10.1007/978-3-031-11203-4_15

the analysis of the current image considers the past. One way to represent this temporal context is in the form of estimated optical flow. However, the classical methodology for its' calculation is an iterative approach [4] too slow for realtime inference. Most recent image registration approaches based on deep learning (e.g. [5]) are computationally too expensive to be executed on mobile devices in the required time. Realtime estimation of optical flow of ultrasound sequences would be advantageous in many practical point-of-care ultrasound (POCUS) applications that are based on intelligent guidance through image analysis. The aim of this work is to train a network, that learns from larger, pre-trained flow estimation networks and is able to accurately propagate relevant information (e.g. segmentations of important anatomies) in ultrasound. Ultrasound images often exhibit ambiguous structure depiction and a network, that employs only 2D convolution without temporal context, is not able to perfectly interpret the image with satisfying accuracy. So instead, utilising the motion of the images can leverage temporal context without requiring access to the whole temporal sequence. A CNN can be trained to estimate the temporal context e.g. by learning to propagate anatomical labels correctly between two images (which is usually coined weakly-supervised registration [6]). Clinically, this is relevant e.g. for the diagnosis of deep vein thrombosis (DVT), for which vessels in the leg need to be labelled.

2   Related Work

2.1   Dynamic Ultrasound Analysis

The use of automated image analysis for ultrasound is constantly increasing both in research and practical clinical translations [7]. The recent MICCAI challenge CLUST [8] has studied the quality of image registration algorithms for tracking ultrasound but without realtime constraints. A Siamese network for respiratory motion estimation on ultrasound images has been proposed by Liu and colleagues [9], which is capable of tracking landmarks through a video sequence.

A system for compression-based DVT examination in ultrasound (US) images was proposed by Tanno and colleagues [10]. The system, named AutoDVT, uses a dual-task network to help make predictions about the patient's VTE status. One of the tasks consists of classifying the compression status of a registered vein as either closed or open. The network itself uses stacked consecutive frames as input to create temporal consistency. The different task networks share the majority of convolutional layers and only separate the two tasks in the last convolutional layer, thus each task regulates the other during training.

To achieve higher temporal consistency and capture a more holistic view of dynamic sequences, optical flow estimation between frames can be leveraged. To ensure fast inference time, it is of importance that the optical flow prediction takes as little time as possible, while still generating accurate estimations.

2.2 Optical Flow Estimation

In recent research in deep learning and optical flow estimation numerous capable network solutions have been proposed, including Flownet [11], its evolution Flownet2 [12], and PWC-Net [1]. Flownet uses CNN feature extractors on two images, correlates these features over a discretised displacement search window (originally 21 × 21 pixels with a stride of 4), and further processes these correlations to predict a flow field. Flownet2 extends the original Flownet approach by employing multiple different and fine-tuned versions of this architecture.

PWC-Net, which was proposed by Sun et al. [1], on the other hand, uses pyramidal images with a combination of a cost-volume layer and a warping layer to estimate the optical flow of the input images.

In the medical domain LapIRN [13] and PDD-Net [2] are two capable networks for estimating large deformations. PDD-Net utilizes deformable convolution layers for feature extraction, which are then correlated. The correlation layer is followed by a min convolution and mean-field inference to predict dense displacement probabilities in volumes.

Some of these networks are larger, with up to 162 million parameters and up to 0.6 s of inference time on an NVIDIA graphics card [11,12]. However, these models are very accurate, which makes them valuable teachers in a student-teacher setting. Other models, such as the PDD-Net, with less parameter counts use little space and computation.

2.3 Knowledge Distillation

Student-teacher learning, also known as knowledge distillation (KD), was proposed by Li et al. [14]. The method uses one (or more) large and accurately trained neural network(s), also called teacher, and tries to teach the output distribution to a smaller network, also called student, by minimizing the KL divergence between the teacher's output and the students' prediction.

Yuan et al. proposed that not only accurate teachers can be used in a knowledge distillation setting. In [15] they found that also insufficiently trained teachers can increase the performance of the students, as they provide a representative distribution of the classes in the classification task. Thus, the teachers not only provide accurate information about the output but also provide regularized soft targets.

In [16] Kim et al. compared the KL divergence as a loss function, which is widely used in knowledge distillation, to a mean squared error loss and found, that the mean squared error loss is superior to the KL divergence, especially, when using a small tau, as the label noise is mitigated.

2.4 Contributions

We utilize the aforementioned knowledge distillation process [14] to train a small and lightweight optical flow estimator network (PDD-Net) for ultrasound motion estimation and vessel segmentation propagation in ultrasound images. We also compare this method to a label loss trained network to evaluate the usage of the distilled knowledge and find an increase in Dice score, as well as a decrease in Hausdorff distance (HD). As segmented medical reference data is scarce, this approach could potentially help increase performances for ultrasound image processing.

We aim at a short inference time of the optical flow to either create an additional input for further image analysis networks or to use the optical flow itself for segmentation propagation on mobile devices, such as tablets or phones. This constrains size and throughput of the network, as computational power on mobile devices differs greatly from stationary setups. Therefore, we use a lightweight version of the aforementioned PDD-Net as student.

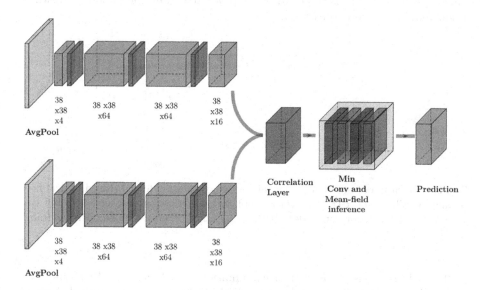

Fig. 1. Overview of the PDD-Network architecture for image registration, which comprises deformable convolutions with batch normalisation and ReLU (red), a correlation layer (blue) and differentiable mean-field inference as regularisation (purple and green). (Color figure online)

We use the PDD-Net [2], which achieved competitive results in the Learn2Reg challenge [17], and was made available[1] in a 2D version (Fig. 1). In this version of the model, an average pooled (yellow) input image is processed by three convolutional layers each followed by batch normalisation and ReLU (red). After the first convolution, we adapt the 2D implementation by applying an Obelisk layer [18], which is then followed by two more convolutional layers. The Obelisk layer is a form of deformable convolution, which uses learnable weights and a gridsampling operator, to increase the receptive field of the next convolutional layer [18].

---

[1] https://www.kaggle.com/mattiaspaul/learn2reg-tutorial.

For a fixed and a moving image, the extracted features are correlated, akin to the correlation used in Flownet-C [11] and then further processed with min convolutions and mean-field inference [19] (gray box).

The whole model yields an inference time of around 2.7 ms on an Nvidia RTX 2060 Ti GPU. When looking at the model (Fig. 1), we can see two feature extractors, which share weights. By processing one fixed frame at time t and keeping this frame as a fixed frame, we only need to process the moving frame at point $t + x$ of the video through the CNN. By reducing the convolutional operations needed during video processing, the network's inference time can be reduced to 1.7 ms. The same optimization can be applied when using different fixed images. In that case the extracted feature map of the moving frame (at time t) can be re-purposed as feature map of fixed frame (at time $t + x$), when a new moving frame is presented.

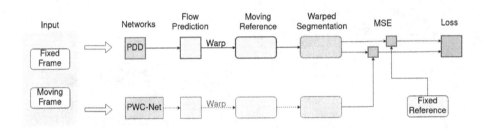

Fig. 2. Illustration of our concept for knowledge distillation for DL-based optical flow estimation. The teacher (PWC-Net) was not trained on ultrasound sequences but can provide a soft target for our student (PDD-Net) based on only a single reference frame segmentation.

The PDD-Net is trained on a combination of soft and hard targets. The hard target loss is calculated as the MSE between the one-hot encoded reference segmentation ("fixed reference" in Fig. 2), and the networks' prediction. The prediction is generated by using the predicted flow field to warp the reference segmentation from the moving frame towards the fixed frame. This warped segmentation is then compared to the reference segmentation of the fixed frame.

We use the established optical flow estimator PWC-Net [1] as a teacher to provide soft targets during training. This is done as shown in Fig. 2. To generate the soft targets, the PWC-Net's optical flow prediction is used to warp the reference segmentation of the moving frame towards the fixed frame. We calculate the MSE loss between the one-hot encoded warped moving reference segmentations of teacher and student networks. The soft and hard target loss are then summed up, where the soft target loss is scaled by 0.5.

Experimental Setup: We train two networks with different methods on the same data. One network is trained solely on hard labels (labeled PDD), as described above. The other network is trained with additional soft target influence (labeled PDD_{KD}). The dataset used for training and evaluation was pro-

Segmented Frame at 15

Segmented frame at 50

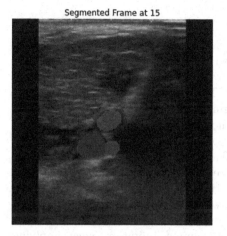
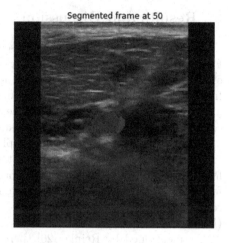

Fig. 3. Exemplary image pair used in the fine tuning data set. Reference segmentations are added for better visualization where the artery is shown in blue and the vein is marked in red. (Color figure online)

vided by ThinkSono GmbH[2]. It contains video sequences of DVT examinations that were annotated by experts. An overlay of these reference segmentations can be seen in Fig. 3. We use 250 video IDs to create two datasets with which we capture two distinct properties. The first dataset is created as a training dataset and contains 1743 image pairs with a fixed frame distance of 6 frames that were randomly sampled. Thus, capturing smaller and larger displacements while also providing heterogenous image quality. The second dataset is created as a fine-tuning dataset. This dataset is created to provide task specific data. For every ID, we select one random frame in the first fifth of the video, or before the onset of the vein compression (whatever came first). We then sample the coming frames with a frame distance of 4 and create various image pairs with the same fixed frame and different moving frames, resulting in 3285 image pairs. In this dataset larger displacements and vein compressions are captured. The evaluation task is to propagate a single reference segmentation of veins and arteries through a video of unseen IDs of about 10 s of a DVT examination.

We proceed to train the PDD-Net adaptation on the training data set with additional soft targets from the PWC-Net (Fig. 2) over 100 epochs with a learning rate of 0.002 and an Adam optimizer. We then trained the distilled network on the fine-tuning data set for 200 epochs with a learning rate of 0.00025. For comparison, we also train one version of the PDD-Net adaptation without additional soft targets in the same manner.

---

[2] https://thinksono.com/.

3   Results and Discussion

We evaluate both networks on 23 unseen videos containing approximately 1600 Frames overall. For each video, we selected one random frame in the first fifth of the video, which we refer to as f_t, for frame at time point t. Each following frame f_{t+x} is used as moving frame input. The estimated optic flow between f_t and f_{t+x} is used to warp the reference segmentation from t to $t + x$, where it is compared to reference segmentation at time pint $t + x$.

This procedure allows us to apply the mentioned runtime optimization towards video processing. By passing the fixed frame once, keeping it in memory for correlation, solely the moving frames need to be passed through the CNN for feature extraction. The reduced inference time per image is about as fast as a reference segmentation network, nnU-Net, which takes 1.6 ms on the same GPU (Nvidia RTX 280Ti).

As mentioned by Reinke [20] there are common limitations when applying only one metric to measure the performance of segmentation masks. Therefore, we evaluate the two networks on Dice score and Hausdorff distance. The dice score is used as a measurement of overlap between the reference and predicted segmentation. It ranges from 0 to 1, where 1 is the best score, which we have denoted by ↑. The HD is used as a measurement of furthest distance between reference and predicted segmentation. We show the absolute values, where lower is better, as denoted by ↓. The mean results over all IDs can be seen in Table 1.

Table 1. Mean Dice ↑ over the test IDs and Mean HD ↓ over the IDs. Comparison between label loss and KD trained PDD-Nets

Score	Registration		Segmentation
	PDD	*PDD$_{KD}$*	nnU-Net
vein Dice %	46.9 ± 4.13	**47.92** ± 4.15	45.93 ± 6.47
artery Dice %	44.48 ± 6.08	*46.67* ± 6.28	**66.80** ± 6.91
overall Dice %	45.69 ± 5.0	*47.3* ± 5.09	**56.36** ± 7.77
vein HD	25.28 ± 166.82	*24.16* ± 159.5	**23.71** ± 366.06
artery HD	28.3 ± 205.19	*27.7* ± 205.54	**26.88** ± 640.51
overall HD	26.79 ± 183.79	*25.93* ± 181.26	**25.33** ± 508.84

We found the distilled network to perform slightly better compared to the label loss trained network over both metrics. When looking at the dice score between the two networks, we found a 2% increase in accuracy over artery segmentation and a 1% increase in vein segmentation. When looking at the HD, we found a similar pattern. The KD trained network outperforms the label loss trained network slightly. We argue that this slight increase is due to the different conceptual representation learned by the distilled network, which would be in line with current research [14,16,21]. The PWC-Net scored at 40.56 ± 3.74 in overall dice and 26.51 ± 160.42 in overall HD on the evaluation videos.

When compared to a 2D segmentation network (nnU-Net [22] Table 1), which was trained on the same image IDs, as the optical flow estimator, we find that the distilled network is performing slightly worse in HD, and worse in Dice score. This result is somewhat expected, since the motion during longer sequences can have significant deformations (compression of veins) and substantial drift. The frame-by-frame segmentation is in principle translation invariant and was trained with a large number of ground truth segmentation annotations. However, when visually looking at estimated segmentations (and quantitatively the variance in HD between the optical flow method and the nnU-Net), we can see that the segmentation network has limited temporal consistency. This suggests that the 2D nnU-Net creates less smooth segmentations over a video, compared to the optical flow method. In the future, we therefore plan to experiment with the optical flow as additional input for a segmentation network. Using a deformation field between two frames, instead of a stacked tensor of all frames, can reduce the computational effort needed for processing, while at the same time containing almost as much information as stacked consecutive frames.

Especially during compression of the vein, this additional information can be leveraged. Figure 4 shows the estimated deformation field between the fixed and the moving frame. The compression is clearly visible as and marked with a black bounding box.

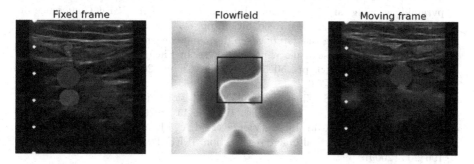

Fig. 4. Visualised deformation field between fixed and moving frame. Segmentation was overlayed for better visibility. The bounding box shows where the compression of the vein (pink) is located and in which direction the vein is compressed. (Color figure online)

4   Conclusion

In this paper, we presented experiments on possible benefits of cross-domain knowledge distillation (from computer vision to medical imaging) for training an optical flow estimator. By using additional teacher-generated soft targets during training, we were able to achieve a small increase in Dice score and a small decrease in Hausdorff distance. This shows that cross-domain KD can have a beneficial effect applied in the training of an image registration network.

We were able to adjust our approach to video inference, such that it is capable of running in realtime, with 1.7 ms per frame pair or more than 500 frames per second. Estimating our approach to use approximately 0.14 GFlops per image, we can calculate an upper limit of roughly 230 frames per second on modern mobile GPUs (Qualcomm Adreno 660).

Performance of segmentation networks still exceeded segmentation via this optical flow based registration of the labels. But we suggest an increase in the segmentation networks' accuracy is possible by combining optical flow information with image features, to add temporal context to the segmentation formation. This was already suggested in previous research in medical video segmentation [23], where improved temporal coherence is reported when optical flow is incorporated. Therefore, we will further investigate the influence of optical flow on vessel segmentation in ultrasound videos.

The results are part of a masters thesis and the code is made available via a git repository[3].

References

1. Sun, D., Yang, X., Liu, M.-Y., Kautz, J.: PWC-Net: CNNs for optical flow using pyramid, warping, and cost volume. In: Proceedings of the IEEE Conference on Computer Vision and Pattern Recognition, pp. 8934–8943 (2018)
2. Heinrich, M.P.: Closing the gap between deep and conventional image registration using probabilistic dense displacement networks. In: Shen, D., et al. (eds.) MICCAI 2019. LNCS, vol. 11769, pp. 50–58. Springer, Cham (2019). https://doi.org/10.1007/978-3-030-32226-7_6
3. Yao, R., Lin, G., Xia, S., Zhao, J., Zhou, Y.: Video object segmentation and tracking: a survey. ACM Trans. Intell. Syst. Technol. **11** (2020)
4. Klein, S., Staring, M., Murphy, K., Viergever, M.A., Pluim, J.P.: Elastix: a toolbox for intensity-based medical image registration. IEEE Trans. Med. Imaging **29**(1), 196–205 (2009)
5. Mok, T.C.W., Chung, A.C.S.: Conditional deformable image registration with convolutional neural network. In: De Bruijne, M., et al. (eds.) MICCAI 2021. LNCS, vol. 12904, pp. 35–45. Springer, Cham (2021). https://doi.org/10.1007/978-3-030-87202-1_4
6. Hu, Y., et al.: Weakly-supervised convolutional neural networks for multimodal image registration. Med. Image Anal. **49**, 1–13 (2018)
7. Noble, J.A.: Reflections on ultrasound image analysis (2016)
8. De Luca, V., et al.: Evaluation of 2d and 3d ultrasound tracking algorithms and impact on ultrasound-guided liver radiotherapy margins. Med. Phys. **45**(11), 4986–5003 (2018)
9. Liu, F., Liu, D., Tian, J., Xie, X., Yang, X., Wang, K.: Cascaded one-shot deformable convolutional neural networks: developing a deep learning model for respiratory motion estimation in ultrasound sequences. Med. Image Anal. **65**, 101793 (2020)

---

[3] https://github.com/TillNicke/KD-for-optical-flow.git

10. Tanno, R.: AutoDVT: joint real-time classification for vein compressibility analysis in deep vein thrombosis ultrasound diagnostics. In: Frangi, A.F., Schnabel, J.A., Davatzikos, C., Alberola-López, C., Fichtinger, G. (eds.) MICCAI 2018. LNCS, vol. 11071, pp. 905–912. Springer, Cham (2018). https://doi.org/10.1007/978-3-030-00934-2_100

11. Dosovitskiy, A., et al.: Flownet: learning optical flow with convolutional networks. In: Proceedings of the IEEE International Conference on Computer Vision, pp. 2758–2766 (2015)

12. Ilg, E., Mayer, N., Saikia, T., Keuper, M., Dosovitskiy, A., Brox, T.: FlowNet 2.0: evolution of optical flow estimation with deep networks. In: Proceedings of the IEEE Conference on Computer Vision and Pattern Recognition, pp. 2462–2470 (2017)

13. Mok, T.C.W., Chung, A.C.S.: Large deformation diffeomorphic image registration with Laplacian pyramid networks. In: Martel, A.L., et al. (eds.) MICCAI 2020. LNCS, vol. 12263, pp. 211–221. Springer, Cham (2020). https://doi.org/10.1007/978-3-030-59716-0_21

14. Li, J., Zhao, R., Huang, J.-T., Gong, Y.: Learning small-size DNN with output-distribution-based criteria. In: Fifteenth Annual Conference of the International Speech Communication Association (2014)

15. Yuan, L., Tay, F.E., Li, G., Wang, T., Feng, J. : Revisiting knowledge distillation via label smoothing regularization. In: Proceedings of the IEEE/CVF Conference on Computer Vision and Pattern Recognition, pp. 3903–3911 (2020)

16. Kim, T., Oh, J., Kim, N., Cho, S., Yun, S.Y.: Comparing kullback-leibler divergence and mean squared error loss in knowledge distillation, arXiv preprint arXiv:2105.08919 (2021)

17. Hering, A., et al.: Learn2Reg: comprehensive multi-task medical image registration challenge, dataset and evaluation in the era of deep learning, arXiv preprint arXiv:2112.04489 (2021)

18. Heinrich, M.P., Oktay, O., Bouteldja, N.: Obelisk-net: fewer layers to solve 3d multi-organ segmentation with sparse deformable convolutions. Med. Image Anal. **54**, 1–9 (2019)

19. Krähenbühl, P., Koltun, V.: Efficient inference in fully connected CRFs with gaussian edge potentials. Adv. Neural Inf. Process. Syst. **24**, 109–117 (2011)

20. Reinke, A., et al.: Common limitations of image processing metrics: a picture story, arXiv preprint arXiv:2104.05642 (2021)

21. Hofstätter, S., Althammer, S., Schröder, M., Sertkan, M., Hanbury, A.: Improving efficient neural ranking models with cross-architecture knowledge distillation, arXiv preprint arXiv:2010.02666 (2020)

22. Isensee, F., Jaeger, P.F., Kohl, S.A., Petersen, J., Maier-Hein, K.H.: nnU-Net: a self-configuring method for deep learning-based biomedical image segmentation. Nat. Methods **18**(2), 203–211 (2021)

23. Yan, W., Wang, Y., Li, Z., van der Geest, R.J., Tao, Q.: Left ventricle segmentation via optical-flow-net from short-axis cine MRI: preserving the temporal coherence of cardiac motion. In: Frangi, A.F., Schnabel, J.A., Davatzikos, C., Alberola-López, C., Fichtinger, G. (eds.) MICCAI 2018. LNCS, vol. 11073, pp. 613–621. Springer, Cham (2018). https://doi.org/10.1007/978-3-030-00937-3_70

Metrics/Losses

Motion Correction in Low SNR MRI Using an Approximate Rician Log-Likelihood

Ivor J. A. Simpson[1(✉)], Balázs Örzsik[2], Neil Harrison[3], Iris Asllani[2,4], and Mara Cercignani[3]

[1] Department of Informatics, University of Sussex, Brighton, UK
i.simpson@sussex.ac.uk
[2] CISC, Brighton and Sussex Medical School, Brighton, UK
[3] CUBRIC, University of Cardiff, Cardiff, UK
[4] Biomedical Engineering, Rochester Institute of Technology, Rochester, USA

Abstract. Certain MRI acquisitions, such as Sodium imaging, produce data with very low signal-to-noise ratio (SNR). One approach to improve SNR is to acquire several images, each of which takes may take more than a minute, and then average these measurements. A consequence of such a lengthy acquisition procedure is subject motion between each image. This work investigates a solution for retrospective motion correction in this scenario, where the high level of Rician noise renders standard registration tools less effective. We employ a simple generative model for the data based on tissue segmentation maps, and provide a differentiable approximation of the Rician log-likelihood to fit the model to the observations. We find that this approach substantially outperforms a Gaussian log-likelihood baseline on synthetic data that has been corrupted by Rician noise of varying degrees. We also provide results of our approach on real Sodium MRI data, and demonstrate that we can reduce the effects of substantial motion compared to a general purpose registration tool.

Keywords: Motion correction · Rician distribution · Low SNR

1 Introduction

Subject motion is a common issue in long MRI acquisition protocols; in situations where several images have been acquired, motion can be retrospectively corrected using image registration. For brain MRI images with reasonable signal-to-noise-ratio (SNR), general purpose linear image registration tools, e.g. [2,3,8,10,13], have been shown to be highly effective. However, in low SNR MRI data, such as acquired with Sodium MRI, traditional cost functions may become less effective. One cause is the noise properties of the analysed data, which consists of the magnitude of the complex signal components. The noise in such data is described using a Rician distribution [6]. When the SNR of the acquired complex signal is high, the resulting noise is approximately Gaussian. Conversely, when the SNR

© The Author(s), under exclusive license to Springer Nature Switzerland AG 2022
A. Hering et al. (Eds.): WBIR 2022, LNCS 13386, pp. 147–155, 2022.
https://doi.org/10.1007/978-3-031-11203-4_16

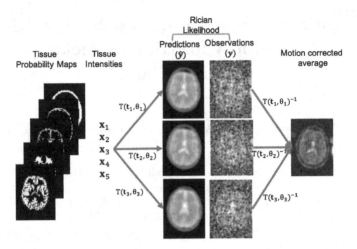

Fig. 1. Illustration of the generative model, which predicts noise free images, $\hat{y}$, parameterised by: T1 estimated tissue segmentation maps, G, multiplied by estimated tissue intensities, $\mathbf{x}$. These are transformed by a translation, $\mathbf{t}$ and rotation θ. The error between the observations, $\mathbf{y}$ and predictions is described using a Rician likelihood, which is used to drive the parameter estimation.

is low the Rician distribution is asymmetric and dissimilar from a Gaussian. This distinction is particularly significant for registration approaches considering cost functions derived from a Gaussian, e.g. sum-of-squared differences.

This paper introduces a linear motion correction model using a simple generative model of the data. This is inspired by the seminal "Unified Segmentation" paper [1]. A diagram of our approach is given in Fig. 1. Our model produces noise-free predictions, which are rigidly aligned to each of the observed images. The novel contribution of this work lies in our approximation of the Rician log-likelihood that enables gradient estimates through automatic differentiation [14]. This is in contrast to previous work using Rician likelihoods for motion correction [16], which required a gradient-free optimisation of the transformations.

We demonstrate how our approach can be used to remove substantial motion from Sodium MRI data. Sodium is an emerging imaging modality, with several potential biomedical applications [11,19]. However, it has poor SNR due to the relatively low concentration and magnetic susceptibility of Sodium, as shown in Fig. 1. Our results illustrate the effectiveness of this approach in removing substantial motion from high noise situations in both real and synthetic datasets.

2   Background: The Rice Distribution

The noise in magnitude MR images is known to follow a Rice distribution [6]:

$$p(\mathbf{y}|\hat{\mathbf{y}}, \sigma) = \text{Rice}(\mathbf{y}; \hat{\mathbf{y}}, \sigma) = \frac{\mathbf{y}}{\sigma^2} \exp\left(\frac{-(\mathbf{y}^2 + \hat{\mathbf{y}}^2)}{2\sigma^2}\right) I_0\left(\frac{\mathbf{y}\hat{\mathbf{y}}}{\sigma^2}\right) \qquad (1)$$

where I_0 is a modified Bessel function of the first kind with order zero (described in Sect. 3.2). Unlike the Gaussian, this distribution: is not symmetric with respect to its first parameter, $\hat{\mathbf{y}}$; does not fulfill any of the algebraic conjugacy properties that enable derivation of closed-form parmeter updates; it also does not provide an obvious cost function for directly comparing two images, as it requires a parameterisation in terms of the clean signal, $\hat{\mathbf{y}}$. Generative models can be used to provide such a parameterisation [1].

3   Method

We consider a generative model for the image data based on 5 probabilistic tissue segmentation maps, G, derived from a T1 image acquired in the same space. We denote G as a matrix of size $N \times 5$, where N corresponds to the number of voxels. The intensity of any voxel can be predicted by matrix multiplication with $\mathbf{x}$, a vector containing the intensity for each tissue class. We consider a geometric transformation associated with each observed image:

$$\hat{\mathbf{y}}_i = \mathrm{P}(\mathrm{T}(G\mathbf{x}, \mathbf{t}_i, \boldsymbol{\theta}_i)) \tag{2}$$

where T provides a rigid transformation of $G\mathbf{x}$, according to translation $\mathbf{t}$ and rotation parameters given by $\boldsymbol{\theta}$. We also include a convolution, P, which corresponds to the point-spread function of the acquisition sequence; this is estimated a-priori from the sequence reconstruction method [18]. The predictions $\hat{\mathbf{y}}_i$ can now be fit to the observed data $\mathbf{y}_i$ using an appropriate likelihood function.

3.1   Priors

In this problem, we are considering the registration of noisy data. Accordingly, the model requires the specification of prior knowledge to enable robust inference. We choose a physiologically based Gaussian prior over the concentration of Sodium, measured in mM, for different tissue types:

$$p(\mathbf{x}) = \mathcal{N}([40, 30, 140, 50, 50], [4, 4, 6, 10, 10]^2) \tag{3}$$

where the means are from [11] and the standard deviations are empirically selected.

The translations have a Normal prior, with a standard deviation specified in mm. The rotations, which are described through an axis-angle representation (in Radians), also employ a Normal prior distribution:

$$p(\mathbf{t}_i) = \mathcal{N}(0, 1.25^2)$$
$$p(\boldsymbol{\theta}_i) = \mathcal{N}(0, 0.025^2)$$

3.2   A Stable Approximation of the Rician Log-Likelihood

Most of the Rician likelihood (Eq. 1) is amenable to efficient calculation in a differentiable manner. However, I_0, corresponds to a modified Bessel function of the first kind with order zero [20], which is an infinite series:

$$I_0(z) = \sum_{k=0}^{\infty} \frac{(\frac{1}{4}z^2)^k}{(k!)^2} \tag{4}$$

The result can be approximated as a sum of the first N_k terms. However, this necessitates a differentiable form for the factorial in the denominator. By noting both that $k! = \Gamma(k+1)$, where Γ is the Gamma function, and that we only require the log probability, we can write an approximation for $\log I_0(z)$ as:

$$\log I_0(z) \approx \text{log-sum-exp}(\mathbf{k}(\log(0.25) + 2*\log(z)) - 2\ln\Gamma(\mathbf{k}+1)) \tag{5}$$

where $\ln\Gamma$ refers to the log Gamma function. $\mathbf{k}$ is a vector containing values from 0 to N_k, which is summed over. log-sum-exp$(\mathbf{z})$ is a numerically stable and convex function [5] for calculating the logarithm of the sum of exponentiated terms, log-sum-exp$(\mathbf{z}) = \log(\sum_i \exp(z_i))$. This implementation is empirically numerically stable, although inefficient in terms of memory as we require multiplying each voxel by N_k values. We found that $N_k = 50$ provided sufficient precision.

3.3   Inference

We perform maximum-a-posteriori (MAP) inference on the model parameters $\Theta = \{\mathbf{x}, \mathbf{t}, \boldsymbol{\theta}, \sigma\}$, with the following cost function:

$$\mathcal{L} = -\sum_{i}^{N} [\log p(\mathbf{y}_i|\mathbf{x}, \mathbf{t}_i, \boldsymbol{\theta}_i, \sigma) + \log p(\mathbf{t}_i) + \log p(\boldsymbol{\theta}_i)] + \log p(\mathbf{x}) \tag{6}$$

Updates alternated between two groups of parameters, those that are shared for all images $\Theta_1 = \{\mathbf{x}, \sigma\}$ and those that vary per image $\Theta_2 = \{\mathbf{t}, \boldsymbol{\theta}\}$. The updates for Θ_1 were calculated using batches of 5 images at a time, and Θ_2 were updated per image. To account for the batching in updating Θ_1, we perform two update steps on these parameters for every step for Θ_2. The Adam [9] optimiser was used to optimise the model parameters, with a fixed learning rate of $2e^{-2}$ for Θ_1 and $1e^{-3}$ for Θ_2 with $\beta_1 = 0.0$ and $\beta_2 = 0.9$. We stopped the inference after 300 rounds of iterations, at which point the model parameters appeared to have converged. This took approximately 3.5 min for 16 images, or 4.5 min for 32 images on an NVIDIA Quadro RTX 6000 with 24 GB of RAM.

4   Experiments

4.1   Synthetic Data

We generate synthetic data by drawing samples from our generative model with random tissue parameters, drawn from Eq. 3, with additional random voxelwise variability with standard deviations $[4, 4, 6, 10, 10]$ mM. These synthetic images were then transformed to simulate random motion, with translations sampled from $\mathcal{N}(0, 5\,mm^2)$ and angles from $\mathcal{N}(0, 0.1^2)$. Each of these images was then corrupted with Rician noise at various levels. We then tried to correct for the simulated motion using our model with either a Gaussian or Rician likelihood.

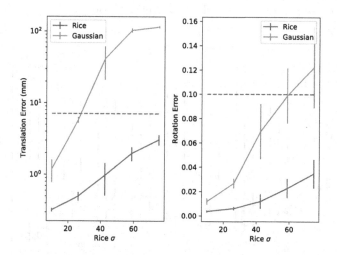

Fig. 2. Synthetic data experiments where the ground truth translation (mean euclidean distance) and rotation error (mean Frobenius norm of the difference of log matrices) are given in the above plots for varying Rician noise level. The dashed line indicates the average initial error. As can be seen, the error when using a Gaussian likelihood rises very quickly, whereas the Rician likelihood is less affected by noise. In this example, $\sigma = 40$ is roughly equivalent to the Sodium MRI data.

Figure 2 illustrates that using the correct likelihood model has a substantial impact on registration performance, particularly in high noise scenarios.

4.2 Real Sodium MRI

^{23}Na MR images were acquired using a dual-tuned, 2-channel (one channel for sodium and one for proton) birdcage ^{23}Na ^{1}H coil developed by RAPID Biomedical GmbH on a 3T Siemens Prisma scanner. Sodium images were acquired using the FLORET spiral sequence [15] with parameters TR = 120 ms, TE = 0.2 ms, FOV = $256 \times 256 \times 256$ mm, flip angle = $80°$, 3 hubs at $22°$, 200 interleaves, pulse duration= 0.5ms and dwell time = 0.01 ms. Each acquisition took 1 min and 10 s, and was repeated either 16 or 32 times. The k-space data were transferred offline and image reconstruction was performed in Matlab using 3D regridding [15] with density compensation [22]. The data was reconstructed with an isotropic resolution of $4\,mm^3$ and an image size of $64 \times 64 \times 64$. Examples slices are shown in Fig. 1. A T1-weighted image ($2\,mm^3$ isotropic) was also acquired using the same coil prior to the Sodium data. This was used for preparing tissue segmentation maps using SPM12.

To enable quantification, a set of 4 Sodium phantoms with known concentrations (30, 50, 70, 120 mM) were attached to the head. We use these to map the tissue specific priors, defined in Eq. 3, to the correct intensity range in each image. This mapping is inferred through linear regression of the median signal for each of these phantoms from the true concentrations.

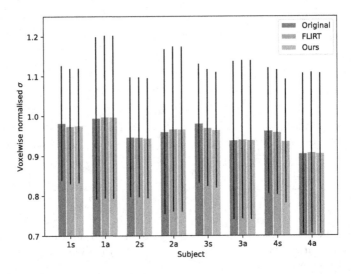

Fig. 3. Bar chart illustrating the mean and std. dev. voxelwise σ, estimated using scipy.stats.Rice, over the motion corrected images for 4 subjects either sleeping (s) or awake (a). These numbers are normalised by the estimated σ in the background.

Using this acquisition protocol, we collected data for 4 subjects either when they are asleep (32 Sodium images) or awake (16 images). The data acquired when sleeping is much more likely to contain motion artefacts due to both the length of the scan and unintentional movements during sleep. Accordingly, we use a more permissive transformation prior (with double the standard deviation for rotation and translations) for these examples.

We experiment with motion correcting the sodium magnitude images using either our proposed approach or "mcflirt" [8], using a cost function of normalized correlation and co-registering to the average image. Nearest neighbour interpolation was used as the final step for both approaches for comparable results without introducing additional smoothness or distortion of noise characteristics.

Validation of the proposed model is complicated by the low SNR exhibited in the motion corrected and averaged images, see Fig. 4 for some examples. Desirable properties of aligned images include: similar values at each voxel over images, and sharp boundaries between regions in the average image. We can measure the first of these by fitting a Rice distribution to each voxel, see Fig. 3.

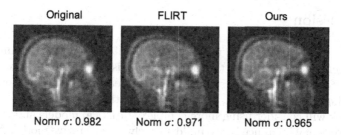

Fig. 4. Example average images calculated by averaging 32 Sodium MRI acquisition. Norm σ refers to the mean voxelwise Rice σ, normalised by an estimate of σ in the image background. In this example, where a lot of motion was detected, our approach leads to a visibly sharper average image.

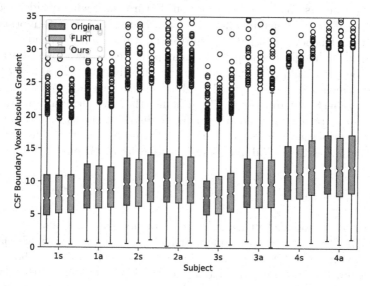

Fig. 5. Boxplot illustrating the absolute gradient of the average image in voxels on the boundary of CSF. Larger values indicate the presence of stronger edges.

We observe that for sleeping acquisitions that are corrupted with visible motion, particularly subject 3 and 4, our approach reduces the mean voxelwise noise compared to other methods. However, in some awake acquisitions, the use of either motion correction approach increases σ; we hypothesise this may be due to interpolation artefacts when correcting sub-voxel motion.

Considering the sharpness of the average image, we can visually observe sharper looking average images in examples with large motion, particularly in subject 3 shown in Fig. 4, where we estimated a mean translation of 7.75 mm (6.49 mm std. dev.) and rotation norm of 0.145 (0.12 std. dev.). To quantify the image sharpness, we examine the distribution of absolute gradient values in voxels that lie on the boundary between CSF and anything else, which should have high contrast. We observe that our motion correction induces stronger edges in most of the acquisitions of sleeping participants.

5   Discussion

The presented approach uses a very simple generative model for the image data, which prevents it overfitting to the high level of noise in the data. However, it also prevents it from making use of strong distinctive features such as the eyes, which contain a high level of Sodium, or the phantoms that are attached to the head. Future work will consider using more complex statistical models and techniques, such as variational inference [7], to build a voxelwise generative model. Amortised inference strategies could also be investigated to improve efficiency [4].

In our experimentation, we observed that in some cases where low motion was observed, our algorithm overestimated the level of movement. We found that this was removed by introducing variable transformation permissiveness based on our prior beliefs on the level of motion. Future work will consider methods for inferring these parameters, and using auto-regressive priors on motion [21].

This work has not investigated preprocessing the data using denoising methods, e.g. [12]; although such approaches may produce cleaner representations for aligning the data, they also manipulate the underlying image statistics being modelled, which may lead to biased results. We also have not compared against the use of robust cost functions [17], although these are generally more suited to heavy tailed rather than asymmetric noise distributions as we have here.

We have published our code on GitHub[1]. The data are not currently available for distribution as the initial analysis of a wider dataset is ongoing.

6   Conclusions

This paper has introduced an algorithm for data modelling and motion correction of low SNR MRI data using a differentiable approximation of the Rician log-likelihood. Our synthetic experiments illustrated the importance of choosing the right cost function for generative models for motion correction, as the Gaussian likelihood performs very poorly where the errors take a different form. On real Sodium MRI data, our results provide support for the use of our method in resolving substantial motion artefacts and creating sharper average images.

Acknowledgments. We acknowledge funding from the University of Sussex used in the data acquisition. We thank Guillaume Madelin for providing the Sodium MRI sequence.

References

1. Ashburner, J., Friston, K.J.: Unified segmentation. Neuroimage **26**(3), 839–851 (2005)
2. Ashburner, J., Neelin, P., Collins, D., Evans, A., Friston, K.: Incorporating prior knowledge into image registration. Neuroimage **6**(4), 344–352 (1997)

---

[1] https://github.com/ivorsimpson/sodium-mri-inference.

3. Avants, B.B., Tustison, N., Song, G., et al.: Advanced normalization tools (ants). Insight j **2**(365), 1–35 (2009)
4. Balakrishnan, G., Zhao, A., Sabuncu, M.R., Guttag, J., Dalca, A.V.: VoxelMorph: a learning framework for deformable medical image registration. IEEE Trans. Med. Imaging **38**(8), 1788–1800 (2019)
5. Boyd, S., Boyd, S.P., Vandenberghe, L.: Convex Optimization. Cambridge University Press, Cambridge (2004)
6. Gudbjartsson, H., Patz, S.: The Rician distribution of noisy MRI data. Magn. Reson. Med. **34**(6), 910–914 (1995)
7. Hoffman, M.D., Blei, D.M., Wang, C., Paisley, J.: Stochastic variational inference. J. Mach. Learn. Res. **14**(40), 1303–1347 (2013). https://jmlr.org/papers/v14/hoffman13a.bib
8. Jenkinson, M., Bannister, P., Brady, M., Smith, S.: Improved optimization for the robust and accurate linear registration and motion correction of brain images. Neuroimage **17**(2), 825–841 (2002)
9. Kingma, D.P., Ba, J.: Adam: a method for stochastic optimization. arXiv preprint arXiv:1412.6980 (2014)
10. Klein, S., Staring, M., Murphy, K., Viergever, M.A., Pluim, J.P.: Elastix: a toolbox for intensity-based medical image registration. IEEE Trans. Med. Imaging **29**(1), 196–205 (2009)
11. Madelin, G., Regatte, R.R.: Biomedical applications of sodium MRI in vivo. J. Magn. Reson. Imaging **38**(3), 511–529 (2013)
12. Manjón, J.V., Carbonell-Caballero, J., Lull, J.J., García-Martí, G., Martí-Bonmatí, L., Robles, M.: MRI denoising using non-local means. Med. Image Anal. **12**(4), 514–523 (2008)
13. Ourselin, S., Roche, A., Subsol, G., Pennec, X., Ayache, N.: Reconstructing a 3D structure from serial histological sections. Image Vis. Comput. **19**(1–2), 25–31 (2001)
14. Paszke, A., et al.: Automatic differentiation in PyTorch (2017)
15. Pipe, J.G., Zwart, N.R., Aboussouan, E.A., Robison, R.K., Devaraj, A., Johnson, K.O.: A new design and rationale for 3D orthogonally oversampled k-space trajectories. Magn. Reson. Med. **66**(5), 1303–1311 (2011)
16. Ramos-Llordén, G., Arnold, J., Van Steenkiste, G., Van Audekerke, J., Verhoye, M., Sijbers, J.: Simultaneous motion correction and t1 estimation in quantitative t1 mapping: an ml restoration approach. In: 2015 IEEE International Conference on Image Processing (ICIP), pp. 3160–3164. IEEE (2015)
17. Reuter, M., Rosas, H.D., Fischl, B.: Highly accurate inverse consistent registration: a robust approach. Neuroimage **53**(4), 1181–1196 (2010)
18. Riemer, F., Solanky, B.S., Stehning, C., Clemence, M., Wheeler-Kingshott, C.A., Golay, X.: Sodium (23Na) ultra-short echo time imaging in the human brain using a 3D-cones trajectory. Magn. Reson. Mater. Phys., Biol. Med. **27**(1), 35–46 (2014)
19. Rose, A.M., Valdes, R., Jr.: Understanding the sodium pump and its relevance to disease. Clin. Chem. **40**(9), 1674–1685 (1994)
20. Wolfram Mathworld: Modified Bessel function of the first kind. https://mathworld.wolfram.com/ModifiedBesselFunctionoftheFirstKind.html
21. Woolrich, M.W., Jenkinson, M., Brady, J.M., Smith, S.M.: Fully Bayesian spatio-temporal modeling of fMRI data. IEEE Trans. Med. Imaging **23**(2), 213–231 (2004)
22. Zwart, N.R., Johnson, K.O., Pipe, J.G.: Efficient sample density estimation by combining gridding and an optimized kernel. Magn. Reson. Med. **67**(3), 701–710 (2012)

Cross-Sim-NGF: FFT-Based Global Rigid Multimodal Alignment of Image Volumes Using Normalized Gradient Fields

Johan Öfverstedt$^{(\boxtimes)}$ ⓘ, Joakim Lindblad ⓘ, and Nataša Sladoje ⓘ

Department of Information Technology, Uppsala University, Uppsala, Sweden
johan.ofverstedt@it.uu.se

Abstract. Multimodal image alignment involves finding spatial correspondences between volumes varying in appearance and structure. Automated alignment methods are often based on local optimization that can be highly sensitive to initialization. We propose a novel efficient algorithm for computing similarity of normalized gradient fields (NGF) in the frequency domain, which we globally optimize to achieve rigid multimodal 3D image alignment. We validate the method experimentally on a dataset comprised of 20 brain volumes acquired in four modalities (T1w, Flair, CT, [18F] FDG PET), synthetically displaced with known transformations. The proposed method exhibits excellent performance on all six possible modality combinations and outperforms the four considered reference methods by a large margin. An important advantage of the method is its speed; global rigid alignment of 3.4 Mvoxel volumes requires approximately 40 s of computation, and the proposed algorithm outperforms a direct algorithm for the same task by more than three orders of magnitude. Open-source code is provided.

Keywords: Image registration · Global · Exhaustive search · NGF · FFT · Matching · GPU implementation

1 Introduction

Multimodal image alignment (also known as registration) involves finding correspondences between images with varying degrees of difference of appearance and structure [18], often applied with the goal of combining the complementary information of each modality via image fusion. Alignment of large displacements is particularly challenging since correspondences to be inferred are far apart and presence of multiple local optima becomes increasingly problematic as the search space grows, thereby often requiring global contextual and spatial information.

A large number of methods exist for multimodal alignment [14], including local optimization methods based on mutual information (MI) [8,17] or normalized gradient fields (NGF) [3,13], and representation extraction techniques

© The Author(s), under exclusive license to Springer Nature Switzerland AG 2022
A. Hering et al. (Eds.): WBIR 2022, LNCS 13386, pp. 156–165, 2022.
https://doi.org/10.1007/978-3-031-11203-4_17

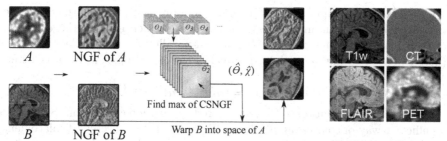

(a) Illustration of the global image volume alignment method.

(b) The modalities included in this study.

Fig. 1. Main steps of one level of the multi-level rigid alignment method (a), and examples of the modalities considered in the evaluation (b) (images from [9]). (a) Two image volumes of modalities A (here [18F] FDG PET), and B (T1 weighted MR), are used as input. For a set of 3D rotations θ, the similarity measure s_{ANGF} between the NGF of A and the NGF of B (rotated), (here shown as RGB images where each color channel represents one component of the 3D vector field $n_1(\cdot; A), n_2(\cdot; A), n_3(\cdot; A)$) is computed for all 3D displacements. The rigid alignment $(\hat{\theta}, \hat{\chi})$ is found by locating the maximum s_{ANGF}.

based on local self-similarities [4] or Deep Feature Learning [5,12]. Most of the (intensity-based) methods are based on some form of local optimization, which usually require a good initial guess to work well. However, several global alignment methods do exist, including [1,6] as well as a recently proposed method based on the cross-mutual information function (CMIF) [10].

We propose a new global alignment method based on NGF that is fast and exhibits excellent performance on a rigid multimodal 3D medical image alignment task. Our evaluation on 6 pairs of modality combinations shows that it outperforms well known methods which rely on local optimization of MI [8,17] and NGF [3] as well as the recently proposed approach based on global optimization of CMIF [10]. Figure 1 illustrates the general idea of the method.

A fast PyTorch-based implementation of the method is shared as open-source at http://github.com/MIDA-group/cross_sim_ngf.

2   Background

The (regularized) normalized gradient field [3], for image A at point x, is

$$n(x; A) = \frac{\nabla A(x)}{\sqrt{\|\nabla A(x)\|_2^2 + \epsilon^2}},$$ (1)

where ϵ is a small constant to reduce the impact of gradients with very small magnitude and avoid division by zero. In this work we use $\epsilon = 10^{-5}$ for $A(x) \in [0, 1]$, selected empirically; higher values yielded more failed alignments and lower values mostly made the measure more noisy.

The main assumption of NGF-based alignment is that parts of images (acquired by different modalities) are in correspondence when the directions of their intensity changes are parallel or anti-parallel. A local similarity of NGF (SNGF) based on the squared dot-product of the elements of the NGF is defined as

$$s_{\mathrm{NGF}}(x; A, B) = \langle \boldsymbol{n}(x; A), \boldsymbol{n}(x; B) \rangle^2. \tag{2}$$

Orientation correlation (OC) and squared orientation correlation (SOC) offer an efficient way of computing SNGF of 2D images for all discrete displacements [1]. In 2D, the vectors $\boldsymbol{n}(\cdot; \cdot)$ are represented as complex numbers. A fast algorithm utilizing log-polar Fourier transform for OC-based alignment w.r.t.rotation and scaling is proposed in [16]. A computationally efficient extension to 3D [2] required a modification of the similarity measure; the authors proposed to, instead of (2), use its unsquared version:

$$s_{\mathrm{US\text{-}NGF}}(x; A, B) = \langle \boldsymbol{n}(x; A), \boldsymbol{n}(x; B) \rangle. \tag{3}$$

By observing three separable components of the (unsquared) dot-product in (3), the authors [2] formulated an algorithm for efficiently computing the measure for all discrete displacements using cross-correlation in the frequency domain. None of the existing work, however, describes a method for computing similarities of NGF using the squared measure (2) efficiently in the frequency domain for 3D volumes, a gap which we aim to fill with this work.

The ability to use the squared measure rather than the unsquared measure is beneficial for multimodal image alignment [1]. Equation (3), similarly to (the unsquared) OC [1], exhibits useful properties such as invariance to changes of contrast and absolute intensity levels, which are suitable for monomodal registration tasks. However, multimodal scenarios are often characterized by the appearance of parts of a sample that are dark in one modality and bright in another; in such cases, aligned samples actually minimize $s_{\mathrm{US\text{-}NGF}}$.

3   Method

Here we define a similarity measure between NGF based on (2), a cross-similarity (c.f.cross-correlation) formulation of the measure, and propose an algorithm for computing it efficiently in the frequency domain for all 3D discrete displacements.

In [3], the point-wise contributions of s_{NGF} (2) are aggregated by summation. A downside of this choice is that it imposes a strong bias towards full overlap of the images which can be especially problematic for global optimization. We instead formulate a scaled similarity measure that is applied to selected regions of the images $A: X_A \to \mathbb{R}$ and $B: X_B \to \mathbb{R}$, defined by indicator functions (masks) $M_A: X_A \to \{0, 1\}$ and $M_B: X_B \to \{0, 1\}$, ignoring the parts of the finite rectangular domains where either M_A or M_B are zero-valued. The average similarity of NGF is

$$s_{\mathrm{ANGF}}(A, B; M_A, M_B) = \frac{1}{\sum_x M_A(x) M_B(x)} \sum_x M_A(x) M_B(x) s_{\mathrm{NGF}}(x; A, B). \tag{4}$$

Based on s_{ANGF}, we define the *Cross Similarity of NGF*

$$\mathrm{CSNGF}(\chi; A, B, M_A, M_B) = \frac{1}{N(\chi)} \sum_x M_A(x) M_B(x+\chi) s_{\mathrm{NGF}}(x; A(x), B(x+\chi)), \quad (5)$$

where $\chi \in S$ is a discrete translation from the set S representing all the considered discrete translations and $N(\chi)$ is the number of overlapping voxels (where M_A and M_B intersect) as a function of χ. $N(\chi)$ can be computed as the cross-correlation between the mask images $N(\chi) = (M_A \star M_B)(\chi)$. An analogous approach is taken in [10] to compute CMIF. Masks are essential for computation of CSNGF, for any choice of S which results in a partial overlap of the images. Figure 1 illustrates CSNGF as a part of a rigid 3D alignment method.

A *direct method* for computing CSNGF for all $\chi \in S$ involves looping over each χ, and compute and aggregate s_{NGF} for all overlapping voxels. If $|S| = O(|X_A|)$, then the run-time complexity of the direct method is $O(|X_A||X_B|)$ which for equisized images A and B gives a quadratic run-time complexity in the size of the images, which is not feasible for volumes of realistic sizes.

We propose a more efficient algorithm for computing CSNGF for all $\chi \in S$ in 3D. By reformulating (2), and expanding the squared dot-product,

$$s_{\mathrm{NGF}}(x; A, B) = \sum_{i=1}^{3} \left(n_i(x;A)^2 n_i(x;B)^2 + 2 \sum_{j=i+1}^{3} n_i(x;A) n_j(x;A) n_i(x;B) n_j(x;B) \right), \quad (6)$$

we express it as 6 separable parts comprising 3 squared components ($i \in \{1, 2, 3\}$), as well as products of 3 pairs of components ($(i,j) \in \{(1,2),(1,3),(2,3)\}$), of the NGF vector fields (see Fig. 1a), which can be computed independently for all χ using cross-correlation. Let n_i^M denote a modified NGF scaled by the associated mask, $n_i^M(x; A) = M_A(x) n_i(x; A)$. The required cross-correlations $((n_i^M(\cdot; A)^2) \star (n_i^M(\cdot; B)^2))$ and $((n_i^M(\cdot; A) n_j^M(\cdot; A)) \star (n_i^M(\cdot; B) n_j^M(\cdot; B)))$ are efficiently computed in the frequency domain; $(n_i^M(\cdot; A)^2 \star n_i^M(\cdot; B)^2) = F^{-1}(\overline{F(n_i^M(\cdot; A)^2)} \odot F(n_i^M(\cdot; B)^2))$, where $F(\cdot)$ denotes the Fourier transform, $\overline{z}$ denotes complex conjugation and $\odot$ denotes element-wise multiplication. For efficiency, the 6 separable parts are aggregated in the Fourier domain. Computing CSNGF involves 14 real-valued FFTs (6 per image plus 1 mask per image) and 2 inverse FFTs. Generalization to nD is straightforward.

3.1   Method for Global 3D Rigid Alignment

The fast algorithm for computing CSNGF for all $\chi \in S$ provides direct means of global optimization of s_{ANGF} w.r.t. axis-aligned shifts. To reach global optimization w.r.t. rigid transformations, we adopt a hybrid approach where the space of 3D rotations $\theta = (\theta_x, \theta_y, \theta_z)$ (represented as Euler angles) is explored via a multi-stage combination of Gaussian pyramids, random search, and global optimization of s_{ANGF}. One stage of this coarse-to-fine method is illustrated in Fig. 1. This multi-stage approach facilitates global search at the lowest considered resolution, followed by more local search to refine the alignment.

Initially, a Gaussian resolution pyramid with m levels is constructed through the application of Gaussian blur and downsampling. For each level $k \in \{1 \ldots m\}$, a random search is performed in a coarse-to-fine sequence, by sampling angles θ either (a) as random rotations from the set of all possible rotations, for the first level ($k = 1$), or (b) as rotations close to one of the p_{k-1} best solutions of the previous level, for levels $k \in \{2, \ldots, m\}$. An angle "close to" is realized by perturbing the previous solution by a change in rotation around axes (x, y, z), sampled from $\mathbb{U}(-u_{k-1}, u_{k-1})$ for each axis. For each θ, the corresponding transformation T_θ is applied to the floating image $B_\theta = B \circ T_\theta$ using trilinear interpolation and its mask $M_{B_\theta} = M_B \circ T_\theta$ using nearest neighbor interpolation. $n(\cdot; B_\theta)$ is computed, followed by computation of $\arg\max_\chi \text{CSNGF}(\chi; A, B_\theta, M_A, M_{B_\theta})$ for all $\chi \in S$, where S is the set of displacements satisfying a user-selected amount of minimum overlap γ. A suitable zero padding scheme is used to enable partial overlaps (following [10]). For $k > 1$, the p_{k-1} best solutions of the previous level are also evaluated unmodified to not risk discarding good solutions. For $k = m$, the best rotation and displacement are taken as the final rigid transformation.

The method is parameterized by blur-levels $\sigma = (\sigma_1, \ldots, \sigma_m)$, downsampling factors $\mathbf{d} = (d_1, \ldots d_m)$, largest allowed steps $\mathbf{u} = (u_1, \ldots u_{m-1})$, number of rotations $\mathbf{a} = (a_1, \ldots a_m)$, and number of kept best solutions $\mathbf{p} = (p_1, \ldots, p_{m-1})$. For all related experiments, $\mathbf{d} = (4, 2, 2, 1)$, $\mathbf{a} = (5000, 3000, 300, 0)$, $\mathbf{u} = (10, 3, 0)$, and $\mathbf{p} = (20, 3, 1)$. We use $\gamma = 0.5$ everywhere in this study.

4   Performance Analysis

The empirical evaluation of the proposed method is based on the CERMEP-IDB-MRXFDG dataset [9], available upon request from the authors. The dataset consists of images of brains of 33 subjects acquired by 4 different modalities: T1 weighted MR, Flair MRI, Computed Tomography (CT), [18F] FDG PET, all mapped to the standard MNI space (see Fig. 1b), thus providing ground-truth for image alignment method evaluation, and a possibility to consider 6 different combinations of modalities, enabling evaluation of the generality of the methods.

4.1   Similarity Landscape of the Average SNGF

First, we perform an empirical analysis of how (4) is affected by spatial transformations of the observed images. The aim is to provide evidence of the relevance of global optimization for multimodal image alignment. We consider two images acquired with the modalities FLAIR and PET and study the similarity landscape as the PET volume is rotated around a single axis of rotation; the result is shown in Fig. 2. We observe that the similarity landscape exhibits characteristics that impede local methods without a good initial guess for all parameters.

4.2   Multimodal Brain Image Volume Alignment

We compare the proposed method with two global and two local alignment methods on the task of recovering rigid transformations of brain image volumes.

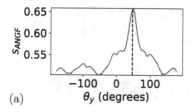 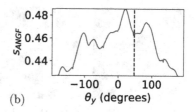

(a)   (b)

Fig. 2. Similarity landscape of s_{ANGF} for a pair of FLAIR and PET images of a brain (blur: $\sigma = 5$), w.r.t.the rotation angle θ_y. Two scenarios are presented: (a) with no additional transformation, *i.e.*, all transformation parameters other than θ_y have their correct values, and (b) when the FLAIR image has been rotated by 5° around a random axis (other than y) and translated by 20 vx in a randomly direction. The vertical dashed lines mark the sought angle. We observe that, (a) even without displacement, the convergence region of the sought angle has a limited size, with local maxima near the global maximum, and that (b) displacements along multiple dimensions make the search using local approaches further challenging; here the sought angle (dashed line) is between local optima.

For each of the twelve (ordered) pairs of modalities (six unordered modality combinations) included in the CERMEP-IDB-MRXFDG dataset, and for each of the first 20 subjects (the last 13 used for parameter tuning), we randomly (uniformly) sample a 3D rotation $\boldsymbol{\theta}$, and an axis-aligned shift $\chi_i \in [-30\ vx, +30\ vx]$ for each axis i. These transformations are applied, using inverse mapping and bicubic interpolation, to the first image volume of each pair. The transformed image is taken as reference image and the untransformed image as floating image in the alignment task. Finally, a block of size $151 \times 151 \times 151\ vx$ (*c.f.*original size $207 \times 243 \times 226$) at the center of the volume is extracted, retaining most of the content of interest, while omitting most of the background and avoiding padding introduced by inverse mapping outside the image domain. This setup enables evaluation of the accuracy of the proposed method w.r.t.alignment of multimodal 3D images by recovering these known transformations. Example slices of pairs from the selected modality combination are shown in Fig. 3.

With the aim to evaluate the benefit of the proposed algorithm, based on the original similarity of NGF (2), compared to the one proposed in [2], we let USNGF refer to an alignment method similar to CSNGF, but with s_{NGF} in (5) replaced by $s_{US\text{-}NGF}$. We evaluate both USNGF and "USNGF-", where the latter denotes USNGF but with an intensity-inverted floating image, to observe the sensitivity of USNGF to the sign of the gradients [1]. We also include the recently proposed CMIF-based global alignment method [10], which has exhibited excellent performance and outperformed several recent Deep Learning methods (including [12]) on multiple biomedical datasets. The selected global optimization methods are implemented in Python/PyTorch [11] with CUDA/GPU-acceleration.

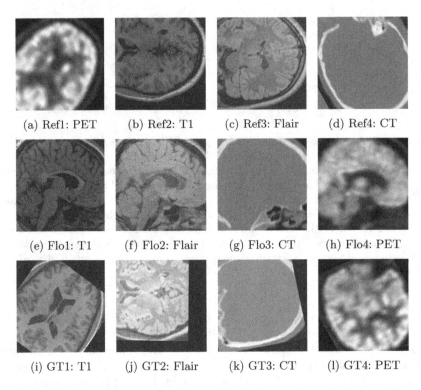

(a) Ref1: PET   (b) Ref2: T1   (c) Ref3: Flair   (d) Ref4: CT

(e) Flo1: T1   (f) Flo2: Flair   (g) Flo3: CT   (h) Flo4: PET

(i) GT1: T1   (j) GT2: Flair   (k) GT3: CT   (l) GT4: PET

Fig. 3. Sample slices of 3D image pairs from the evaluation dataset generated from the CERMEP-IDB-MRXFDG dataset [9]. (a-d) the reference (transformed) images and (e-h) the floating images. Image (e) is to be registered to (a); (f) to (b), (g) to (c) and (h) to (d). The bottom row shows the ground-truth (GT) of each floating (Flo) image aligned to the corresponding reference (Ref) image.

We also compare with local optimization-based methods using MI and NGF as objective functions, relying on open-source implementations Elastix [8] and AIRLab [15] respectively.

We use the mean Euclidean distance between the corresponding corner points of the extracted block before and after the performed (recovered) alignment as a displacement measure, denoted d_E. We consider an alignment successful if $d_E < 5\ vx$, which is approximately 2% of the length of the diagonal of the blocks. For CMIF we use $k = 16$ (for the k-means clustering), and $\sigma = (3.0, 1.5, 1.0, 0.0)$. For NGF, USNGF (and USNGF-), we use $\sigma = (5.0, 3.0, 2.0, 1.5)$. For local optimization MI (LO-MI) [8,17], we use 6 pyramid levels, the Adaptive Stochastic Gradient Descent optimizer [7], 4096 maximum iterations for each level. For local optimization NGF (LO-NGF) [3], we use 5 pyramid levels, ADAM optimizer, iteration counts according to the schedule (4096, 4096, 1024, 100, 50), with downsampling factors (16, 8, 4, 2, 1) and Gaussian smoothing parameters (15.0, 9.0, 5.0, 3.0, 1.0), with learning-rate 0.01. Trilinear interpolation is used.

Results. The results of the evaluation of the 6 considered methods on the multimodal brain image dataset are presented in Table 1. The proposed method provides overall excellent performance, and is the best choice for all observed modality combinations. Most of the competitors show generally poor performance, completely failing on one or more modality combinations. Near-successes are also of interest, since those solutions may be refined with a local optimization method; therefore, we plot the distribution up to the threshold $d_E < 20$ as Fig. 4.

Table 1. Image alignment performance presented in terms of success-rate, where the threshold of success is set to 5 vx. The modality names are abbreviated in the headings (T: T1, F: Flair, C: CT, P: [18F] FDG PET).

Method	Modalities					
	T/F	T/C	T/P	F/C	F/P	C/P
LO-MI	0.05	0.025	0.075	0.025	0.1	0.075
LO-NGF	0.025	0.00	0.00	0.00	0.00	0.00
CMIF	0.675	0.30	0.325	0.80	0.85	0.525
USNGF	0.225	0.00	0.00	0.00	**0.925**	0.10
USNGF-	0.00	0.275	0.00	0.00	0.00	0.00
CSNGF	**1.00**	**0.95**	**0.925**	**0.90**	**0.925**	**0.95**

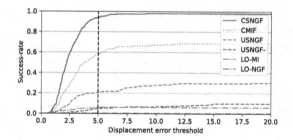

Fig. 4. Success-rate of each considered method as a function of the acceptable displacement error t (fraction of the 240 alignments where $d_E < t$); the results for all modality combinations are aggregated. Up and to the left is better.

4.3   Time Analysis

We compare the run-times of the global rigid registration methods, as well as the run-times of the novel Cross-Sim-NGF algorithm with a direct (not FFT-based) approach. The reported results are obtained on a Nvidia GeForce RTX 3090.

Both the FFT-based algorithm and the direct method are implemented in Python/PyTorch using GPU-acceleration; the direct method consists of a loop over all axis-aligned shifts $\chi \in S$, and computation of the squared dot-products.

The average run-times of the methods CMIF, USNGF, and CSNGF are 569 s, 33 s, and 41 s, respectively. Comparison of the run-times of the FFT-based algorithm and the direct method, as a function of image size, is presented in Table 2. We observe that for size 128, the here proposed algorithm is approximately 6275 times (more than three orders of magnitude) faster.

Table 2. Run-time (s) comparison of FFT-based CSNGF and a direct algorithm for computing CSNGF, for all $\chi \in S$ where the overlap is 50% or higher, on cube image volumes of increasing size (expressed as side-length).

Method	Size				
	8	16	32	64	128
Direct algorithm	0.129	0.557	3.537	27.07	502.4
FFT-based alg.	**0.002**	**0.002**	**0.002**	**0.008**	**0.088**

5   Conclusion

We propose a novel NGF-based method for global rigid 3D multimodal alignment, which extends a well-performing method for 2D image alignment, outperforming a previous extension that relies on an unsquared version of the similarity measure. We confirm both its great performance and its high efficiency. Through the comparison with CMIF-based alignment [10], the method is indirectly compared with several approaches based on deep learning while leaving a more comprehensive comparative study as future work. The method does not use any training (data), which is a large advantage for (bio)medical applications [5].

Acknowledgments. We thank Ines Mérida and team for the CERMEP-IDB-MRXFDG dataset. We acknowledge financial support by Vinnova, MedTech4Health Grant 2017-02447.

References

1. Fitch, A., Kadyrov, A., Christmas, W., Kittler, J.: Orientation correlation. In: Proceedings of British Machine Vision Conference, pp. 133–142 (2002)
2. Fotin, S.V., et al.: Normalized gradient fields cross-correlation for automated detection of prostate in magnetic resonance images. In: Medical Imaging 2012: Image Processing, vol. 8314, p. 83140V. International Society for Optics and Photonics (2012)
3. Haber, E., Modersitzki, J.: Intensity gradient based registration and fusion of multi-modal images. In: Larsen, R., Nielsen, M., Sporring, J. (eds.) MICCAI 2006. LNCS, vol. 4191, pp. 726–733. Springer, Heidelberg (2006). https://doi.org/10.1007/11866763_89
4. Heinrich, M.P., et al.: MIND: modality independent neighbourhood descriptor for multi-modal deformable registration. Med. Image Anal. 16(7), 1423–1435 (2012)

5. Islam, K.T., Wijewickrema, S., O'Leary, S.: A deep learning based framework for the registration of three dimensional multi-modal medical images of the head. Sci. Rep. **11**(1), 1–13 (2021)
6. Jenkinson, M., Smith, S.: A global optimisation method for robust affine registration of brain images. Med. Image Anal. **5**(2), 143–156 (2001)
7. Klein, S., Pluim, J.P.W., Staring, M., Viergever, M.A.: Adaptive stochastic gradient descent optimisation for image registration. Int. J. Comput. Vis. **81**(3), 227 (2008)
8. Klein, S., Staring, M., Murphy, K., Viergever, M.A., Pluim, J.P.W.: Elastix: a toolbox for intensity-based medical image registration. IEEE Trans. Med. Imaging **29**(1), 196–205 (2010)
9. Mérida, I., et al.: CERMEP-IDB-MRXFDG: a database of 37 normal adult human brain [18F] FDG PET, T1 and FLAIR MRI, and CT images available for research. EJNMMI Res. **11**(1), 1–10 (2021)
10. Öfverstedt, J., Lindblad, J., Sladoje, N.: Fast computation of mutual information in the frequency domain with applications to global multimodal image alignment. arXiv preprint arXiv:2106.14699 (2021)
11. Paszke, A., Gross, S., et al.: PyTorch: an imperative style, high-performance deep learning library. Adv. Neural Inf. Process. Syst. **32**, 8026–8037 (2019)
12. Pielawski, N., et al.: CoMIR: contrastive multimodal image representation for registration. In: Neural Information Processing System, vol. 33, pp. 18433–18444 (2020)
13. Pluim, J., Maintz, J., Viergever, M.: Image registration by maximization of combined mutual information and gradient information. IEEE Trans. Med. Imaging **19**(8), 809–814 (2000)
14. Saiti, E., Theoharis, T.: An application independent review of multimodal 3D registration methods. Comput. Graph. **91**, 153–178 (2020)
15. Sandkühler, R., Jud, C., Andermatt, S., Cattin, P.C.: Airlab: autograd image registration laboratory. arXiv preprint arXiv:1806.09907 (2018)
16. Tzimiropoulos, G., Argyriou, V., Zafeiriou, S., Stathaki, T.: Robust FFT-based scale-invariant image registration with image gradients. IEEE Trans. Pattern Anal. Mach. Intell. **32**(10), 1899–1906 (2010)
17. Viola, P., Wells, W.M., III.: Alignment by maximization of mutual information. Int. J. Comput. Vis. **24**(2), 137–154 (1997)
18. Zitova, B., Flusser, J.: Image registration methods: a survey. Image Vis. Comput. **21**(11), 977–1000 (2003)

Identifying Partial Mouse Brain Microscopy Images from the Allen Reference Atlas Using a Contrastively Learned Semantic Space

Justinas Antanavicius[1], Roberto Leiras[2], and Raghavendra Selvan[1,2(✉)]

[1] Department of Computer Science, University of Copenhagen,
Copenhagen, Denmark
raghav@di.ku.dk

[2] Department of Neuroscience, University of Copenhagen, Copenhagen, Denmark

Abstract. Registering mouse brain microscopy images to a reference atlas is crucial to determine the locations of anatomical structures in the brain, which is an essential step for understanding the function of brain circuits. Most existing registration pipelines assume the identity of the reference plate – to which the image slice is to be registered – is known beforehand. This might not always be the case due to three main challenges in microscopy image data: missing image regions (partial data), different cutting angles compared to the atlas plates and a large number of high-resolution images to be identified. Manual identification of reference plates as an initial step requires highly experienced personnel and can be biased, tedious and resource intensive. On the other hand, registering images to all atlas plates can be slow, limiting the application of automated registration methods when dealing with high-resolution image data. This work proposes to perform the image identification by learning a *low-dimensional* space that captures the similarity between microscopy images and the reference atlas plates. We employ Convolutional Neural Networks (CNNs), in the *Siamese* network configuration, to first obtain low-dimensional embeddings of microscopy image data and atlas plates. These embeddings are contrasted with positive and negative examples in order to learn a semantically meaningful space that can be used for identifying corresponding 2D atlas plates. At inference, atlas plates that are closest to the microscopy image data in the learned embedding space are presented as candidates for registration. Our method achieved TOP-3 and TOP-5 accuracy of 83.3% and 100%, respectively, compared to the SimpleElastix-based baseline which obtained 25% in both the Top-3 and Top-5 accuracy (Source code is available at https://github.com/Justinas256/2d-mouse-brain-identification).

Keywords: Image registration · Mouse brain · Partial data · Deep learning

© The Author(s), under exclusive license to Springer Nature Switzerland AG 2022
A. Hering et al. (Eds.): WBIR 2022, LNCS 13386, pp. 166–176, 2022.
https://doi.org/10.1007/978-3-031-11203-4_18

1   Introduction

Determining the location of anatomical structures in a mouse brain is an essential step for analyzing and understanding the architecture and function of brain circuits, and of the overall whole-brain activity [4]. Structures of interest can be located using standardized anatomical reference atlases, usually taking a two-step approach:

1. Identification: The input brain slice has to be identified, i.e., the corresponding 2D atlas plate has to be found.

2. Registration: The identified slice is registered to the corresponding atlas plate. Anatomical structures are determined based on the registered annotated plate.

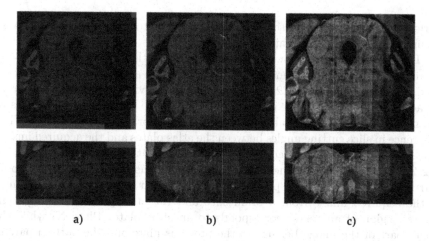

a)   b)   c)

Fig. 1. a) Two typical high-resolution microscopy images showing the cross-sectional view of a mouse spinal cord in pseudo-color. The size of the input images in this work varied between 17408 × 10240 and 25600 × 20480 pixels. Notice the artefacts due to low contrast, tiling and missing regions, which make them challenging to process. b) Input images after gray scale conversion c) Pre-processed images with histogram equalization

In most cases, the acquired microscopy images of brain slices often suffer from artefacts due to missing tissue regions, irregular staining, titling errors, air bubbles and tissue wrinkles [15], as shown in Fig. 1. This is further aggravated due to additional variations in the images depending on the experimental procedures, instrumentation noise, etc. This makes it difficult to identify and register mouse brain images. For these reasons, practitioners usually resort to manually comparing image slices to 2D atlas plates which can be very time-consuming.

Compared to the registration of mouse brain images, the first part of identification has received far less attention from the brain imaging community. At the outset, wrong identification of brain slices could lead to incorrect determination of anatomical structures regardless of how well the image registration itself is

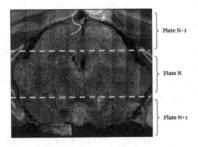

Fig. 2. Due to the difference in cutting angles compared to the atlas plates, no single ground truth plate can be registered to the input images. In this illustration, we point this out where the expert user usually would register different, usually consecutive, plates to different regions of the image.

performed. Therefore, precise determination of anatomical structures requires accurate identification of brain slices as a precursor.

The correspondence between brain slices and atlas plates could be found by reconstructing a 3D volume from the brain slices and then registering them to the 3D reference atlas [12]. However, it is not always possible to construct an accurate brain volume, e.g. when brain slices are cut at different angles or when only few brain slices are available or partial brain images are used. The difference in slice cutting angles between the atlas plates and the acquired images is a common challenge affecting the usefulness of atlas-based registration. In Fig. 2 we illustrate an instance where different regions of the same image could correspond to different atlas plates due to a mismatch between the cutting angle of the acquired image from a brain slice and the atlas plates. This way, the central region of an image corresponds to an atlas plate (Plate N) while the upper part of the image belongs to the previous plate and the bottom part to the next atlas plate. Another approach could be based on content-based image retrieval where images are queried based on some underlying image or sub-image feature descriptions [11,13].

In this study, we investigate the problem of identifying the atlas plates corresponding to mouse brain slices, when the image data are partial and/or acquired at different cutting angles. The brain slices are identified by finding the corresponding 2D coronal plates in the Allen Mouse Brain Atlas [9]. The proposed approach has similarities to some of the ideas explored within the domain of content-based image retrieval [1,13]. The brain slice identification is achieved by using convolutional neural networks (CNNs), used in the Siamese Network configuration [2,8], to obtain low-dimensional representations of the image data. These low-dimensional embeddings are contrasted with positive and negative pairs to learn a semantically meaningful space where the correspondence between brain slices and atlas plates can be determined. The image identification method is compared to SimpleElastix, which is based on the widely used tool Elastix [10], in terms of accuracy and speed.

2 Methods

Siamese Networks: In this work, CNNs are used to identify brain slices by matching them to their corresponding atlas plates. The network architecture is comprised of identical CNNs in the Siamese Network configuration [8], as shown in Fig. 3. The CNN, $S_\theta(\cdot)$, takes an image I (of height H and width W) as input and outputs a low-dimensional feature vector (embedding), h, i.e., $S_\theta(\cdot) : I \in \mathbb{R}^{H \times W} \mapsto h \in \mathbb{R}^L$, where L is the size of the embedding space and θ are the learnable network parameters. In the pairwise setting, two *sister* neural networks with shared parameters are used (Fig. 3-b).

The embeddings for brain slices, treated as the fixed image, are obtained as $h_F = S_\theta(I_F) \in \mathbb{R}^L$. The embeddings for the atlas plates, treated as the moving image, are obtained in a similar manner, $h_M = S_\theta(I_M) \in \mathbb{R}^L$. After obtaining the embeddings of the fixed and moving images, their similarity is determined based on the Euclidean distance between these embeddings, $d(h_M, h_F)$. The reference atlas plate with the lowest distance is then predicted to be the corresponding atlas plate for a given brain slice.

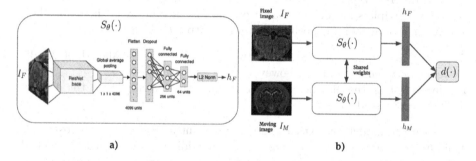

Fig. 3. a) Network architecture of the model used comprising of a ResNet-backbone and a multi-layered perceptron, $S_\theta(\cdot)$, used to obtain low-dimensional embeddings of the brain slices and atlas plates. b) Computing the similarity between brain slices and atlas plates with CNNs based on the low-dimensional representations corresponding to the moving and fixed images obtained from the identical CNNs, in a Siamese network layout, which are further used to compute their pairwise similarity, $d(\cdot)$.

Metric Learning: The distance between the embeddings of more similar images should be smaller than that between dissimilar images for the low-dimensional embedding space to be meaningful. This is achieved in this work using weakly supervised metric learning [3]. The Siamese networks for brain slice identification are trained to learn the representation of images such that corresponding brain slices and atlas plates would be closer to each other in the embedding space. We compare the embedding space learned based on training the networks with two different loss functions:

1. Contrastive loss [5], given as:

$$L = \begin{cases} \frac{1}{2}d(h_F, h_M)^2, & \text{if positive pair} \\ \frac{1}{2}\max(0, m - d(h_F, h_M))^2, & \text{if negative pair} \end{cases} \tag{1}$$

where the positive pair is comprised of the microscopy image, I_F, and the corresponding ground truth atlas plate, I_M, and the negative pair can consist of any non-ground truth atlas plate. The parameter $m \in \mathbb{R}_+$ is the margin used to control the contribution from negative pairs.

2. Triplet loss [14], given as:

$$L = \max(d(h_A, h_P) - d(h_A, h_N) + m, 0) \tag{2}$$

where h_A, h_P, h_N are the embeddings of anchor- (I_A), positive- (I_P) and negative- (I_N) images, respectively. Note that in case of triplet loss, a third sister network with shared weights is included to obtain feature embeddings.

Two different types of triplets (I_A, I_P, I_N) are sampled to calculate the triplet loss. These triplets are defined based on the distance between the embeddings h_A, h_P, h_N of anchor I_A, positive I_P and negative I_N images:

i) Semi-hard triplets: the distance between h_A and h_P is smaller than the distance between h_A and h_N, however, the loss is still positive.
ii) Hard triplets: the distance between h_A and h_N is smaller than the distance between h_A and h_P.

When the models are either trained with contrastive- or triplet- losses, the training process enforces structure to the embedding space so that the embeddings of similar images are pulled closer, whereas embeddings of dissimilar images are pushed away from each other. At inference, new microscopy images are ideally closer to their corresponding atlas plates in the embedding space. An overview of CNNs in Siamese network configuration for atlas plate prediction with moving and fixed images is shown in Fig. 3.

3   Data and Experiments

3.1   Data

Microscopy Data: Eighty-four high-resolution microscopy images of mouse brain slices were acquired using a 10x objective in a Zeiss LSM 900 confocal microscope from four animals. The size of the images varied between 17408×10240 px and 25600×20480 px. Most of the images were partial as they were not capturing the entire brain slice. For instance, the cortex or the cerebellar cortex were captured partially or, in some images, were not captured at all as seen in first column of Fig. 4. The images of brain slices were preprocessed, cropped and equalized using Contrast Limited Adaptive Histogram Equalization

(CLAHE) to reduce some artefacts, as shown in Fig. 1. The dataset was split into four sets: training (50 images), validation-1 (12 images), validation-2 (10 images) and test (12 images).

Ground Truth: The Allen Mouse Brain Atlas [9] was used as the reference atlas. It consisted of 132 Nissl-stained coronal plates spaced at 100 μm, seen in the second column of Fig. 4. The ground truth in these experiments were the atlas plate numbers which were provided by a neuroscientist with expertise in manual registration of these images. For a given brain slice, there could be several matching plates due to the difference in cutting angles, as shown in Fig. 2. However, the domain expert marked a single plate to be the ground truth depending on whichever plate best described specific regions of interest. This is to say, in most applications involving these data there are no hard ground truths as each slice could correspond to several consecutive atlas plates due to the difference in cutting angles.

Data Augmentation: To capture variations in the microscopy data beyond the limited training set extensive data augmentation (affine transformation, cropping and padding, pepper noise) was applied to the training dataset. Data augmentation was performed on all the 50 training set brain slices and also the 132 atlas plates. In order to reduce computations, the high resolution images were resized to square inputs of size 1024^2, 512^2 or 224^2 depending on the experiment.

3.2   Experiments

Experiments: The performance of our CNN-based slice identification method was compared with a baseline SimpleElastix-based algorithm that identifies brain slices based on mutual information (MI). The baseline method affinely registers each brain slice with every atlas plate and picks the atlas plate with the highest

Table 1. Mean Absolute Error (MAE) on the *validation-2* dataset for identifying brain slices with our method. The lowest MAE is achieved by the network with ResNet50v2 base, trained with semi-hard triplet loss and using 1024^2 images. B is the training batch size.

Loss	B	ResNet50v2			ResNet101v2		
		224^2	448^2	1024^2	224^2	448^2	1024^2
Triplet (semi-hard)	32	2.5	2.2	2.8	1.9	3.1	3.1
	16	2.0	3.7	**1.8**	2.6	2.1	2.7
Triplet (hard)	32	2.4	3.0	3.0	2.8	3.7	2.7
	16	3.1	2.8	2.6	2.0	2.7	2.8
Contrastive	32	3.6	2.1	3.4	4.2	2.5	5.6

MI. In total, 100 random hyperparameters from the SimpleElastix affine parameter map were tested. The results of the best performing baseline model (with 7 resolutions using recursive image pyramid and random sample region, 2800 iterations in each resolution level and disabled automatic parameter estimation) are used for comparison.

Metrics: The methods were evaluated based on three metrics: Mean Absolute Error (MAE), TOP-N accuracy and inference time. MAE measured the accuracy of predictions. For each brain slice all 132 atlas plates were ranked (starting from zero) based on the similarity score (the Euclidean distance or MI, depending on the method). Then MAE was computed as $MAE = (\sum_{i=0}^{N} y_i)/N$, where N is the number of brain slices, y_i is the position of ranked ground truth atlas plate for a given brain slice i. With 132 atlas plates used, MAE can have values in the range $[0, 131]$. If all brain slices are identified correctly, MAE is equal to 0. To account for the inherent ambiguity in ground truth we report Top-3, Top-5 and Top-10 accuracy.

Hyperparameters: Fig. 3-a) shows the architecture of the Siamese Networks with the embedding space feature dimension $L = 64$. The base of network consists of a CNN-backbone implemented as ResNet network [6] pre-trained on the ImageNet dataset. The CNN backbone is followed by a multi-layered perceptron that outputs the embedding. While training the networks, all layers of the ResNets were *frozen* except the last ones starting with the prefix *conv5*. The networks were trained on the *training* dataset for a maximum of 10 k iterations using the Adam optimizer [7] with an initial learning rate of 10^{-4}. The experiments were performed on Nvidia GeForce RTX 3090 GPU, i7-10700F CPU and 32 GB memory. The training was stopped if MAE on the *validation-1* dataset was not decreasing for more than 2 k iterations.

Results: The converged models based on *validation-1* set were evaluated on the *validation-2* dataset, and the MAE performance for two ResNet backbones (ResNet50, ResNet101), the various loss functions, input- and batch- sizes are reported in Table 1. The best performing configuration is the ResNet50 backbone network trained with batch size (B) of 16 using input size 1024^2 with the semi-hard triplet loss with MAE=1.8. This best performing model was further evaluated on the *test* dataset and compared with the SimpleElastix-based approach, reported in Table 2. We notice that the MAE on test set for our method is 1.42 compared to 60.4 for the baseline. Our method obtained Top-3 accuracy of obtaining 83.3% compared to 25% for the baseline. A similar trend is observed for Top-5 and Top-10 accuracy, where our method achieves 100% accuracy. The total inference time on the *test* set for the two methods are also reported in Table 2 where we observe that the baseline method takes orders of magnitude more time than the trained CNN model.

Finally, the Top-5 predicted atlas plates on a subset of the *test* dataset are reported in Table 3. In all the cases, the ground truth plate is within the Top-5 predictions highlighted in bold. Examples of the predicted atlas plates by our method that have the highest similarity are visualized in Fig. 4.

Table 2. Performance of our CNN-based method compared to the SimpleElastix-based approach on the *test* dataset for identifying brain slices reported as Top-N accuracy. Our method trained with semi-hard triplet loss outperforms SimpleElastix-based approach by a large margin in all the evaluated metrics. Inference time measures the time taken to identify all 12 brain slices from the *test* dataset.

	MAE	TOP-1	TOP-3	TOP-5	TOP-10	Infer. time
SimpleElastix	60.4	16.7%	25%	25%	25%	12h 25 m
Siamese Networks	**1.42**	**25%**	**83.3%**	**100%**	**100%**	**7.2 s**

Table 3. Identifying brain slices from the subset of the *test* dataset: the labels of ground truth and Top-5 predicted atlas plates by our CNN-based method. Even though some predictions are incorrect, all of them are close to the ground truth labels. Labels define the position of atlas plates in the reference atlas.

Ground truth	Top-5 predictions
91	92, **91**, 93, 90, 94
130	129, 128, **130**, 131, 127
86	87, 88, **86**, 85, 89
63	62, 61, 60, **63**, 59
108	109, 110, 111, 112, **108**

4   Discussions and Conclusions

Our CNN-based method in the Siamese network configuration used to identify brain slices have shown impressive results, i.e. in finding corresponding coronal 2D atlas plates. Our method performed well even when most images were missing image regions, and some images belonging to different classes (plate numbers) looked very similar to each other, thus making the identification task even more complex. Training with contrastive- and triplet- losses solve this issue by using margin, i.e., dissimilar images are not pushed away if the distance between them is larger than the margin.

The identification accuracy (MAE) had no clear correlation with the batch size (16 and 32), the image resolution (224×224, 448×448, 1024×1024) and the type of the base for the Siamese network (ResNet50v2 and ResNet101v2), as seen in Table 1. However, using images with lower resolution and networks with fewer parameters could further improve the inference time. We did not observe the performance of our method to be highly influenced by the choice of loss functions. The models trained with triplet loss rather than contrastive loss, on average, achieved higher accuracy, however, the difference is not significant.

Evaluating the performance of the method using ambiguous ground truth data due to variations in cutting angle was another challenge. This was overcome by evaluating the methods using Top-N accuracy instead of only predicting the most similar atlas plate. We observe that our method achieved TOP-5 accuracy

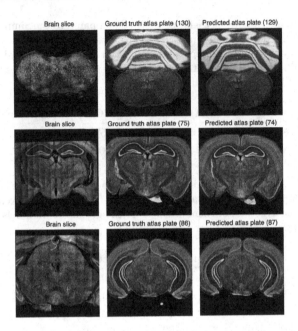

Fig. 4. Examples of the predicted (most similar) atlas plates by our method. Note that in all cases the ground truth plates are predicted within the Top-5 candidates in Table 3. Columns: **(1)** brain slices from the test dataset; **(2)** ground truth atlas plates; **(3)** predicted atlas plates. The number in parentheses shows the label of the atlas plate, i.e. the position of the atlas plate in the reference atlas.

of 100% meaning that the actual corresponding atlas plate always falls in the top 5 predicted atlas plates, as seen in Table 2. Further, the variations within the Top-5 predictions for all five cases reported in Table 3 could be plausible, as most of the predictions are neighbouring atlas plates of the ground truth. We also report the Top-1 accuracy and notice a drop in performance for both methods due to the inherent ambiguity in the ground truth. The inherent ambiguity of the ground truth makes our method more useful as practitioners can explore several likely candidate atlas plates to register to.

In conclusion, we proposed to use CNNs in Siamese Network configuration trained with contrastive- and triplet- losses as a method for identifying correspondence between complete and partial mice brain slices. Challenges such as partial/missing data and variations in cutting angles were overcome by learning a semantically meaningful embedding space. Our method has shown large performance improvements in both accuracy and inference times compared to the SimpleElastix-based baseline. With this work, we have we demonstrated the usefulness of this approach with a 2D reference atlas. We hypothesize that the same method can also be applied to a 3D reference atlas for further improved precision in the slice identification task.

5   Discussions and Conclusions

The Siamese networks used to identify brain slices has shown impressive results, i.e. in finding corresponding coronal 2D atlas plates. It achieved TOP-5 accuracy of 100% meaning that the actual corresponding atlas plate always falls in the top 5 predicted atlas plates. The identification accuracy (MAE) had no clear correlation with the batch size (16 and 32), the image resolution (224×224, 448×448, 1024×1024) and the type of the base for the Siamese network (ResNet50v2 and ResNet101v2). However, using images with lower resolution and networks with fewer parameters could improve the inference time. We did not observe that the performance of the Siamese network would be highly influenced by the loss function, namely contrastive and triplet losses. The models trained with triplet loss rather than contrastive loss, on average, achieved higher accuracy, however, the difference is not significant.

The Siamese networks produced impressive results even though some images of different classes looked very similar to each other, thus making the identification task even more complex. The distance between such images should be lower than the distance between two completely dissimilar images. Maximizing the distance between all images of different classes would make it difficult for networks to learn representations of these classes. Contrastive and triplet losses solve this issue by using margin, i.e., dissimilar images are not pushed away if the distance between them is larger than the margin.

In this study, we proposed Siamese Networks as a method for identifying complete and partial mouse brain slices, i.e. finding the corresponding 2D atlas plates. The networks have shown a high precision and significantly improved inference time compared to the baseline. While we demonstrated this with a 2D reference atlas, the same method can also be applied to a 3D reference atlas for even higher identification precision.

6   Conclusions

In this study, we proposed Siamese Networks as a method for identifying complete and partial mouse brain slices, i.e. finding the corresponding 2D atlas plates. The networks have shown a high precision and significantly improved inference time compared to the baseline. While we demonstrated this with a 2D reference atlas, the same method can also be applied to a 3D reference atlas for even higher identification precision.

Acknowledgments. The authors thank Kiehn Lab (University of Copenhagen, Denmark) for providing access to the microscopy images and the hardware used to train the models. They also acknowledge the Core Facility for Integrated Microscopy (CFIM) at the Faculty of Health and Medical Sciences for support with image acquisition.

Compliance with Ethical Standards. All animal experiments and procedures were carried according to the EU Directive 2010/63/EU and approved by the Danish Animal Experiments Inspectorate (Dyreforsøgstilsynet) and the Local Ethics Committee at the University of Copenhagen.

References

1. Breznik, E., Wetzer, E., Lindblad, J., Sladoje, N.: Cross-modality sub-image retrieval using contrastive multimodal image representations. arXiv preprint arXiv:2201.03597 (2022)
2. Bromley, J., Guyon, I., LeCun, Y., Säckinger, E., Shah, R.: Signature verification using a Siamese time delay neural network. Adv. Neural Inf. Process. Syst. **6**, 737–744 (1993)
3. Chopra, S., Hadsell, R., LeCun, Y.: Learning a similarity metric discriminatively, with application to face verification. In: 2005 IEEE Computer Society Conference on Computer Vision and Pattern Recognition (CVPR 2005), vol. 1, pp. 539–546. IEEE (2005)
4. Furth, D., et al.: An interactive framework for whole-brain maps at cellular resolution. Nat. Neurosci. **21**(1), 139–149 (2017). https://doi.org/10.1038/s41593-017-0027-7
5. Hadsell, R., Chopra, S., LeCun, Y.: Dimensionality reduction by learning an invariant mapping. In: 2006 IEEE Computer Society Conference on Computer Vision and Pattern Recognition (CVPR 2006), vol. 2, pp. 1735–1742 (2006). https://doi.org/10.1109/CVPR.2006.100
6. He, K., Zhang, X., Ren, S., Sun, J.: Deep residual learning for image recognition. In: Proceedings of the IEEE Conference on Computer Vision and Pattern Recognition, pp. 770–778 (2016)
7. Kingma, D.P., Ba, J.: Adam: a method for stochastic optimization (2017)
8. Koch, G., Zemel, R., Salakhutdinov, R.: Siamese neural networks for one-shot image recognition (2015)
9. Lein, E.S., Hawrylycz, M.J., Ao, N.: Genome-wide atlas of gene expression in the adult mouse brain. Nature **445**(7124), 168–176 (2007). https://doi.org/10.1038/nature05453, https://www.nature.com/articles/nature05453
10. Marstal, K., Berendsen, F., Staring, M., Klein, S.: SimpleElastix: a user-friendly, multi-lingual library for medical image registration. In: 2016 IEEE Conference on Computer Vision and Pattern Recognition Workshops (CVPRW), pp. 574–582 (2016). https://doi.org/10.1109/CVPRW.2016.78
11. Müller, H., Michoux, N., Bandon, D., Geissbuhler, A.: A review of content-based image retrieval systems in medical applications-clinical benefits and future directions. Int. J. Med. Inform. **73**(1), 1–23 (2004)
12. Pichat, J., Iglesias, J.E., Yousry, T., Ourselin, S., Modat, M.: A survey of methods for 3D histology reconstruction. Med. Image Anal. **46**, 73–105 (2018). https://doi.org/10.1016/j.media.2018.02.004
13. Qayyum, A., Anwar, S.M., Awais, M., Majid, M.: Medical image retrieval using deep convolutional neural network. Neurocomputing **266**, 8–20 (2017)
14. Schroff, F., Kalenichenko, D., Philbin, J.: FaceNet: a unified embedding for face recognition and clustering. In: 2015 IEEE Conference on Computer Vision and Pattern Recognition (CVPR), June 2015. https://doi.org/10.1109/cvpr.2015.7298682, http://dx.doi.org/10.1109/CVPR.2015.7298682
15. Xiong, J., Ren, J., Luo, L., Horowitz, M.: Mapping histological slice sequences to the Allen mouse brain atlas without 3D reconstruction. Front. Neuroinform. **12**, 93 (2018). https://doi.org/10.3389/fninf.2018.00093, https://www.frontiersin.org/article/10.3389/fninf.2018.00093

Transformed Grid Distance Loss
for Supervised Image Registration

Xinrui Song[1], Hanqing Chao[1], Sheng Xu[2], Baris Turkbey[3],
Bradford J. Wood[2], Ge Wang[1], and Pingkun Yan[1(✉)]

[1] Department of Biomedical Engineering and Center for Biotechnology
and Interdisciplinary Studies, Rensselaer Polytechnic Institute, Troy, NY 12180, USA
yanp2@rpi.edu
[2] Center for Interventional Oncology, Radiology and Imaging Sciences,
National Institutes of Health, Bethesda, MD 20892, USA
[3] Molecular Imaging Program, National Cancer Institute, National Institutes
of Health, Bethesda, MD 20892, USA

Abstract. Many deep learning image registration tasks, such as volume-to-volume registration, frame-to-volume registration, and frame-to-volume reconstruction, rely on six transformation parameters or quaternions to supervise the learning-based methods. However, these parameters can be very abstract for neural networks to comprehend. During the optimization process, ill-considered representations of rotation may even trap the objective function at local minima. This paper aims to expose these issues and propose the Transformed Grid Distance loss as a solution. The proposed method not only solves the problem of rotation representation but unites the gap between translation and rotation. We test our methods both with synthetic and clinically relevant medical image datasets. We demonstrate superior performance in comparison with conventional losses while requiring no alteration to the network input, output, or network structure at all.

1 Introduction

Existing deep learning-based image registration methods have explored many types of supervision. Unsupervised methods such as [1,4,11] relies on image intensity-based similarity metrics to supervise the network. These methods, however, are limited to single-modality registration tasks, or multi-modal images with very similar content and texture. Weakly supervised registration [2,7] incorporated weak labels such as organ segmentation to guide the training process.

In contrast, supervised methods require the ground truth annotations of registration for training [3,5]. For deformable image registration, providing such annotations can be unrealistically difficult. However, for tasks in which no significant differences were found between rigid and deformable registrations [10], using rigid registration reduces the annotation cost significantly. For example, in image-fusion guided prostate cancer biopsies, the manual registration between the MR and ultrasound images has been a routine for the clinical procedure.

A. Hering et al. (Eds.): WBIR 2022, LNCS 13386, pp. 177–181, 2022.
https://doi.org/10.1007/978-3-031-11203-4_19

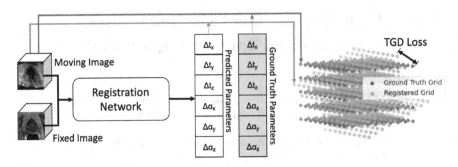

Fig. 1. Illustration of the transformed grid distance (TGD) loss.

Requesting these manual registration labels for training come at no additional cost to the clinicians. In other scenarios where deformable registration is preferred, a rigid transformation is also often required to pre-align the images before any deformable registration can be performed. For the above reasons, supervised deep learning based rigid image registration has been intensively studied, and will be the focus of this work.

Labels and loss function are critical components of supervised image registration. Since 3D rigid transformation is commonly represented by six transformation parameters, including three rotation angles and a 3D translation vector, a straightforward option is to use the distance between the ground truth and estimated transformation parameters as the loss to train the image registration network. However, numerous works [6,8,12] pointed out that the Euler angle representation is problematic for loss computation. In some cases, quaternion angles are used instead. In this paper, we argue that neither of them is the optimal choice for being used directly in a loss function. Instead, these abstract mathematical expressions should be first converted into more physically intuitive values. We propose a new loss – the Transformed Grid Distance (TGD) loss for network training.

2   Transformed Grid Distance

In supervised rigid registration, transformation parameters are often used as the label for network supervision. Compact transformation parameters, either in Euler or quaternion representation, can be difficult for neural networks to learn through conventional loss functions (e.g. L1 and MSE loss).

Instead of directly supervising the transformation parameters themselves, we apply the estimated transformation on the moving image grid, and supervise the distance between the transformed points and their corresponding points in the ground truth grid as illustrated in Fig. 1. Let $G \in \mathbb{R}^{m \times n \times l}$ denote a 3D moving image grid. TGD loss is computed as

$$L_{TGD} = \|T_\theta(G) - T_{gt}(G)\|_2 , \tag{1}$$

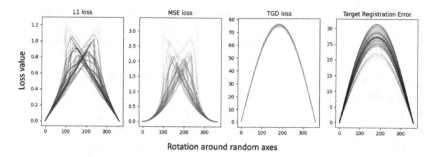

Fig. 2. We generated the rotations in this figure by randomly selecting 27 unit vectors and ranging the rotation amplitude from 0 to 360 °C. The Target Registration Error serves as the evaluation metric, as in many registration tasks.

where T denotes a 3D transformation matrix converted from θ. The key difference here is to convert the abstract representation θ into a dense and intuitive representation, which guides the network optimization process through circumventing any non-linear transformation conversions that the network would otherwise have to figure out.

The proposed TGD loss elegantly unifies both rotation and transformation into point-wise distance, which results in a smooth loss landscape that guides the network learning process. Had more meaningful points been acquired (*i.e.* anatomical landmarks), the loss can be simply adapted into Target Registration Error (TRE) by replacing the grid with those clinically relevant points. One major weakness of the Euler angles is that they must be applied in a fixed order, which is not reflected at all with L1 or MSE loss. During training, each line from Fig. 2 can be regarded as a training sample. The loss curves for either L1 or MSE loss on Euler angles vary wildly from sample to sample, while the proposed TGD loss stays consistent with the Target Registration Error (TRE).

The quaternion system seems to be a better solution than the Euler angles. However, due to the fact that the quaternion expression is divided into two intertwined parts, it is hard to guarantee that the direction of optimization is at all correct. For example, slight error in the rotation axis would result in a large TRE regardless of the rotation angle.

3   Experiments

In this section, we present both a synthetic and a clinically relevant experiment. Our dataset consists of 528 manually labeled cases of MR-transrectal ultrasound (TRUS) volume pair for training, 66 cases for validation, and 68 cases for testing.

In the first experiment, we use an MR volume as the fixed image, and its own perturbed result as the moving image. We have also included the result of TRE-TGD loss, which is another version of the proposed method that replaces the regular grid points in TGD loss with the target prostate surface points. The quaternion loss, on the other hand, failed to converge in this experiment where

Table 1. Performance of different loss functions in MR-MR registration.

Method	Mean TRE (mm)	Percentiles [25th, 50th, 75th, 95th]
Initial	12.66 ± 7.30	[6.39, 12.83, 18.79, 23.88]
Quaternion loss	12.95 ± 7.36	[6.64, 12.93, 19.00, 24.67]
MSE Euler angle loss	2.68 ± 2.31	[1.19, 2.04, 3.43, 6.82]
L1 Euler angle loss	2.80 ± 2.68	[1.04, 2.09, 3.72, 7.48]
TGD loss	$\mathbf{1.51 \pm 1.45}$	[0.62, 1.15, 1.93, 3.85]
TRE-TGD loss	$\mathbf{1.50 \pm 1.53}$	[0.65, 1.14, 1.87, 3.83]

Table 2. Performance of different loss functions in MR-TRUS registration.

Method	Mean TRE (mm)	Percentiles [25th, 50th, 75th, 90th]
Initial	9.93 ± 5.87	[4.89, 9.82, 14.89, 19.10]
MSE Euler angle loss	5.57 ± 2.86	[3.47, 5.06, 7.07, 10.98]
SRE-TGD loss	$\mathbf{4.40 \pm 2.49}$	[2.57, 3.97, 5.77, 8.88]

large rotation errors are concerned. Results in Table 1 show that simply through 'rephrasing' the transformation parameters into physical distance between grid points, the network was guided to converge at a lower minimum.

The second experiment treats the TRUS volume as the moving image, and the corresponding MR volume as the fixed image. This is a use case, where an accurate alignment between the transrectal ultrasound (TRUS) and MR volume greatly benefits the ultrasound-guided prostate cancer biopsy [9]. For each pair of MR and TRUS volume, we are provided with the manual label for rigid registration from TRUS to MR, as well as the prostate surface points in MR. Similar to the TRE-TGD loss in the first experiment, the SRE-TRD loss also calculates the distance between corresponding points, thereby a subset to the proposed TGD loss. Table 2 compares the result of multi-modal registration between the conventional MSE loss and SRE-TGD loss. With the same network architecture and other settings, the proposed loss function results in a significant ($p < 0.001$ under paired t-test) improvement over the conventional MSE loss.

4   Discussions and Conclusion

In this paper, we revealed the limitation of directly using abstract transformation parameters for loss computation in supervised training of image registration networks. With such insight, we introduced a simple yet effective tool to boost the performance of supervised rigid volume registration. Although the analysis and experiments are mainly conducted in a rigid setting, this idea can be easily adapted for a non-rigid affine registration task.

References

1. Balakrishnan, G., Zhao, A., Sabuncu, M.R., Guttag, J., Dalca, A.V.: VoxelMorph: a learning framework for deformable medical image registration. IEEE Trans. Med. Imaging **38**(8), 1788–1800 (2019)
2. Baum, Z.M.C., Hu, Y., Barratt, D.C.: Multimodality biomedical image registration using free point transformer networks. In: Hu, Y., et al. (eds.) ASMUS/PIPPI - 2020. LNCS, vol. 12437, pp. 116–125. Springer, Cham (2020). https://doi.org/10.1007/978-3-030-60334-2_12
3. Guo, H., Kruger, M., Xu, S., Wood, B.J., Yan, P.: Deep adaptive registration of multi-modal prostate images. Comput. Med. Imaging Graph. **84**, 101769 (2020)
4. Hansen, L., Heinrich, M.P.: GraphRegNet: deep graph regularisation networks on sparse keypoints for dense registration of 3D lung CTs. IEEE Trans. Med. Imaging **40**(9), 2246–2257 (2021)
5. Haskins, G., et al.: Learning deep similarity metric for 3D MR-TRUS image registration. Int. J. Comput. Assist. Radiol. Surg. **14**(3), 417–425 (2019)
6. Hou, B., et al.: 3-D reconstruction in canonical co-ordinate space from arbitrarily oriented 2-D images. IEEE Trans. Med. Imaging **37**(8), 1737–1750 (2018)
7. Hu, Y., et al.: Weakly-supervised convolutional neural networks for multimodal image registration. Med. Image Anal. **49**, 1–13 (2018)
8. Kendall, A., Grimes, M., Cipolla, R.: PoseNet: a convolutional network for real-time 6-DOF camera relocalization. In: Proceedings of the IEEE International Conference on Computer Vision, pp. 2938–2946 (2015)
9. Song, X., et al.: Cross-modal attention for MRI and ultrasound volume registration. In: de Bruijne, M., et al. (eds.) MICCAI 2021. LNCS, vol. 12904, pp. 66–75. Springer, Cham (2021). https://doi.org/10.1007/978-3-030-87202-1_7
10. Venderink, W., de Rooij, M., Sedelaar, J.M., Huisman, H.J., Fütterer, J.J.: Elastic versus rigid image registration in magnetic resonance imaging-transrectal ultrasound fusion prostate biopsy: a systematic review and meta-analysis. Eur. Urol. Focus **4**(2), 219–227 (2018)
11. de Vos, B.D., Berendsen, F.F., Viergever, M.A., Staring, M., Išgum, I.: End-to-End unsupervised deformable image registration with a convolutional neural network. In: Cardoso, M.J., et al. (eds.) DLMIA/ML-CDS -2017. LNCS, vol. 10553, pp. 204–212. Springer, Cham (2017). https://doi.org/10.1007/978-3-319-67558-9_24
12. Wei, W., Haishan, X., Alpers, J., Rak, M., Hansen, C.: A deep learning approach for 2D ultrasound and 3D CT/MR image registration in liver tumor ablation. Comput. Methods Programs Biomed. **206**, 106117 (2021)

Efficiency

Deformable Lung CT Registration by Decomposing Large Deformation

Jing Zou[1], Lihao Liu[2], Youyi Song[1], Kup-Sze Choi[1], and Jing Qin[1(✉)]

[1] Centre for Smart Health, School of Nursing, The Hong Kong Polytechnic University, Hong Kong, China
harry.qin@polyu.edu.hk
[2] Centre for Mathematical Sciences, University of Cambridge, Cambridge, UK

Abstract. Deformable lung CT registration plays an important role in image-guided navigation systems, especially in the situation with organ motion. Recent progress has been made in image registration by utilizing neural networks for end-to-end inference of a deformation field. However, there are still difficulties to learn the irregular and large deformation caused by organ motion. In this paper, we propose a patient-specific lung CT image registration method. We first decompose the large deformation between the source image and the target image into several continuous intermediate fields. Then we compose these fields to form a spatio-temporal motion field and refine it through an attention layer by aggregating information along motion trajectories. The proposed method can utilize the temporal information in a respiratory circle and can generate intermediate images which are helpful in image-guided systems for tumor tracking. Extensive experiments were performed on a public dataset, showing the validity of the proposed methods.

Keywords: Image registration · Lung CT · Organ movement · Deformation field decomposition · Attention layer

1 Introduction

Image-guided navigation systems have greatly enhanced the therapeutic efficiency of complicated interventions [1]. However, in such systems, organ motions caused by respiration is a major challenge of accurate lesion targeting. In current practice, this challenge is often handled by asking the patients to hold their breath and scanning repeated CTs. This will either cause distress to patients or increase the radiation exposure. To the end, registration is a promising technique to correct the position offset of the targeting organ or tumor.

Recently, deep networks have been applied to address deformable registration problems and achieved remarkable success [3,6–10]. However, it is still difficult to accurately estimate the large deformation due to respiration (tumors and sensitive structures in the thorax can move more than 20 mm [12]). In this paper, we propose a lung CT registration method that utilizes temporal information

© The Author(s), under exclusive license to Springer Nature Switzerland AG 2022
A. Hering et al. (Eds.): WBIR 2022, LNCS 13386, pp. 185–189, 2022.
https://doi.org/10.1007/978-3-031-11203-4_20

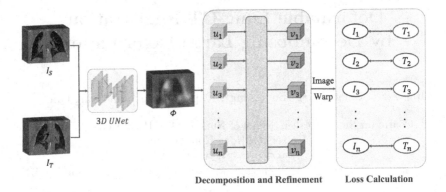

Fig. 1. The illustrative pipeline of our method.

during respiration. During training, the images at extreme phases , as well as the intermediate images, are employed as training data. Once the network is trained, it can infer a deformation field without the intermediate images.

2   Methodology

Let I_S and $I_T \in \mathbb{R}^{H \times W \times C}$ be the source and target lung CT images, respectively. Our aim is to figure out the deformation filed $\Phi \in \mathbb{R}^{H \times W \times C \times 3}$ that stores the coordinate offset between I_S and I_T. We employ a deep network f that takes I_S and I_T as the input to predict Φ by solving the below problem:

$$\underset{f \in \mathcal{F}}{\arg\min} \, \ell(I_T, I_S \circ \Phi_f) + \lambda \mathcal{R}(\Phi_f), \tag{1}$$

where $\mathcal{F}$ denotes the function space of f and Φ_f stands for Φ with f given the input (I_S, I_T). $I_S \circ \Phi_f$ represents I_S warped by Φ_f, and ℓ is the loss function to measure the discrepancy between I_T and $I_S \circ \Phi_f$. $\mathcal{R}(\Phi_f)$ stands for the regularization term with the hyperparameter λ to balance its importance.

This training paradigm can work well when the deformation of lungs is small [4], but fail for a large and irregular deformation, in which pixels are dramatically deformed, diminishing the accuracy of the registration. Our method aims to solve this issue by decomposing the deformation field into several ones with small deformations and gradually refining them through an attention layer. An overview of the proposed method is shown in Fig. 1.

Decomposition. We first decompose the deformation field Φ. This field describes the directions and the distances for all voxels moving from I_S to I_T. Considering the progressive movement of the lung, the deformation field can be decomposed by incremental steps to obtain intermediate deformation fields u_i.

We assume that each voxel deforms along a straight line [11]. Thus the decomposition can be achieved by linear interpolation: $u_i = \Phi/n$, where n denotes the phases in a respiratory circle.

Refinement. Above mentioned linear interpolation of the deformation field relies on the assumption that the displacement of each voxel is homogeneous. However, in practice, the deformation may be irregular. So we refine these small deformation fields by firstly concatenating them to form a spatio-temporal motion field U, which contains spatial and temporal information during respiration. Then we input the motion field U to a self-attention layer, and output the refined field V. At last, V is decomposed again to obtain refined intermediate fields v_i, with which I_S are warped to generate intermediate images I_n that are used to calculate loss with ground truth intermediate images T_n. Finally, our decomposition method aims to train the deep network f for deformable registration by solving the following problem:

$$\underset{f \in \mathcal{F}}{\mathrm{argmin}} \sum_{t=1}^{n} \ell_t(T_t, I_S \circ (t\Phi_f/n)) + \lambda \mathcal{R}(\Phi_f). \tag{2}$$

3 Experiments

Experimental Setup: Our method was evaluated on a public dataset [5], which has ten thoracic 4D CTs obtained at ten different respiratory phases in a respiratory cycle. In each 4D CT, 300 anatomical landmarks were manually annotated at two extreme phases. We evaluate our method with target registration error (TRE), which is formulated as the average Euclidean distance between the fixed landmarks and the warped moving landmarks. We implemented our method with Pytorch on an NVIDIA RTX 3090 GPU.

Experimental Results: We compare our method with five competitive methods: BL [2] (CVPR 2018), IL [6] (MedIA 2019), VM [3] (TMI 2019), MAC [7] (MedIA 2021), and CM [8] (MedIA 2021), denoting the baseline and existing methods that use iterative learning strategy, lung masks as the supervision, landmarks as the supervision, and the cycle consistency, respectively. For a fair comparison, we employed the same backbone network (3D UNet) and the same learning setting.

The results via cross-validation are reported in Table 1. We can see that our method achieved the best performance of the average *TRE* (denoted as *Ave.* in the table). It improves the performance over the second-best method (VM) with 8.0%. This demonstrates the validity of our method. We also can see that our method works consistently well in the best and worst cases (denoted as *Best* and *Worst*). Moreover, the performance of our method is less diverse than others as we have the lowest *Std.* (1.06). These evidences suggest that our algorithm is more reliable and effective. We also checked the statistical significance of the performance improvement by paired t-test. We can see that, expect VM (whose p-value is 0.053), other p-values are less than 0.05, which implies that our method significantly improves the registration performance.

Table 1. The *TRE* (*mm*) results of our algorithm and compared methods.

	BL	IL	VM	MAC	CM	Ours
Ave.	3.53	3.85	3.38	3.53	3.56	**3.11**
Std.	1.38	1.25	1.17	1.25	1.56	**1.06**
Best	1.97	2.19	2.19	2.19	1.97	**1.75**
Worst	6.02	5.93	5.79	6.27	6.77	**4.8**
p-value	0.015	0.001	0.053	0.022	0.045	—

4   Conclusion

In this paper, we have investigated a simple and effective method to learn the large deformation field in lung CT image registration, which is helpful in image-guided navigation systems. This method decomposes the large deformation field into small fields, and then composes these small fields and refines them by attention layer. The experimental results show that our method works better than existing methods.

Acknowledgement. The work described in this paper is supported by a grant of Hong Kong Research Grants Council under General Research Fund (no. 15205919).

References

1. Anzidei, M., et al.: Preliminary clinical experience with a dedicated interventional robotic system for CT-guided biopsies of lung lesions: a comparison with the conventional manual technique. Eur. Radiol. **25**(5), 1310–1316 (2015)
2. Balakrishnan, G., Zhao, A., Sabuncu, M.R., Guttag, J., Dalca, A.V.: An unsupervised learning model for deformable medical image registration. In: Proceedings of the IEEE Conference on Computer Vision and Pattern Recognition, pp. 9252–9260 (2018)
3. Balakrishnan, G., Zhao, A., Sabuncu, M.R., Guttag, J., Dalca, A.V.: VoxelMorph: a learning framework for deformable medical image registration. IEEE Trans. Med. Imaging **38**(8), 1788–1800 (2019)
4. Beg, M.F., Miller, M.I., Trouvé, A., Younes, L.: Computing large deformation metric mappings via geodesic flows of diffeomorphisms. Int. J. Comput. Vis. **61**(2), 139–157 (2005)
5. Castillo, R., et al.: A framework for evaluation of deformable image registration spatial accuracy using large landmark point sets. Phys. Med. Biol. **54**(7), 1849 (2009)
6. De Vos, B.D., Berendsen, F.F., Viergever, M.A., Sokooti, H., Staring, M., Išgum, I.: A deep learning framework for unsupervised affine and deformable image registration. Med. Image Anal. **52**, 128–143 (2019)
7. Hering, A., Häger, S., Moltz, J., Lessmann, N., Heldmann, S., van Ginneken, B.: CNN-based lung CT registration with multiple anatomical constraints. Med. Image Anal. **72**, 102139 (2021)

8. Kim, B., Kim, D.H., Park, S.H., Kim, J., Lee, J.G., Ye, J.C.: CycleMorph: cycle consistent unsupervised deformable image registration. Med. Image Anal. **71**, 102036 (2021)

9. Liu, L., Aviles-Rivero, A.I., Schönlieb, C.B.: Contrastive registration for unsupervised medical image segmentation. arXiv preprint arXiv:2011.08894 (2020)

10. Liu, L., Huang, Z., Liò, P., Schönlieb, C.B., Aviles-Rivero, A.I.: Pc-SwinMorph: patch representation for unsupervised medical image registration and segmentation. arXiv preprint arXiv:2203.05684 (2022)

11. Sarrut, D., Boldea, V., Miguet, S., Ginestet, C.: Simulation of four-dimensional CT images from deformable registration between inhale and exhale breath-hold CT scans. Med. Phys. **33**(3), 605–617 (2006)

12. Schreibmann, E., Chen, G.T., Xing, L.: Image interpolation in 4D CT using a BSpline deformable registration model. Int. J. Radiat. Oncol. Biol. Phys. **64**(5), 1537–1550 (2006)

You only Look at Patches: A Patch-wise Framework for 3D Unsupervised Medical Image Registration

Lihao Liu[✉], Zhening Huang, Pietro Liò, Carola-Bibiane Schönlieb, and Angelica I. Aviles-Rivero

University of Cambridge, Cambridge, UK
11610@cam.ac.uk

Abstract. Medical image registration is a fundamental task for a wide range of clinical procedures. Automatic systems have been developed for image registration, where the majority of solutions are supervised techniques. However, those techniques rely on a large and well-representative corpus of ground truth, which is a strong assumption in the medical domain. To address this challenge, we propose a novel unified unsupervised framework for image registration and segmentation. The highlight of our framework is that patch-based representation is key for performance gain. We first propose a patch-based contrastive strategy that enforces locality conditions and richer feature representation. Secondly, we propose a patch stitching strategy to eliminate artifacts. We demonstrate, through our experiments, that our technique outperforms current state-of-the-art unsupervised techniques.

1 Introduction

Image registration seeks to find a mapping that aligns an unaligned image to a reference one. The estimated spatial mapping aims to best align the anatomical structure of interest. Majority of existing works have been investigated from the classic perspective. Whilst promising performance has been reported, those techniques build upon costly optimisation schemes, which limits their efficiency when using a large volume of data. This limitation has encouraged the fast development of deep learning techniques for medical image registration. A set of techniques have been reported based on supervised learning. However, the need for a well-representative and high-quality ground truth is a strong assumption and hard to obtain in the medical domain. Another set of techniques have been devoted to explore unsupervised techniques e.g. [1–4,7,8]. Existing techniques have proposed several network mechanisms and explicit regularisers, to accommodate a certain level, with the lack of prior knowledge. However, the performance is still limited due to the lack of high-quality prior knowledge.

Our work is motivated by the aforementioned limitation. We argue that better quality prior can be estimated from patches rather than the full image. Medical images have complex anatomical structures, which impose a challenge when

© The Author(s), under exclusive license to Springer Nature Switzerland AG 2022
A. Hering et al. (Eds.): WBIR 2022, LNCS 13386, pp. 190–193, 2022.
https://doi.org/10.1007/978-3-031-11203-4_21

estimating an image-to-image mapping. Therefore, our modeling hypothesis is that patch embeddings are more meaningful representation for performance gain. In this work, we introduce a novel unified framework for unsupervised image registration and segmentation, which we call PC-SwinMorph (**P**atch **C**ontrastive Strategy with **S**hifted-**win**dow multi-head self-attention based on Voxel**Morph**). We underline two major highlights of our framework. Firstly, we introduce a patchwise contrastive registration strategy for richer feature representation. Secondly, we propose a patch stitching strategy to address the splitting effect caused by the image patch-based partition. We evaluate our framework using the benchmark dataset LPBA40. We demonstrate through our experimental result that our two patch-based strategies lead to better performance than the state-of-the-art techniques for unsupervised registration and segmentation.

2 Proposed Framework

In this section, we describe the overall workflow of our proposed framework.

Overview Workflow. In Fig. 1, our PC-SwinMorph first take the moving and fixed images as inputs. We then generate non-overlap patches from the two input images, and perform patch-level contrastive learning to refine the features (Patch-based Strategy I from Fig. 1). Then the contrasted features are fed into two weight-shared CNN encoders. Followed by a decoder, the features are recursively concatenated and enlarged with skip connec-

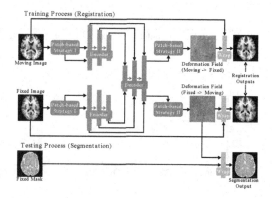

Fig. 1. Workflow of our proposed framework.

tions to reconstruct two sets of deformation field patches. We then use a 3D W-MSA and a 3D SW-MSA module [6] to refine and stitch the deformation field patches to obtain the full deformation field (Patch-based Strategy II from Fig. 1). Finally, we wrap the moving image → fixed image, and the fixed image → moving image. After the training registration process, we also adopt the full deformation field to transfer the segmentation mask for fixed masks to obtain the segmentation mask of the moving image. We underline that no masks are used in the training registration process, and they are only used in the testing segmentation stage. Hence, our framework is a unified unsupervised registration and segmentation network.

3   Experimental Results

In this section, we detail the experimental setup and experimental results to validate our proposed unified unsupervised registration and segmentation framework.

Experimental Setup. We evaluate our framework on the publicly available LONI Probabilistic Brain Atlas (LPBA40) dataset[1] using Dice evaluation metrics. For the implementation details regarding the network architecture, data pre-processing, and training and testing schemes, we refer to the reader to [5]. Our code will be publicly available upon the acceptance of this work.

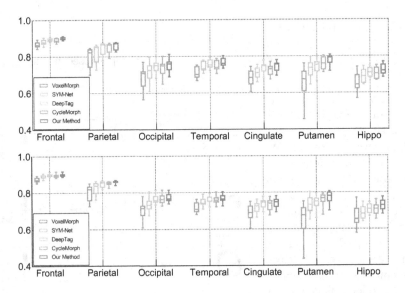

Fig. 2. Boxplots in terms of Dice, per anatomical region, for registration (top) and segmentation (bottom) tasks. The comparison displays our Method (PC-SwinMorph) against SOTA techniques.

Comparison to the State-of-the-Art Techniques. We compared our technique with recent unsupervised brain segmentation methods, including Voxel-Morph [1], DeepTag [8], SYMNet [7], CycleMorph [2]. For a fair comparison, all models use the same backbone, VoxelMorph, which has been fine-tuned to achieve optimal performance. In Fig. 2, the boxplots summarise performance-wise, in terms of the *Dice* coefficient, the compared SOTA methods, and our PC-SwinMorph. In a closer look at the boxplots, we observe that our method

---

[1] https://loni.usc.edu/research/atlases.

outperforms all other SOTA methods by a large margin on all seven majority anatomical regions for both registration and segmentation tasks. Particularly, for both registration and segmentation tasks, our results report an improvement of 5.9% compared to VoxelMorph on the average Dice results, and 3.9–4.3% against the other compared SOTA techniques on the average Dice score.

4   Conclusion

We introduced a novel unified unsupervised framework for image registration and segmentation. We showed that patches are crucial for obtaining richer features and preserving anatomical details. Our intuition behind the performance gain of our technique, is that patches can capture not only global but also local spatial structures (more meaningful embeddings). We demonstrated, that at this point in time, our technique reported SOTA performance for both tasks.

References

1. Balakrishnan, G., et al.: Voxelmorph: a learning framework for deformable medical image registration. IEEE Trans. Med. Imaging **38**(8), 1788–1800 (2019)
2. Kim, B., et al.: Cyclemorph: cycle consistent unsupervised deformable image registration. Med. Image Anal. **71**, 102036 (2021)
3. Liu, L., Aviles-Rivero, A.I., Schönlieb, C.B.: Contrastive registration for unsupervised medical image segmentation. arXiv preprint arXiv:2011.08894 (2020)
4. Liu, L., Hu, X., Zhu, L., Heng, P.-A.: Probabilistic multilayer regularization network for unsupervised 3D brain image registration. In: Shen, D., et al. (eds.) MICCAI 2019. LNCS, vol. 11765, pp. 346–354. Springer, Cham (2019). https://doi.org/10.1007/978-3-030-32245-8_39
5. Liu, L., Huang, Z., Liò, P., Schönlieb, C.B., Aviles-Rivero, A.I.: Pc-swinmorph: patch representation for unsupervised medical image registration and segmentation. arXiv preprint arXiv:2203.05684 (2022)
6. Liu, Z., et al.: Swin transformer: hierarchical vision transformer using shifted windows. arXiv preprint arXiv:2103.14030 (2021)
7. Mok, T.C., Chung, A.: Fast symmetric diffeomorphic image registration with convolutional neural networks. In: CVPR, pp. 4644–4653 (2020)
8. Ye, M., et al.: Deeptag: an unsupervised deep learning method for motion tracking on cardiac tagging magnetic resonance images. In: CVPR, pp. 7261–7271 (2021)

Recent Developments of an Optimal Control Approach to Nonrigid Image Registration

Zicong Zhou[1]([✉])([iD]) and Guojun Liao[2]

[1] Institute of Natural Sciences, Shanghai Jiao Tong University, Shanghai, China
zicongzhou818@sjtu.edu.cn
[2] Math Department, University of Texas at Arlington, Arlington, TX, USA

Abstract. The Variational Principle (VP) forms diffeomorphisms (non-folding grids) with prescribed Jacobian determinant (JD) and curl under an optimal control set-up, which satisfies the properties of a Lie group. To take advantage of that, it is meaningful to regularize the resulting deformations of the image registration problem into the solution pool of VP. In this research note, (1) we provide an optimal control formulation of the image registration problem under a similar optimal control set-up as is VP; (2) numerical examples demonstrate the confirmation of diffeomorphic solutions as expected.

Keywords: Diffeomorphic image registration · Computational diffeomorphism · Jacobian determinant · Curl · *Green*'s identities

1 Our Approach to Image Registration

This work connects the resulting registration deformations to the solution pool of VP in [1], which achieves a recent progression in describing non-folding grids in a diffeomorphism group. Hence, to restrict the image registration method built in [3] satisfying the constraint of VP, it is reformulated and proposed as follows: let I_m be a ***moving*** image is to be registered to a ***fixed*** image I_f on the fixed and bounded domain ($\boldsymbol{\omega} = < x, y, z > \in)\Omega \subset \mathbb{R}^3$, the energy function $Loss$ is minimized over the form $\boldsymbol{\phi} = \boldsymbol{id} + \boldsymbol{u}$ on Ω with $\boldsymbol{u} = \boldsymbol{0}$ on $\partial\Omega$,

$$Loss(\boldsymbol{\phi}) = \frac{1}{2}\int_\Omega [I_m(\boldsymbol{\phi}) - I_f]^2 d\boldsymbol{\omega} \quad \text{subjects to } \Delta\boldsymbol{\phi} = \boldsymbol{F}(f, g) \text{ in } \Omega, \quad (1)$$

where the scalar-valued f and the vector-valued $\boldsymbol{g}$ are the control functions in the sense of VP that mimic the prescribed JD and curl, respectively.

A. Hering et al. (Eds.): WBIR 2022, LNCS 13386, pp. 194–197, 2022.
https://doi.org/10.1007/978-3-031-11203-4_22

1.1   Gradient with Respect to Control F

The variational gradient of (1) with respect to $\delta\Delta\phi = \delta\Delta u = \delta F$ is derived. For all δF vanishing on $\partial\Omega$ and by *Green's* identities with fixed boundary condition,

$$\delta Loss(\phi) = \delta(\frac{1}{2}\int_\Omega [I_m(\phi) - I_f]^2 d\omega) = \int_\Omega [(I_m(\phi) - I_f)\nabla I_m(\phi) \cdot \delta\phi]d\omega$$

$$= \int_\Omega [\Delta b \cdot \delta\phi]d\omega = \int_\Omega [b \cdot \delta\Delta\phi]d\omega = \int_\Omega [b \cdot \delta F]d\omega \quad \Rightarrow \quad \frac{\partial Loss}{\partial F} = b, \tag{2}$$

where $\Delta b = (I_m(\phi) - I_f)\nabla I_m(\phi)$, so, a gradient-based algorithm can be formed.

1.2   Hessian Matrix with Respect to Control Function F

In case of a Newton optimizing scheme is applicable, from (2), one can derive the Hessian matrix H of (1) with respect to F as follows,

$$\delta^2 Loss(\phi) := \delta(\delta Loss(\phi)) = \delta(\int_\Omega [(I_m(\phi) - I_f)\nabla I_m(\phi) \cdot \delta\phi]d\omega) = \int_\Omega [\delta\phi^\top K\delta\phi]d\omega,$$

$$\text{where } \Delta^2 H = K = \nabla I_m(\phi)[\nabla I_m(\phi)]^\top + (I_m(\phi) - I_f)\nabla^2 I_m(\phi),$$

$$\text{and } \nabla^2 I_m(\phi) = \begin{pmatrix} I_m(\phi)_{xx} & I_m(\phi)_{xy} & I_m(\phi)_{xz} \\ I_m(\phi)_{yx} & I_m(\phi)_{yy} & I_m(\phi)_{yz} \\ I_m(\phi)_{zx} & I_m(\phi)_{zy} & I_m(\phi)_{zz} \end{pmatrix},$$

$$\text{so, } \delta^2 Loss(\phi) = \int_\Omega [\delta\phi^\top \Delta^2 H\delta\phi]d\omega = \int_\Omega [\delta\Delta\phi^\top H\delta\Delta\phi]d\omega \Rightarrow \frac{\partial^2 Loss}{(\partial F)^2} = H. \tag{3}$$

A necessary condition that ensures a Newton scheme works is to show such Hessian H must be of Semi-Positive Definite matrix. This is left for future study.

1.3   Partial Gradients with Respect to Control Functions $\hat{f}$ and g

To ensure (1) producing diffeomorphic solutions that is controlled by $J_{min} \in (0,1)$, instead of optimizing along F by (2), it can be set that $f := J_{min} + \hat{f}^2$ in (1). Since it is known $\delta\Delta u = \delta F = \delta(\nabla f - \nabla \times g)$, then, it carries to,

$$\delta Loss(\phi) = \int_\Omega [b \cdot \delta\Delta\phi]d\omega = \int_\Omega [b \cdot \delta F]d\omega = \int_\Omega [b \cdot \delta(\nabla f - \nabla \times g)]d\omega$$

$$= \int_\Omega [b \cdot (\nabla\delta(J_{min} + \hat{f}^2))]d\omega + \int_\Omega [-b \cdot \nabla \times \delta g]d\omega$$

$$= \int_\Omega [b \cdot (2\hat{f}\nabla\delta\hat{f})]d\omega + \int_\Omega [-b \cdot \nabla \times \delta g]d\omega = \int_\Omega [-2\hat{f}\nabla \cdot b\delta\hat{f}]d\omega + \int_\Omega [-\nabla \times b \cdot \delta g]d\omega$$

$$\Rightarrow \quad \frac{\partial Loss}{\partial\hat{f}} = -2\hat{f}\nabla \cdot b \quad \text{and} \quad \frac{\partial Loss}{\partial g} = -\nabla \times b. \tag{4}$$

2   Numerical Examples

In our algorithms, $J_{min} = 0.5$ is artificially set. It is desirable to design a mechanism that yields optimal values of J_{min}. The gradient-based algorithms can be structured with (1) the coarse-to-fine **multiresolution** technique, which fits better in large deformation problems over binary images, as it did in [2]; and (2) the function composition **regriding** technique, which divides the problem difficulty and prevent non-diffeomorphic solutions on medical image registrations. These observations are demonstrated by the next example.

2.1   A Large Deformation Test and a MRI Registration Test

The J-to-V part of this example is done with **multiresolution** and the Brain Morph part is done with **regriding**. In Fig. 1(c, j), ϕ is the diffeomorphic solution found by the proposed method; Fig. 1(d, k), $I_m(\phi)$ is the registered image that is close to I_f, Fig. 1(b, i). Next, ϕ_{vp}^{-1} is the inverse of ϕ that constructed by VP. In Fig. 1(f,m), ϕ is composed by ϕ^{-1}, in Red grid, and superposed on Black grid *id* but the Black grid barely shows. This shows the composition $T = \phi_{vp}^{-1} \circ \phi$ is very close to *id*. Therefore, ϕ_{vp}^{-1} can be treated as the inverse to ϕ and they are of the same diffeomorphism group which VP focuses (Fig. 1).

Table 1. Evaluation of the proposed image registration

e.g.	Ω	$ratio = Loss(\phi)/Loss(id)$	$\min(\det \nabla \phi)$	JSC	DICE
J-to-V	$[1,128]^2$	0.0034	0.2191	0.9337	0.9657
Brain Morph	$[1,128]^2$	0.0605	0.2540	0.9849	0.9924

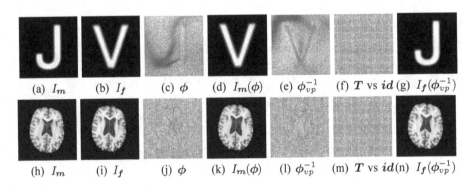

(a) I_m (b) I_f (c) ϕ (d) $I_m(\phi)$ (e) ϕ_{vp}^{-1} (f) T vs id (g) $I_f(\phi_{vp}^{-1})$

(h) I_m (i) I_f (j) ϕ (k) $I_m(\phi)$ (l) ϕ_{vp}^{-1} (m) T vs id(n) $I_f(\phi_{vp}^{-1})$

Fig. 1. Resulting registration deformations and their inverses by VP

The question is whether ϕ_{vp}^{-1} is also a valid inverse registration deformation that moves I_f back to I_m. The answer is YES, at least in our tested examples. $I_f(\phi_{vp}^{-1})$

is indeed close to I_m. That means ϕ_{vp}^{-1} can be treated as a valid registration deformation from I_f to I_m, as it is confirmed by the Table 2 records.

Table 2. Evaluation of ϕ_{vp}^{-1} by VP in the sense of Image Registration

e.g.	$ratio$ (of $Loss$ from $I_f(\phi_{vp}^{-1})$ to I_m)	$\min(\det\nabla\phi_{vp}^{-1})$	JSC	DICE
J-to-V	0.0029	0.1520	0.9195	0.9581
Brain morph	0.0657	0.3212	0.9832	0.9915

3   Discussion

This note provides the analytic description with simple demonstration of the proposed method. A full paper with extensive experiments will be available soon.

References

1. Zhou, Z., Liao, G.: Construction of diffeomorphisms with prescribed jacobian determinant and curl. In: International Conference on Geometry and Graphics, Proceedings (2022). (in press)
2. Zhou, Z., Liao, G.: A novel approach to form Normal Distribution of Medical Image Segmentation based on multiple doctors' annotations. In: Proceedings of SPIE 12032, Medical Imaging 2022: Image Processing, p. 1203237 (2022). https://doi.org/10.1117/12.2611973
3. Zhou, Z.: Image Analysis Based on Differential Operators with Applications to Brain MRIs, Ph.D. Dissertation, University of Texas at Arlington (2019)

2D/3D Quasi-Intramodal Registration of Quantitative Magnetic Resonance Images

Batool Abbas[1]($\boxtimes$), Riccardo Lattanzi[2], Catherine Petchprapa[2],
and Guido Gerig[1]

[1] Computer Science and Engineering, New York University Tandon School
of Engineering, New York, NY, USA
{batool.abbas,gerig}@nyu.edu
[2] Department of Radiology, New York University Grossman School of Medicine,
New York, NY, USA
riccardo.Lattanzi@nyulangone.org, catherine.petchprapa@nyumc.org

Abstract. Quantitative Magnetic Resonance Imaging (qMRI) is backed
by extensive validation in research literature but has seen limited use in
clinical practice because of long acquisition times, lack of standardization
and no statistical models for analysis. Our research focuses on developing a novel quasi-intermodal 2D slice to 3D volumetric pipeline for an
emerging qMR technology that aims to bridge the gap between research
and practice. The two-part method first initializes the registration using
a 3D reconstruction technique then refines it using a 3D to 2D projection technique. Intermediate results promise feasibility and efficacy of
our proposed method.

1 Introduction

Biochemical changes often precede observable changes in morphologyand insight
into these earlier asymptomatic deviations can help inform clinical strategy. Magnetic Resonance Imaging (MRI) has been traditionally used to acquire visual
insight into the anatomy, morphology and physiology of living organisms. Quantitative MRI (qMRI) can capture and express the biochemical composition of
the imaged structures as quantitative, calibrated physical units [8]. Despite a his-
torically large body of research evidence providing validation for qMRI [3,17], it
has seen limited integration into routine clinical practice due to obstacles such as
infeasible acquisition time, insufficient standardization and a lack of statistical
models for computational analysis [7].

One particularly promising approach for clinical integration of qMR enables
rapid high-resolution and simultaneous mapping of multiple parameters in six 2D
sections oriented around a central axis of rotation [10]. This method drastically
cuts down acquisition time and has been proven to be highly reproducible [4]
but it lacks normative models to perform comparative, population-based and
longitudinal analysis. This is partly owing to the novelty of the data but also

A. Hering et al. (Eds.): WBIR 2022, LNCS 13386, pp. 198–205, 2022.
https://doi.org/10.1007/978-3-031-11203-4_23

because spatial normalization necessitates an effective 2D slice to 3D volume registration technique which continues to be an open problem today [5,12].

At first glance, this seems like a straightforward intermodal problem because the acquisition principle for both images is the same. However, the differences in protocol and parameters result in widely differing intensity distributions. The 3D qualitative volume comprises weighted intensity values while the 2D quantitative slices record raw un-weighted measurements. This difference categorizes this as a quasi-intra-modal registration problem [12] and adds another facet to its complexity. Thus the novelty of our proposed technology is rooted in both the originality of our data and in the research question that it aims to address.

1.1   Clinical Motivation

Symptomatic hip osteoarthritis (OA) is a degenerative joint disease that severely hinders functional mobility and impacts quality of life. It is one of the most common joint disorders in the United States [18] and the leading indication for primary total hip replacement surgeries [9]. The development of effective preventative and treatment measures necessitates the study of its causative factors.

1.2   Clinical Data

Each volunteer was scanned to collect a 3D qualitative scan of the hip and six 2D quantitative data scans acquired via incremental 30° rotations around a central axis passing through the femur bone as shown in Fig. 1. The specifics of the target 3D volume itself are less relevant since 3D/3D volume registration has several well-established and effective solutions that can be used to transform a template to a scan and vice versa [2]. For this reason, the 3D qualitative scan serves as our fixed volume. To start the process, a 'localizer' plane is maneuvered over the opening of the acetabulum by the MRI technician. The axis of rotation passes through this plane meaning that all acquired 2D scans are normal to this

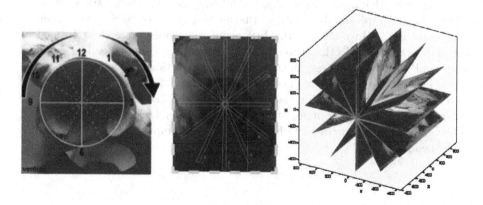

Fig. 1. Radial scans orientation(from left to right): i) superimposed over hip socket, ii) superimposed over a 2D MRI scan, iii) visualized in 3D space

plane. The images in Fig. 1i) and ii) are parallel to this localizer. The mechanics of the acquisition technology [11] mean that only the center and normal vector of the localizer are accessible in the resulting DICOM images of all six qMR scans. The scans are expected roughly correlate to the diagram in Fig. 1i) but there is no guarantee with respect to the order or the directions of the final images. For comparison, a sequence of real scans from a volunteer can be compared to the expected orientations in the Appendix Fig. 4 and Fig. 5 respectively.

2   Method

Given the complexity of the anatomy imaged in these scans, we rely on initial segmentations of the femur and acetabulum to initialize our registrations. Different tissues express themselves differently in the modalities but the bony structures are consistently identifiable across all scans. We use a combination of random forests trained on samples from three different modalities of the 2D scans and a neural network pre-trained on a much larger dataset of shoulder joints to segment out the 2D and 3D bones respectively. Our proposed registration method can be broken down into two main steps. The first part includes recreating a visual hull [16] of the femur bone from the 2D slices that can then be registered to the 3D volume to estimate a reasonable initialization. The second step requires fine-tuning this registration through an iterative process of manipulation the 3D volume to 'emulate' the 2D slices, comparing these emulations to the real scans using a feature-based intermodal similarity metric such as mutual information [15] and updating the locations accordingly.

2.1   2D to 3D Reconstruction Using SFS

Shape-From-Silhouette (SFS) is a 3D reconstruction technique that uses images of 2D silhouettes to produce an output termed the visual or convex hull [16]. Traditional SFS problems are posed as a 3D object surrounded by cameras that capture 2D images of the object's silhouettes from their various point of views [6]. A simple model showing the moving parts of SFS can be seen in Appendix Fig. 3. In our case, the MR acquisition system can be reframed as an orthographic projection extending a polyhedral prism instead of a visual cone and with the bone segmentations as cross-sections. The six intersecting slices can be re-imagined as having been produced from similarly positioned external cameras surrounding the hip such that the 3D volume lies within the intersection of the visual prisms associated with the silhouette of the hip bones. This adaptation is illustrated in Fig. 2 and can be compared to the original SFS in Appendix Fig. 3.

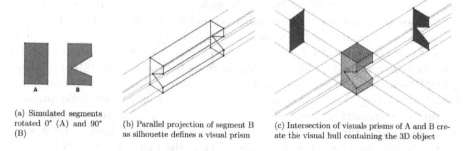

(a) Simulated segments rotated 0° (A) and 90° (B)

(b) Parallel projection of segment B as silhouette defines a visual prism

(c) Intersection of visuals prisms of A and B create the visual hull containing the 3D object

Fig. 2. SFS reconstruction adapted to qMRI acquisition framework

Our exact reconstruction algorithm is still under construction, but we expect to share its details along with preliminary results in the next iteration of our publication. Additionally, this technique is based on the assumption that the relative locations of the 2D slices with respect to each other is accurate and known. This, as confirmed by Figs. 4 and 5, is not true for our case. To address this, we preprocessed our slices by comparing them to the expected orientations and manually aligned them into their appropriate positions. This process is also expected to be automated in the near future.

2.2   3D to 2D Projection Using Binary Search

Once the location of the 2D slices has been initialized, a set of 2D scans are 'emulated' via projection of the 3D volume onto the planes where the slices intersect. A second set of emulated scans are acquired after rotating these planes clockwise and anti-clockwise by an angle of 15°. The reason for this choice of angle is to explore the space of possibilities using a binary search in logarithmic time instead of an exhaustive linear search. All the actual 2D slices are then compared to these emulated slices and cumulatively vote to move the search space to one of the two sub-regions then repeat the process using 7° rotations and so on. Results from the binary search based optimization technique are pending the finalization of the initialization procedure, but results of the emulated scans from the 3D volumes can be seen in Appendix Figs. 6 and 7 respectively, visualized using 3DSlicer [1, 13, 14]

3   Discussion

Quantitative MRI technology allows earlier insight into asymptomatic morphological abnormalities that may improve the likelihood of positive prognoses. Despite extensive validation in research literature, qMRI has not translated into routine clinical practice for reasons including long acquisition times, lack of standardization and an absence of statistical models for analysis. Our research aims to enable a particularly promising new qMRI technology that has reduced acquisition times to a clinically feasible range and proven to be highly reproducible over time and scanners. Our contribution aims to enable registration of these 2D qMR slices to a normative 3D volumetric space to allow performing comparative and longitudinal analysis in larger scale or longer studies. We propose an initialization method using the 2D slices to create a 3D reconstruction and a followup optimization technique that emulates the qMRI acquisition process by capturing 2D slices from the 3D volume. While the method is currently under development, we have included intermediate results from its various sub-methods that show a lot of promise for the efficacy of our final 2D/3D multi-dimensional quasi-intermodal registration process.

Appendix

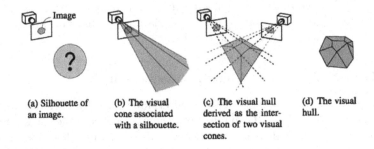

(a) Silhouette of an image.   (b) The visual cone associated with a silhouette.   (c) The visual hull derived as the intersection of two visual cones.   (d) The visual hull.

Fig. 3. Typical SFS reconstruction procedure as illustrated in [6] (a) Camera captures a silhouette (b) The silhouette defines a visual cone. (c) The intersection of two visual cones contains an object. (d) A visual hull of an object is the intersection of many visual cones

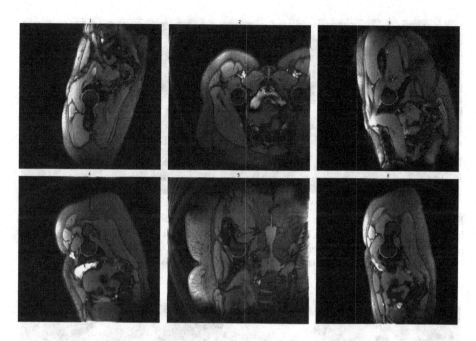

Fig. 4. Actual appearance of the six 2D qMR scans acquired from a volunteer

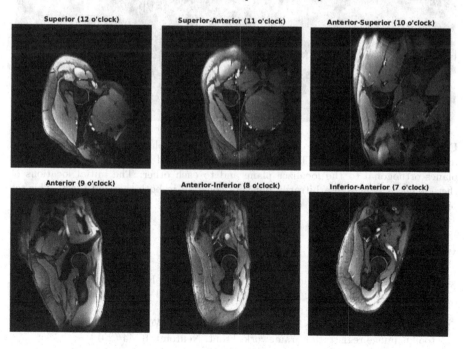

Fig. 5. Expected appearance of the six 2D qMR scans

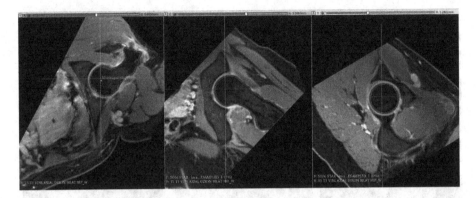

Fig. 6. The image on the far right depicts the localizer plane and so it is unchanging in both sets of emulations. The images on the center and left are captured using two planes orthogonal to the localizer plane and to each other. The initial locations of these planes was chosen arbitrarily but kept constant in both this and Fig. 7. This set produced via rotation by 90° clockwise

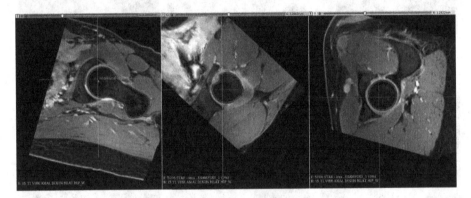

Fig. 7. The image on the far right depicts the localizer plane and so it is unchanging in both sets of emulations. The images on the center and left are captured using two planes orthogonal to the localizer plane and to each other. The initial locations of these planes was chosen arbitrarily but kept constant in both this and Fig. 6. This set produced via rotation by 45° counter-clockwise

References

1. Fedorov, A., et al.: 3D Slicer as an image computing platform for the Quantitative Imaging Network. Magn. Reson. Imaging **30**(9), 1323–1341 (2012). https://www.slicer.org/
2. Avants, B.B., Tustison, N.J., Stauffer, M., Song, G., Wu, B., Gee, J.C.: The insight toolkit image registration framework. Front. Neuroinf. **8**, 44 (2014)
3. Bashir, A., Gray, M.L., Burstein, D.: Gd-dtpa2- as a measure of cartilage degradation. Magn. Reson. Med. **36**, 665–673 (1996)

4. Cloos, M.A., Assländer, J., Abbas, B., Fishbaugh, J., Babb, J.S., Gerig, G., Lattanzi, R.: Rapid radial t1 and t2 mapping of the hip articular cartilage with magnetic resonance fingerprinting. J. Magn. Reson. Imaging **50**(3), 810–815 (2019)
5. Ferrante, E., Paragios, N.: Slice-to-volume medical image registration: a survey. Med. Image Anal. **39**, 101–123 (2017)
6. Imiya, A., Sato, K.: Shape from silhouettes in discrete space. In: Gagalowicz, A., Philips, W. (eds.) CAIP 2005. LNCS, vol. 3691, pp. 296–303. Springer, Heidelberg (2005). https://doi.org/10.1007/11556121_37
7. Jazrawi, L.M., Alaia, M.J., Chang, G., Fitzgerald, E.F., Recht, M.P.: Advances in magnetic resonance imaging of articular cartilage. J. Am. Acad. Orthopaedic Surg. **19**, 420–429 (2011)
8. Jazrawi, L.M., Bansal, A.: Biochemical-based MRI in diagnosis of early osteoarthritis. Imaging Med. **4**(1), 01 (2012)
9. Katz, J.N., et al.: Association between hospital and surgeon procedure volume and outcomes of total hip replacement in the united states medicare population. JBJS **83**(11), 1622–1629 (2001)
10. Lattanzi, R., et al.: Detection of cartilage damage in femoroacetabular impingement with standardized dgemric at 3 t. Osteoarthritis Cartilage **22**(3), 447–456 (2014)
11. Ma, D., et al.: Magnetic resonance fingerprinting. Nature **495**(7440), 187 (2013)
12. Markelj, P., Tomaževič, D., Likar, B., Pernuš, F.: A review of 3D/2D registration methods for image-guided interventions. Med. Image Anal. **16**(3), 642–661 (2012). https://doi.org/10.1016/j.media.2010.03.005, https://www.sciencedirect.com/science/article/pii/S1361841510000368, computer Assisted Interventions
13. Pieper, S., Halle, M., Kikinis, R.: 3D slicer. In: 2004 2nd IEEE International Symposium on Biomedical Imaging: Nano to Macro (IEEE Cat No. 04EX821), pp. 632–635. IEEE (2004)
14. Pieper, S., Lorensen, B., Schroeder, W., Kikinis, R.: The NA-MIC kit: ITK, VTK, pipelines, grids and 3D slicer as an open platform for the medical image computing community. In: 3rd IEEE International Symposium on Biomedical Imaging: Nano to Macro, 2006, pp. 698–701. IEEE (2006)
15. Pluim, J.P., Maintz, J.A., Viergever, M.A.: Mutual-information-based registration of medical images: a survey. IEEE Trans. Med. Imaging **22**(8), 986–1004 (2003)
16. Schneider, D.C.: Shape from silhouette, pp. 725–726. Springer, Boston (2014). https://doi.org/10.1007/978-0-387-31439-6_206
17. Venn, M., Maroudas, A.: Chemical composition and swelling of normal and osteoarthrotic femoral head cartilage. I. Chemical composition. Ann. Rheumatic Dis. **36**, 121–129 (1977)
18. Zhang, Y., Jordan, J.M.: Epidemiology of osteoarthritis. Clin. Geriatric Med. **26**(3), 355–369 (2010)

Deep Learning-Based Longitudinal Intra-subject Registration of Pediatric Brain MR Images

Andjela Dimitrijevic[1,2(✉)], Vincent Noblet[3(✉)],
and Benjamin De Leener[1,2,4(✉)]

[1] NeuroPoly Lab, Institute of Biomedical Engineering, Polytechnique Montréal,
Montréal, QC, Canada
{andjela.dimitrijevic,benjamin.de-leener}@polymtl.ca

[2] Research Center, Ste-Justine Hospital University Centre, Montréal, QC, Canada

[3] ICube-UMR 7357, Université de Strasbourg, CNRS, Strasbourg, France
vincent.noblet@unistra.fr

[4] Computer Engineering and Software Engineering, Polytechnique Montréal,
Montréal, QC, Canada

Abstract. Deep learning (DL) techniques have the potential of allowing fast deformable registration tasks. Studies around registration often focus on adult populations, while there is a need for pediatric research where less data and studies are being produced. In this work, we investigate the potential of unsupervised DL-based registration in the context of longitudinal intra-subject registration on 434 pairs of publicly available Calgary Preschool dataset of children aged 2–7 years. This deformable registration task was implemented using the DeepReg toolkit. It was tested in terms of input spatial image resolution (1.5 vs 2.0 mm isotropic) and three pre-alignement strategies: without (NR), with rigid (RR) and with rigid-affine (RAR) initializations. The evaluation compares regions of overlap between warped and original tissue segmentations using the Dice score. As expected, RAR with an input spatial resolution of 1.5 mm shows the best performances. Indeed, RAR has an average Dice score of of 0.937 ± 0.034 for white matter (WM) and 0.959 ± 0.020 for gray matter (GM) as well as showing small median percentages of negative Jacobian determinant (JD) values. Hence, this shows promising performances in the pediatric context including potential neurodevelopmental studies.

Keywords: Learning-based image registration · Pediatric · MRI

1 Introduction

Registration consists of bringing a pair of images into spatial correspondence. There are hardly any registration methods dedicated to the pediatric brain,

Supported by Polytechnique Montréal, by the Canada First Research Excellence Fund, and by the TransMedTech Institute.

A. Hering et al. (Eds.): WBIR 2022, LNCS 13386, pp. 206–210, 2022.
https://doi.org/10.1007/978-3-031-11203-4_24

mainly because of the difficulties arising from major changes that occur during neurodevelopment [5]. Conventional deformable registration involves estimating a deformation field through an iterative optimization problem. This process is time consuming, but provides accurate results. Convolutional neural networks (CNN) can allow faster registrations by applying a learning-based approach [4]. Hence, applying DL methods to pediatric brain scans could improve registration and future diagnostics for medical applications. Ultimately, it would be relevant to validate the potential use of DL-based frameworks for pediatric populations.

The general objective of this study is to validate a DL framework which allows fast intra-subject deformable registrations after training on pediatric MRI scans. To do so, different initial conditions are considered by fragmenting the non-rigid transformation into its simpler parts. Pre-network rigid registration, RigidReg (RR) and rigid-affine registration, RigidAffineReg (RAR) are performed on each intra-subject pair using ANTs [1] in order to determine their respective impact on the network's performance. Also, a third method called NoReg (NR) is investigated where no pre-alignment task is done. These three methods are then trained using a U-Net like CNN architecture implemented via the DeepReg toolkit [3]. The robustness of these DL techniques is assessed by using different input resolutions (1.5 vs 2.0 mm isotropic) for the same network architectures.

2 Methodology

Preprocessing Pipeline. Each image was corrected for bias field inhomogeneity using N4 algorithm. Both rigid and rigid-affine pre-alignments were performed with ANTs registration framework. The Mattes similarity metric was used.

Unsupervised Deformable Registration Framework. The U-Net architecture used to generate the deformation field consists of a 3-layer encoder and decoder with 8, 16 and 32 channels each. As for the loss function, it is composed of a local normalized cross-correlation similarity measure and an L2-norm gradient regularization factor to ensure realistic physical deformation fields. Local normalized cross-correlation is chosen for its robustness to local variations of intensities. The ADAM optimizer is used with a learning rate set to 1.0e-4. Finally, the network was trained on a GeForce RTX 2080 Ti GPU.

3 Experiments

Data. 434 pairs of moving/fixed 3D images were extracted from the longitudinal Calgary Preschool dataset [6] containing 247 T1-weighted images from 64 children aged 2–7 years old. The average time interval between consecutive scans is of 1.15 ± 0.68 years. The original images have a native resolution of $0.4492 \times 0.4492 \times 0.9$ mm^3. The resized images of 1.5 mm as well as 2.0 mm isotropic resolution have respectively a matrix size of $153 \times 153 \times 125$ and $114 \times 114 \times 94$.

Evaluation. To acquire white matter (WM), gray matter (GM) and cerebrospinal fluid (CSF) segmentations for evaluation purposes, each image was non-linearly registered to the MNI pediatric template for children 4.5–8.5 years old [2]. This also allowed obtaining skull-stripped images via the available mask in the template space. Unsupervised networks are then evaluated using the Dice score as a performance metric. In addition, the generated deformation fields are evaluated using the percentage of negative JD values indicating unwanted local foldings. To compare the impact from the three initialization methods or input resolutions, one-sided Wilcoxon signed-rank tests were performed.

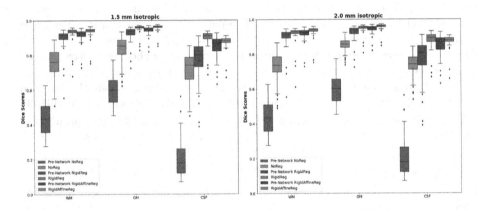

Fig. 1. Dice scores results for different input resolutions obtained for each method compared to their pre-network Dice scores represented as boxplots. The Dice scores are calculated for all subjects and WM, GM and CSF regions using the test set.

Table 1. Average Dice scores per resolution calculated over all segmented regions and subjects using the test set for the three studied methods. Median percentages of negative JD values are given because of highly right-skewed distributed data. ANTs pre-registration tasks are performed on the native resolution of $0.4492 \times 0.4492 \times 0.9$ mm^3 and using the available CPU implementation.

Methods	1.5 mm isotropic				2.0 mm isotropic				Native resolution
	Dice score	% of JD<0	Train time/epoch	Test time/pair	Dice score	% of JD<0	Train time/epoch	Test time/pair	ANTs pre-reg time/pair
NR	0.764 ± 0.105	1.11e−1	189.3 s	4.88 s	0.770 ± 0.088	1.33e−1	74.8 s	1.87 s	0 s
RR	0.929 ± 0.045	1.86e−4	137.7 s	3.55 s	0.916 ± 0.051	0	78.1 s	1.88 s	168.6 s
RAR	0.924 ± 0.047	0	177.5 s	4.13 s	0.922 ± 0.047	0	75.8 s	1.86 s	365.8 s

4   Results

A 85/15 % split was respectively done for train and test sets. Then, a three-fold cross-validation technique is employed to train and evaluate each method containing 123 pairs per fold. In total, 65 pairs are used for test purposes. Above, presented results for the two evaluated resolutions in Table 1 and Fig. 1 come from this unseen test set. Figure 2 shows the differences of obtained predicted

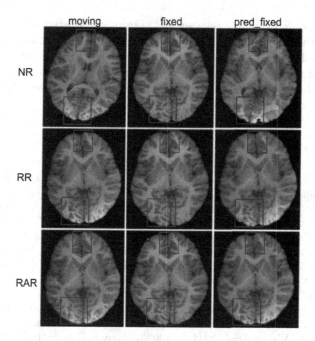

Fig. 2. Resulting images for a specific pair (age interval of 3.37 years) for the three pre-alignment strategies using an input spatial image resolution of 1.5 mm isotropic.

fixed images for all the three considered initialization methods. Also, two one-sided Wilcoxon tests were conducted comparing, first, the median Dice scores differences between RR and RAR (RAR-RR) for both resolutions as well as between 2.0 and 1.5 (1.5–2.0) for all initialization methods. The first test allowed rejecting the null hypothesis only for WM and GM (p<1.06e–10) showing higher median Dice scores for RAR compared to RR for both resolutions. The second test shows that 1.5 mm isotropic resolution, at the cost of longer train and test times, yields slightly, but statistically significant better performances than 2.0 for RR and RAR methods for all segmented regions (p<1.51e–4). This improvement is not significant for GM and CSF regions for the NR method.

5 Discussion and Conclusion

In this study, we demonstrated that DL-based deformable registration succeeds to improve registration accuracy regardless of the initialization method and for both tested resolutions (see Fig. 1). RAR demonstrated higher Dice scores compared to RR for WM (0.937 ± 0.034 vs 0.930 ± 0.046) and GM (0.959 ± 0.020 vs 0.955 ± 0.025). Differing results for CSF may be due to its thin surface and errors arising from the skull-stripping process. Both RR and RAR reached high registration quality, while NR shows lower registration performance due to its incapacity to extract both global and local transformations simultaneously, shown in

Fig. 2. However, NR remains relevant as no prior registration is needed. Future work will evaluate the capacity of a neural network to decompose the global and local transformations. Considering all studied combinations of pre-alignment strategies and input resolutions, RAR provides better Dice scores with 1.5 mm isotropic resolution images, which could help to perceive neurodevelopmental changes from a large age range of pediatric data.

References

1. Avants, B.B., Tustison, N.J., Song, G., Cook, P.A., Klein, A., Gee, J.C.: A reproducible evaluation of ants similarity metric performance in brain image registration. NeuroImage **54**(3), 2033–2044 (2011). https://doi.org/10.1016/j.neuroimage.2010.09.025, https://www.sciencedirect.com/science/article/pii/S1053811910012061
2. Fonov, V., Evans, A.C., Botteron, K., Almli, C.R., McKinstry, R.C., Collins, D.L.: Unbiased average age-appropriate atlases for pediatric studies. NeuroImage **54**(1), 313–327 (2011). https://doi.org/10.1016/j.neuroimage.2010.07.033, https://www.sciencedirect.com/science/article/pii/S1053811910010062
3. Fu, Y., et al.: Deepreg: a deep learning toolkit for medical image registration. J. Open Source Softw. **5**(55) (2020). https://doi.org/10.21105/joss.02705
4. Haskins, G., Kruger, U., Yan, P.: Deep learning in medical image registration: a survey. Mach. Vision Appl. **31**(1–2) (2020). https://doi.org/10.1007/s00138-020-01060-x
5. Phan, T.V., Smeets, D., Talcott, J.B., Vandermosten, M.: Processing of structural neuroimaging data in young children: bridging the gap between current practice and state-of-the-art methods. Dev. Cogn. Neurosci. **33**, 206–223 (2018). https://doi.org/10.1016/j.dcn.2017.08.009
6. Reynolds, J.E., Long, X., Paniukov, D., Bagshawe, M., Lebel, C.: Calgary preschool magnetic resonance imaging (MRI) dataset. Data Brief **29**, 105224 (2020). https://doi.org/10.1016/j.dib.2020.105224, https://www.ncbi.nlm.nih.gov/pubmed/32071993

Real-Time Alignment for Connectomics

Neha Goyal[✉], Yahiya Hussain, Gianna G. Yang, and Daniel Haehn

University of Massachusetts - Boston, Boston, MA 02125, USA
sneh.goyal.22@gmail.com

Abstract. In Connectomics, researchers are creating the brain's wiring diagram at nanometer resolution. As part of this processing workflow, 2D electron microscopy (EM) images must be aligned to 3D volumes. However, existing alignment methods are computationally expensive and can take a long time. We hypothesize that adding biological features improve and accelerate the alignment procedure. Since especially mitochondria can be detected accurately and fast, we propose a new alignment method, MITO, that uses these structures as landmark points. With MITO, we can decrease the alignment time by 27%, and our experiments indicate a throughput of 33 Megapixels/s, which is faster than the acquisition speed of current microscopes. We can align an image volume of $1268 \times 1524 \times 160$ voxels in less than 12 s. We compare our method to the following feature generators: ORB, BRISK, FAST, and FREAK.

Keywords: Image alignment · Registration · Feature matching

1 Introduction

Connectomics studies the functional and structural connections of a brain to understand the correlation between the physiology of the brain and its behavior. This correlation will help better treatment solutions, design new drugs for mental pathologies, construct custom neural prostheses, etc. Therefore, a registration process is required to map every synaptic connection to build a computer-generated brain wiring diagram. When needed, the image registration process is necessary to map the similarities between images acquired at different times or across other subjects by various sensors. Moreover, image registration is a crucial processing step in various other bio-medical image applications. In this study, we used diamond-knife-sliced electron microscopy (EM) images that provide high resolution such that individual synaptic connections between neurons are visible. We hypothesize to align these images by adding biological features can improve state-of-the-art registration methods. We have used a feature extraction model that follows four steps: feature detection, feature extraction, feature matching, and estimating the transformation matrix. Using the biological features, we get faster real-time alignment performance.

© The Author(s), under exclusive license to Springer Nature Switzerland AG 2022
A. Hering et al. (Eds.): WBIR 2022, LNCS 13386, pp. 211–214, 2022.
https://doi.org/10.1007/978-3-031-11203-4_25

2   Methods

We used unaligned two-dimensional EM images with nanometer resolution, and the corresponding mitochondria mask data as labeled data. The original dataset is called Lucchi++ and was the result of the study 'Fast Mitochondria Detection for Connectomics [1].' This dataset included two stacks: image and mask of 160 tiles, each having 768 × 1024 px. We created the unaligned dataset from the original by rotating each image tile and its corresponding mask tile at an arbitrary angle between $(-\pi, +\pi)$ and added a pad size of 250 px on all the sides to prevent information loss at the time of rotation. The new unaligned dataset has two stacks: image and mask, with 160 tiles and dimensions 1268 × 1524 px.

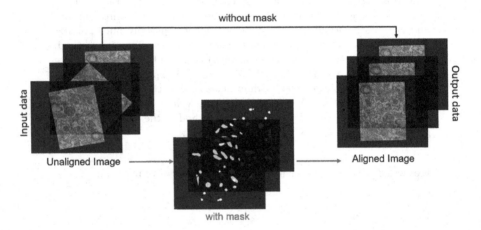

Fig. 1. Mapping of input images with and without adding the biological features. The unaligned input EM images (left) were mapped in real-time with and without adding the biological features (mask data). We generated a stack of aligned images (right) as output in both the cases to draw comparisons.

We performed an automatic registration on the unaligned EM images using a custom-build interactive program that runs the feature extraction model and calculates alignment score, execution time, and throughput for the entire dataset. This model used existing computer vision algorithms such as FAST [6], ORB [2], BRISK [3] to learn the features or patterns from the input dataset. We propose a new feature detector mechanism called **MITO** that detects the keypoints in EM images using mitochondria from mask images as a region of interest (ROI). In this feature detection step, we introduced mask images as additional biological features to improve the alignment performance. In the feature description step, the model uses ORB, BRISK, and FREAK [4] algorithms to create descriptors that are unique and could be referred to as a keypoint's numerical fingerprint. In the next step, we used feature matching algorithms such as BF [8] and FLANN [9] matcher to map (x_i, y_i) of the source image to (x_i', y_i') of the target image. Finally, with the help of the homography matrix, the model transforms the source image and outputs the aligned image. We generated two stacks of registered images with and without the help of mitochondria masks for comparisons (see Fig. 1).

3 Results

We perform experiments on the unaligned Lucchi++ dataset to measure timing and alignment accuracy. When we combine biological features using the MITO method with the BF and FLANN matchers, we observe a maximum execution time of 9.49 ($\pm$0.37) seconds for the whole stack. When comparing the accuracy, we measure a dice score of over 0.89 for both BF and FLANN, indicating quality alignment. The average throughput with MITO is at least 33 Megapixels/s which is faster than the acquisition speed of modern electron microscopes (11 Megapixels/s). Our findings indicate that MITO can be used to align connectomics image data in real-time during image acquisition. Table 1 shows the full evaluation.

Table 1. Alignment Results on Lucchi++. We compare the BF and FLANN matchers with a variety of feature descriptors. When using the MITO detector, we measure the throughput of at least 33 Megapixels/s, indicating real-time performance.

Matcher	Detector + Descriptor	Mask	Dice score	Execution time (sec.)	Stack throughput (MP/s)
BF	BRISK	✓	0.9354	47.0052(±1.5173)	6.7879(±0.2170)
			0.8569	**19.3020(±0.2625)**	**16.5210(±0.2256)**
	ORB	✓	0.7529	19.4427(±1.8462)	16.4941(±1.4953)
			0.8226	20.4218(±0.5493)	15.6208(±0.4259)
	FAST + BRISK	✓	0.9184	2419.9270(±99.9857)	0.1319(±0.0053)
			0.8762	**28.4635(±1.2776)**	**11.2167(±0.4908)**
	ORB + BRISK	✓	0.6291	16.3020(±1.4923)	19.6693(±1.8124)
			0.7935	16.9687(±1.6858)	18.9180(±1.9290)
	FAST + FREAK	✓	0.9405	2391.9479(±137.7484)	0.1335(±0.0074)
			0.9140	**25.1302(±0.5)**	**12.6912(±0.2498)**
	ORB + FREAK	✓	0.8320	16.6458(±1.8088)	19.2979(±1.9733)
			0.7637	16.8072(±0.1365)	18.9718(±0.1545)
	MITO(ours) + BRISK	✓	**0.9142**	**7.7708(±0.0888)**	**41.035(±0.4713)**
	MITO(ours) + FREAK	✓	**0.8963**	**8.3697(±0.0888)**	**38.0983(±0.4027)**
FLANN	BRISK	✓	0.9344	40.1145(±0.9393)	7.9514(±0.1887)
			0.8338	**19(±2.4111)**	**16.9513(±2.0058)**
	ORB	✓	0.8069	19.3802(±1.2145)	16.4941(±0.9979)
			0.8280	20.6875(±1.1149)	15.4417(±0.8082)
	FAST + BRISK	✓	0.9338	3082.2343(±130.2627)	0.1035(±0.0043)
			0.8784	**29.6041(±0.2350)**	**10.7709(±0.0856)**
	ORB + BRISK	✓	0.6297	16.9322(±1.7772)	18.9655(±1.9261)
			0.7648	15.2031(±1.1735)	21.0579(±1.6571)
	FAST + FREAK	✓	0.9450	2628.3229(±32.5343)	0.1213(±0.0015)
			0.9091	**31.4166(±4.7502)**	**10.2940(±1.4380)**
	ORB + FREAK	✓	0.8285	16.2812(±0.0563)	19.5841(±0.0676)
			0.7402	17.2083(±1.2107)	18.5882(±1.2665)
	MITO(ours) + BRISK	✓	**0.9062**	**9.2239(±0.7265)**	**34.7050(±2.6154)**
	MITO(ours) + FREAK	✓	**0.8928**	**9.4843(±0.3694)**	**33.6528(±1.3213)**

4 Conclusion

Fast registration is crucial to creating 3D volumetric connectomics datasets from unaligned EM images. This process can be computationally expensive. Based on our studies, adding biological features to register these images results in faster alignment. Specifically, we include mitochondria masks as part of our MITO feature detector. With MITO, the overall dice score is higher than 0.80, and the throughput is faster than 11 Megapixels/s. These measurements indicate the possibility of real-time alignment during the image acquisition with modern electron microscopes.

References

1. Casser, V., Kang, K., Pfister, H., Haehn, D.: Fast mitochondria detection for connectomics. In: Proceedings of the Third Conference on Medical Imaging with Deep Learning, PMLR, pp. 111–120 (2020)
2. Rublee, E., Rabaud, V., Konolige, K., Bradski, G.R.: ORB: an efficient alternative to SIFT or SURF. In: 2011 International Conference on Computer Vision, pp. 2564–2571. IEEE (2011)
3. Leutenegge, S., Chli, M., Siegwart, R.Y.: BRISK: binary robust invariant scalable keypoints. In: 2011 International Conference on Computer Vision, pp. 2548–2555. IEEE (2011)
4. Alahi, A., Ortiz, R., Vandergheynst, P.: Freak: fast retina keypoint. In: 2012 IEEE Conference on Computer Vision and Pattern Recognition (CVPR), pp. 510–517 (2012)
5. Calonder, M., Lepetit, V., Strecha, C., Fua, P.: BRIEF: binary robust independent elementary features. In: Daniilidis, K., Maragos, P., Paragios, N. (eds.) ECCV 2010. LNCS, vol. 6314, pp. 778–792. Springer, Heidelberg (2010). https://doi.org/10.1007/978-3-642-15561-1_56
6. Rosten, E., Drummond, T.: Machine learning for high-speed corner detection. In: Leonardis, A., Bischof, H., Pinz, A. (eds.) ECCV 2006. LNCS, vol. 3951, pp. 430–443. Springer, Heidelberg (2006). https://doi.org/10.1007/11744023_34
7. Fischler, M.A., Bolles, R.C.: Random sample consensus: a paradigm for model fitting with applications to image analysis and automated cartography. Commun. ACM **24**, 381–395 (1981)
8. OpenCV modules. http://docs.opencv.org/3.1.0. Accessed 19 Apr 2017
9. Muja, M., Lowe, D.G.: Fast approximate nearest neighbors with automatic algorithm configuration. In: VISAPP (2009)
10. Khachikian, S., Emadi, M.: Applying fast & freak algorithms in selected object tracking. Int. J. Adv. Res. Electr. Electron. Instrument. Eng. **5**, 5829–5839 (2016)
11. Phan, D., Oh, C.-M., Kim, S.-H., Na, I.-S., Lee, C.-W.: Object recognition by combining binary local invariant features and color histogram. In: 2013 2nd IAPR Asian Conference on Pattern Recognition (ACPR), pp. 466–470 (2013)
12. Rosten, E., Porter, R., Drummond, T.: Faster and better: a machine learning approach to corner detection. IEEE Trans. Pattern Anal. Mach. Intell. **32**, 105–119 (2010)
13. Szeliski, R.: Computer Vision: Algorithms and Applications. Springer-Verlag, London (2010). https://doi.org/10.1007/978-1-84882-935-0
14. Li, X., Zhanyi, H.: Rejecting mismatches by correspondence function. Int. J. Comput. Vision **89**, 1–17 (2010)

Weak Bounding Box Supervision
for Image Registration Networks

Mona Schumacher[1,2(✉)], Hanna Siebert[1], Ragnar Bade[2], Andreas Genz[2],
and Mattias Heinrich[1]

[1] Institute of Medical Informatics, University of Luebeck, Luebeck, Germany
[2] MeVis Medical Solutions AG, Bremen, Germany
mona.schumacher@mevis.de

Abstract. Image registration is a fundamental task in medical image
analysis. Many deep learning based methods use multi-label image segmentations during training to reach the performance of conventional
algorithms. But the creation of detailed annotations is very time-
consuming and expert knowledge is essential. To avoid this, we propose
a weakly supervised learning scheme for deformable image registration
that uses bounding boxes during training. By calculating the loss function based on these bounding box labels, we are able to perform an image
registration with large deformations without using densely labeled annotations. The performance of the registration of inter-patient 3D Abdominal CT images can be enhanced by approximately 10% only with little annotation effort in comparison to unsupervised learning methods.
Taken into account this annotation effort, the performance also exceeds
the performance of the label supervised training.

Keywords: Deformable image registration · Weak supervision ·
Bounding box supervision

1 Introduction

Medical image registration is the process of the alignment of the anatomical
structures of two or more images in order to be able to do follow up studies,
image-guidance or to plan a treatment. Deep learning methods have become
increasingly important. They have demonstrated low computation times and
are promising to enable real time registration approaches. For the case of brain
image registration [1], which only require small deformation, already satisfac-
tory results could be achieved. The registration of images of highly deformable
body regions, such as the abdominal region or thorax are, due to the respiration
or digestion, more complex and still often solved with conventional algorithms
[2,3]. Deep learning methods have started to address the challenge of handling
large deformations (for example in the Learn2Reg Challenge, cf. learn2reg.grand-
challenge.org) [4,5]. Mok et al. [6] use Laplacian pyramids to solve the registration in a coarse-to-fine scheme inspired by classical algorithms. They show that

© The Author(s), under exclusive license to Springer Nature Switzerland AG 2022
A. Hering et al. (Eds.): WBIR 2022, LNCS 13386, pp. 215–219, 2022.
https://doi.org/10.1007/978-3-031-11203-4_26

label supervision substantially increases the registration accuracy, which is also shown by Siebert et al. [7]. In image segmentation, weak label supervision has already gained interest. Rajchl et al. [8], for example, use an extension of the GrabCut algorithm and learn segmentation from bounding box annotations. In this paper, our aim is to close the gap between supervised and unsupervised registration methods and propose a weakly supervised learning scheme for deformable image registration including large deformations and introduce a loss function based on 3D bounding boxes to decrease the effort of the labeling process. We use inter-patient 3D Abdominal CT images and are able to increase the overlap of organs by approximately 10% in comparison to unsupervised image registration methods. If the time of the labeling process is taken into account, the performance of supervised algorithms can also be exceeded.

2   Methods

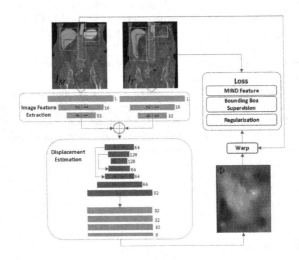

Fig. 1. Architecture of proposed method: Image features are extracted for I_F and I_M separately in two decoders (shared weights). The concatenated features are passed through a U-Net-like architecture and are finally used to estimate a displacement Φ to warp I_M. The loss consists of three parts: MIND features, regularization and the proposed bounding box supervision. The resolution in relation to the input resolution of the different steps are displayed in the layers.

The network consists of two parts: an image feature extraction part and a displacement estimation part. An overview of the architecture is shown in Fig. 1. The image feature extraction part extracts the low level features of the input images in two streams (with shared weights for monomodal registration). The displacement estimation part uses the concatenated low level features and estimates the displacement field. The 32 concatenated feature maps of I_F and I_M are used as input to extract 32 joint feature maps with a U-Net-like network with three encoder and four decoder blocks. Three additional sequences are added to estimate the displacement field. The final displacement field is generated by reducing the 32 feature maps to the three displacement dimensions with a $1 \times 1 \times 1$ convolution and transformed to normalized sampling voxel locations (value range from -1 to 1) with the *tanh* activation function to match the PyTorch grid definition. The deformation has the same size as the input images.

To train the network, weak label supervision is used. Instead of using detailed labels for the calculation of the loss function, bounding boxes are used. The advantage of this method is that a significant reduction in time can be achieved and the variance between raters is also lower. A combination of three loss functions is used: the modality independent neighbourhood descriptor (MIND) with self-similar context (SSC) [10], a diffusion regularization and the mean squared error for the bounding boxes. The bounding box loss is multiplied by a factor of two.

To generate the final registration result including large deformations, we apply the network twice. The first input images are I_F and I_M. Then, I_M is warped with the first displacement field. The resulting warped moving image is used as second input.

3   Experiments

To train and evaluate our method, we use the publicly available Learn2Reg challenge dataset (Task3, 2020). This dataset contains 30 abdominal CT scans with thirteen manually labeled abdominal organs [4,5]. For training and testing, we use the split and validation pairs as in the official challenge. The data is already preprocessed to same voxel sizes and spatial dimensions. We downsample the images for the experiments to a size of $144 \times 112 \times 144$ due to GPU memory requirements. For all labels, tight bounding boxes as well as a bounding box with a random error of $\pm 5\%$ are generated. The network is trained using Adam optimizer with a learning rate of 0.001 for 7500 iterations.

We train our network three times: unsupervised (not using the label loss), with the proposed bounding box loss, and with the voxelwise manually labeled organ segmentations. To establish comparability between training with label and weak label loss, we perform additional runs of supervised training with less training data. In this way, we simulate manual generation of labels or bounding boxes that takes the same amount of time. In total, we have five experiments: unsupervised, tight-weakly-supervised, weakly-supervised, supervised and supervised_50%. Tight-weakly refers to perfect bounding boxes, weakly refers to bounding boxes with an additional error of $\pm 5\%$ and supervised_50% refers to the experiment with less labeled data.

4   Results

In Table 1 the average Dice scores for all organs are listed for the different trainings. In comparison to the initial overlap of the organs, the overlap can

Table 1. Dice scores [%] for spleen ■, right kidney ■, left kidney ■, gall bladder ■, esophagus ■, liver ■, stomach ■, aorta ■, inferior vena cava ■, portal and splenic vein ■, pancreas ■, left adrenal gland ■, and right adrenal gland ■.

	■	■	■	■	■	■	■	■	■	■	■	■	■	avg ± std
initial	42	34	35	2	23	62	24	33	36	5	15	8	9	25 ± 13
unsupervised	67	57	61	5	33	81	35	54	50	15	21	18	14	39 ± 14
tight-weakly-supervised	70	67	69	7	33	86	41	53	56	20	27	25	17	44 ± 13
weakly-supervised	67	64	64	6	32	83	40	54	56	18	28	24	16	43 ± 13
supervised	81	73	78	8	43	86	50	67	61	17	25	21	16	48 ± 11
supervised-_50%	67	55	59	6	38	81	39	51	42	10	18	23	9	38 ± 13

be increased by approximately 14%. For the tight bounding box training, the overlap can be increased by approximately 19% and 18% for the bounding box training with random error. The label supervised trained network increased the overlap by approximately 22%. The standard deviation of the Jacobian determinant as well as the proportion of negative values are comparable for all trainings. It can be shown that a higher Dice score can be obtained for larger organs or for organs that initially already have a high overlap. The largest organ, the liver, for example, has the highest initial Dice overlap of 62%, and also the highest Dice overlap after registration for all variants (in a range of 81–85%). Organs with a small initial overlap, e.g. left adrenal gland (initial overlap 8%), also have a relatively low overlap after registration for all methods (in a range of 18–25%). For these organs, however, the Dice of weakly-supervised is higher than for supervised (e.g. left adrenal gland: 25% for weakly-supervised and 21% for supervised).

5   Discussion and Conclusion

We presented a deep-learning-based method for deformable image registration with weak bounding box supervision. We compared our method with an unsupervised and a label supervised training. The resulting registration of our method shows an improvement of about 5% for the Dice overlap in comparison to the unsupervised training. To simulate a realistic annotation of bounding boxes, we added an inter-observer-error of 5% per bounding box side, and showed that the quality of the result does not change significantly (approximately 1%) compared to tight bounding boxes. Organs with small initial overlap show the highest Dice score after the registration with the weak bounding box supervised network.

If the time for the labeling process was taken into account, so that less labels are available than bounding boxes, the accuracy of the label supervised training is less than for our bounding box supervision. Hence, for the purpose of medical image registration the proposed weak supervision strategy (labeling more images with lower effort) is beneficial.

References

1. de Vos, B.D., Berendsen, F.F., Viergever, M.A., Sokooti, H., Staring, M., Isgum, I.: A deep learning framework for unsupervised affine and deformable image registration. Med. Image Anal. **52**, 128–143 (2019)
2. Eppenhof, K.A., Pluim, J.P.: Pulmonary CT registration through supervised learning with convolutional neural networks. IEEE Trans. Med. Imag. **38**(5), 1097–1105 (2018)
3. Sentker, T., Madesta, F., Werner, R.: GDL-FIRE^{4D}: deep learning-based fast 4D CT image registration. In: Frangi, A.F., Schnabel, J.A., Davatzikos, C., Alberola-López, C., Fichtinger, G. (eds.) MICCAI 2018. LNCS, vol. 11070, pp. 765–773. Springer, Cham (2018). https://doi.org/10.1007/978-3-030-00928-1_86
4. Hansen, L., Hering, A., Heinrich, M.P., et al.: Learn2Reg: 2020 MICCAI registration challenge (2020). https://learn2reg.grand-challenge.org
5. Xu, Z., Lee, C.P., Heinrich, M.P., et al.: Evaluation of six registration methods for the human abdomen on clinically acquired CT. IEEE Trans. Biomed. Eng. **63**(8), 1563–1572 (2016)
6. Mok, T.C.W., Chung, A.C.S.: Large deformation diffeomorphic image registration with Laplacian pyramid networks. In: Martel, A.L., et al. (eds.) MICCAI 2020. LNCS, vol. 12263, pp. 211–221. Springer, Cham (2020). https://doi.org/10.1007/978-3-030-59716-0_21
7. Siebert, H., Hansen, L., Heinrich, M.P.: Evaluating design choices for deep learning registration networks. In: Bildverarbeitung für die Medizin 2021. I, pp. 111–116. Springer, Wiesbaden (2021). https://doi.org/10.1007/978-3-658-33198-6_26
8. Rajchl, M., et al.: DeepCut: object segmentation from bounding box annotations using convolutional neural networks. IEEE Trans. Med. Imag. **36**(2), 674–683 (2016)
9. Hering, A., Kuckertz, S., Heldmann, S., Heinrich, M.P.: Memory-efficient 2.5 D convolutional transformer networks for multi-modal deformable registration with weak label supervision applied to whole-heart CT and MRI scans. Int. J. Comput. Assist. Radiol. Surg. **14**(11), 1901–1912 (2019)
10. Heinrich, M.P., Jenkinson, M., Papież, B.W., Brady, S.M., Schnabel, J.A.: Towards realtime multimodal fusion for image-guided interventions using self-similarities. In: Mori, K., Sakuma, I., Sato, Y., Barillot, C., Navab, N. (eds.) MICCAI 2013. LNCS, vol. 8149, pp. 187–194. Springer, Heidelberg (2013). https://doi.org/10.1007/978-3-642-40811-3_24

Author Index

Abbas, Batool 198
Albertini, F. 57
Andresen, Julia 3
Antanavicius, Justinas 166
Asllani, Iris 147
Audigier, Chloe 75
Aviles-Rivero, Angelica I. 190

Bade, Ragnar 215
Balbastre, Yaël 103
Bastiaansen, Wietske A. P. 29
Bigalke, Alexander 37
Bloch, Isabelle 8

Cercignani, Mara 147
Cetin, Oezdemir 124
Chao, Hanqing 177
Chen, Junyu 96
Choi, Kup-Sze 185

Dalca, Adrian V. 103
De Leener, Benjamin 206
Dimitrijevic, Andjela 206
Dorent, Reuben 75
Du, Yong 96

Ehrhardt, Jan 3

Fischl, Bruce 103
Flinner, Nadine 124
François, Anton 8
Frey, Eric C. 96
Frintrop, Simone 134

Genz, Andreas 215
Gerig, Guido 198
Glaunès, Joan 8
Gori, Pietro 8
Goyal, Neha 211
Graf, Laura 134

Haehn, Daniel 211
Handels, Heinz 3
Hansen, Lasse 85

Harrison, Neil 147
Heinrich, Mattias 215
Heinrich, Mattias P. 37, 85, 119, 134
Hempe, Hellena 37
Hernandez, Monica 18
Huang, Zhening 190
Hussain, Yahiya 211

Iglesias, Juan Eugenio 103

Joutard, Samuel 75

Kepp, Timo 3
Klein, Stefan 29
Koeppl, Heinz 124
Koning, Anton H. J. 29
Krishnan, Venkateswaran P. 47
Kruggel, Frithjof 67
Kruse, Christian N. 37

Lattanzi, Riccardo 198
Lauri, Mikko 134
Leiras, Roberto 166
Liao, Guojun 194
Lindblad, Joakim 156
Liò, Pietro 190
Liu, Lihao 185, 190
Lomax, T. 57

Maillard, Matthis 8
Mansi, Tommaso 75
Mayordomo, Elvira 18
Mischkewitz, Sven 134
Modat, Marc 75

Nicke, Till 134
Niessen, Wiro J. 29
Noblet, Vincent 206

Öfverstedt, Johan 156
Oppenheim, Catherine 8
Örzsik, Balázs 147

Pallud, Johan 8
Petchprapa, Catherine 198

Pheiffer, Thomas   75
Piat, Sebastien   75

Qin, Jing   185

Ramon, Ubaldo   18
Roider, Johann   3
Rousian, Melek   29

Schönlieb, Carola-Bibiane   190
Schumacher, Mona   215
Selvan, Raghavendra   166
Shu, Yiran   124
Siebert, Hanna   119, 215
Simpson, Ivor J. A.   147
Sivaswamy, Jayanthi   47
Sladoje, Nataša   156
Smolders, A.   57
Song, Xinrui   177
Song, Youyi   185
Steegers-Theunissen, Régine P. M.   29

Thottupattu, Alphin J.   47
Turkbey, Baris   177

Vercauteren, Tom   75
von der Burchard, Claus   3

Wang, Ge   177
Weber, D. C.   57
Weihsbach, Christian   37
Wells, William M.   103
Wild, Peter   124
Wohlfahrt, Patrick   75
Wood, Bradford J.   177

Xu, Sheng   177

Yan, Pingkun   177
Yang, Gianna G.   211
Young, Sean I.   103

Zhou, Zicong   194
Ziegler, Paul   124
Zou, Jing   185

Printed in the United States
by Baker & Taylor Publisher Services

Printed in the United States
by Baker & Taylor Publisher Services